Gondwana Six: Structure, Tectonics, and Geophysics

Geophysical Monograph Series

Including
Maurice Ewing Volumes
Mineral Physics Volumes

GEOPHYSICAL MONOGRAPH SERIES

Geophysical Monograph Volumes

1 **Antarctica in the International Geophysical Year** *A. P. Crary, L. M. Gould, E. O. Hulburt, Hugh Odishaw, and Waldo E. Smith (Eds.)*

2 **Geophysics and the IGY** *Hugh Odishaw and Stanley Ruttenberg (Eds.)*

3 **Atmospheric Chemistry of Chlorine and Sulfur Compounds** *James P. Lodge, Jr. (Ed.)*

4 **Contemporary Geodesy** *Charles A. Whitten and Kenneth H. Drummond (Eds.)*

5 **Physics of Precipitation** *Helmut Weickmann (Ed.)*

6 **The Crust of the Pacific Basin** *Gordon A. Macdonald and Hisashi Kuno (Eds.)*

7 **Antarctic Research: The Matthew Fontaine Maury Memorial Symposium** *H. Wexler, M. J. Rubin, and J. E. Caskey, Jr. (Eds.)*

8 **Terrestrial Heat Flow** *William H. K. Lee (Ed.)*

9 **Gravity Anomalies: Unsurveyed Areas** *Hyman Orlin (Ed.)*

10 **The Earth Beneath the Continents: A Volume of Geophysical Studies in Honor of Merle A. Tuve** *John S. Steinhart and T. Jefferson Smith (Eds.)*

11 **Isotope Techniques in the Hydrologic Cycle** *Glenn E. Stout (Ed.)*

12 **The Crust and Upper Mantle of the Pacific Area** *Leon Knopoff, Charles L. Drake, and Pembroke J. Hart (Eds.)*

13 **The Earth's Crust and Upper Mantle** *Pembroke J. Hart (Ed.)*

14 **The Structure and Physical Properties of the Earth's Crust** *John G. Heacock (Ed.)*

15 **The Use of Artificial Satellites for Geodesy** *Soren W. Henriksen, Armando Mancini, and Bernard H. Chovitz (Eds.)*

16 **Flow and Fracture of Rocks** *H. C. Heard, I. Y. Borg, N. L. Carter, and C. B. Raleigh (Eds.)*

17 **Man-Made Lakes: Their Problems and Environmental Effects** *William C. Ackermann, Gilbert F. White, and E. B. Worthington (Eds.)*

18 **The Upper Atmosphere in Motion: A Selection of Papers With Annotation** *C. O. Hines and Colleagues*

19 **The Geophysics of the Pacific Ocean Basin and Its Margin: A Volume in Honor of George P. Woollard** *George H. Sutton, Murli H. Manghnani, and Ralph Moberly (Eds.)*

20 **The Earth's Crust: Its Nature and Physical Properties** *John G. Heacock (Ed.)*

21 **Quantitative Modeling of Magnetospheric Processes** *W. P. Olson (Ed.)*

22 **Derivation, Meaning, and Use of Geomagnetic Indices** *P. N. Mayaud*

23 **The Tectonic and Geologic Evolution of Southeast Asian Seas and Islands** *Dennis E. Hayes (Ed.)*

24 **Mechanical Behavior of Crustal Rocks: The Handin Volume** *N. L. Carter, M. Friedman, J. M. Logan, and D. W. Stearns (Eds.)*

25 **Physics of Auroral Arc Formation** *S.-I. Akasofu and J. R. Kan (Eds.)*

26 **Heterogeneous Atmospheric Chemistry** *David R. Schryer (Ed.)*

27 **The Tectonic and Geologic Evolution of Southeast Asian Seas and Islands: Part 2** *Dennis E. Hayes (Ed.)*

28 **Magnetospheric Currents** *Thomas A. Potemra (Ed.)*

29 **Climate Processes and Climate Sensitivity (Maurice Ewing Volume 5)** *James E. Hansen and Taro Takahashi (Eds.)*

30 **Magnetic Reconnection in Space and Laboratory Plasmas** *Edward W. Hones, Jr. (Ed.)*

31 **Point Defects in Minerals (Mineral Physics Volume 1)** *Robert N. Schock (Ed.)*

32 **The Carbon Cycle and Atmospheric CO_2: Natural Variations Archean to Present** *E. T. Sundquist and W. S. Broecker (Eds.)*

33 **Greenland Ice Core: Geophysics, Geochemistry, and the Environment** *C. C. Langway, Jr., H. Oeschger, and W. Dansgaard (Eds.)*

34 **Collisionless Shocks in the Heliosphere: A Tutorial Review** *Robert G. Stone and Bruce T. Tsurutani (Eds.)*

35 **Collisionless Shocks in the Heliosphere: Reviews of Current Research** *Bruce T. Tsurutani and Robert G. Stone (Eds.)*

36 **Mineral and Rock Deformation: Laboratory Studies—The Paterson Volume** *B. E. Hobbs and H. C. Heard (Eds.)*

37 **Earthquake Source Mechanics (Maurice Ewing Volume 6)** *Shamita Das, John Boatwright, and Christopher H. Scholz (Eds.)*

38 **Ion Acceleration in the Magnetosphere and Ionosphere** *Tom Chang (Ed.)*

39 **High Pressure Research in Mineral Physics (Mineral Physics Volume 2)** *Murli H. Manghnani and Yasuhiko Syono (Eds.)*

Maurice Ewing Volumes

1 **Island Arcs, Deep Sea Trenches, and Back-Arc Basins** *Manik Talwani and Walter C. Pitman III (Eds.)*

2 **Deep Drilling Results in the Atlantic Ocean: Ocean Crust** *Manik Talwani, Christopher G. Harrison, and Dennis E. Hayes (Eds.)*

3 **Deep Drilling Results in the Atlantic Ocean: Continental Margins and Paleoenvironment** *Manik Talwani, William Hay, and William B. F. Ryan (Eds.)*

4 **Earthquake Prediction—An International Review** *David W. Simpson and Paul G. Richards (Eds.)*

5 **Climate Processes and Climate Sensitivity** *James E. Hansen and Taro Takahashi (Eds.)*

6 **Earthquake Source Mechanics** *Shamita Das, John Boatwright, and Christopher H. Scholz (Eds.)*

Mineral Physics Volumes

1 **Point Defects in Minerals** *Robert N. Schock (Ed.)*

2 **High Pressure Research in Mineral Physics** *Murli H. Manghnani and Yasuhiko Syono (Eds.)*

Geophysical Monograph 40

Gondwana Six: Structure, Tectonics, and Geophysics

Garry D. McKenzie
Editor

American Geophysical Union
Washington, D.C.
1987

Published under the aegis of AGU Geophysical Monograph Board.

Library of Congress Cataloging-in-Publication Data

Gondwana six : structure, tectonics, and geophysics.

(Geophysical monograph, ISSN 0065-8448 ; 40)
"Papers presented at the Sixth International Gondwana Symposium held at the Institute of Polar Studies, the Ohio State University, Columbus, Ohio, 19–23 August, 1985"—Pref.
 1. Gondwana (Geology)—Congresses. 2. Geology, Structural—Congresses. I. McKenzie, Garry D.
II. International Gondwana Symposium (6th : 1985 : Institute of Polar Studies, Ohio State University)
III. Ohio State University. Institute of Polar Studies.
IV. Series.
QE511.5.G66 1987   551.7   87-11408
ISBN 0-87590-064-X
ISSN 0065-8448

Copyright 1987 by the American Geophysical Union, 2000 Florida Avenue, NW, Washington, DC 20009

Figures, tables, and short excerpts may be reprinted in scientific books and journals if the source is properly cited.

Authorization to photocopy items for internal or personal use, or the internal or personal use of specific clients, is granted by the American Geophysical Union for libraries and other users registered with the Copyright Clearance Center (CCC) Transactional Reporting Service, provided that the base fee of $1.00 per copy plus $0.10 per page is paid directly to CCC, 21 Congress Street, Salem, MA 10970. 0065-8448/87/$01. + .10.
 This consent does not extend to other kinds of copying, such as copying for creating new collective works or for resale. The reproduction of multiple copies and the use of full articles or the use of extracts, including figures and tables, for commercial purposes requires permission from AGU.

Printed in the United States of America.

DEDICATION

John W. Cosgriff, Jr., was born November 10, 1931, in Denver, Colorado. He was awarded a B.A. in anthropology from the University of Arizona in 1953 and an M.A. and Ph.D. in vertebrate paleontology from the University of California, Berkeley, in 1960 and 1963, respectively. He was a Senior Research Fellow at the Department of Geology, University of Tasmania, from 1964 to 1967 and served as Assistant Professor, Associate Professor, and Professor in the Department of Biological Sciences, Wayne State University, Detroit, Michigan, until his death on April 28, 1985.

Affectionately called "Cos" by his students, John was a dedicated teacher known for his warm-hearted nature and generosity. His friends and colleagues knew him as a mild-mannered scholar with a tremendous breadth of knowledge.

John's interests in Triassic vertebrate paleontology took him to field and museum work in Europe, South Africa, India, Australia, and Antarctica. In the austral summer of 1977-1978, he led a team of vertebrate paleontologists to the Cumulus Hills of the central Transantarctic Mountains, where they collected a large number of Lystrosaurus Zone vertebrate fossils from the Lower Triassic Fremouw Formation. But for his untimely death, he would have returned to the central Transantarctic Mountains in the 1985-1986 field season. John published more than 25 papers on Triassic labyrinthodonts and reptiles.

One of John's final wishes was to return some of the kindnesses shown him by his Indian colleagues on his recent sabbatical leave at the Indian Statistical Institute in Calcutta. His contribution of financial support for this conference helped to sponsor several Indian scientists. We are grateful to his wife, Bette, and his two sons, Kevin and Ethan, for carrying out this request.

CONTENTS

Dedication   vii

Preface   xi

Acknowledgments   xii

Seismic Refraction Measurements of Crustal Structure in West Antarctica   *Sean T. Rooney, Donald D. Blankenship, and Charles R. Bentley*   1

Satellite Magnetic Anomalies and Continental Reconstructions   *R. R. B. von Frese, W. J. Hinze, R. Olivier, and C. R. Bentley*   9

A Revised Reconstruction of Gondwanaland   *Lawrence A. Lawver and Christopher R. Scotese*   17

The Handler Formation, a New Unit of the Robertson Bay Group, Northern Victoria Land, Antarctica   *Thomas O. Wright and Colin Brodie*   25

Radiometric Ages of Pre-Mesozoic Rocks From Northern Victoria Land, Antarctica   *H. Kreuzer, A. Höhndorf, H. Lenz, P. Müller, and U. Vetter*   31

A Review of the Problems Important for Interpretation of the Cambro-Ordovician Paleogeography of Northern Victoria Land (Antarctica), Tasmania, and New Zealand   *R. H. Findlay*   49

Paleozoic Magmatism and Associated Tectonic Problems of Northern Victoria Land, Antarctica   *S. G. Borg and E. Stump*   67

Construction of the Pacific Margin of Gondwana During the Pannotios Cycle   *Edmund Stump*   77

Early Paleozoic Westward Directed Subduction at the Pacific Margin of Antarctica   *G. Kleinschmidt and F. Tessensohn*   89

Joint U.K.-U.S. West Antarctic Tectonics Project: An Introduction   *Ian W. D. Dalziel and Robert J. Pankhurst*   107

Crustal Structure of the Area Around Haag Nunataks, West Antarctica: New Aeromagnetic and Bedrock Elevation Data   *S. W. Garrett, L. D. B. Herrod, and D. R. Mantripp*   109

Outline of the Structural and Tectonic History of the Ellsworth Mountains-Thiel Mountains Ridge, West Antarctica   *B. C. Storey and I. W. D. Dalziel*   117

Correlation of Gabbroic and Diabasic Rocks From the Ellsworth Mountains, Hart Hills, and Thiel Mountains, West Antarctica   *Walter R. Vennum and Bryan C. Storey*   129

Petrology, Geochemistry, and Tectonic Setting of Granitic Rocks From the Ellsworth-Whitmore Mountains Crustal Block and Thiel Mountains, West Antarctica   *Walter R. Vennum and Bryan C. Storey*   139

Rb-Sr Geochronology of the Region Between the Antarctic Peninsula and the Transantarctic Mountains: Haag Nunataks and Mesozoic Granitoids   *I. L. Millar and R. J. Pankhurst*   151

Ellsworth-Whitmore Mountains Crustal Block, Western Antarctica: New Paleomagnetic Results and Their Tectonic Significance   *A. M. Grunow, I. W. D. Dalziel, and D. V. Kent*   161

The Ellsworth-Whitmore Mountains Crustal Block: Its Role in the Tectonic Evolution of West Antarctica   *I. W. D. Dalziel, S. W. Garrett, A. M. Grunow, R. J. Pankhurst, B. C. Storey, and W. R. Vennum*   173

Sedimentary Rocks of the English Coast, Eastern Ellsworth Land, Antarctica   *T. S. Laudon, D. J. Lidke, T. Delevoryas, and C. T. Gee*   183

The Gondwanian Orogeny Within the Antarctic Peninsula: A Discussion   *B. C. Storey, M. R. A. Thomson, and A. W. Meneilly*   191

Sandstone Detrital Modes and Basinal Setting of the Trinity Peninsula Group, Northern Graham Land, Antarctic Peninsula: A Preliminary Survey   *J. L. Smellie*   199

Structural Evolution of the Magmatic Arc in Northern Palmer Land, Antarctic Peninsula   *A. W. Meneilly, S. M. Harrison, B. A. Piercy, and B. C. Storey*   209

Late Paleozoic Accretionary Complexes on the Gondwana Margin of Southern Chile: Evidence From the Chonos Archipelago   *John Davidson, Constantino Mpodozis, Estanislao Godoy, Francisco Hervé, Robert Pankhurst, and Maureen Brook*   221

Early Paleozoic Structural Development in the NW Argentine Basement of the Andes and Its Implication for Geodynamic Reconstructions   *A. P. Willner, U. S. Lottner, and H. Miller*   229

Paleomagnetism of Permian and Triassic Rocks, Central Chilean Andes   *Randall D. Forsythe, Dennis V. Kent, Constantino Mpodozis, and John Davidson*   241

Late Paleozoic Pseudoalbaillellid Radiolarians From Southernmost Chile and Their Geological Significance   *Hsin Yi Ling and Randall D. Forsythe*   253

The Late Paleozoic Evolution of the Gondwanaland Continental Margin in Northern Chile   *C. M. Bell*   261

Permian to Late Cenozoic Evolution of Northern Patagonia: Main Tectonic Events, Magmatic Activity, and Depositional Trends   *M. A. Uliana and K. T. Biddle*   271

Petrology and Facies Analysis of Turbiditic Sedimentary Rocks of the Puncoviscana Trough (Upper Precambrian-Lower Cambrian) in the Basement of the NW Argentine Andes   *P. Ježek and H. Miller*   287

Aspects of the Structural Evolution and Magmatism in Western New Schwabenland, Antarctica   *G. Spaeth*   295

Plate Tectonic Development of Late Proterozoic Paired Metamorphic Complexes in Eastern Queen Maud Land, East Antarctica   *Kazuyuki Shiraishi, Yoshikuni Hiroi, Yoichi Motoyoshi, and Keizo Yanai*   309

Tectonic Position of Karoo Basalts, Western Zambia   *Raphael Unrug*   319

Symposium Participants   323

PREFACE

This volume contains many of the papers presented at the Sixth International Gondwana Symposium, held at the Institute of Polar Studies, The Ohio State University, Columbus, Ohio, August 19-23, 1985. The symposium was the first held outside the Gondwanaland continents; other symposia were held in Buenos Aires, Argentina, 1967; Cape Town and Johannesburg, South Africa, 1970; Canberra, Australia, 1973; Calcutta, India, 1977; and Wellington, New Zealand, 1980.

The Columbus symposium attracted 150 scientists from 19 countries to five days of technical sessions, six field trips, commission and working group meetings, and workshops. Topics covered in the technical sessions were generally similar to those of earlier meetings and included reconstruction of Gondwanaland, vertebrate and invertebrate paleontology, biogeography, glacial geology, Gondwana stratigraphy, economic geology, and tectonics and sedimentation at plate margins. A notable difference was in geographic coverage. As might be expected at a meeting co-hosted by the Institute of Polar Studies and the Department of Geology and Mineralogy at The Ohio State University, the focus of the meetings was on Antarctica, with 45% of the 102 papers covering the Ross Sea sector, West Antarctica, and northern Victoria Land.

The 56 papers are presented in two volumes: Gondwana Six: Structure, Tectonics, and Geophysics, and Gondwana Six: Stratigraphy, Sedimentology, and Paleontology; there is some overlap of topics.

The papers in this volume include 20 that focus on Antarctica, seven on South America, three on Gondwanaland in general, and one on southern Africa. Most of these address problems related to the Pacific margin of Gondwanaland. Papers on the joint U.K.-U.S. West Antarctic tectonics project form an important section. The authors present results of recent fieldwork and the geological, geophysical, and geochemical data that provide support for an improved understanding of the sedimentary basins, magmatism, tectonics, and paleogeography of this region. Many papers on similar topics but other areas, for instance north Victoria Land, Antarctica, and by scientists from other nations are included. The results reported are important for an improved reconstruction of Gondwanaland and have implications for the paleoceanographic history of the southern hemisphere.

The papers in the second volume deal with paleontological and biostratigraphical topics and cover many regions, including Australia, New Zealand, Asia, India, Africa, and Madagascar, often exploring the biogeographical connections between them. These papers range from comprehensive reviews of paleontological groups, for example, Devonian vertebrates and the Trematosauridae, to the description of a new dinosaur. Two examples of stratigraphic and sedimentologic topics are the comparison of Gondwana sequences (Tasmania-Antarctica, Africa-South America) and facies analysis of glacigenic deposits.

All papers in these volumes report on basic research, but two also address a topic of economic interest (uranium). However, with the current concern for global resource availability, all papers, and particularly those on Antarctica, have resource implications because they provide an improved understanding of the geology and geologic history of Gondwanaland.

Although the editorial committee was firm in the editorial format, some leeway was given for individual and national writing styles. Further, the committee was influenced by the arguments of A. M. C. Sengör (Geologische Rundschau, 72, 397-400, 1983) on the use of "Gondwana" vs. "Gondwanaland" and decided to leave the decision to the authors.

The symposium was organized by James W. Collinson and David H. Elliot (Co-chairmen); Peter J. Anderson, Garry D. McKenzie, and Peter N. Webb, The Ohio State University; and J. M. Dickins, Bureau of Mineral Resources, Canberra, Australia.

The technical sessions were preceded and followed by field trips. The field trip guidebooks (Geology of the Southern Appalachians, Glacial Geology of Central Ohio, Lower Carboniferous Clastic Sequence of Central Ohio, Carboniferous of Eastern Kentucky, and Quaternary and Proterozoic Glacial Deposits) and the abstracts of papers presented, were published by the Institute of Polar Studies.

Social events included a welcoming party, an ox roast, and a symposium banquet at which Campbell Craddock, University of Wisconsin, discussed the stages of development of American views on Gondwanaland from rejection (1920s), reconsideration (1950s), acceptance (1960s), to refinement (1970 to present).

Preceding the Symposium, on August 16-17, 1985, a workshop on Cenozoic geology of southern high latitudes was hosted by P. N. Webb, Department of Geology and Mineralogy, The Ohio State University, and was attended by 30 scientists, most of whom took part in the Symposium.

J. W. Collinson
D. H. Elliot
S. M. Haban
G. D. McKenzie

Editorial Committee

ACKNOWLEDGMENTS

The editorial and organizing committees thank the many organizations and individuals who contributed to the success of the symposium and the preparation of this volume. Without the cosponsors and contributors of financial support, the meeting would not have been possible. Stephanie Haban played a major role in the planning and daily activities of the symposium, and as technical editor of this volume spent many hours transforming manuscripts into papers. Not enough praise can be given for the editorial assistance offered by the reviewers. Without their expertise and time, the quality of the papers certainly would not have been sustained.

Lynn Lay, librarian at the Institute of Polar Studies, spent many hours completing and upgrading references for the author's manuscripts. We also thank Robert Tope for his preparation of illustrations for many of the papers.

Thanks are due to the students, staff, and faculty (and their spouses) of the Institute of Polar Studies and the Department of Geology and Mineralogy for their support of symposium activities. We are especially grateful to those who organized and prepared field excursions and guidebooks: A. C. Rocha-Campos (University of São Paulo, Brazil); D. E. Pride, R. O. Utgard, I. M. Whillans, G. D. McKenzie, S. M. Bergstrom, R. H. Blodgett, T. N. Taylor, and E. M. Smoot (The Ohio State University); K. B. Bork and R. J. Malcuit (Denison University, Ohio); J. R. Chaplin (Oklahoma Geological Survey); B. L. Lowry-Chaplin (University of Texas-Arlington); C. E. Mason (Morehead State University, Kentucky); R. T. Lierman (The George Washington University, Washington, D.C.); D. R. Sharpe (Geological Survey of Canada); and G. M. Young, (University of Western Ontario, Canada).

Major funding was provided by U.S. National Science Foundation grant EAR-8407780, awarded to The Ohio State University. Other support was received from the International Union of Geological Sciences (IUGS), The Ohio State University, and the General Electric Company. The symposium was co-sponsored by the IUGS Subcommission on Gondwana Stratigraphy, the Geological Society of America, and the American Geophysical Union. At the request of her late husband, John W. Cosgriff of Wayne State University, Bette Cosgriff supported the travel of several Indian scientists. We are grateful for all of these contributions.

Publication of this volume is partially supported by National Science Foundation Grant No. EAR-8407780; however, any opinions, findings, conclusions, or recommendations expressed herein are those of the authors and do not necessarily reflect the views of the Foundation. Publication is also supported by The Ohio State University.

SEISMIC REFRACTION MEASUREMENTS OF CRUSTAL STRUCTURE IN WEST ANTARCTICA

Sean T. Rooney, Donald D. Blankenship, and Charles R. Bentley

Geophysical and Polar Research Center, University of Wisconsin-Madison
Madison, Wisconsin 53706

Abstract

Two seismic refraction profiles yield new information about the crustal structure of West Antarctica. One profile on the Ross Ice Shelf grid northwest of Crary Ice Rise shows basement rock with a compressional wave velocity (v_p) of 5.9 ± 0.2 km s^{-1} underlying a layer of presumed sediment about 400 m thick. The irregular basement surface lies about 1.0 km below sea level and has an average apparent dip of 1° to the grid north. The second profile, on the Siple Coast, reveals 1 to 2 km of sediment overlying a basement ($v_p = 5.4 \pm 0.2$ km s^{-1}) that has an apparent dip of 4° to the grid southeast. An abrupt 2-km decrease in depth to basement within the profile is interpreted as being due to faulting. Deep reflections were observed on one record at 50 km, but data are insufficient for identification of their source.

Introduction

Two seismic refraction profiles were obtained during the 1983-1984 and 1984-1985 Antarctic field seasons, one on the Ross Ice Shelf (station CIR, see Figure 1) and the other on the Siple Coast (station UPB). Both profiles give information about crustal structure beneath the ice.

At CIR, shot distances were obtained (to an accuracy of a small fraction of a meter) with an electronic distance measuring device (AGA geodimeter), and absolute positions were determined by satellite positioning equipment (Magnavox geoceiver). Shot distances at station UPB were found by using several methods, including measurement by geodimeter, satellite positioning (good to a few meters), and direct P waves through the ice (accurate to better than 1%). Ice thickness values along the profile lines were determined by electromagnetic sounding (S. Shabtaie, personal communication, 1985). Ten geophone arrays, separated by 60 or 90 m over a total spread length of 690 m, were used at each recording station. Each array comprised six 8-Hz geophones planted in a rectangular pattern. Explosive charges ranging from 90 to 500 kg were used as sources at both CIR and UPB. Data were recorded by a high-speed digital recording system (developed at the Geophysical and Polar Research Center) with a 0.5-ms sampling interval on each channel. Time breaks were recorded for all shots of both profiles.

Ross Ice Shelf Profile

An unreversed profile consisting of five shots spaced roughly 3 km apart was obtained at CIR with shot-to-spread distances ranging from 14 to 27 km. The spread location remained fixed at grid position 6.16°S, 1.45°W (Figure 1). Refracted arrivals through the seafloor were observed on all five records. To analyze these data, travel times were plotted, and arrivals that could be traced across all records were identified. The apparent velocity of each arrival on a single record (the "forward velocity") was measured as well as the apparent velocity of each arrival as recorded on the same channel for all five records (the "reverse velocity"). The forward velocity (v_f) is a function of the dip under the receivers, whereas the reverse velocity (v_r) is a function of dip between the shots (Figure 2). If an assumption of constant dip from the shots to the receivers is made, these two apparent velocities can be used to calculate the true velocity and apparent dip of the refractor.

A reduced-time record section (Figure 3) was made by using the two measured apparent velocities. (Reduced times are travel times adjusted, at some assumed reducing velocity, to some standard distance.) For Figure 3, reducing velocities of $v_f = 6.18$ km s^{-1} and $v_r = 5.67$ km s^{-1} were used. Thus a persistent arrival with these apparent velocities should form a straight line across all records. In fact, for shot-to-spread distances greater than 20 km, the first arrivals (at 3.05 s in Figure 3) do line up well. At 17.7 km and 14.3 km, however, the arrivals were delayed by 0.03 s and 0.11 s, respectively. In principle, the delay could be due to changes in water depth, thickness of low-velocity sediments overlying the refractor, or both. In fact, we believe from radar sounding studies that the ice is intermittently grounded in this region [Shabtaie and Bentley, 1986], so we attribute the delay to a thicker sedimentary column under those two stations.

Using the travel times and apparent velocities from the more distant shots and assuming a veloci-

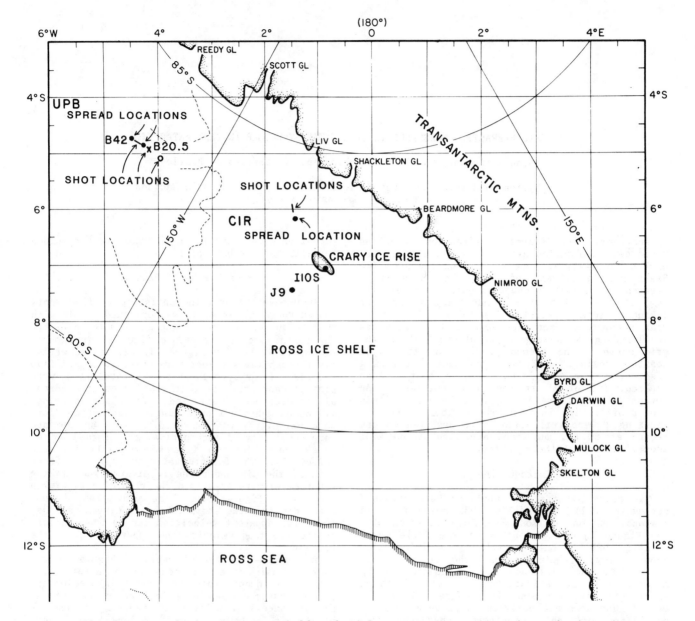

Fig. 1. Map of the Ross Ice Shelf and Siple Coast, West Antarctica, showing the locations of the two seismic refraction profiles described in the text. At CIR shot points were spaced equally along the short line. At UPB shot points were placed along the short line and also at each of the points marked by a cross and a circle. The spreads were at sites B42 and B20.5, both marked with solid circles. In the rectangular grid coordinate system, 0° grid longitude lies along the Greenwich and 180° meridians with grid north toward Greenwich, and the grid equator passes through the geographic South Pole.

ty of 2.4 km s^{-1} in a sedimentary layer (velocities less than that in the ice, 3.8 km s^{-1}, cannot be measured), we find a "true" velocity of 5.9 km s^{-1}, a mean depth to the layer of 1 km below sea level, and a grid northerly dip of 1.1°. The dip beneath the two shots nearest the spread, between which the apparent velocity is 6.5 km s^{-1}, is 2.3° to the grid south. Thus an undulating basement with a relief of a few hundred meters is implied (shown schematically in Figure 4).

This, of course, violates the assumption of constant dip over the entire profile, which was in any case an extreme one, not to be taken literally. However, with apparent velocities that differ only by 0.4 km s^{-1}, it is unlikely that a major error in true velocity will be made. Furthermore, a large dip beneath a layer that averages only about a kilometer in thickness can exist only locally, so that an apparent velocity over a 6-km stretch cannot differ greatly from the true vel-

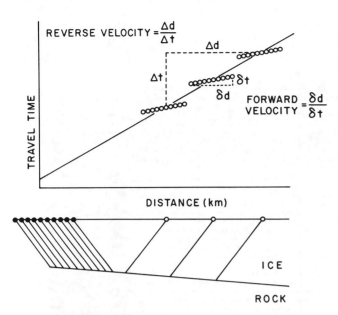

Fig. 2. Diagram showing the arrangement of shots and receivers (not to scale), and the meaning of the forward and reverse velocities.

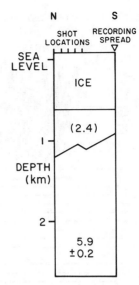

Fig. 4. Upper-crustal structure deduced from the CIR refraction profile. The numbers in the column are P-wave velocities in kilometers per second. Parentheses denote an assumed value. N and S are grid directions (cf. Figure 1).

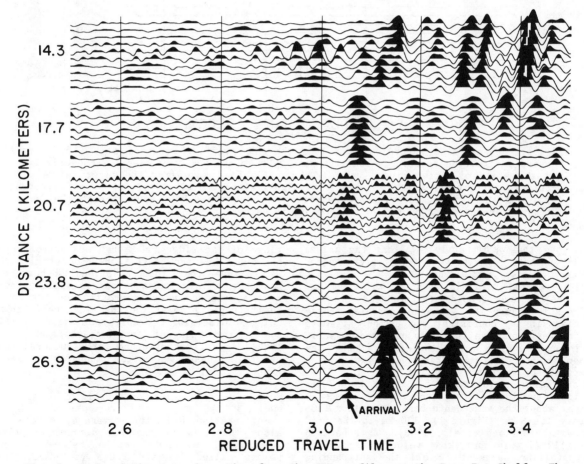

Fig. 3. Reduced-time record section from the CIR profile, on the Ross Ice Shelf. The forward and reverse reducing velocities were 6.18 km s^{-1} and 5.67 km s^{-1}, respectively, and the standard distance used was 14 km. The head wave from the first refractor occurs at a corrected time of 3.05 s.

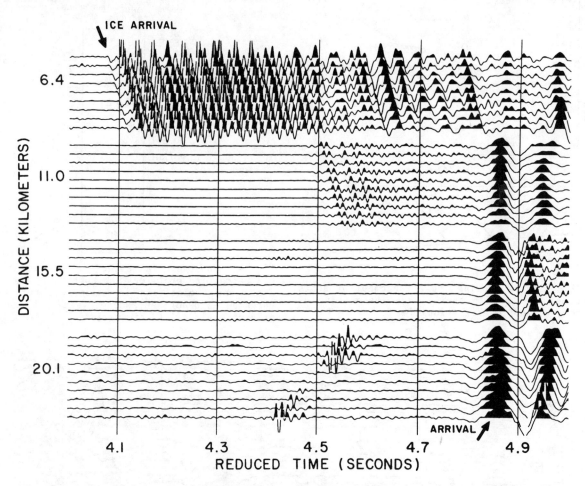

Fig. 5. Reduced-time record section from the UPB profile, recorded at B20.5. The forward and reverse reducing velocities were 5.05 km s^{-1} and 5.80 km s^{-1}, respectively, and the standard distance used was 20.2 km. The head wave from the first refractor occurs at a corrected time of 4.8 s. The high-frequency arrivals at 4.1 s, 4.5 s, and 4.95 s on the first three records, respectively, are the direct wave through the ice. The event around 4.5 s on the 20.1-km record is noise.

ocity. We thus arbitrarily say that the actual velocity probably lies between the apparent velocities, i.e., v = 5.9 ± 0.2 km s^{-1}.

Strong energy arrivals with v_f = 6.7 km s^{-1} occur on all our records (at reduced time of 3.1-3.4 s in Figure 3). We cannot correlate these arrivals with any consistent V_r, and our attempts to interpret them quantitatively have led to serious inconsistencies. We therefore cannot say whether they represent refraction energy from a deeper, higher velocity layer.

The basement velocity of 5.9 ± 0.2 km s^{-1} is not significantly different from that (5.7 km s^{-1}) found at about the same depth beneath station I10S on Crary Ice Rise (Figure 1), 130 km to the grid southeast [Robertson et al., 1982]. As Robertson et al. [1982] point out, that velocity is typical of crystalline basement in all the surrounding geological provinces, so it is difficult to make any specific association. However, since both their profiles and ours were located on the same ridge of the ridge-and-trough sea-bottom topography under the ice shelf [Robertson et al., 1982], we believe that this additional occurrence of basement rock at a depth that is shallow compared to those found on other profiles on the Ross Ice Shelf (e.g., 2 km at the Ross Ice Shelf Geophysical and Glaciological Survey, station J9 (Figure 1) in the adjacent trough [Robertson et al., 1982]) is further evidence for structural control of the linear sea-bottom topographic features [Robertson et al., 1982].

Siple Coast Profile

At station UPB, the coverage, more or less along the axis of the ice stream, comprised a 20-km-long reversed profile between recording points called B42 and B20.5 (solid circles in Figure 1), plus two extended shots, one recorded at B42 and set off 28 km to the grid southeast (cross in Figure 1), and the other recorded at

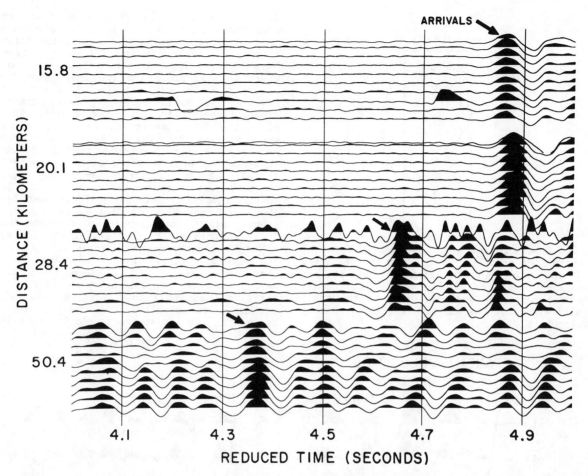

Fig. 6. Reduced-time record section from the UPB profile. The 16-km, 20-km, and 28-km records were recorded at B42; the 50-km record was recorded at B20.5. The forward and reverse reducing velocities were 5.80 km s^{-1} and 5.05 km s^{-1}, respectively, and the standard distance used was 20.2 km. The head wave from the first refractor occurs at corrected times of 4.85, 4.85, 4.65, and 4.35 s.

B20.5 and fired 50 km to the grid southeast (open circle in Figure 1). The records from these two shots are both included in Figure 6, the other records from shots being set off grid southeast of the recording point.

Two reduced-time record sections (Figures 5 and 6) were prepared by using the same pair of reducing velocities (v_f on one leg equals v_r on the other). The reversed profile included shots at 6, 11, 16, and 20 km recorded at B20.5 (Figure 5) and at 16 and 20 km recorded at B42 (top two records in Figure 6). Five of the six records of the reversed profile show a strong coherent arrival with excellent alignment, which indicates that the dip of the refractor is steady across the profile; the calculated true velocity is 5.4 ± 0.2 km s^{-1}.

To calculate depths, we assume that the layer between the ice and the refractor is composed of lithified sediments, and that the corresponding wave velocity is 4 km s^{-1}, as is commonly indicated elsewhere in West Antarctica [Bentley and Clough, 1972]. Then the depth of the 5.4 km s^{-1} layer under B42 is 2.4 km and the dip is 4° to the grid southeast (Figure 7).

The shot 50 km southeast of the spread (B20.5) yielded a first arrival at 10.2 s (Figure 8) that has an apparent velocity of 4.6 km s^{-1}, which is lower than the apparent velocity of the first arrivals on all the closer shots. (Note that the total travel time constrains the average speed along the ray path beneath the ice to greater than 5 km s^{-1}). The strong arrival with an apparent velocity of 5.8 km s^{-1} at around 11 s in Figure 8 can also be seen at a reduced time of 4.35 s in Figure 6. Its apparent velocity is close to that for the head wave from the 5.4 km s^{-1} layer, but it arrives 0.5 s earlier than predicted by extension of the single-planar-layer model derived from the reversed profile described above to the southeast. The first arrival on the 28-km shot recorded at B42 also arrives earlier (at reduced time of 4.65 s on Figure 6) than expected for a head wave from the 5.4 km s^{-1} layer. We interpret all these occurrences as being due to a decrease in basement

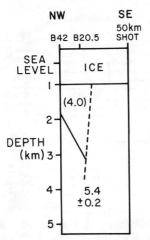

Fig. 7. Upper-crustal structure deduced from the UPB refraction profile. The numbers in the column are P-wave velocities in kilometers per second. Parentheses denote an assumed value. The dashed line is a possible fault discussed in the text; see also Figure 9. NW and SE are grid directions (cf. Figure 1).

depth southeast of the UPB camp, perhaps caused by faulting. The 4.6 km s^{-1} arrival was successfully modeled by ray tracing as a refraction through an interface with an apparent dip of 7° (Figure 9). This suggests that there is a fault located somewhere between B20.5 and the shot point for the record at 28 km. A fault throw of about 2.2 km is required to match observed travel times with travel times obtained from ray tracing. The model of Figure 9 is also consistent with the observed early arrivals from the 5.4 km s^{-1} layer (apparent velocity 5.8 km s^{-1}) recorded from the 50-km shot.

Since the profile was shot nearly parallel to the axis of the ice stream, the 7° dip represents the component of dip normal to that axis. Because the subglacial topography [Albert and Bentley, 1986; map reproduced in Bentley and Ježek, 1981] shows a grain continuing inland from the Ross Ice Shelf that is almost parallel to the ice stream, we might expect basement structures to trend in the same direction. The 7° apparent dip could, therefore, result from a steep fault striking nearly parallel to the seismic profile. (The limiting case would be a vertical fault with a strike 7° off parallel with the profile). The facts that the low-apparent-velocity arrival does

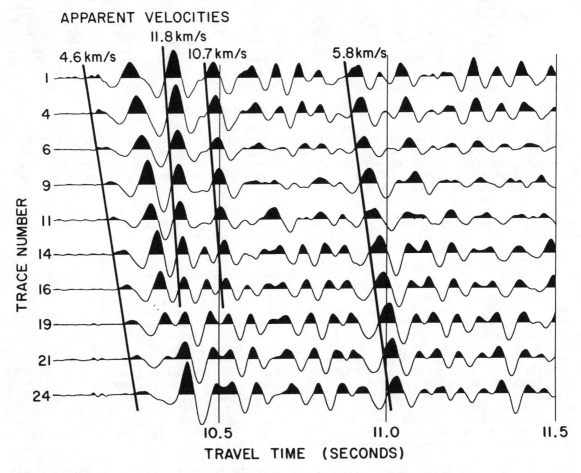

Fig. 8. Seismogram from the 50-km shot at UPB. Note the compressed time scale. Lines denote refracted arrivals with apparent velocities of 4.6 km s^{-1} and 5.8 km s^{-1}, and reflections with apparent velocities of 10.7 km s^{-1} and 11.8 km s^{-1}.

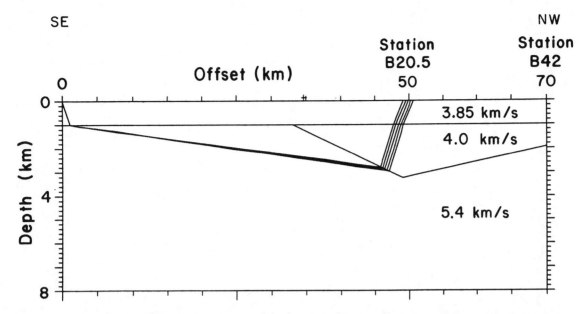

Fig. 9. Crustal model derived to produce an arrival with an apparent velocity of 4.6 km s^{-1} at 10.1 s from the 50-km shot, recorded at B20.5. Refraction through a fault face with an apparent dip of 7° is indicated. SE and NW are grid directions.

not occur on the 28-km shot, which is laterally offset by about 1.9 km, and that the first arrival (from the 5.4 km s^{-1} layer) occurs earlier than expected, are consistent with such a model.

The two arrivals recorded on the 50-km shot at 10.4 and 10.5 s (Figure 8) have apparent velocities, 11.8 km s^{-1} and 10.7 km s^{-1}, respectively, characteristic of reflection events. These reflections cannot be from irregularities in the ice rock boundary because the travel time is less than the travel time of the direct wave in the ice. In fact, when the velocity model derived from the reversed profile is used and a velocity of 6.5 km s^{-1} at depths greater than 6 km is assumed, the reflection travel times indicate that the reflector must be at least 8 km, but not more than 20 km, deep and must lie within 10 km laterally. The reflector must be quite large (the radius of the first Fresnel zone is ~2 km) in order to result in the large reflection amplitude. The arrival could be a reflection from Moho at 20 km, but the large dip (>18°) indicated by the apparent velocities makes this identification questionable. Because we observe these reflections clearly on only one record, it is impossible for us to delineate their source any more precisely.

Acknowledgments. This work was supported by National Science Foundation grant DPP-8412404. We wish to thank B. R. Weertman, J. E. Nyquist, R. Flanders, S. Shabtaie, D. A. Holland, J. Dallman, and D. R. MacAyeal for help with field measurements, P. B. Dombrowski for figure drafting, and A. N. Mares for manuscript preparation. This is contribution number 440 of the Geophysical and Polar Research Center, University of Wisconsin-Madison.

References

Albert, D.G., and C. R. Bentley, Seismic investigations on the Ross Ice Shelf, Antarctica, in The Ross Ice Shelf: Glaciology and Geophysics, Antarct. Res. Ser., vol 42, edited by C. R. Bentley and D. E. Hayes, AGU, Washington, D.C., in press, 1986.

Bentley, C. R., and J. W. Clough, Seismic refraction shooting in Ellsworth and Queen Maud Lands, in Antarctic Geology and Geophysics, edited by R. J. Adie, pp. 169-172, Universitetsforlaget, Oslo, 1972.

Bentley, C. R., and K. C. Ježek, RISS, RISP, and RIGGS: Post-IGY glaciological investigations of the Ross Ice Shelf in the U.S. program, J. R. Soc. N. Z., 11(4), 355-372, 1981.

Robertson, J. D., C. R. Bentley, J. W. Clough, and L. L. Greischar, Sea-bottom topography and crustal structure below the Ross Ice Shelf, Antarctica, in Antarctic Geoscience, edited by C. Craddock, pp. 1083-1090, University of Wisconsin Press, Madison, 1982.

Shabtaie, S., and C. R. Bentley, West Antarctic ice streams draining into the Ross Ice Shelf: Configuration and mass balance, J. Geophys. Res., in press, 1986.

Copyright 1987 by the American Geophysical Union.

SATELLITE MAGNETIC ANOMALIES AND CONTINENTAL RECONSTRUCTIONS

R. R. B. von Frese,[1] W. J. Hinze,[2] R. Olivier,[3] and C. R. Bentley[4]

Abstract. Regional magnetic anomalies observed by NASA's magnetic satellite mission over the eastern Pacific Ocean, North and South America, the Atlantic Ocean, Europe, Africa, India, Australia, and Antarctica are adjusted to a fixed elevation of 400 km and differentially reduced to the radial pole of intensity of 60,000 nT. Having been normalized for differential inclination, declination, and intensity effects of the core field, these radially polarized anomalies in principle directly map the geometric and magnetic property variations of sources within the lithosphere. Continental satellite magnetic data show a sharp truncation and even parallelism of anomalies along the active edges of the North and South American plates, whereas across passive plate continental margins the truncation of anomalies is less distinct, and possibly reflects subsided continental crust or the tracks of hotspots in the ocean basins. When plotted on an Early Cambrian reconstruction of Gondwanaland, many of the radially polarized anomalies of the continents demonstrate detailed correlation across the continental boundaries to verify the prerift origin of their sources. Accordingly, these anomalies provide fundamental constraints on the geologic evolution of the continents and their reconstructions.

Introduction

Magnetic measurements collected by NASA's Polar Orbiting Geophysical Observatory (POGO) and magnetic satellite mission (MAGSAT) have yielded fundamental and unique information concerning not only the core field but also anomalies derived from the lithosphere. Quantitative geologic analyses of these anomalies are limited by several factors including anomaly distortion introduced by the variable attitude and intensity attributes of the core field. This distortion, which is severe when it affects tectonic analyses on global or regional scales, is demonstrated by the regional magnetic anomalies mapped by the MAGSAT mission in Figure 1a. Satellite magnetic anomalies for the eastern Pacific Ocean, North and South America, the Atlantic Ocean, Europe, Africa, India, Australia, and Antarctica are differentially reduced to the pole by an equivalent point dipole inversion procedure to minimize their contamination by the core field. Finally, the geological significance of these adjusted anomalies is considered for studying the evolution and dynamics of the continents and oceans.

Enhancement of Satellite Magnetic Anomalies for Geologic Analysis

Satellite magnetic anomalies implemented in this study are based on areal averages of MAGSAT scalar observations obtained during periods of low temporal magnetic activity. A model developed by NASA-Goddard Space Flight Center was used to remove the geomagnetic core field from the observations. Quadratic functions were then fitted by least squares and subtracted from the orbital magnetic profiles to account for external field effects and to enhance data consistency. Removal of these quadratic functions tends to enhance east-west anomaly components over the north-south components as MAGSAT was essentially a polar-orbiting satellite [Langel et al., 1982]. The resulting data were then averaged within 2° bins to produce the anomalies plotted in Figure 1a.

The 2°-averaged MAGSAT scalar magnetic anomalies cannot be readily compared in terms of the magnetic properties and geometries of lithospheric sources because the data are registered at altitudes that range 120 km about a mean elevation of 400 km over the study area. Furthermore, the anomalies are contaminated by differential inclination, declination, and intensity variations of the geomagnetic field. The attributes of the United Kingdom Institute of Geological Sciences (IGS) 1975 geomagnetic reference field updated to 1980 over the surface of the study area are plotted in Figure 2 to illustrate their variability. Inclinations of the reference field in this region vary from 0° at the magnetic equator to ±90° at the poles, whereas declinations vary from about 51° to -66°. These attitudinal characteristics operate over the geometries of the lithospheric sources and at the observation points to vary the attributes of magnetic anomalies continuously from

---

[1] Department of Geology and Mineralogy and Institute of Polar Studies, The Ohio State University, Columbus, Ohio 43210.
[2] Department of Geosciences, Purdue University, West Lafayette, Indiana 47907.
[3] Institut de Géophysique, Université de Lausanne, 1005 Lausanne, Switzerland.
[4] Department of Geology and Geophysics, University of Wisconsin-Madison, Madison, Wisconsin 53706.

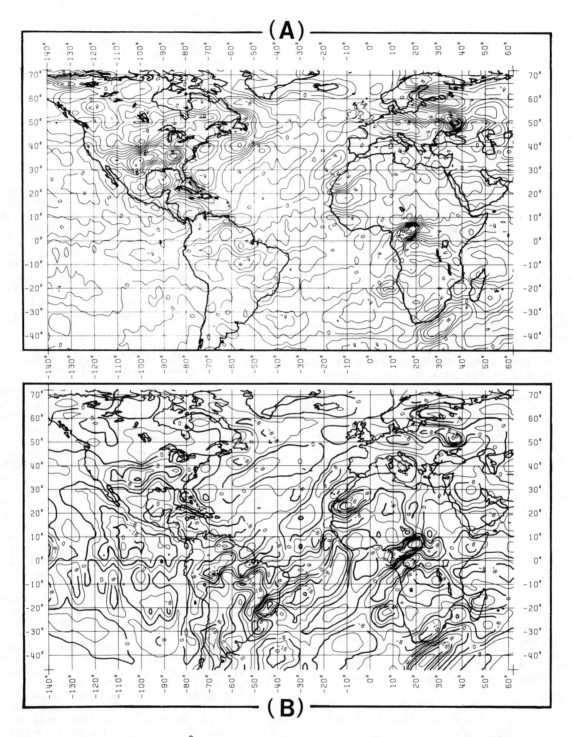

Fig. 1. (a) MAGSAT scalar 2°-averaged magnetic anomalies for the eastern Pacific Ocean, North and South America, the Atlantic Ocean, and Euro-Africa. These total field anomalies are contoured at 2-nT intervals. (b) MAGSAT scalar magnetic anomalies adjusted to a fixed elevation of 400 km and differentially reduced to the pole of 60,000-nT intensity. The radially polarized anomalies are contoured at 4-nT intervals.

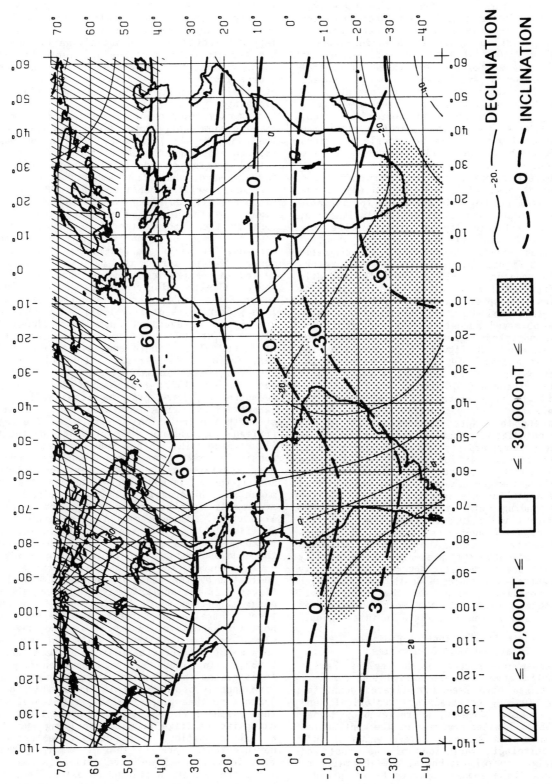

Fig. 2. Attributes of the core field over the surface of the region considered in Figure 1.

the poles to the equator where inversion of the anomaly signs occurs. Assuming induced magnetization, the anomaly amplitudes are a function of the susceptibility of the sources and the ambient field strength which varies as shown in Figure 2 from roughly 23,000 nT to 62,000 nT. Thus, an inductively magnetized source located in central southeastern South America produces an anomaly which is only about one third to one half as strong as the anomaly produced by the same source at high magnetic latitudes. In general, the sign, shape, and strength of the anomalies over the study area in Figure 1a are not simple functions of the magnetic and geometric properties of the lithospheric sources.

An equivalent point dipole inversion scheme [von Frese et al., 1981] was used to adjust the regional satellite magnetic anomalies for variable elevation effects and differential inclination, declination, and intensity variations of the core field. Application of the procedure involved least-squares matrix inversion of the magnetic anomalies of Figure 1a, using the IGS 1975 reference field updated to 1980 to determine magnetic susceptibilities for an array of point dipoles constrained to a spherical surface and spaced on a 4° grid. This produced stable point sources which model the observed anomalies with negligible error. The adjusted magnetic anomalies were then obtained as shown in Figure 1b by computing the anomalies at a fixed elevation (400 km) from the equivalent point dipoles assuming a radial field of constant strength (60,000 nT).

The effects of differential reduction to the radial pole are readily appreciated by comparing Figure 1a and 1b. Note the marked shift of the radially polarized anomalies relative to their total field counterparts along isogonic lines toward the poles. At low geomagnetic latitudes this phase shift is as great as several degrees, whereas along the geomagnetic equator the radially polarized anomalies are reversed in sign relative to corresponding total field anomalies. These effects have a major impact on relating satellite elevation magnetic anomalies with regional lithospheric features. Also, the radially polarized anomalies indicate the presence of relatively strong lithospheric magnetic sources, particularly in east-central and southern South America, which are not readily apparent in the total field anomaly data because of the weak polarization characteristics of the core field in this region.

Geologic Implications of Radially Polarized MAGSAT Anomalies

The lithospheric sources of the regional satellite magnetic anomalies are varied and complex, and only a few have been investigated quantitatively by using constraining geological and geophysical data. However, as reviewed by Mayhew et al. [1985], limited analyses of the continental magnetic anomalies suggest sources which include regional petrologic variations of the crust and upper mantle, crustal thickness, and thermal perturbations. These anomaly sources behave predominantly as inductively magnetized features and include sources whose magnetization may be dominated by viscoremanence. Accordingly, the radially polarized anomalies of the continents are in principle centered over their sources, and source geometric and magnetic property variations are mapped directly by these anomaly variations. This has important implications for tectonic analyses on a continental or global basis, as geologic source regions may be compared directly in terms of their radially polarized magnetic anomaly signatures.

For example, previous investigators have noted, using predominantly total field POGO [Frey et al., 1983] and MAGSAT [Galdeano, 1983] anomalies, the existence of magnetically disturbed regions on the rifted margins of continents which match except for the sign, shape, and amplitude of the anomalies on prerift reconstructions. However, when differentially adjusted to the radial pole as in Figure 1b, the anomalies are sufficiently consistent in these attributes that they may be contoured across or along the rifted continental margins. In fact, the continuity of lithospheric source regions across the rifted margins is suggested in remarkable detail as shown in Figure 3, where radially polarized MAGSAT data are plotted on an Early Cambrian reconstruction of Gondwanaland [after Smith et al., 1981]. Included in Figure 3 are 2°-averaged anomalies reduced to an elevation of 400 km and to a radial pole strength of 60,000 nT for South America, Africa, Madagascar, India, and Australia. For Antarctica, 3°-averaged scalar anomalies, derived by Ritzwoller and Bentley [1983] from MAGSAT vector component measurements, have also been adjusted to these parameters of fixed elevation and radial pole strength for inclusion in Figure 3. The adjacent (rifted) margins of the present continents as illustrated in this Cambrian reconstruction consist primarily of Precambrian age rocks with only spatially restricted superimposed intrusive and extrusive igneous rocks and essentially nonmagnetic Phanerozoic sedimentary rocks which presumably overlie Precambrian basement crystalline rocks. Phanerozoic crust or crust strongly modified by orogenic-related tectonic and thermal processes occurs primarily along the margins of the Gondwanaland supercontinent shown in Figure 3.

The correlation of magnetic source regions indicated by the radially polarized anomalies in Figure 3 is particularly striking along the rifted margins between South America and Africa and between Australia and Antarctica. The most intense positive anomaly of Figure 1b is the Bangui anomaly of west-central Africa which has been related to a major intracrustal feature by Regan and Marsh [1982]. In Figure 3, the Bangui anomaly correlates across the Atlantic rift margin with a positive anomaly of the Sao Luiz Craton which projects farther northeastward (in the grid coordinates of the cylindrical equidistant projection) as an extensive positive anomaly overlying the Central Brazilian Shield. This positive feature extends northwestward back to the rift margin where it ties into a broad positive African anomaly overlying the Cubango Basin and an extensive region of Precambrian rocks. The details of the last correlation are better illustrated in Figure 1b where

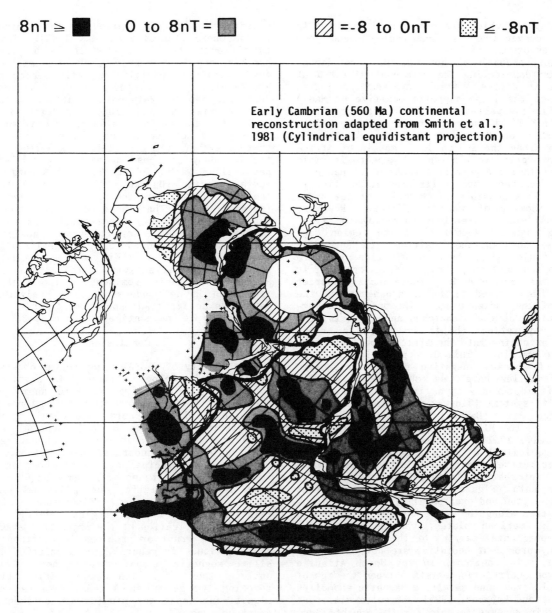

Fig. 3. Radially polarized MAGSAT anomalies plotted on an Aldanian reconstruction of Gondwanaland. The coastlines of Antarctica and Africa are highlighted to facilitate identifying the continental components of Gondwana.

the data are contoured at a finer interval than was used in Figure 3.

Adjacent to the Bangui anomaly on the north is an intense magnetic low (Figure 3) which reaches a minimum at its westernmost limit in Africa over the Zaire Basin. This feature correlates across the Atlantic rift margin with a comparable low overlying the Sao Franciso Craton of southern Brazil. Additional striking associations include the positive anomaly overlying Archean-Proterozoic cratonic blocks in south-central and western Australia which correlates with a pronounced high over Wilkes Land in Antarctica. Also, the magnetic low flanking the large Australian positive anomaly on the north and overlying the Adelaide and Tasman orogens correlates with an Antarctic minimum over the Ross Sea Embayment and Transantarctic Mountains.

In contrast to these associations is the overall poor correlation of magnetic anomalies for the Africa-Madagascar-India-Antarctica fit in Figure 3. The anomaly data suggest problems with the adopted reconstruction if it can be assumed that the anomaly sources actually predate rifting. Verifying this proviso is clearly difficult for regional-scale deep-seated magnetic features of the lithosphere which are not well understood in terms of conventional surface geologic and geophysical evidence.

The interpretational complexities are demonstrated by considering the prominent mismatch of anomalies in Figure 3 which involves the juxtaposition of the large magnetic positive of Madagascar with the well-defined minimum at the African margin. If the sources of the radially polarized anomaly of Madagascar and the continental anomalies along the Indian Ocean margin of Africa were formed prior to breakup, it appears feasible to suggest that the prerift attachment of Madagascar to Africa was close to its present position in Figure 1b as a possible continuation of the broad positive anomaly of southern Africa. However, recent analysis of sea-floor spreading and fracture-zone magnetic anomalies of the region supports the Gondwanaland position of Madagascar which is indicated in the reconstruction of Figure 3 [Martin and Hartnady, 1986]. Accordingly, it is possible that one or both of these anomaly sources may have been formed during or subsequent to the opening of the Indian Ocean. The pronounced magnetic minimum of the mismatched anomaly pair, for example, may reflect the failed arm of a triple junction extending into the African continent from the ocean margin. This interpretation is suggested only because negative radially polarized MAGSAT anomalies have been observed to overlie other rift features such as the Amazon River and Takatu rift systems [Ridgway and Hinze, 1986], the Rio Grande rift [Mayhew, 1985], and the rift structure of the Mississippi River embayment [von Frese et al., 1982]. However, no other geophysical or geological evidence appears available to support or further constrain this hypothesis.

Other interesting features of the radially polarized data (Figure 1b) include an apparent sharp truncation and even parallelism of continental anomalies along the western edges of the North and South American plates, whereas across the passive continental margins of the South Atlantic Ocean many prominent anomalies are not distinctly truncated. In contrast, in the North Atlantic Ocean a good correlation exists between the age of the oceanic crust and satellite magnetic anomalies [LaBrecque and Raymond, 1985]. A quite different relationship appears to hold for the South Atlantic Ocean where the radially polarized anomalies transect the central ridge.

The source of passive continental margin anomalies, which for the South Atlantic Ocean can be traced into the South American continent and Africa for considerable distances, is not immediately obvious. They may be related to external geomagnetic electrojet effects, especially in the region between ±30° inclination of Figure 2. However, they are replicated in hundreds of orbits [Ridgway and Hinze, 1986] to yield the temporally static signatures which typify anomalies of magnetic sources within the lithosphere. In their geologic context, on the other hand, passive continental margin anomalies may be associated with oceanic rises which have continental affinities. A possible example involves the area of the southern tip of Africa where seismic refraction and dredging results indicate continental crust beneath the Agulhas Plateau in the southwestern Indian Ocean [Tucholke et al., 1981]. This area is characterized in Figure 1b by a pronounced positive magnetic anomaly which is consistent with the anomalously positive magnetic source regions geographically associated with the West Antarctic Peninsula and the Patagonian Platform of southern South America as reconstructed in Figure 3.

In the central South Atlantic Ocean, the anomalies trend northeast, but closer to the African shoreline they turn generally to a north-northeast trend. Many of these anomalies show a striking parallelism with the tracks of hotspots in the South Atlantic [Duncan, 1981; Morgan, 1983] and their extensions into the sub-African upper mantle [Morgan, 1983]. That these tracks have no consistent magnetic anomaly sign complicates their interpretation. However, it seems plausible that the hotspots may indeed leave a magnetic signature imprint by virtue of their related thermal aureoles and magmatic activity. The origin of these anomalies and their prominence in the South Atlantic Ocean is not understood, but analyzing them may prove an important source of further clues on the history of the continents and oceans.

Conclusions

The utility of MAGSAT magnetometer observations for regional geologic analysis is significantly enhanced by normalizing magnetic anomalies for global attitude and intensity variations of the magnetic field of the core. The resulting radially polarized anomalies of a reconstructed Gondwanaland show remarkable correlation across present continental boundaries. This strongly suggests that a principal source of these anomalies is the prerift terranes which have acquired their magnetic characteristics during Precambrian tectonic and thermal events. Discrepancies in correlation across rifted margins reflect rift or postrift modification of the magnetic characteristics of the crust or problems in continental reconstruction. In general, the resolution of satellite magnetic anomalies and the capacity to analyze them for lithospheric information is reaching the stage where these data can provide significant input for studies of continental reconstructions.

Acknowledgments. Financial support for this investigation was provided by the Goddard Space Flight Center under NASA contract NAG5-304 and by grant DPP-8313071 from the National Science Foundation. This is contribution number 567 of the Institute of Polar Studies, Ohio State University, Columbus, Ohio.

References

Duncan, R. A., Hotspots of the southern oceans--An absolute frame of reference for motion of the Gondwana continents, Tectonophysics, 74, 29-43, 1981.
Frey, H., R. A. Langel, G. Mead, and K. Brown, POGO and Pangea, Tectonophysics, 95, 181-189, 1983.
Galdeano, A., Acquisition of long wavelength mag-

netic anomalies predates continental drift, Phys. Earth Planet. Inter., 32, 289-292, 1983.

LaBrecque, J. L., and C. A. Raymond, Sea-floor spreading anomalies in the Magsat field of the North Atlantic, J. Geophys. Res., 90, 2565-2575, 1985.

Langel, R. A., J. D. Phillips, and R. J. Horner, Initial scalar magnetic anomaly map from MAGSAT, Geophys. Res. Lett., 9, 269-272, 1982.

Martin, A. K., and C. J. H. Hartnady, Plate tectonic development of the southwest Indian Ocean: A revised reconstruction of East Antarctica and Africa, J. Geophys. Res., 91, 4767-4786, 1986.

Mayhew, M. A., Curie isotherm surfaces inferred from high-altitude magnetic anomaly data, J. Geophys. Res., 90, 2647-2654, 1985.

Mayhew, M. A., B. D. Johnson, and P. J. Wasilewski, A review of problems and progress in studies of satellite magnetic anomalies, J. Geophys. Res., 90, 2511-2522, 1985.

Morgan, W. J., Hotspot tracks and the early rifting of the Atlantic, Tectonophysics, 94, 123-139, 1983.

Regan, R. D., and B. D. Marsh, The Bangui magnetic anomaly: Its geological origin, J. Geophys. Res., 87, 1107-1120, 1982.

Ridgway, J. R., and W. J. Hinze, MAGSAT scalar anomaly map of South America, Geophysics, 51, 1472-1479, 1986.

Ritzwoller, M. H., and C. R. Bentley, Magnetic anomalies over Antarctica measured from MAGSAT, in Antarctic Earth Science, edited by R. L. Oliver, P. R. James, and J. B. Jago, pp. 504-507, Australian Academy of Science, Canberra, 1983.

Smith, A. G., A. M. Hurley, and J. C. Briden, Phanerozoic Paleocontinental World Maps, 102 pp., Cambridge University Press, New York, 1981.

Tucholke, B. E., R. E. Houtz, and D. M. Barrett, Continental crust beneath the Agulhas Plateau, southwest Indian Ocean, J. Geophys. Res., 86, 3791-3806, 1981.

von Frese, R. R. B., W. J. Hinze, and L. W. Braile, Spherical earth gravity and magnetic anomaly analysis by equivalent point source inversion, Earth Planet. Sci. Lett., 53, 69-83, 1981.

von Frese, R. R. B., W. J. Hinze, and L. W. Braile, Regional North American gravity and magnetic anomaly correlations, Geophys. J. R. Astron. Soc., 69, 745-761, 1982.

Copyright 1987 by the American Geophysical Union.

A REVISED RECONSTRUCTION OF GONDWANALAND

Lawrence A. Lawver and Christopher R. Scotese

Institute for Geophysics, University of Texas, Austin, Texas 78751

Abstract. A revised reconstruction of Gondwana is presented. It is based on previous reconstructions that were geologically well constrained and utilizes marine magnetic anomalies and recent paleomagnetic results as additional constraints. Three areas of conflict, namely, the position of Madagascar, the location of India-Sri Lanka, and the overlap of the Antarctic Peninsula with the Falkland Plateau are discussed. An interactive graphics terminal was used to minimize continental overlap while reducing obvious gaps between pre-breakup components. Poles and angles of rotation were determined directly from the terminal for optimal geometric fit.

Reconstruction

Du Toit [1937] proposed a reasonable model for the reconstruction of the southern continents based on geological lineations. As our knowledge of the earth has increased, aided by the discovery and correlation of marine magnetic anomalies, models of Gondwana have evolved. In our reconstruction, we have used marine magnetic anomalies to position the major pieces of Gondwana and have utilized an interactive graphics terminal to then produce the best fit and to generate the poles of rotation and angles of rotation. We have relied on previous reconstructions for those parts of the southern supercontinent that can be well constrained both geologically and with magnetic anomaly data. We have also taken into consideration continental stretching that precedes initial rifting to produce a closer fit. Our revised reconstruction bears a striking resemblance to that presented by du Toit [1937].

Three major areas have produced controversy regarding reconstructions of Gondwana. These are the location of Madagascar with respect to Africa, the fit of Sri Lanka and India with East Antarctica, and the location of the Antarctic Peninsula with respect to the Falkland Plateau (both recognized continental fragments). Recent identifications of marine magnetic anomalies in the Somali Basin have confirmed the northern fit of Madagascar with the Somali-Kenya coast of northeast Africa [Segoufin and Patriat, 1980; Rabinowitz et al., 1983]. Rotation of Sri Lanka closed with southern India [Katz, 1978] produces a prebreakup configuration of India-Sri Lanka that fits tightly into the Lützow-Holm Bay region of Enderby Land and a remarkable fit of the Indian coastline with Enderby Land, East Antarctica (Figure 1). The closures of South America with Africa [Ladd, 1974; Rabinowitz and LaBrecque, 1979], and Australia with East Antarctica, have generated only minor controversy and are used with only slight modifications in our reconstruction.

A reasonable fit of Africa-Madagascar-India-Sri Lanka with East Antarctica is produced by using the marine magnetic anomalies identified in the Somali Basin, in the Mozambique Basin [Segoufin, 1978; Simpson et al., 1979], and off the coast of East Antarctica [Bergh, 1977] as constraints on the closure of Africa-Madagascar to East Antarctica, and suggested lineaments between Madagascar and India [Katz and Premoli, 1979] as well as the above mentioned marine magnetic anomalies in the Indian Ocean [McKenzie and Sclater, 1971]. Further evidence for the position of India-Sri Lanka in Gondwana is given by the juxtaposition of charnockite localities in southern India and Enderby Land [Grew, 1982a, b]. Grindley and Davey [1982] presented a comprehensive reconstruction of New Zealand-Australia-Antarctica which we have used with only minor modifications. Consequently, in the closure of South America-Africa-Madagascar with India-Sri Lanka-East Antarctica-Australia and the fit of New Zealand-Australia-Antarctica, the only circumpolar pieces of Gondwana that are not well constrained are the pieces of West Antarctica.

Our pole of rotation for Australia fitted to Antarctica is different from those of Norton and Molnar [1977] and Griffiths [1974] but not quite as extreme as the McKenzie and Sclater [1971] pole that simply relied on a computer-generated least-squares fit of the coastlines. Griffiths [1974] attempted to align the Bowers group of north Victoria Land with the Dundas Trough of Tasmania, but more recent workers [Cooper et al., 1983] have concluded that there is insufficient geological evidence to require an exact paleoalignment. By shifting the Australia fit slightly clockwise with respect to Antarctica, the Tasman Rise can be treated as a relatively rigid piece of the Australian plate and does not have to be rotated with respect to Australia as it does in the Norton and Molnar [1977] and Griffiths [1974] models.

Reasonable assumptions regarding the partial

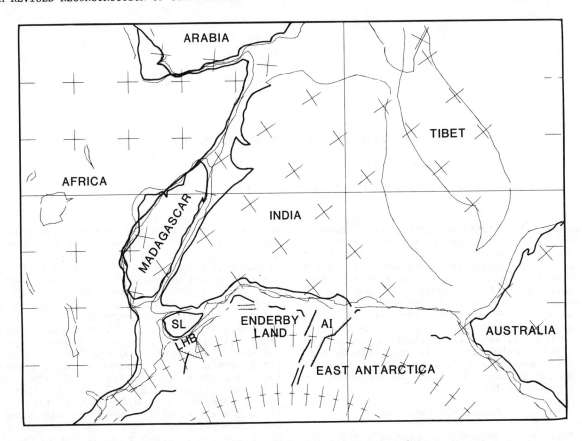

Fig. 1. Reconstruction of Africa-Madagascar-India-Sri Lanka-East Antarctica. Africa is held fixed in a present-day reference frame. On continental fragments other than East Antarctica, a heavy line is the present-day coastline. On East Antarctica, the heavy line represents the northernmost extent of basement outcrops or the boundary of the Transantarctic Mountains taken from Drewry and Jordan [1983]. The light line is a 2000-m contour, and 5- by 5-degree grid marks are prerotation present-day latitude and longitude. Where the present-day extent of continental block is unknown (i.e., the northern boundary of India), the continental outline is left as a light line. SL is Sri Lanka; LHB, Lützow-Holm Bay; AI, Amery Ice Shelf.

closure of Marie Byrd Land (West Antarctica with East Antarctica) can be made on the basis of assumed crustal stretching in such areas as the Marie Byrd Basin and the Ross Sea area. The bedrock surface map of Antarctica [Drewry and Jordan, 1983], as well as the Deep Sea Drilling Project results [Hayes, et al., 1975], both give the impression that the area beneath the Ross Sea and the Ross Ice Shelf and extending to the Ellsworth Mountains region is an area of horsts and grabens vaguely similar in appearance to the Basin and Range province of the western United States. Doake et al. [1983] and Garrett et al. [this volume] present data that strongly suggest that the subsurface ice topography is produced tectonically and not by ice movement. We assume that the crust between East Antarctica and Marie Byrd Land has been stretched to about 40-50%, keeping Marie Byrd Land in the same general relationship with East Antarctica that it presently is.

Identified marine magnetic lineations in the Tasman Sea were used to rotate Lord Howe Rise-North Island of New Zealand closed to Australia [Weissel et al., 1977]. Anomalies in the southwest Pacific can be used to close Campbell Plateau-South Island of New Zealand with Marie Byrd Land [Molnar et al., 1975]. By assuming the partial closure of the Ross Sea, the revised position of Australia with respect to Antarctica, and the closure of North New Zealand-Lord Howe Rise with Australia, the South Island New Zealand-Chatham Rise-Campbell Plateau fits into the space available (Figure 2). To restore the Campbell magnetic anomaly on the Campbell Plateau to the Stokes magnetic anomaly on South Island requires the closure of the Bounty Trough between the Chatham Rise and the Campbell Plateau as Grindley and Davey [1982] have suggested. Such a closure produces a tight fit between the Chatham Rise-Campbell Plateau and Marie Byrd Land similar to that proposed by Grindley and Davey [1982].

The above constraints on the reconstructions of Gondwana leave only a small area into which the remaining pieces of West Antarctica must fit,

ka, and East Antarctica have remained in
roximately the same configuration. The fit of
tralia to East Antarctica and South America to
ica are easily recognizable. It is interesting
t the reconstruction of Gondwana has not been
atly revised in the past 50 years.
Our revised reconstruction of Gondwana places
straints on the area in which the pieces of
t Antarctica can be placed. The paleomagnetic
ults of Grunow et al. [this volume] indicate
t there may be a major break between the pen-
ula and Marie Byrd Land, since our reconstruc-
n indicates that Marie Byrd Land cannot have
n rotated in the simple fashion that is sug-
ted for the peninsula and the Ellsworth and
tmore Mountains. Further work, particularly in
region between the Antarctic Peninsula and
ie Byrd Land, is needed.

Acknowledgments. This work was supported by
grant 84-05968 to the Institute for Geophysics
by contributions to the Paleoceanographic
ping Project. This is the University of Texas,
titute for Geophysics Contribution number 666.
wish to thank I. Dalziel, D. Sandwell, J.
ater, and T. Shipley as well as one anonymous
iewer for reading and commenting on this
uscript.

References

ker, P. F., et al., Evolution of the southwestern Atlantic Ocean basin: Results of Leg 36,
eep Sea Drilling Project, Initial Rep. Deep Sea
rill. Proj. 1969, 36, 993-1014, 1976.
ron, E. J., C. G. A. Harrison, and W. W. Hay, A
evised reconstruction of the southern conti-
ents, EOS Trans. AGU, 59, 436-449, 1978.
gh, H. W., Mesozoic seafloor off Dronning Maud
and, Antarctica, Nature, 269, 686-687, 1977.
oper, R. A., J. B. Jago, A. J. Rowell, and P.
raddock, Age and correlation of the Cambrian-
rdovician Bowers Supergroup, north Victoria
and, Antarctic Earth Sciences, edited by R. L.
liver, P. R. James, and J. B. Jago, pp. 128-
31, Australian Academy of Sciences, Canberra,
983.
lziel, I. W. D., Comments on Mesozoic evolution
f the Antarctic Peninsula and the southern
ndes, Geology, 8, 260-261, 1980.
lziel, I. W. D., and D. H. Elliot, West Antarc-
ica: Problem child of Gondwanaland, Tectonics,
, 3-19, 1982.
Wit, M. J., The evolution of the Scotia Arc as
key to the reconstruction of southwestern
ondwanaland, Tectonophysics, 37, 53-81, 1977.
etz, R. S., and J. B. Holden, Reconstruction of
angea: Breakup and dispersion of continents,
ermian to Present, J. Geophys. Res., 75, 4939-
956, 1970.
ake, C. S. M., R. D. Crabtree, and I. W. D.
alziel, Subglacial morphology between Ellsworth
ountains and Antarctic Peninsula: New data and
ectonic significance, in Antarctic Earth Sci-
nces, edited by R. L. Oliver, P. R. James,
nd J. B. Jago, pp. 270-273, Australian Academy
f Sciences, Canberra, 1983.

Drewry, D. J., and S. R. Jordan, The bedrock surface geology of Antarctica, in Sheet 3 of Antarctica: Glaciological and Geophysical Folio,
edited by D. J. Drewry, Scott Polar Research
Institute, Cambridge, England, 1983.
du Toit, A. L., Our Wandering Continents, an Hypothesis of Continental Drifting, 366 pp., Oliver
and Boyd, Edinburgh, 1937.
Garrett, S. W., L. D. B. Herrod, and D. R. Mantripp, Crustal structure of the area around Haag
Nunataks, West Antarctica: New aeromagnetic and
bedrock elevation data, this volume.
Grew, E., The Antarctic margin, in The Ocean
Basins and Margins, edited by A. E. M. Nairn
and F. G. Stehli, pp. 697-775, Plenum, New York,
1982a.
Grew, E., Sapphirine-bearing rocks in Antarctica,
south India, Madagascar, and southern Africa
(abstract), Volume of Abstracts, for Fourth
International Symposium on Antarctic Earth Sciences, pp. 73, Adelaide University, Adelaide,
Australia, 1982b.
Griffiths, J. R., Revised continental fit of Australia and Antarctica, Nature, 249, 336-338,
1974.
Grindley, G. W., and F. J. Davey, The reconstruction of New Zealand, Australia, and Antarctica,
in Antarctic Geoscience, edited by C. Craddock,
pp. 15-29, University of Wisconsin Press, Madison, 1982.
Grunow, A. M., I. W. D. Dalziel, and D. V. Kent,
Ellsworth-Whitmore Mountains crustal block, West
Antarctica: New paleomagnetic results and their
tectonic significance, this volume.
Hayes, D. E., et al., Initial Rep. Deep Sea
Drill. Proj. 1969, 28, pp. 3-942, 1975.
Katz, M. B., Sri Lanka in Gondwanaland and the
evolution of the Indian Ocean, Geol. Mag., 115,
237-244, 1978.
Katz, M. B., and C. Premoli, India and Madagascar
in Gondwanaland based on matching Precambrian
lineaments, Nature, 297, 312-315, 1979.
Ladd, J. W., South Atlantic sea-floor spreading
and Caribbean tectonics, Ph.D. thesis, 251 pp.,
Columbia University, New York, 1974.
Lawver, L. A., J. G. Sclater, and L. Meinke,
Mesozoic and Cenozoic reconstructions of the
South Atlantic, Tectonophysics, 114, 233-254,
1985.
Longshaw, S. K., and D. H. Griffiths, A paleomag-
netic study of Jurassic rocks from the Antarctic
Peninsula and its implications, J. Geol. Soc.
London, 140, 945-954, 1983.
McKenzie, D., and J. G. Sclater, The evolution of
the Indian Ocean since the Late Cretaceous,
Geophys. J. R. Astron. Soc., 25, 437-528, 1971.
Molnar, P., T. Atwater, J. Mammerickx, and S. M.
Smith, Magnetic anomalies, bathymetry, and the
tectonic evolution of the South Pacific since
the Late Cretaceous, Geophys. J. R. Astron.
Soc., 40, 383-420, 1975.
Norton, I. O., and P. Molnar, Implications of a
revised fit between Australia and Antarctica for
the evolution of the eastern Indian Ocean,
Nature, 267, 338-339, 1977.
Norton, I. O., and J. G. Sclater, A model for the
evolution of the Indian Ocean and the breakup of

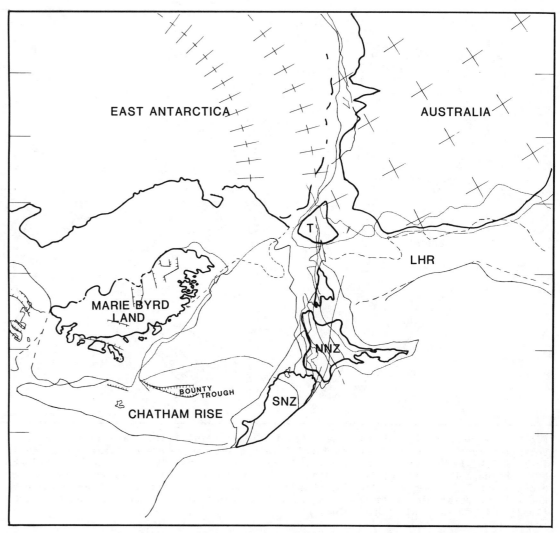

Fig. 2. Reconstruction of New Zealand-Marie Byrd Land-East Antarctica-Australia
modified from Grindley and Davey [1982]. Continental outlines are the same as in Figure
1. The dashed light line is the presumed continental part of Lord Howe Rise (LHR). T
is Tasmania; SNZ, southern New Zealand; NNZ, northern New Zealand.

assuming that they are not allochthonous. Previous reconstructions of Gondwana have resorted to
three different solutions with regard to the fit
of the pieces of West Antarctica. The first solution has been to ignore the problem and to simply
show the Antarctic Peninsula overlapping the Falkland Plateau [Dietz and Holden, 1970; Smith and
Hallam, 1970; Norton and Sclater, 1979]. An alternative solution has been to place the Antarctic
Peninsula on the Pacific side of South America
[Barron et al., 1978]. While there are no marine
magnetic data to dispute this conclusion, geological data seem to imply that the westward facing
tip of South America fronted an active subduction
zone [Dalziel, 1980] during the period 120-165 Ma,
the period for which an overlap must be considered
a problem [Lawver et al., 1985]. The third solution has involved movement of the peninsula with
respect to East Antarctica [de Wit, 1977; Barker
et al., 1976; Dalziel and Elliot, 1982]. From
data presented by Longshaw and Griffiths [1983]
and Grunow et al. [this volume] concerning the
paleomagnetic position for the pieces of West
Antarctica, constraints can be placed on the
amount of rotation as well as the paleolatitude
that both the Ellsworth-Whitmore Mountains crustal
block and the Antarctic Peninsula could have
undergone if the paleomagnetic results are strictly adhered to. While our revised version of
Gondwana does not use the new poles determined by
Grunow et al. [this volume], we do use their conclusion that the Antarctic Peninsula and the Ellsworth Mountains-Whitmore Mountains were roughly in
the same position relative to each other during
the Jurassic as they are now. We have moved the
Ellsworth Mountains block and the Whitmore Moun-

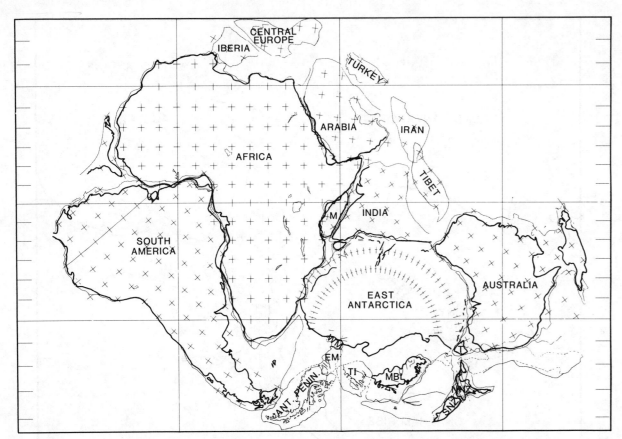

Fig. 3. Reconstruction of Gondwana. Continental outline is the same as in Figure 1. Poles for reconstruction to Africa fixed in a present-day reference frame are given in Table 1. WM is Whitmore Mountains; EM, Ellsworth Mountains; TI, Thurston Island block; NNZ, northern New Zealand; SNZ, southern New Zealand attached to Campbell Plateau; M, Madagascar; T, Tasmania. The rotation of the Antarctic Peninsula/Ellsworth-Whitmore Mountains crustal blocks is based on the paleomagnetic work of Longshaw and Griffiths [1983] and Grunow et al. [this volume].

tains block slightly closer together on the basis of the recent work of Doake et al. [1983] and Garrett et al. [this volume]. The prebreakup locations of the pieces of West Antarctica with respect to Gondwana, with the exception of Marie Byrd Land, are tentative, so they are shown with a light line to distinguish them from the pieces that are constrained by marine magnetic anomalies. Other pieces, particularly those north of the main body of Gondwana, are also shown with a light line, since marine magnetic anomaly identifications are again not available to constrain them in a prebreakup configuration.

Our revised reconstruction is shown in Figure 3. We have kept Africa fixed and plotted the reconstructed pieces in a Mercator projection. The advantages of plotting the pieces fixed to Africa and using a Mercator projection are twofold. First, the continents remain recognizable to those accustomed to seeing the world plotted in a Mercator projection. Second, the pieces that are not controversial are plotted without much enlargement, while the more poorly positioned pieces of New Zealand and West Antarctica are enlarged so that some of the controversial areas are easier to decipher. We feel that the major problems in our revised reconstruction of Gondwana are the placement of the pieces of West Antarctica and the relative position of the whole assemblage. Table 1 lists the poles of rotation that we used to produce the reconstruction shown in Figure 3.

Since we choose to have Africa remain fixed, the poles are either listed as rotations of the pieces to Africa or are listed as being rotated to another plate, fixed to that plate, and then the package of plates rotated to Africa. To change the location of Gondwana with respect to the present reference frame, it would only be necessary to change the rotation pole for Africa.

Du Toit's [1937] reconstruction of Gondwana (Figure 4) stands up remarkably well with time; the only obvious change is the location of Antarctica with respect to Africa. Otherwise, the location of Madagascar with respect to Africa has not changed, although its location has produced debate during the last 20 years, and India, Sri

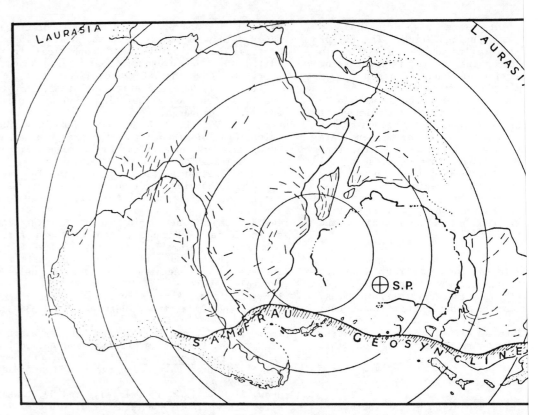

Fig. 4. Reproduction of Figure 7 in Our Wandering Continents [du Toit, 1937], the reassembly of Gondwana during the Paleozoic era. The space between the portions was then mostly land. Short lines indicate the Precambrian or Early "grain." Diagonal rules show the "Samfrau" Geosyncline of the late Paleozoic. marks indicate our regions of Late Cretaceous and Tertiary compression, acco Lambert's equal area polar projection.

TABLE 1. Rotation Poles for Jurassic Gondwana Fit

Plate	Latitude	Longitude	Angle	With Respect to Plate
Africa	90.0	0.0	0.0	Held Fixed
South America	45.5	-32.2	58.2	Africa
India	-4.44	16.74	-92.77	East Antarctica
Sri Lanka	-13.67	31.11	-107.14	East Antarctica
Australia	-1.58	39.02	-31.29	East Antarctica
Madagascar	-3.41	-81.70	19.73	Africa
Antarctic Peninsula	73.87	108.59	-41.79	East Antarctica
Ellsworth Mountains	72.64	100.37	-37.44	East Antarctica
Whitmore Mountains	72.55	97.64	-39.73	East Antarctica
Thurston Island	62.27	21.84	13.27	East Antarctica
Marie Byrd Land	62.27	21.84	13.27	East Antarcitca
North New Zealand	24.19	-19.91	44.61	Australia
South New Zealand	65.14	-52.00	62.38	Marie Byrd Land
Chatham Rise	41.00	-15.90	7.47	South New Zealand
East Antarctica	-7.78	-31.42	58.0	Africa
Florida	57.00	-20.80	88.90	Africa
Central Europe	64.3	147.3	20.6	Africa
Iberia	50.0	3.3	-27.0	Central Europe
Arabia	26.5	21.5	-7.6	Africa
Iran/Turkey	2.02	36.9	-24.6	Arabia
Tibet	-43.0	0.5	8.0	India

Gondwanaland, J. Geophys. Res., 84, 6803-6830, 1979.

Rabinowitz, P. D., and J. L. LaBrecque, The Mesozoic south Atlantic Ocean and evolution of its continental margins, J. Geophys. Res., 84, 5973-6002, 1979.

Rabinowitz, P. D., M. F. Coffin, and D. Falvey, The separation of Madagascar and Africa, Science, 220, 67-69, 1983.

Segoufin, J., Anomalies magnetiques mesozoiques dans le bassin de Mozambique, C. R. Sceances Acad. Sci. Ser. 2., 287D, 109-112, 1978.

Segoufin, J., and P. Patriat, Existence d'anomalies mesozoiques dans le bassin de Somalie, Implications pour les relations Afrique-Antarctique-Madagascar, C. R. Sceances Acad. Sci. Ser. 2, 291B, 85-88, 1980.

Simpson, E. S. W., J. G. Sclater, B. Parsons, I. O. Norton, and L. Meinke, Mesozoic magnetic lineations in the Mozambique Basin, Earth Planet. Sci. Lett., 43, 260-264, 1979.

Smith, A. G., and A. Hallam, The fit of the southern continents, Nature, 225, 139-144, 1970.

Weissel, J. K., D. E. Hayes, and E. M. Herron, Plate tectonics synthesis: The displacements between Australia, New Zealand, and Antarctica since the Late Cretaceous, Mar. Geol., 25, 231-277, 1977.

Copyright 1987 by the American Geophysical Union.

THE HANDLER FORMATION, A NEW UNIT OF THE ROBERTSON BAY GROUP, NORTHERN VICTORIA LAND, ANTARCTICA

Thomas O. Wright

Earth Sciences Division, National Science Foundation, Washington, D.C. 20550

Colin Brodie

Geology Department, Otago University, Dunedin, New Zealand

Abstract. The geology of northern Victoria Land provides a critical link between Australia and the rest of the Paleozoic margin of Gondwanaland. The Robertson Bay Group occupies a large area along the Ross Sea, and new information allows the stratigraphic subdivision of the group. The Handler Formation is defined as a sequence of quartzose sandstones, gray slate with minor red slate, and pebbly mudstone and megaconglomerate that constitutes the upper stratigraphic part of the Robertson Bay Group. Its lower contact is gradational with the lower part of the Robertson Bay Group, which remains undivided, and is distinguished from it primarily by the presence of red slate and pebbly mudstone beds. At the type section, Handler Ridge (72°30'S, 167°E), the top of the sequence is truncated by a thrust fault that places the Bowers Supergroup on top structurally. The age of the Handler Formation is constrained by the presence of fossils of youngest Cambrian/oldest Ordovician age in limestone olistoliths. This maximum age applies to the upper part of the Handler Formation; the age of the lower part and the underlying Robertson Bay Group is not well constrained paleontologically. The Handler Formation rocks are strongly folded and cleaved; total thickness is estimated at several thousand meters. The Handler Formation also occurs in a strip along the east side of the Victory Mountains for approximately 70 km north of Handler Ridge and in the Mirabito and Admiralty ranges.

Introduction

Robertson Bay rocks were originally described by Rastall and Priestley [1921]. They visited the Robertson Bay area where folded graywacke and slate beds occur (Figure 1). Harrington et al. [1964] formally established the group and proposed a type section near Cape Hallett. Sturm and Carryer [1970] extended the group's known location westward to several ranges, and this distribution has been further refined by more recent work [e.g., Tessensohn et al., 1981; Stump et al., 1983]. These, and other pre-Devonian rocks of northern Victoria Land, were folded and cleaved during the Ross Orogeny, thought to be of Cambro-Ordovician age, and by younger events.

Wright [1981] and Field and Findlay [1983] studied the sedimentology of the Robertson Bay Group and concluded that most of the sequence was basin plain to outer fan facies turbidites and hemipelagic slate. The presence of red beds and coarser sediments has been noted, but the Robertson Bay Group rocks are predominantly gray medium- to fine-grained sandstones and dark gray slates. The sequence has been interpreted as being derived from continental sources [Wright, 1985], based on the high percentage of quartzose metamorphic rock fragment clasts and the lack of volcanic rock fragments.

The age of the Robertson Bay Group is poorly constrained because of a lack of fossils, except for trace feeding burrows that imply latest Precambrian age or younger. During the 1982/1983 season, fossils were found in a sequence of rocks of uncertain affinity on Handler Ridge. The fossils came from limestone blocks within a debris-flow deposit and gave a maximum age of the limestone of uppermost Cambrian/lower Ordovician [Wright et al., 1984; Burrett and Findlay, 1984]. A whole-rock phyllite K-Ar age from the fossil site of 477 Ma [Adams and Kreuzer, 1984] provides a minimum age for the matrix slate, assuming no excess argon is present in detrital grains or from other sources. However, K-Ar whole-rock phyllite ages from Robertson Bay Group rocks show scatter between 455 and 505 Ma. With this 50 m.y. spread, it is likely that these are composite ages, and the possibility of inherited or acquired excess argon cannot be discounted. Wright and Findlay [1984] discussed the possible affinities of these rocks and concluded that they may be the upper part of the Robertson Bay Group and may or may not be correlative with rocks of the same age in the Bowers Supergroup, which occurs to the west of the Robertson Bay Group across the Leap Year Fault. They favored correlation to the Bowers rocks, but due to the reconnaissance nature of the work, these conclusions remained tentative.

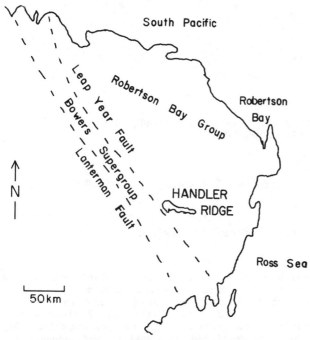

Fig. 1. Sketch map of northern Victoria Land showing location of Handler Ridge.

During the 1984/1985 season, Handler Ridge (Figure 2) was studied in detail. Additional fossils were found, and structures were mapped in order to resolve some of the uncertainties surrounding these rocks. These data are reported here and lead to the conclusion that the sequence is the upper part of the Robertson Bay Group and that they do not correlate with any of the Bowers Supergroup rocks. It is proposed that a new formation, the Handler Formation, be erected with the type section at Handler Ridge.

The Handler Formation rocks are folded about northwest axes, and folds have shallow plunges except near thrust faults. Axial planar cleavage is well developed, yet primary sedimentary features are preserved. For approximately 8 km from the eastern end of Handler Ridge (Figure 2) the folds are west-vergent asymmetric folds with overturned west limbs. Folds expose higher stratigraphic sections from east to west, with the fossil-containing unit preserved in the core of the westernmost major syncline in the sequence. To the west of this syncline, rocks face east and are on the western limb of the syncline. Within 2 km west of the fossil site, minor folds in the Handler sequence increase in angle of plunge from less than 20° to near vertical. Stratal disruption occurs in zones in which graywacke beds are boudinaged and cleavage intensity increases. Four kilometers west of the fossil site the sequence is abruptly truncated by a thrust fault, marked by a shear zone at least 100 m wide. Stretching lineations and C and S bands indicate reverse dip-slip movement, with a west over east sense.

Using these structural relationships to help determine original stratigraphy, a clear pattern emerges based on superposition, despite a lack of fossil control. Structurally lowest in Handler Ridge are the rocks exposed near the eastern tip. These are primarily quartzose sandstones about 10-50 cm thick alternating with dark gray slate. The beds are medium-fine grained, commonly show grading and do not contain coarse material other than occasional rip-up clasts. Beds are laterally persistent, and little or no bed amalgamation was noted. Red slate beds are present, but are rare. Tracks and trails were not observed. Except for the presence of red beds, these rocks are identical to the Robertson Bay rocks that widely occur to the east and north. Stratigraphically above these beds, several gradational lithologic changes occur. Red-colored fine-grained beds become more prominent and thicker. The grain size of these red beds varies from siltstone to very fine slate. In many red beds a lateral change in color to light green was observed, sometimes accompanied by a small grain size change. Soft sediment slump folds were observed, particularly in the finest-grain-sized beds. In the same stratigraphic unit, a distinctive type of deposit appears and becomes more prominent upward. This is the presence of pea-sized and larger well-rounded clasts of white quartz and, less commonly, quartz-rich rock fragments. These appear as beds up to several meters in thickness or as discontinuous streaks in sandstone. Most commonly, these pebbles and cobbles are matrix supported by a fine- to medium-grained sandy matrix. While not well sorted, the clasts are exceptionally well rounded. These are rarely clast-supported conglomerates, and volumetrically, the gravel-sized clasts are never dominant, yet their presence is quite distinctive.

At the top of the section a thick unit occurs that is exposed in the core of the westernmost syncline. This unit is primarily dark gray slate and gray sandstone, but bedding is not clearly developed. The most distinctive feature is the chaotic occurrence of blocks of limestone, quartz pebble conglomerate, and medium to fine quartz sandstone. Blocks range in size from a few centimeters to 7 m in diameter. Sorting is poor, except that the largest blocks appear to co-occur in patches, with smaller boulders and small pebbles occurring between. The matrix slate has many small-scale folds with variable plunge. Detached fold hinges are common. Wright and Findlay [1984] interpreted this deposit as a submarine debris flow, probably in a marine channel. They saw only the edge of this deposit, however, and did not observe the full range of features.

The composition of these deposits indicates that they were derived from quartz-rich continental rocks and their sedimentary cover. No significant volcanic rock fragments were noted. Even after considerable searching, no plutonic or volcanic blocks were found in the channel deposit, and thin sections of the fine-sand matrix and gray sandstone beds yielded low percentages of volcanic clasts.

Thickness estimates for these units forming the Handler Formation are based on incomplete sections separated by either snow or rubble-covered areas

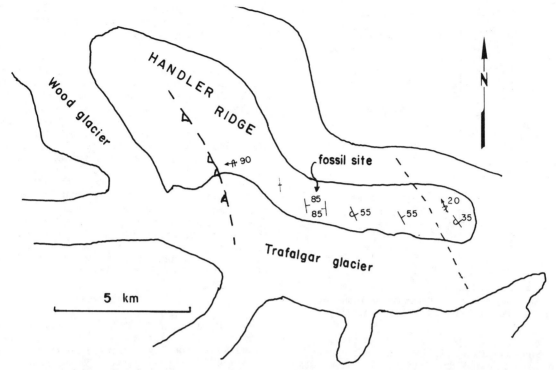

Fig. 2. Index map of Handler Ridge. A thrust fault separates Bowers Supergroup rocks (upper plate) from the Handler Formation east of the Wood Glacier. The dashed line on the east end of Handler Ridge indicates gradational contact between the Handler Formation and undivided lower Robertson Bay Group rocks.

and complicated by structure. The lower Robertson Bay-like unit is approximately 1000-1500 m thick, the red slate and quartz pebble-containing unit is 500-1000 m thick, and the channel deposit is 500+ m thick. These characteristics are summarized and presented in a schematic stratigraphic column in Figure 3.

The Handler Formation also occurs in ridges to the north, at least as far as Turret Ridge. Red slate beds very similar to the ones at Handler Ridge have been reported from the Mirabito and Admiralty ranges. Thus it is likely that the Handler Formation extends considerably further than presently mapped.

Discussion

The characteristics of the Handler sequence provide strong evidence supporting the conclusion that it is a part of the Robertson Bay Group. No fault separating the sequence from the Robertson Bay rocks was found, and the development, style, and orientation of folds and cleavage are identical to that found in outcrops of the Robertson Bay Group in ranges to the north and east of Handler Ridge. The sedimentary features, including the clast composition of the sandstone beds, also are entirely compatible with this conclusion.

The stratigraphy developed for the Handler Formation defines an upward coarsening sequence, with the introduction of quartz gravel and, at the top of the exposed section, large olistoliths of various lithologies. The matrix material, however, does not change significantly from the lower Robertson Bay; it is quartzose, medium-grained gray sandstone and dark gray slate. While macrofossils are not present in the matrix of any Robertson Bay Group rocks including the Handler Formation, trace fossils, especially bed-parallel branching burrows, increase in abundance upward in the Handler Formation. The coarse material appears to be in debris flows; therefore the upward coarsening does not necessarily imply shallowing.

The simultaneous appearance of the well-rounded quartz pebbles, in the form of individual grains and as blocks of conglomerate, limestone blocks and red slate beds, may imply a common origin. The limestone blocks show clear evidence of shallow-water derivation. In addition to abundant fossils, they contain alga, oolites, rip-up clasts, and other features that leave little doubt that these originally were deposited on a shallow stable shelf, free of first-cycle debris. One large limestone boulder contained a solution cavity filled with well-rounded coarse quartz conglomerate, identical to the blocks of conglomerate in the same debris flow. This evidence of uplift and karst development provides a model to explain the origin of both the conglomerate and the red slates. Uplift of a shallow carbonate tract would result in karst and development of a terra rosa soil horizon. The rounding of the quartz pebbles

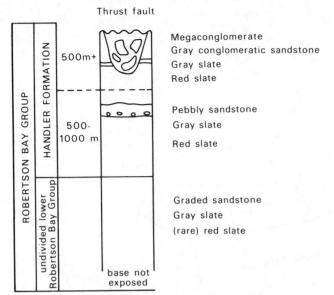

Fig. 3. Generalized stratigraphic column for the Handler Ridge Formation at the type locality.

in the conglomerates implies several previous cycles of erosion and deposition, an interpretation supported by the extreme mineralogic maturity. Transport of elements of this stable, probably low-topography material over the shelf edge to deeper water would account for the observed phenomena. The blocks of medium- to fine-grained gray sandstone in the debris flow are interpreted as blocks of previously deposited Robertson Bay sandstones exposed higher on the continental slope.

The significance of these new observations are as follows:

1. The Robertson Bay Group is at least as young as the Cambrian/Ordovician boundary and may be younger. This suggests the need to reevaluate the age of the Ross Orogeny in northern Victoria Land, which has been thought to be of this age.

2. Because of stratigraphic continuity, a significant part of the group may be Cambrian in age, although the older limit is unconstrained. The presence of trace fossils in most of the Robertson Bay Group may indicate that these rocks are no older than Vendian.

3. The Handler Formation, the upper part of the group, is defined on the appearance of red slate beds and well-rounded pea-sized quartz gravel, either as separate pebbles in sandstone beds or as blocks of conglomerate. Limestone blocks also occur but are rare or absent in the lower part.

4. The Handler Formation includes debris-flow deposits that contain transported elements of a tectonically stable, limestone-polycyclic quartz terrain. The clasts are of youngest Cambrian/ oldest Ordovician age and show little evidence of tectonic activity.

5. The Handler Formation likely occurs in several other ranges, based on reports of red slate in the literature, but it has not been mapped.

Acknowledgments. The field portion of this work was supported by the West German Bundesanstalt für Geowissenschaften und Rohstoffe (project GANOVEX) during the 1982/1983 season, by the Antarctic Division of the New Zealand Department of Scientific and Industrial Research during the 1984/1985 season, and by the Divisions of Earth Sciences and Polar Programs, United States National Science Foundation. Discussions and joint fieldwork on Handler Ridge with John Bradshaw, Franz Tessensohn, John Begg, and Werner Buggisch sharpened and improved the concepts presented. The outstanding spirit of cooperation and friendship that prevailed during this project stand as testimony to international cooperation in its best sense. The manuscript was improved by the reviews of Albert J. Rowell and anonymous reviewers.

References

Adams, C. J., and H. Kreuzer, Potassium-argon age studies of slates and phyllites from the Bowers and Robertson Bay terranes, north Victoria Land, Antarctica, Geol. Jahrb., Reihe B, 60, 265-288, 1984.

Burrett, C. F., and R. H. Findlay, Cambrian and Ordovician conodonts from the Robertson Bay Group, Antarctica and their tectonic significance, Nature, 307, 723-725, 1984.

Field, B. D., and R. H. Findlay, The sedimentology of the Robertson Bay Group, north Victoria Land, in Antarctic Earth Science, edited by R. L. Oliver, P. R. James, and J. B. Jago, pp. 102-106, Australian Academy of Science, Canberra, 1983.

Harrington, H. J., B. L. Wood, I. C. McKellar, and G. J. Lensen, The geology of Cape Hallett - Tucker Glacier district, in Antarctic Geology, edited by R. J. Adie, pp. 220-228, North Holland, Amsterdam, 1964.

Rastall R. H., and R. E. Priestley, The slate-graywacke formation of Robertson Bay: Br. Antarct. Terra Nova Exped. 1910, Nat. Hist. Rep. Geol., 1, 121-129, 1921.

Stump, E., M. G. Laird, J. D. Bradshaw, J. R. Holloway, G. S. Borg, and K. E. Lapham, Bowers graben and associated tectonic features cross northern Victoria Land, Antarctica, Nature, 304, 334-336, 1983.

Sturm, A., and S. J. Carryer, Geology of the region between Matusevitch and Tucker glaciers, north Victoria Land, Antarctica, N.Z. J. Geol. Geophys., 13, 408-435, 1970.

Tessensohn, F., K. Duphorn, H. Jordan, G. Kleinschmidt, D. N. B. Skinner, U. Vetter, T. O. Wright, and D. Wyborn, Geological comparison of basement units in north Victoria Land, Antarctica, Geol. Jahrb., Reihe B, 41, 31-88, 1981.

Wright, T. O., Sedimentology of the Robertson Bay Group, north Victoria Land, Antarctica, Geol. Jahrb., Reihe B, 41, 127-138, 1981.

Wright, T. O., Late Precambrian and early Paleozoic tectonism and associated sedimentation in northern Victoria Land, Antarctica, Geol. Soc. Am. Bull, 96, 1332-1339, 1985.

Wright, T. O., and R. H. Findlay, Relationships between the Robertson Bay Group and the Bowers Supergroup - New progress and complications from the Victory Mountains, north Victoria Land, Geol. Jahrb., Reihe B, 60, 105-116, 1984.

Wright, T. O., R. J. Ross, Jr., and J. E. Repetski, Newly discovered youngest Cambrian or oldest Ordovician fossils from the Robertson Bay terrane (formerly Precambrian), northern Victoria Land, Antarctica, Geology, 12, 301-305, 1984.

Copyright 1987 by the American Geophysical Union.

RADIOMETRIC AGES OF PRE-MESOZOIC ROCKS FROM NORTHERN VICTORIA LAND, ANTARCTICA

H. Kreuzer, A. Höhndorf, H. Lenz, P. Müller, and U. Vetter

Bundesanstalt für Geowissenschaften und Rohstoffe, D-3000 Hannover 51,
Federal Republic of Germany

Abstract. Northern Victoria Land forms the Pacific end of the Transantarctic Mountains. Three major thrust-bounded terranes are distinguished. The Wilson Terrane (WT), with low- to high-grade partly polyphase metamorphics, is in the west; the Robertson Bay Terrane (RBT), with low-grade metaturbidites, is in the east; and squeezed between these two, with indications of a suture zone at the margin of the western terrane, is a third, the narrow Bowers Terrane (BT), with a low-grade metamorphic regressive Cambrian sequence of volcanics and marine to fluviatile sediments. All three terranes were affected by early Ordovician events. Models for the relationships of the three terranes have to consider the following observations. Rb-Sr whole-rock isochron ages of about 600 and 530 Ma are suggested for the WT as well as for the northern margin of the RBT. The Ordovician Granite Harbour Intrusives are restricted to the WT, but similar Ordovician ages of 500 to 460 Ma are determined for all three terranes on micas or schists and slates. Hence, a common history is possible for all three terranes since the Ordovician. The atectonic Devonian to Carboniferous Admiralty Intrusives (370-350 Ma) intrude mainly the eastern terrane but probably cut the bounding thrust zones of the terranes. They show a southwestward decreasing crustal contamination and have characteristics of postcollision I-type granitoids, allowing the speculation of a juxtaposition of the three terranes in the Devonian. Permian Beacon sediments and Jurassic Ferrar Dolerites in the western, the Bowers, and the Wilson terranes indicate that the three terranes were thrust together at least in the late Paleozoic.

Introduction

Northern Victoria Land (NVL) forms part of the western border of mobile "West Antarctica" (here in the east) against the Antarctic shield (Figure 1). Because of its position opposite Tasmania-Australia, NVL is a key area for correlations of the Antarctic and Australian shield areas in reconstructions of Gondwana. This paper summarizes the results of age determinations in NVL in order to provide constraints for models of its evolution, which presently range from juxtaposition of suspect terranes by strike-slip faulting [Weaver et al., 1984] to an almost continuous evolution at a convergent margin [e.g., Kleinschmidt and Tessensohn, this volume].

Regional Geology

Faults and glaciers divide NVL into north-northwest/south-southeast trending zones (Figure 2). The major structural units are thrust bounded and are, from west to east, the Wilson, the Bowers, and the Robertson Bay terranes [Weaver et al., 1984].

The Wilson Terrane

The Wilson Terrane, situated at the east edge of the polar plateau, can be subdivided into at least two zones, separated by the Rennick and Aviator glaciers.

The Wilson zone. The western part of the Wilson Terrane mainly consists of low- to high-temperature low-pressure metamorphic first-cycle psammites and migmatites, the Rennick Schists, and the Wilson Gneisses, with up to four phases of deformation [Kleinschmidt, 1981; Kleinschmidt and Skinner, 1981; Plummer et al., 1983] in the early Ordovician intruded by synkinematic postkinematic, mainly S-type granitoids of the Granite Harbour Intrusives [Vetter et al., 1983; Wyborn, 1983; Borg and Stump, this volume].

The Lanterman zone. The eastern part of the Wilson Terrane comprises the thrust-bounded Lanterman, Salamander, and Mountaineer ranges, with polymetamorphic paragneisses [Kleinschmidt et al., 1984; Roland et al., 1984], intruded by syntectonic I-type granitoids. Meta-ultramafics [Grew et al., 1984] indicate a suture line.

The Bowers Terrane

The narrow Bowers Terrane is thrust bounded to the west by the Rennick Graben and the Lanterman Fault and to the east by the Leap Year Fault zone. It is underlain by the Bowers Supergroup, a tectonized sequence comprising the Middle Cambrian [Cooper et al., 1983] intraoceanic arc volcanics, which interdigitate with marine sediments and terrestrial input from the southwest [Weaver et al., 1984], a Middle to Late Cambrian regressive

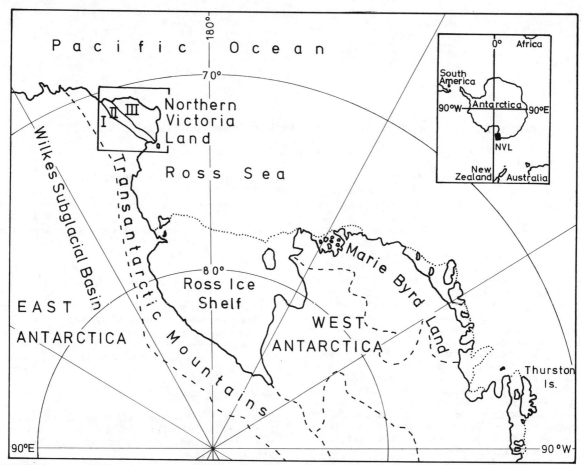

Fig. 1. Sketch map of the Pacific end of the Transantarctic Mountains [from Adams and Kreuzer, 1984]. I is Wilson Terrane; II, Bowers Terrane; III, Robertson Bay Terrane.

shallow marine sequence, both sequences overlain, with erosional contacts, by conglomerates and fluviatile sandstones [Laird and Bradshaw, 1983].

The Robertson Bay Terrane

The wide Robertson Bay Terrane is bordered by the sea from the north to the southeast. It is underlain by the Robertson Bay Group of probable second-cycle turbidites which were shed in a northerly direction [Wright, 1981; Wyborn, 1983]. At Handler Ridge (Victory Mountains, southwestern Robertson Bay Terrane), the Robertson Bay Group seems to pass into a sequence which is similar in age and lithology to the upper part of the shallow marine sediments of the Bowers Supergroup [Wright and Findlay, 1984].

All three terranes yielded Early Ordovician radiometric ages [Adams et al., 1982b]. In the central Transantarctic Mountains, Cambro-Ordovician events are subsumed under the term "Ross Orogeny" and distinguished from a late Proterozoic "Beardmore Orogeny" [Adams et al., 1982a].

The Bowers and Robertson Bay terranes have

Fig. 2. (Opposite) Geological sketch map of northern Victoria Land, Antarctica [after Laird and Bradshaw, 1983]; Kleinschmidt et al., 1984]. In the Wilson Terrane (I), A is Mount Anakiwa; BR, Black Ridge; C, Mount Camelot; D, Daniels Range; GH, Gallipoli Heights; GI, Gerlache Inlet; J, Jupiter Amphitheatre (Morozumi Range); M, Mountaineer Range; K, Kavrayskiy Hills; L, Lanterman Range; MD, Mount Dickason; MO, Morozumi Range; R, Reninie Rocks; S, Salamander Range; SS, Schroeder Spur; ST, Mount Staley; TS, Thompson Spur; U, Unconformity Valley (Morozumi Range). In the Bowers Terrane (II), LP is Lawrence Peaks; MS, Mount Supernal; TG, Tiger Gabbro; and Z, Znamenskiy Island. In the Robertson Bay Terrane (III), B is Black Prince; DP, Daniell Peninsula; CH, Champness area; CO, Zykov Glacier, Cooper Bluffs, and Cooper Spur; E, Everett Range; F, Football Saddle; G, Gregory Bluffs; MP, Malta Plateau; P, Platypus Ridge; SP, Sputnik Island; TI, Tucker Inlet; and Y, Yule Bay, Yule Batholith, and Surgeon Island. For the faults, LF is Lanterman Fault; RF, Rennick Fault; and LYF, Leap Year Fault.

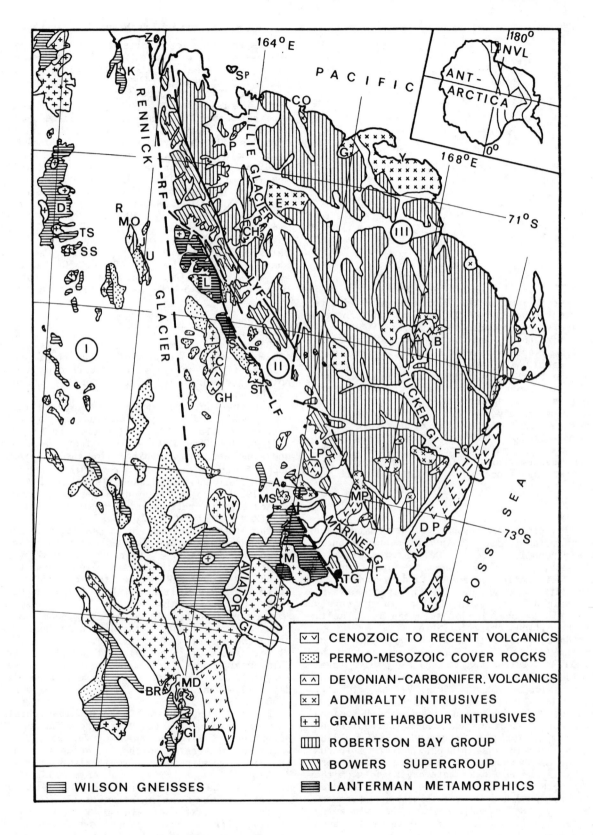

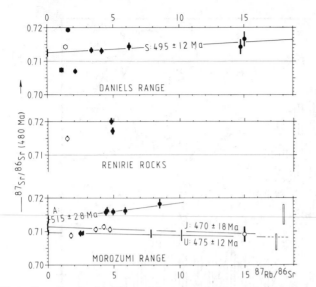

Fig. 3. Strontium isotopic composition of Granite Harbour Intrusives at 480 Ma versus $^{87}Rb/^{86}Sr$ and tentative isochrons. In this plot, a 480 Ma isochron is horizontal, and the ordinate of each data point closely approximates the Sr initial ratio of the rock sample. For the Daniels Range the solid circles are for Schroeder Spur, and the other symbols are for Thompson Spur, namely, solid square is tonalite, solid hexagon is migmatite, and open hexagon is metasediment. For Renirie Rocks, solid circles are granites, and the open circle is an aplite dike. For the Morozumi Range the solid circles are the main adamellite massif of the northern Morozumi Range (A), the open circles are the Jupiter Amphitheatre (J), and the bars are the Unconformity Valley (U).

undergone a single major phase of deformation and low-grade metamorphism [Kleinschmidt and Skinner, 1981; Kleinschmidt, 1983]. Only at the dividing thrust zone, the Millen zone, is more than one phase of deformation observed [Findlay and Field, 1983]. But Jordan et al. [1984] discuss a syngenetic origin of the main cleavage in the Bowers Terrane, of the shearing of the Leap Year Fault, and of the Millen Schist foliation.

In Middle Devonian to Early Carboniferous time, mainly the Robertson Bay Terrane was intruded by the Admiralty Intrusives, a suite of high-level granitoids with chemical and mineralogical I-type characteristics [Wyborn, 1981; Vetter et al., 1983, 1984]. Three distinct episodes have been postulated, Middle Devonian, around the Devonian-Carboniferous boundary, and middle Carboniferous [Adams et al., 1986]. Tessensohn [1984, p. 378] concludes that in some places, Admiralty Intrusives cut the boundary faults of the three terranes. This and Permo-Mesozoic cover rocks of the Beacon Supergroup, present in the Wilson [Grindley and Oliver, 1983] and the Bowers terranes [Tessensohn et al., 1981, p. 82], suggest that since the Carboniferous, at the latest, the three terranes had a common history.

Age Records of Northern Victoria Land

All radiometric results including quoted ones are calculated with the constants recommended by the International Union of Geological Scientists [Steiger and Jäger, 1977]. Error estimates are given as 95% confidence intervals of the analytical precision. For Rb-Sr isochron data and replicates with a mean of the squared weighted deviations (MSWD) larger than 1, the analytical error is enhanced by the square root of the MSWD. K-Ar whole-rock dates on whole-rock samples of granitoids (dates of Ravich and Krylov [1964] and the four dates of M. G. Ravich and A. J. Krylov, quoted by Gair et al. [1969]) (without mentioning the constant used) should be disregarded, because such dates can be considerably too early [Dalrymple and Lanphere, 1969, Figure 10-3]. Still vague is the stratigraphic meaning of Cambrian dates. The age of the top of the Cambrian is sharply bracketed between 495 and 505 Ma, but the age of its base is debatable, but is at least between 540 and 590 Ma [e.g., Jenkins, 1984].

Pre-Ordovician Isotopic Ages

Beardmore equivalent ages. Two ages of about 600 Ma are indicated for NVL from Rb-Sr scatterchrons (linear arrays with a scatter of the data points too large to be confidently regarded as isochrons).

For the Lanterman zone of the Wilson Terrane, C. J. Adams and H. Lenz (unpublished data, 1984) determined 610 ± 26 Ma (initial ratio (IR) = 0.7132 ± 0.0008, MSWD = 4.0) on four granodioritic gneisses and one amphibolitic gneiss from the upper Zenith Glacier and from Husky Pass in the southeastern Lanterman Range (samples of Adams et al. [1982b]). The same data result if only the granodioritic gneisses are considered (597 ± 32 Ma, IR = 0.7137 ± 0.0012, MSWD = 3.4).

The other indication for a Beardmore equivalent age comes from the northeastern margin of the Robertson Bay Terrane. Vetter et al. [1984] (disregarding two of a total of six data points) reported an age of 599 ± 21 Ma for the foliated granodiorite of Surgeon Island (IR = 0.7590 ± 0.0014, MSWD = 0.25).

Cambrian scatterchron dates around 530 Ma. These ages are found in the Wilson Terrane as well as in the Robertson Bay Terrane.

For nine biotite schists and one slate from the Salamander Range (Lanterman zone of the Wilson Terrane) C. J. Adams and H. Lenz (unpublished data, 1984) calculated 550 ± 26 Ma (IR = 0.7158 ± 0.0008, MSWD = 5.3).

For two lithologies of the Rennick Schists of the Daniels Range in the Wilson zone, Adams [1986] determined a weighted mean age of 532 ± 14 Ma, which he interpreted as the age of metamorphism of the pelitic and psammitic sediments.

A similar age is reported from 350 km to the south in the Wilson zone for five samples of "Terra Nova Granites" [Vetter et al., 1984] from localities scattered up to 40 km apart, but comprising different rock types [see Skinner, 1983, Figure 1]. Restricted to the two or possibly

TABLE 1. Rb-Sr Data of Pre-Devonian Granitoids From Northern Victoria Land

Field No.[c]	Lab. AL No.	Rock Type	Lat. S.	Long. E.	^{87}Rb (ppm)	^{86}Sr (ppm)	$\dfrac{^{87}\text{Rb}^a}{^{86}\text{Sr}}$	$\dfrac{^{87}\text{Sr}^a}{^{86}\text{Sr}}$	$\dfrac{^{87}\text{Sr}^b}{^{86}\text{Sr}}$ (480 Ma)
Schroeder Spur, Daniels Range									
DA 05	1781	granite	71°39'	160°21'	61.17	17.99	3.361	0.73626	0.7133 ± 9
DA 11	1782	granite	71°37'	160°06'	70.91	11.27	6.220	0.75696	0.7144 ± 11
DA 12	1783	granite	71°40'	160°29'	61.18	14.60	4.142	0.74141	0.7131 ± 9
DA 10	1786	pegmatite	71°38'	160°20'	115.9	7.76	14.76	0.81529	0.7143 ± 22
DA 03	1784	leucogranite	71°37'	160°21'	124.5	8.18	15.05	0.81958	0.7166 ± 22
DA 04	1785	granodiorite	71°38'	160°21'	48.48	22.39	2.140	0.72161	0.7070 ± 8
Thompson Spur, Daniels Range									
DA 08	1791	migmatite	71°32'	160°16'	28.90	17.66	1.618	0.73032	0.7193 ± 8
DA 09	1787	tonalite	71°37'	160°20'	27.19	24.74	1.086	0.71485	0.7074 ± 7
DA 13	1793	metasediment	71°32'	160°17'	39.10	27.67	1.397	0.72388	0.7143 ± 7
Renirie Rocks									
RE 01	1774	granite	71°19'	161°19'	69.92	14.07	4.911	0.75078	0.7172 ± 10
TES115/1	1775	granite	71°19'	161°19'	73.39	15.05	4.819	0.75303	0.7201 ± 10
RE 02	1777	aplite dyke	71°19'	161°19'	38.62	26.03	1.466	0.72490	0.7149 ± 8
Morozumi Adamellite Massif, Morozumi Range									
MO 01	1762	adamellite	71°31'	161°47'	79.35	13.36	5.871	0.75633	0.7162 ± 11
MO 02	1763	adamellite	71°33'	161°45'	69.43	15.55	4.414	0.74602	0.7158 ± 10
MO 03	1764	adamellite	71°34'	161°43'	73.26	16.28	4.448	0.74676	0.7163 ± 10
MO 05	1765	adamellite	71°34'	161°40'	71.82	14.35	4.947	0.74977	0.7159 ± 10
MO 04	1767	leucogranite	71°34'	161°40'	62.20	7.25	8.477	0.77623	0.7183 ± 14
MO 06	1766	adamellite	71°30'	161°41'	65.14	26.20	2.458	0.72623	0.7094 ± 8
Jupiter Amphitheatre, Morozumi Range									
MO 08	1769	leucogranite	71°35'	161°51'	52.78	29.84	1.748	0.72078	0.7088 ± 8
4181	1979	adamellite	71°35'	161°54'	55.52	15.12	3.630	0.73539	0.7106 ± 7
4182	1980	adamellite	71°35'	161°54'	58.21	12.23	4.704	0.74281	0.7106 ± 8
4184	1982	adamellite	71°35'	161°49'	60.97	14.16	4.255	0.74054	0.7114 ± 7
4185	1983	leucogranite	71°35'	161°35'	106.1	6.84	15.34	0.81405	0.7091 ± 22
Unconformity Valley, Morozumi Range									
MO 07	1768	leucogranite	71°39'	162°03'	65.36	6.38	10.12	0.77798	0.7088 ± 16
4190	1984	adamellite	71°39'	162°02'	65.45	24.19	2.675	0.72790	0.7096 ± 6
4191	1985	granite	71°39'	162°02'	64.76	24.27	2.638	0.72749	0.7094 ± 6
4192	1986	leucogranite	71°39'	162°02'	70.65	8.94	7.810	0.76268	0.7093 ± 12
4193	1987	adamellite, greisenized	71°39'	162°02'	70.63	4.01	17.41	0.82582	0.7067 ± 24
4194	1988	granite, greisenized	71°39'	162°02'	67.83	3.72	18.00	0.83760	0.7145 ± 25

[a] Analytical precision at the level of 95% confidence: $d(^{87}\text{Rb}/^{86}\text{Sr}) = d(x) = 2\%$, and $d(^{87}\text{Sr}/^{86}\text{Sr}) = d(y) = 0.1\%$ and 0.06% for Laboratory AL No. smaller and larger than 1900, respectively.
[b] $^{87}\text{Sr}/^{86}\text{Sr}$ ratio at 480 a before present $(y(480) = y - f \cdot x$, with $f = (\exp(480 \cdot 1.42 \cdot 10^{-5}) - 1))$, i.e., the $^{87}\text{Sr}/^{86}\text{Sr}$ ratio at the time of emplacement of the Granite Harbour Intrusives, which for all rocks except the migmatite and the metasediment represents the "initial" ratio, within the analytical uncertainties. The given digits for the error estimate correspond to the last digits of given $y(480)$ value. They include the error of x: $d(y(480))^2 = d(y)^2 + f^2 \cdot d(x)^2$.
[c] Samples collected by U. Vetter, TES115/1 by F. Tessensohn.

TABLE 2a. K-Ar Mineral Data, Predominantly of Granitoids: Pre-Devonian Granitoids of the Wilson Terrane

Field No.	Lab. AL No.	Rock Type Lat. S. Long. E.	K-Ar Age (Ma)		K-Ar Dates of the Sieve Fractions (Ma)		K (wt%)	±	Rad. STP (nl/g)	±	Atm. STP (nl/g)
					Schroeder Spur, Daniels Range						
DA 11	1782	granite 71°37' 160°06'	M	477 ± 2.5			8.84	4	187.5	9	2.3
			-		B_c	468.5 ± 2.5	7.72	4	160.5	6	2.9
					B_f	465 ± 2.5	7.69	4	158.4	6	2.9
DA 12	1783	granite 71°40' 160°29'	M	476.5 ± 2	M_c	476 ± 3	8.83	5	186.9	10	2.2
					M_f	477 ± 3	8.78	5	186.3	10	2.5
			B	473.5 ± 2	B_c	474 ± 3	7.80	4	164.4	9	2.9
					B_f	473 ± 3	7.80	4	163.8	9	3.0
DA 03	1784	leucogranite 71°37' 160°21'	M	473.5 ± 3	*M_c	473.5 ± 3.5	8.77	5	184.6	10	1.9
					M_f	473 ± 3.5	8.81	5	185.2	10	3.2
			B	471.5 ± 2	*B_c	472 ± 2.5	7.43	3	155.8	6	2.5
					B_f	470.5 ± 3	7.33	4	153.1	6	2.3
DA 04	1785	granodiorite 71°38' 160°21'	B	471 ± 3	B_c	470.5 ± 3.5	7.14	5	149.2	9	2.3
					B_f	471.5 ± 4	7.12	5	149.1	9	2.3
DA 10	1786	pegmatite 71°38' 160°20'	M	474.5 ± 2	M_c	473.5 ± 3	8.83	5	185.8	10	2.2
					M_f	475.5 ± 3	8.81	5	186.2	10	2.2
			B	472 ± 4	B_c	470 ± 3	7.62	4	159.2	9	2.7
					B_f	474 ± 3	7.61	4	160.2	9	2.9
DA 09	1787	tonalite 71°37' 160°20'	H	>485 ± 5	H_c	480.5 ± 4	0.610	4	13.05	7	0.38
					H_f	485 ± 5	0.582	6	12.59	7	0.39
			B	479 ± 3	*B_c	478.5 ± 3	7.68	3	163.6	10	2.2
					B_f	482 ± 5	7.66	3	164.5	17	1.9
					Thompson Spur, Daniels Range						
DA 01	1789	diorite 71°35' 160°07'	H	>483 ± 6	$H+B_c$	478 ± 6	1.034	12	21.97	16	0.31
					$H+B_f$	483 ± 6	0.982	12	21.15	18	0.32
			B	470.5 ± 3	B_c	471 ± 3.5	7.67	5	160.5	10	1.1
					B_f	470.5 ± 3.5	7.62	5	159.2	10	1.0
DA 07	1790	granodiorite 71°32' 160°20'	B	473.5 ± 2	B_c	472.5 ± 3	7.72	4	162.2	9	1.8
					B_f	474 ± 3	7.69	4	162.1	9	1.4
DA 08	1791	migmatite 71°32' 160°16'	*M	476 ± 2.5			8.58	4	181.7	7	2.4
			B	472 ± 2	*B_c	472 ± 3.5	7.75	5	162.4	9	1.6
					B_f	472.5 ± 3	7.62	4	159.8	6	1.7
					Renirie Rocks						
RE 01	1774	granite 71°19' 161°19'	M	479.5 ± 2.5			8.72	3	186.1	7	1.8
			B	478 ± 2	B_c	478 ± 2.5	7.35	3	156.3	7	1.5
					B_f	478 ± 3.5	7.27	5	154.6	7	1.6
TES 115/1	1775	granite 71°19' 161°19'	M_c	479.5 ± 2.5			8.84	4	188.6	7	1.9
			-		B_c	476.5 ± 3.5	7.09	4	150.3	8	1.4
					B_f	469.5 ± 3.5	6.99	4	145.7	8	1.9
TES 115/2	1776	granite 71°19' 161°19'	B	471 ± 2.5	B_c	471.5 ± 3.5	7.38	5	154.6	8	1.9
					B_f	470 ± 3.5	7.32	5	152.7	8	1.9
RE 02	1777	aplite dyke 71°19' 161°19'	-		M_c	483 ± 3	8.72	4	187.6	10	2.6
					M_f	478 ± 5	8.72	4	185.5	18	1.8
			B_c	475 ± 3.5			6.89	4	145.6	9	1.4
					Morozumi Adamellite Massif, Northern Morozumi Range						
MO 01	1762	adamellite 71°31' 161°47'	B	476 ± 2	*B_c	476.5 ± 2.5	7.60	3	161.2	6	2.0
					B_f	475 ± 3	7.55	3	159.4	10	1.8
MO 02	1763	adamellite 71°33' 161°45'	B	468 ± 2.5	B_c	469.5 ± 3	7.08	4	147.5	8	1.8
					B_f	467 ± 3	7.06	4	146.6	6	1.6
MO 05	1765	adamellite 71°34' 161°40'	B	478.5 ± 3	B_c	479.5 ± 3	6.78	4	144.9	8	2.0
					B_f	477 ± 3	6.69	4	142.1	8	1.8
MO 06	1766	adamellite 71°30' 161°41'	M	478 ± 2	*M_c	478 ± 3	8.93	6	189.9	7	2.4
					*M_f	479.5 ± 3	8.94	6	190.6	7	2.7
			B	478.5 ± 2.5	*B_c	480 ± 3	7.79	5	166.5	6	1.8
					*B_f	477.5 ± 3	7.80	5	165.6	6	1.9

TABLE 2a. (continued)

Field No.	Lab. AL No.	Rock Type Lat. S. Long. E.	K-Ar Age (Ma)		K-Ar Dates of the Sieve Fractions (Ma)		K (wt%)	±	Rad. Ar STP (nl/g)	±	Atm. STP (nl/g)

Jupiter Amphitheatre, Morozumi Range

Field No.	Lab. AL No.	Rock Type Lat. S. Long. E.		K-Ar Age (Ma)		K-Ar Sieve Frac (Ma)	K (wt%)	±	Rad. Ar STP (nl/g)	±	Atm. STP (nl/g)
MO 08	1769	leucogranite 71°35' 161°50'	M	482.5 ± 3			8.60	4	185.0	11	1.7
			B	476 ± 2.5	B_c	476 ± 4	6.34	5	134.2	7	1.4
					B_f	476 ± 3	6.00	3	127.1	7	1.0
4181	1979	adamellite 71°35' 161°53'	M	486 ± 3	M_c	486 ± 3.5	8.79	5	190.6	11	1.7
					M_f	486 ± 4	8.84	6	191.6	14	2.4
				–	B_c	458 ± 2.5	5.21	2	105.5	4	1.5
					B_f	462 ± 3	4.68	2	95.8	6	2.0
4182	1980	adamellite 71°35' 161°53'	M	483 ± 3	M_c	482 ± 3.5	8.86	6	190.2	11	2.2
					M_f	484.5 ± 4	8.83	6	190.8	14	1.9
			B	469 ± 3	B_c	467 ± 3.5	6.03	4	125.0	8	1.7
					B_f	470.5 ± 4	5.65	3	118.2	10	2.1
4183	1981	adamellite 71°35' 161°52'		–	M_c	485 ± 4	8.76	5	189.2	13	2.1
					M_f	480 ± 4	8.58	5	183.4	12	1.7
			B	380.5 ± 2.5	B_c	381.5 ± 3	3.73	2	61.7	5	1.1
					B_f	379 ± 4	3.68	2	60.3	6	1.9
4184	1982	adamellite 71°35' 161°49'		–	M_c	483 ± 3.5	8.64	5	185.9	11	1.6
					M_f	487 ± 3.5	8.80	5	191.2	11	1.6
			B	463.5 ± 2.5	B_c	464.5 ± 3.5	5.61	4	115.5	6	1.0
					B_f	462.5 ± 3.5	4.98	3	104.9	8	1.3
4185	1983	leucogranite 71°35' 161°49'	M	483.5 ± 3	M_c	482 ± 3.5	8.88	6	190.8	11	2.7
					M_f	485 ± 4	8.82	6	190.8	14	2.4
				–	B_c	476.5 ± 3.5	7.20	3	152.5	12	1.5
					B_f	469.5 ± 3.5	7.06	3	147.7	9	1.4

Unconformity Valley, Morozumi Range

Field No.	Lab. AL No.	Rock Type Lat. S. Long. E.		K-Ar Age (Ma)		Sieve Frac (Ma)	K (wt%)	±	Rad. Ar STP (nl/g)	±	Atm. STP (nl/g)
4190	1984	adamellite 71°39' 162°02'	M	488 ± 3.5			8.21	5	178.9	11	1.3
				–	B_c	487 ± 2.5	7.22	3	157.0	7	1.6
					B_f	475.5 ± 3	7.30	5	154.4	7	1.9
4191	1985	granite 71°39' 162°02'	M	482 ± 3.5			8.52	5	182.8	11	2.4
			B	482 ± 3	B_c	481.5 ± 3.5	7.35	5	157.6	10	1.8
					B_f	483 ± 4	7.28	5	156.6	11	2.1
4192	1986	leucogranite 71°39' 162°02'	M_f	486 ± 4			8.92	6	193.4	12	2.8
			B	478.5 ± 3.5			6.93	4	147.4	9	1.9
4193	1987	adamellite, greisenized 71°39' 162°02'	M	483 ± 3	M_c	484.5 ± 4	8.76	5	189.1	14	2.0
					M_f	481.5 ± 4	8.67	5	186.0	14	3.0
4194	1988	granite, greisenized 71°39' 162°02'	M	489.5 ± 3	M_c	488 ± 4	8.74	5	190.3	14	1.9
					M_f	491 ± 4	8.66	5	189.8	14	1.9

Lanterman Range

Field No.	Lab. AL No.	Rock Type Lat. S. Long. E.		K-Ar Age (Ma)		Sieve Frac (Ma)	K (wt%)	±	Rad. Ar STP (nl/g)	±	Atm. STP (nl/g)
LA 01	1778	granodiorite 3 km SSW Carnes Crag 71°29' 162°39'	M	484 ± 3	M_c	483 ± 3.5	8.00	5	172.3	10	1.5
					M_f	485.5 ± 4	8.38	5	181.5	14	1.1
				–	B_c	483 ± 3.5	5.87	4	126.5	7	1.7
					B_f	477.5 ± 4	5.58	4	118.6	9	1.2
LA 02	1779	amphibole-rich schlieren 71°29' 162°39'	H	488 ± 7	H_c	488 ± 9	0.181	4	3.950	25	0.25
					H_f	488 ± 10	0.167	4	3.635	25	0.39
			B	480 ± 2.5	*B_c	479 ± 3	7.16	3	152.7	9	3.1
					B_f	481 ± 3.5	7.15	4	153.0	9	2.2
LA 03	1780	diorite S Hoshka Glacier 71°50' 163°24'	H	502 ± 4	H_c	504 ± 7	0.796	11	17.99	13	0.37
					H_f	501 ± 6	0.770	8	17.31	13	0.42
			B	489 ± 2.5	*B_c	489 ± 3	7.78	4	169.7	10	1.9
					B_f	489 ± 4	7.72	5	168.3	10	2.1
2122	2303	Orr Granite, 5 km SSE Mt. Moody, 71°34' 162°54' (recrystallized under amphibolite facies conditions)	B	487.5 ± 2.5	B_c	488 ± 3	7.35	4	159.9	6	1.3
					B_f	487 ± 3.5	7.38	4	160.3	12	1.3

TABLE 2a. (continued)

Field No.	Lab. AL No.	Rock Type Lat. S. Long. E.	K-Ar Age (Ma)	K-Ar Dates of the Sieve Fractions (Ma)	K (wt%)	±	Rad. STP (nl/g)	Ar ±	Atm. STP (nl/g)
		Terra Nova Area							
4250	1989	granite, 6 km NE Mt. Dickason, 74°21' 164°06'	B 458.5 ± 2	B_c 459.5 ± 2.5 B_f 457.5 ± 2.5	7.67 7.65	3 3	156.0 154.8	7 7	1.9 2.0
4251	1990	ditto	B 456.5 ± 2.5	B_c 458 ± 3.5 B_f 455.5 ± 3.5	7.69 7.64	5 5	155.7 153.9	9 9	1.4 1.6
4252	1991	granite, 4 km W Mt. Dickason, 74°24' 163°52'	B 462 ± 2.5	B_c 460.5 ± 3.5 B_f 463 ± 3.5	7.52 7.47	5 5	153.4 153.2	9 9	1.6 1.7
4255	1992	granite, ridge S of Gerlache Inlet, 74°42' 164°08'	B 455.5 ± 2.5	B_c 457 ± 3.5 B_f 454 ± 3.5	7.11 7.00	4 4	143.7 140.6	9 8	2.4 2.3
4256	1993	granodiorite, S sporn of Black Ridge, 74°28' 163°32'	B 475 ± 2.5	B_c 474 ± 3 B_c 476.5 ± 3.5	7.63 7.58	3 5	160.7 160.7	11 11	2.0 2.2

three samples of Dickason Granite, the age is 548 ± 18 Ma or 548 ± 36 Ma (IR = 0.7094 ± 0.0013, MSWD = 4.1), respectively.

At the northern margin of the Robertson Bay Terrane, four samples of the Cooper Spur Granite yielded an isochron age of 525 ± 15 Ma (IR = 0.71240 ± 0.00063, MSWD = 0.5) [Vetter et al., 1984].

Rb-Sr Whole-Rock Dating of Ordovician Events

Wilson Gneisses. Whole-rock analyses of Wilson Gneisses from the northern Daniels Range yielded an isochron age of 490 + 33, - 5 Ma (IR = 0.7205 ± 0.0016, MSWD = 5.8). The initial Sr ratio is the same as that for the psammitic Rennick Schists from which they derived [Adams, 1986]. The younger age limit is given by U-Pb monazite ages of 484-488 Ma for the same rock samples.

The Granite Harbour Intrusives. These syntectonic to posttectonic discordant granitoids have lower initial Sr ratios than the Wilson Gneisses: about 0.716 to 0.710 for S-type granitoids (Table 1 and Figure 3) [Adams, 1986]. Isochron data are based on groups of granitoids with largely different $^{87}Rb/^{86}Sr$ ratios and, hence, possibly different origins and ages. But they cannot be grossly in error because ages for isochrons and micas from different outcrops are similar.

For the southern Daniels Range (Figure 3, upper part), 495 ± 12 Ma (IR = 0.7125 ± 0.0011, MSWD = 1.2) has been suggested [Kreuzer et al., 1981; Vetter et al., 1983].

For the Morozumi Range (Figure 3, lower part), the same authors suggested an isochron age of 515 ± 28 Ma (IR = 0.7136 ± 0.0022, MSWD = 0.9) for the main adamellite massif in the north and a younger age and a significantly lower initial Sr ratio for three samples from dispersed outcrops, meanwhile confirmed in two of them (Tables 1 and 2): Jupiter Amphitheatre (disregarding one of the five analyzed samples) yielded 470 ± 18 Ma (IR = 0.7114 ± 0.0013, MSWD = 2.0). The K-Ar muscovite dates range from 489 to 479 Ma and suggest a slightly older age of 483.7 ± 1.2 Ma (MSWD = 2.1). For Unconformity Valley the four ungreisenized samples yield 475 ± 12 Ma (IR = 0.7097 ± 0.0007, MSWD = 0.15). The K-Ar muscovite dates of the granites (488 to 482 Ma) and of the greisen (491 to 482 Ma) are slightly discordant and also suggest a slightly older age.

TABLE 2b. K-Ar Mineral Data, Predominantly of Granitoids: Tiger Gabbro, Apostrophe Island, Southernmost Bowers Terrane

Field No.	Lab. AL No.	Rock Type Lat. S. Long. E.	K-Ar Age (Ma)	K-Ar Dates of the Sieve Fractions (Ma)	K (wt%)	±	Rad. STP (nl/g)	Ar ±	Atm. STP (nl/g)
		metasomatic hornblende shear zone in layered gabbro 73°31' 167°26'	H 521 ± 10		0.240	5	5.64	5	0.45

TABLE 2c. K-Ar Mineral Data, Predominantly of Granitoids: Admiralty Intrusives From the Robertson Bay Terrane

Field No.	Lab. AL No.	Rock Type Lat. S. Long. E.	K-Ar Age (Ma)	K-Ar Dates of the Sieve Fractions (Ma)	K (wt%)	±	Rad. Ar STP (nl/g)	±	Atm. STP (nl/g)
		Yule Batholith							
TES102	1738	Novosad Island 70°42', 167°30'	*B 361 ± 1.5	(= TS 102, erroneously quoted as Birthday Ridge)					
4010	1848	S of O'Hara Glacier 70°49', 166°42'	B 365.5 ± 2	B$_c$ 365.5 ± 2.5	7.48	3	117.7	7	2.0
				B$_f$ 365.5 ± 3.5	7.44	7	117.3	7	2.1
4012	1850	Tapsell Foreland 70°48', 167°04'	B 364.5 ± 2.5	B$_c$ 365 ± 4	6.85	6	107.7	7	1.2
				B$_f$ 364 ± 3	6.69	4	104.9	7	1.5
4014	1852	Tapsell Foreland 70°49', 167°13'	B 361.5 ± 2	B$_c$ 360.5 ± 3	7.43	5	115.2	7	1.6
				B$_f$ 362.5 ± 3	7.46	5	116.4	7	1.9
4015	1853	Birthday Ridge 70°47', 166°59'	B 360 ± 2	B$_c$ 360.5 ± 3	7.11	4	110.3	7	2.1
				B$_f$ 359 ± 3	7.08	4	109.4	7	2.3
4016	1854	fine-grained bi-hbl-pl-xenolith, loc. as 4015	B 362 ± 2	B$_c$ 361.5 ± 3	7.28	5	113.2	6	1.3
				B$_f$ 362.5 ± 2	7.35	3	114.7	5	1.4
4018	1856	ditto, Birthday Ridge 70°47', 166°59'	B 363.5 ± 2	B$_c$ 362.5 ± 3	7.68	5	119.8	7	1.4
				B$_f$ 364.5 ± 3	7.68	5	120.6	7	1.5
		Gregory Bluffs							
4120	1969	Athurson Bluff 70°45', 166°04'	B 363 ± 2	B$_c$ 363.5 ± 3	7.15	4	111.8	7	1.5
				B$_f$ 362.5 ± 3	7.11	3	111.0	7	1.8
4123	1971	4 km SSW Athurson Bluff 70°47', 166°02'	B 365.5 ± 1.5	B$_c$ 365.5 ± 1.5	7.38	3	116.2	4	1.6
				B$_f$ 366 ± 2	7.19	3	113.5	4	1.8
4127	1972	4 km E Mt. Harwood 70°45', 165°56'	M 364 ± 1.5	M$_c$ 363 ± 2.5	8.69	4	135.9	7	1.4
				M$_f$ 364.5 ± 2.5	8.65	5	135.7	6	1.4
			B 359 ± 2	B$_c$ 358.5 ± 3	7.01	4	108.1	7	1.4
				B$_f$ 359.5 ± 3	6.86	4	106.0	6	1.3
4128	1973	4 km E Mt. Harwood 70°45', 165°56'	M 370.5 ± 2	M$_c$ 370 ± 3	8.50	5	135.5	8	1.6
				M$_f$ 371.5 ± 3	8.25	5	132.3	8	1.8
				(B$_c$) 350 ± 5)	4.64	5	69.6	9	1.0
			-	(B$_f$) 342 ± 5)	4.56	6	66.7	5	1.0
		Everett Range							
EV 06	1748	"810-Ridge" 71°10', 164°29'	H 366 ± 2.5	H$_c$ 366 ± 2.5	0.607	4	9.57	4	0.27
				*H$_f$ 365.5 ± 5	0.584	4	9.19	12	0.35
EV 08	1749	6 km N Pt. 1670 71°07', 164°48'	H 360 ± 3	H$_c$ 361 ± 6	0.522	9	8.10	5	0.25
				*H$_f$ 360 ± 3	0.447	4	6.93	4	0.30
EV 11	1752	2 km NW Pt. 1775 71°08', 164°38'	H 359 ± 4	H$_c$ 358.5 ± 6	0.608	11	9.37	4	0.33
				*H$_f$ 359.5 ± 6	0.532	10	8.23	3	0.36
EV 14	1753	NW Mt. Dockery 71°13', 164°31'	-	B$_c$ 360.5 ± 2	7.04	3	109.1	4	2.0
				B$_f$ 356.5 ± 1.5	6.98	2	106.8	4	2.1

TABLE 2c. (continued)

Field No.	Lab. AL No.	Rock Type Lat. S. Long. E.	K-Ar Age (Ma)	K-Ar Dates of the Sieve Fractions (Ma)	K (wt%)	±	Rad. STP (nl/g)	±	Ar	Atm. STP (nl/g)
\multicolumn{11}{l}{Granitoids West of the Lillie Glacier}										
TES110	1773	Platypus Ridge Lower Lillie Glacier 70°47' 163°35'	H 362 ± 4 B 363 ± 2	H$_c$ 363 ± 6 H$_f$ 360 ± 6 B$_c$ 363 ± 3 B$_f$ 363 ± 3	0.324 0.308 7.22 7.06	5 5 4 4	5.07 4.76 112.8 110.3	5 3 7 7		0.20 0.24 1.5 1.5
\multicolumn{11}{l}{Area of the Champness Glacier}										
4050	1863	Griffith Ridge 71°22' 164°22'	B 354 ± 2	B$_c$ 353.5 ± 2.5 B$_f$ 354.5 ± 2.5	6.96 6.76	4 4	105.6 102.9	6 6		2.0 2.4
4053	1865	Granite Ridge 71°28' 164°14'	—	(B$_c$ 344.5 ± 4.5) (B$_f$ 348.5 ± 2)	4.71 4.23	6 2	69.5 63.3	4 3		1.2 1.0
4054	1866	Griffith Ridge 71°21' 164°27'	B 354.5 ± 2	B$_c$ 355 ± 2.5 B$_f$ 353.5 ± 2.5	6.75 6.49	4 3	102.9 98.5	6 5		2.1 2.1
4055	1867	biotite schlieren loc. as 4054	B 359.5 ± 2	B$_c$ 359 ± 2 B$_f$ 362 ± 4	7.12 6.94	3 7	109.8 108.1	7 7		2.0 1.1
4056	1868	Radspinner Ridge 71°25' 164°36'	H 356 ± 4	H$_c$ 355 ± 5 H$_f$ 356 ± 4	0.502 0.425	9 4	7.65 6.50	8 5		1.1 0.9
4059	1870	Radspinner Ridge 71°23' 164°35'	B 359.5 ± 2 H 348.5 ± 3	B$_c$ 359.5 ± 2 B$_f$ 359 ± 3 H$_c$ 349 ± 5 H$_f$ 348 ± 5	7.19 7.14 0.632 0.465	3 3 5 6	111.1 110.2 9.45 6.93	5 9 7 7		2.1 2.1 1.0 0.8
4060	1871	Copperstain Ridge 71°26' 164°24'	B 359 ± 2.5 B 358 ± 2	B$_c$ 358.5 ± 3.5 B$_f$ 359.5 ± 3.5 B$_c$ 357 ± 3.5 B$_f$ 355.5 ± 2.5	7.37 7.24 7.10 6.85	5 5 5 4	113.5 111.8 108.8 104.5	7 9 10 4		1.9 1.7 1.4 1.6
\multicolumn{11}{l}{Area of the Tucker Inlet (TI)}										
TES 01 1771 (=TS 1)*		Football Saddle N of TI 72°31' 169°42'	B 355.5 ± 2	*B$_c$ 357 ± 3 B$_f$ 354.5 ± 3	7.71 7.73	5 5	118.2 117.7	7 7		2.3 2.6
TU 01	1772	2 km E of Crater Cirque, S side of Lower Tucker Glacier, W of the Tucker Inlet, 72°37' 169°18'	H 363.5 ± 2.5 B 356 ± 2	H$_c$ 364 ± 4.5 H$_f$ 363 ± 3 *B$_c$ 355.5 ± 2.5 B$_f$ 356.5 ± 3	0.982 0.906 7.71 7.66	12 7 3 5	15.40 14.16 117.8 117.3	8 8 7 7		0.39 0.24 1.7 2.2
\multicolumn{11}{l}{Daniell Peninsula, outcrops of bi-hbl and bi granitoids within the Hallett Volcanics}										
3236/ 4412	2002	bi-hbl granite, 5 km N of Cape Phillips, 73°02' 169°38'	—	(B$_c$ 275.5 ± 2.5) (B$_f$ 270.5 ± 2.5)	7.59 7.60	5 5	87.9 86.1	5 6		2.0 2.3
3238/ 4413	2007	bi granite, 1 km N of Cape Phillips, 73°03' 169°38'	—	(B$_c$ 281.5 ± 3) (B$_f$ 274.5 ± 3)	7.22 7.20	5 5	85.4 82.8	8 10		1.8 1.8
\multicolumn{11}{l}{Granites from the Upper Hand Glacier and from the Malta Plateau}										
3228 4411	2001	3 km NNW of Mt. Alberts 73°00' 167°51'	B 350.5 ± 2	B$_c$ 351 ± 2.5 B$_f$ 350 ± 2.5	7.22 7.22	3 3	**108.8** 108.4	7 7		1.3 1.5
3273 4416	2005	S side Upper Hand Gl. 72°54' 167°30'	B 348 ± 2.5	B$_c$ 349 ± 2.5 **B$_f$** 346.5 ± 4	5.10 5.29	2 4	76.3 78.5	5 8		1.2 1.3

3279	2006	3 km NNE Mt. Hussey 72°44', 167°34'	H	363 ± 3.5	H_c H_f	362 ± 5 365 ± 5	0.477 5 0.366 5	7.43 6 5.75 5	0.35 0.34
4417			B	368.5 ± 2.5	B_c B_f	365.5 ± 4 370.5 ± 3	6.43 3 6.13 4	101.4 12 97.9 7	1.8 1.5

Argon in nanoliter per gram at standard conditions ((nl/g) STP) by total fusion and static mass-spectrometer isotope dilution analysis (MAT CH4 and VG MM1200 spectrometers), potassium by flame photometry. Generally, two sieve fractions were analyzed, a coarser one (index c) in size ranges of 630 to 315 μm and a finer one (index f) in ranges of 400 to 160 μm and in the size fractions 160-125 and 125-90 μm, respectively, for muscovites (M) and biotites (B), for hornblendes (H). Data or revised data of Kreuzer et al. [1981] are labelled with an asterisk. All analytical data are corrected for the mean values of short period of blank analyses. Mean relative standard deviations of radiogenic argon (MAT) and of potassium are 0.3% for muscovites and biotites, and 0.5% for hornblendes. Argon analyses with the VG spectrometer have mean standard deviations of 0.7 and 1.5%, respectively. Quoted errors refer to the last digit or the last two digits of the analytical values. They are based on the maximum of general and individual standard deviations and represent the intralaboratory analytical uncertainty at a 95% confidence level. The K-Ar ages are calculated with the IUGS-recommended constants [Steiger and Jäger, 1977]. Our K-Ar date of the standard glaucony Gl-0 is 1% younger than the mean value of the compilation of Odin [1982].

Samples were collected by U. Vetter (DA, RE, MO, LA, EV, and numbers with 4000); F. Tessensohn (TES); M. Schmidt-Thomé (numbers with 3000); N. Roland (2122); D. Wyborn (TU); and G. Delisle (Apostrophe Island).

TS is Thompson Spur; Rad. Ar, radiogenic argon; and Atm., atmosphere.

TABLE 3. Rb-Sr Data of Micas From Intrusives of the Robertson Bay Terrane, Northern Victoria Land

Field UV No.	Lab. AL No.	Rock Type, Mineral	Lat. S. Long. E.	^{87}Rb (ppm)	^{86}Sr (ppm)	^{87}Rb/^{86}Sr[a]	^{87}Sr/^{86}Sr	WR Mineral Age (Ma)	Isochron[b] IR
			Cooper Spur (WR isochron age 525 ± 26 Ma)						
4160	1975	biotite	70°37', 165°02'	231.7	0.4380	522.9	3.3951	359 ± 7	0.7194 ± 6
4161	1976	biotite	70°37', 165°02'	343.9	0.9324	364.6	2.6008	362 ± 7	0.7238 ± 8
		muscovite		176.2	1.209	144.1	1.4609	359 ± 7	0.7240 ± 8
			Gregory Bluffs (WR isochron age 388 ± 13 Ma)						
4120	1969	biotite	70°45', 166°04'	238.2	0.4161	565.9	3.5994	358 ± 7	0.7163 ± 8
4123	1971	biotite	70°47', 166°02'	265.9	0.5794	453.6	3.0262	358 ± 7	0.7154 ± 8
4127	1972	biotite	70°45', 165°56'	441.3	2.168	201.2	1.7394	356 ± 7	0.7191 ± 18
		muscovite		276.5	0.7437	367.5	2.5870	357 ± 7	0.7190 ± 18
4128	1973	biotite	70°45', 165°56'	275.3	2.210	123.1	1.3359	351 ± 7	0.7214 ± 14
		muscovite		267.4	1.208	218.8	1.8383	359 ± 7	0.7204 ± 14

[a]Analytical precision derived from replicates on a standard feldspar with a low Rb/Sr ratio is at the level of 95% confidence: d(^{87}Rb/^{86}Sr) = 2%, and d(^{87}Sr/^{86}Sr) = 0.06%, but it becomes obvious from the much smaller scatter of the age results that the precision in the high Rb/Sr ratios of the micas is about 0.5%, and this error estimate is used in the text.
[b]WR = whole rock; WR analyses from Vetter et al. [1984, Table 1]. IR = initial ^{87}Sr/^{86}Sr ratio; error estimates of the initial ratios correspond to the last digits given.

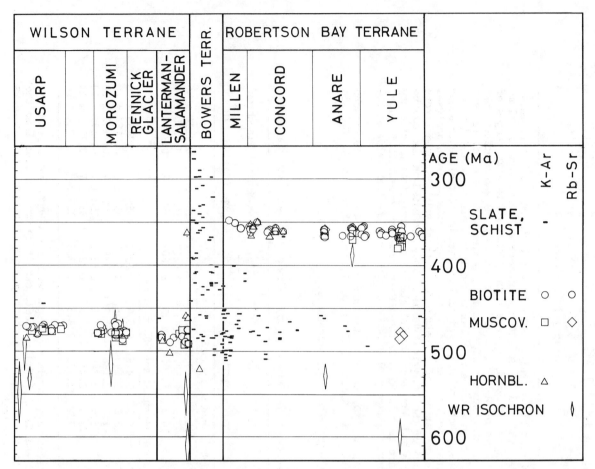

Fig. 4. General WSW-ENE variation of K-Ar and Rb-Sr dates [after Adams and Kreuzer, 1984]. The dates are located according to nine zones which correspond to prominent NW-SE to NNW-SSE trending physiographic features for which names have been arbitrarily chosen. Omitted are the few scattered dates for presumed Admiralty Intrusives from Bowers and Wilson terranes, the dates from the Terra Nova area, a few too young dates for low-potassium biotites, and K-Ar whole-rock dates on granitoids.

K-Ar Dating of Metapelites and K-Ar and Rb-Sr Dating of Minerals

The dominant role of Early Ordovician events in all three terranes is demonstrated by Figure 4. The patterns of K-Ar dates for the three terranes are similar, but the ranges of about 300 to 500 Ma prove the existence of more than one episode of influences.

Wilson Terrane. In the Lanterman metamorphics of the southern Lanterman Range, the maximum ages of 500 and 495 Ma of hornblendes and muscovites provide the same minimum age for the last deformation of regional scale [Adams et al., 1982a]. The diorites to tonalites of the Lanterman Range are foliated, and the unfoliated Orr granite (formerly mapped as an Admiralty Intrusive [Roland et al., 1984]) was reequilibrated at amphibolite facies conditions. It appears that these granitoids alongside the Rennick Glacier are an early phase of the Granite Harbour Intrusives.

In the north of the Lanterman zone and in the Wilson zone of the Wilson Terrane, the maximum mineral ages are around 480 Ma. The existence of pairs of muscovite and biotite with concordant ages, but different ages for closely related locations, suggests a regional uplift and cooling as early as 500 to 480 Ma and local influxes of heat by several pulses of postkinematic Granite Harbour Intrusives from 480 to 470 Ma [Adams et al., 1982a; Kreuzer et al., 1981].

Bowers and Robertson Bay terranes. The oldest dates of 510-500 Ma are scarce [Adams et al., 1982b; Adams and Kreuzer, 1984], but not restricted to the polyphase deformed Millen Schists. The latter is confirmed by recent $^{40}Ar/^{39}Ar$ incremental heating results (T. O. Wright et al., personal communication, 1986) which "have a clear plateau age of 500-505 m.y. for the cleavage formation in both Robertson Bay and Bowers, which is only a few m.y. younger than the fossil ages from Handler Ridge." Hence Robertson Bay and Bowers terranes

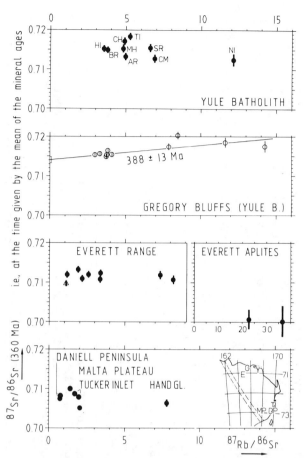

Fig. 5. Strontium isotopic composition of Admiralty Intrusives at 360 Ma versus $^{87}Rb/^{86}Sr$. For abbreviations in the Yule Batholith, see Table 4. For the Everett Range, porphyry is indicated by a triangle.

were deformed right at the Cambrian-Ordovician boundary.

Post-Ordovician Influences on the K-Ar Isotopic Systems

Western and eastern terranes. Figure 4 demonstrates that on a regional scale there were no post-Ordovician influences. In the Wilson Terrane the single young mineral date (415 Ma on a glacial erratic of the Dickason Granite of the southern Wilson zone [Hulston and McCabe, 1972; New Zealand potassium-argon age (NZKA) 25] is outweighed by fairly consistent (MSWD = 1.9) K-Ar ages of 458 ± 3 Ma for biotites from two in situ samples (Table 2). In the Robertson Bay Terrane, outside clear contact metamorphic influences, there are only a few much younger dates, two dates of 419 and 415 Ma from Mount Zykov [Ravich and Krylov, 1964] and a single date of 400 Ma from the Millen Range [Adams and Kreuzer, 1984]. But T. O. Wright et al. (personal communication, 1986) conclude from their $^{40}Ar/^{39}Ar$ work that "all age spectra have been disturbed, some extensively, by later heating events, mostly during the Mesozoic it seems."

Bowers Terrane. In contrast with the other terranes, many younger dates are observed, nearly evenly covering the interval from the Ordovician to less than 300 Ma (Figure 4). Partial resetting due to burial was assumed by Adams and Kreuzer [1984] because of a roughly inverse correlation of the dates with the present-day altitudes of the sampling locations, compatible with a slow uplift since the Mesozoic which is inferred from fission track dating in NVL [Fitzgerald and Gleadow, 1985]. However, paleomagnetic rejuvenations in Mesozoic volcanics on both sides of the Rennick Graben [Delisle and Fromm, 1984] point to a Cretaceous thermal event.

The Devonian/Carboniferous Admiralty Intrusives

The presumed Early Devonian phase. This age is suggested by only three age determinations, two of which are given without analytical data. A single K-Ar whole-rock analysis of the granodiorite of the Upper Tucker Glacier (389 Ma or 375 Ma, depending on the constants used) (M. G. Ravich and A. J. Krylow, personal communication, in the work of Gair et al. [1969]), and "Rb-Sr dating of samples of Mount Adam" (Black Prince area) indicate "closure to isotopic exchange at 370-380 Ma" (E. Stump, personal communication, in the work of Findlay and Jordan [1984]).

Also, the Rb-Sr whole-rock age for the Gregory Bluffs pluton of 388 ± 13 Ma (IR = 0.7141 ± 0.0008, MSWD = 1.7) [Vetter et al., 1984] is not yet convincing, because of a practically three-point isochron (Figure 5).

A former fourth example, the porphyric biotite adamellite from Mount Camelot, Alamein Range, Lanterman zone, with a K-Ar biotite date of 391 ± 10 Ma [Adams, 1975, NZKA 108; Dow and Neall, 1974], is recognized as a Granite Harbour Intrusive (S. G. Borg, et al., unpublished data, 1985).

The phase of Admiralty Intrusives around the Devonian/Carboniferous boundary. This age had been indicated by early biotite dating [Webb et al., 1964; Hulston and McCabe, 1972] and is now convincingly demonstrated by a narrow interval from 370 to 350 Ma for nearly all hornblende and mica ages, representing most of the major granitoid massifs (Figure 4). It is conspicuous that with few exceptions even the minerals of the granitoids with older isochron ages were reset to the age of the younger phase of the Admiralty Intrusives (Table 3).

A middle Carboniferous pulse of Admiralty Intrusives. It has been suggested by Adams et al. [1986], but the radiometric dates in support of it are poor. These dates, mainly from the Bowers Terrane and vicinity, include the 327 ± 10 Ma hornblende date of the Supernal Granite which is in contrast to a biotite date of 354 ± 10 Ma on the same rock specimen [Hulston and McCabe, 1972, NZKA 23; Nathan, 1971], two K-Ar whole-rock dates, that means minimum age estimates of 323 and 316 Ma on "biotite-plagiogranites" from Znamenskiy and Sputnik islands off the Pacific coast [Ravich and

TABLE 4. Rb-Sr Whole-Rock Data of Admiralty Intrusives From Northern Victoria Land

Field No.[d]	Lab. AL No.	Rock Type, Locality[c]	Lat. S.	Long. E.	^{87}Rb (ppm)	^{86}Sr (ppm)	$\frac{^{87}\text{Rb}}{^{86}\text{Sr}}$[a]	$\frac{^{87}\text{Sr}}{^{86}\text{Sr}}$[a]	$\frac{^{87}\text{Sr}}{^{86}\text{Sr}}$[b] (360 Ma)
\multicolumn{10}{l}{Granites of the Yule Batholith}
TES 2	1736	8 km N Cape Moore	70°53'	167°51'	65.14	9.36	6.876	0.74803	0.7128 ± 10
TES 100	1737	Birthday Ridge	70°47'	167°00'	51.20	13.54	3.738	0.73416	0.7150 ± 8
TES 102	1738	Novosad Island	70°42'	167°30'	47.25	3.85	12.13	0.77471	0.7125 ± 15
TES 103	1739	Hughes Island	70°43'	167°39'	49.37	13.83	3.528	0.73335	0.7153 ± 8
TES 104	1740	Sentry Rocks	70°45'	167°24'	68.24	10.25	6.582	0.74905	0.7153 ± 10
TES 105	1741	Ackroyd Ridge	70°50'	166°44'	56.05	11.18	4.955	0.73870	0.7133 ± 9
TES 106	1742	Missen Head	70°40'	166°49'	55.63	11.43	4.811	0.73995	0.7153 ± 9
TES 107	1743	Cape Hooker	70°38'	166°44'	58.57	11.86	4.882	0.74223	0.7172 ± 9
Th 1	1745	Thala Island	70°37'	166°06'	66.49	13.02	5.048	0.74427	0.7184 ± 9
\multicolumn{10}{l}{Magmatic Rocks from the Everett Range}
Ev 01	1746	granite	71°13'	164°28'	48.58	22.10	2.173	0.72210	0.7110 ± 8
Ev 02	1747	granite	71°14'	164°31'	66.14	19.26	3.395	0.72986	0.7125 ± 8
Ev 06	1748	granite	71°10'	164°29'	46.11	38.95	1.170	0.71795	0.7120 ± 7
Ev 08	1749	granite	71°07'	164°48'	55.34	28.75	1.903	0.72304	0.7133 ± 7
Ev 10	1751	granite	71°17'	164°48'	57.00	16.68	3.378	0.72808	0.7108 ± 8
Ev 14	1753	granite	71°13'	164°31'	52.15	19.92	2.588	0.72530	0.7120 ± 8
Ev 05	1755	leucogranite	71°12'	164°31'	67.92	8.16	8.225	0.75285	0.7107 ± 11
Ev 07	1758	pegmatite	71°10'	164°28'	71.36	9.61	7.339	0.74952	0.7119 ± 11
Ev 12	1756	aplite	71°26'	164°24'	75.99	3.31	22.69	0.81695	0.7007 ± 25
Ev 13	1757	aplite	71°12'	164°31'	74.46	2.06	36.64	0.88803	0.7002 ± 39
Ev 03	1759	porphyry	71°14'	164°31'	32.88	28.69	1.133	0.71544	0.7096 ± 7
\multicolumn{10}{l}{Single Granite Samples from the Southern Robertson Bay Terrane}
TES 1	1771	Football Saddle, N of TI	72°31'	169°42'	38.78	22.67	1.691	0.71739	0.7087 ± 5
TU 1	1772	2 km E of Crater Cirque, W of TI	72°37'	169°18'	39.73	19.95	1.969	0.71799	0.7079 ± 5
3228/4411	2001	3 km NNW of Mt. Alberts, W of MP	73°00'	167°51'	34.78	16.75	2.052	0.71562	0.7051 ± 5
3236/4412	2005	5 km N of Cape Phillips, DP	73°02'	169°38'	22.41	31.44	0.7046	0.71113	0.7075 ± 4
3273/4416	2005	S side upper Hand Glacier	72°54'	167°30'	63.59	8.03	7.830	0.74638	0.7063 ± 9
3279/4417	2006	3 km NNE Mt. Hussey, W of MP	72°44'	167°34'	31.11	21.97	1.400	0.71709	0.7099 ± 5
3238/4413	2007	1 km N of Cape Phillips, DP	73°03'	169°38'	22.53	29.61	0.752	0.71205	0.7082 ± 4

[a] Analytical precision at the level of 95% confidence: $d(^{87}\text{Rb}/^{86}\text{Sr}) = d(x) = 2\%$, and $d(^{87}\text{Sr}/^{86}\text{Sr}) = d(y) = 0.1\%$ and 0.06% for Laboratory AL No. smaller and larger than 1770, respectively.
[b] $^{87}\text{Sr}/^{86}\text{Sr}$ ratio at 360 a before present $(y(360) = y - f \cdot x$, with $f = (\exp(360 \cdot 1.42 \cdot 10^{-5}) - 1))$. The given digits for the error estimate correspond to the last digits of given $y(360)$ value. They include the error of x: $d(y_c(360))^2 = d(y)^2 + f^2 \cdot d(x)^2$.
[c] For geographical abbreviations see Figure 2.
[d] Samples collected by F. Tessensohn (TES), U. Vetter (Th, EV), D. Wyborn (TU), M. Schmidt-Thomé (32..).

Krylov, 1964], and one date of 337 ± 14 Ma (R. L. Armstrong, personal communication, in the work of Laird et al. [1974]) for a low-potassium biotite (5.9 wt % K) from Mount Staley. However, Mount Staley belongs to the Salamander Granite Complex, which should not be directly correlated with the Admiralty Intrusives [Borg, 1984].

The initial strontium ratios of the Admiralty Intrusives from the Robertson Bay Terrane. The initial ratios, which strongly resemble the $^{87}Sr/^{86}Sr$ ratios calculated for 360 Ma, seem to decrease from the northeast toward the southwest. Specifically, values of around 0.715 are found in the Yule batholith, values of about 0.712 in the Everett Range, and values of around 0.708 in the southern Robertson Bay Terrane (Figure 5 and Table 4) [Borg and Stump, this volume].

Several of the discriminating features by which Pitcher [1982, Table 2] characterizes Caledonian I-type granitoids apply to the Admiralty Intrusives and would relate them to a Caledonian-type postclosure uplift. These features include high initial Sr ratios; the prevalence of biotite granodiorites and granites; mixed xenolith populations; dispersed isolated complexes of multiple plutons, sometimes associated with basalt-andesite lavas; short sustained plutonism, and they are rarely strongly mineralized. But features not observed include hints for retrograde metamorphism, for postcollision uplift (the Black Prince area is an exception [Findlay and Jordan, 1984]), and for large-scale faulting.

We speculate that the lack of orogenic features is the result of an oblique-motion collision of the Robertson Bay, and probably the Bowers terranes, with the Wilson Terrane during the Devonian.

Devonian to Carboniferous Volcanics and Dikes

Silicic and mafic volcanics are known from the Gallipoli Heights and Mount Anakiwa in the eastern Wilson Terrane, from Lawrence Peaks in the Bowers Terrane, and from the Black Prince area in the central Robertson Bay Terrane [Adams et al., 1986]. Presumably associated with these are dikes of orogenic basalts, andesites, and dacites [Findlay and Jordan, 1984].

For the Gallipoli Rhyolites, Faure and Gair [1970] determined a Rb-Sr isochron age of 370 ± 40 Ma, which is questionable because of only two samples and problematic material; Borg [1984] regards the Gallipoli Rhyolites as a distinct Carboniferous silicic magmatism in the Wilson Terrane.

At Mount Black Prince, the volcanics overlie the Admiralty Intrusives (cooling age of 370-380 Ma) and are intercalated with plant-bearing sediments of a possible age range from Givetian to middle Carboniferous [Findlay and Jordan, 1984]. K-Ar whole-rock dates on (altered) volcanics and dikes range from 380 Ma to less than 300 Ma [Adams et al., 1986]. Findlay and Jordan [1984] and Adams et al. [1986] relate the volcanics to the Devonian-Carboniferous second phase of the Admiralty Intrusives.

Summary

Late Precambrian (600 Ma), Cambrian (550 Ma), and Ordovician phases of metamorphism and granitoid intrusions (500-450 Ma) have been established in the the Wilson Terrane of NVL. They can be related to the Beardmore and Ross orogenies of the central Transantarctic Mountains and confirm the view of the Wilson Terrane as a continuation of the Ross Fold Belt of the Transantarctic Mountains.

However, single examples of similar ages are indicated also for granitoids of the Robertson Bay Terrane. The Beardmore equivalent granodiorite of Surgeon Island may represent a suspect terrane, faulted against the northern margin of the Robertson Bay Terrane.

Ross-equivalent Ordovician ages of 505 to 500 Ma on slates and schists date the deformation of the Robertson Bay and the Bowers terranes at the Cambrian-Ordovician boundary. No comparable lithologies are known from the central Transantarctic Mountains. It is possible that during the Ordovician phase of the Ross Orogeny, the polyphase metamorphic rocks of the Wilson Terrane, the Cambrian intraoceanic volcanic arc of the Bowers Terrane with its regressive cover, and the flysch-like sediments of the Robertson Bay Terrane were juxtaposed.

The Devonian-Carboniferous high-level granitoids of the Admiralty Intrusives are abundant in the Robertson Bay Terrane and die out toward the west. They have many characteristics of Caledonian I-type granitoids and hence point to a post-collision origin. The lack of orogenic features in the Robertson Bay Terrane can be explained by an oblique collision. An allochthonous landmass, including the Robertson Bay and Bowers terranes and extending toward the northeast as suggested by the increasing initial strontium ratios (probably tracing increasing crustal thickness), collided, probably with the Wilson Terrane. This collision occurred before the formation of the Devonian-Carboniferous Admiralty Intrusives, which were themselves generated by the collision.

To overcome the conflicting observations, chemical and isotopic investigations should be focused on the granitoids within and near the borders of the Bowers Terrane, e.g., Supernal Granite, Salamander complex, and the Sputnik and Znamenskiy islands, in order to decide whether they belong to the Admiralty Intrusives.

Acknowledgments. We are indebted to C. J. Adams and S. G. Borg who improved an earlier version of the manuscript. F. Tessensohn, S. G. Borg, R. Burmester, H. Jordan, and N. Roland contributed through valuable discussions. We thank C. Besang and H. Franke for mineral separation, B. Eichmann and H. Schyrocki for the Rb-Sr analyses, and H. Klappert, M. Metz, L. Thiesswald, and D. Uebersohn for the K-Ar analyses.

References

Adams, C. J., New Zealand potassium-argon age list-2, N. Z. J. Geol. Geophys., 18, 443-467, 1975.

Adams, C. J., Age and ancestry of the metamorphic rocks of the Daniels Range, Usarp Mountains, Antarctica, in Geological Investigations in Northern Victoria Land, Antarct. Res. Ser., edited by E. Stump, AGU, Washington, D. C., in press, 1986.

Adams, C. J., and H. Kreuzer, Potassium-argon age studies of slates and phyllites from the Bowers and Robertson Bay terranes, north Victoria Land, Antarctica, Geol. Jahrb., Reihe B, 60, 265-288, 1984.

Adams, C. J., J. Gabites, A. Wodzicki, M. G. Laird, and J. D. Bradshaw, Potassium-argon geochronology of the Precambrian-Cambrian Wilson and Robertson Bay groups and Bowers Supergroup, northern Victoria Land, Antarctica, in Antarctic Geoscience, edited by C. Craddock, pp. 543-548, University of Wisconsin Press, Madison, 1982a.

Adams, C. J., J. Gabites, and G. W. Grindley, Orogenic history of the central Transantarctic Mountains: New K-Ar age data on the Precambrian-early Paleozoic basement, in Antarctic Geoscience, edited by C. Craddock, pp. 817-826, University of Wisconsin Press, Madison, 1982b.

Adams, C. J., P. F. Whitla, R. H. Findlay, and B. F. Field, Age of the Black Prince Volcanics in the central Admiralty Mountains and possibly related hypabyssal rocks in the Millen Range, northern Victoria Land, Antarctica, in Geological Investigations in Northern Victoria Land, Antarct. Res. Ser., edited by E. Stump, AGU., Washington, D. C., in press, 1986.

Borg, S. G., Granitoids of northern Victoria Land, Antarctica, Ph.D. dissertation, 355 pp., Arizona State Univ., Tempe, 1984.

Borg, S. G., and E. Stump, Paleozoic magmatism and associated tectonic problems of northern Victoria Land, Antarctica, this volume.

Cooper, R. A., J. B. Jago, A. J. Rowell, and P. Braddock, Age and correlation of the Cambrian-Ordovician Bowers Supergroup, northern Victoria Land, in Antarctic Earth Science, edited by R. L. Oliver, P. R. James, and J. B. Jago, pp. 128131, Australian Academy of Science, Canberra, 1983.

Dalrymple, G. B., and M. A. Lanphere, Potassium-Argon Dating, 258 pp., W. H. Freeman, San Francisco, 1969.

Delisle, G., and K. Fromm, Paleomagnetic investigations of Ferrar Supergroup rocks, north Victoria Land, Antarctica, Geol. Jahrb., Reihe B, 60, 41-55, 1984.

Dow, J. A. S., and V. E. Neall, Geology of the Rennick Glacier, northern Victoria Land, Antarctica, N. Z. J. Geol. Geophys., 17, 659-714, 1974.

Faure, G., and H. S. Gair, Age determinations of rocks from northern Victoria Land, Antarctica, N. Z. J. Geol. Geophys., 13, 1024-1026, 1970.

Findlay, R. H., and B. D. Field, Tectonic significance of deformations affecting the Robertson Bay Group and associated rocks, northern Victoria Land, Antarctica, in Antarctic Earth Science, edited by R. L. Oliver, P. R. James, and J. B. Jago, pp. 107-112, Australian Academy of Science, Canberra, 1983.

Findlay, R. H., and H. Jordan, The volcanic rocks of Mt. Black Prince and Lawrence Peaks, north Victoria Land, Antarctica, Geol. Jahrb., Reihe B, 60, 143-153, 1984.

Fitzgerald, P. G., and A. J. W. Gleadow, Uplift history of the Transantarctic Mountains, Victoria Land, Antarctica, in Sixth Gondwana Symposium Abstracts, Misc. Publ. 231, p. 42, Institute of Polar Studies, Ohio State University, Columbus, 1985.

Gair, H. S., A. Sturm, S. J. Carrier, and G. W. Grindley, The geology of northern Victoria Land, in Geologic Maps of Antarctica, Antarct. Map Folio Ser., edited by V. C. Bushnell and C. Craddock, folio 12, sheet 13, Am. Geogr. Soc., New York, 1969.

Grew, E. S., G. Kleinschmidt, and W. Schubert, Contrasting metamorphic belts in north Victoria Land, Antarctica, Geol. Jahrb., Reihe B, 60, 253-263, 1984.

Grindley, G. W., and P. J. Oliver, Post-Ross Orogeny cratonization of northern Victoria Land, in Antarctic Earth Science, edited by R. L. Oliver, P. R. James, and J. B. Jago, pp. 133-139, Australian Academy of Science, Canberra, 1983.

Hulston, J. R., and W. J. McCabe, New Zealand potassium-argon age list 1, N. Z. J. Geol. Geophys., 15, 406-432, 1972.

Jenkins, R. J. F., Ediacaran events: Boundary relationships and correlation of key sections, especially in "Amorica," Geol. Mag., 121, 635-643, 1984.

Jordan, H., R. Findlay, G. Mortimer, M. Schmidt-Thomé, A. Crawford, and P. Miller, Geology of the northern Bowers Mountains, north Victoria Land, Antarctica, Geol. Jahrb., Reihe B, 60, 57-81, 1984.

Kleinschmidt, G., Regional metamorphism in the Robertson Bay Group area and in the southern Daniels Range, north Victoria Land, Antarctica: A preliminary comparison, Geol. Jahrb., Reihe B, 41, 201-228, 1981.

Kleinschmidt, G., Trends in regional metamorphism and deformation in northern Victoria Land, Antarctica, in Antarctic Earth Science, edited by R. L. Oliver, P. R. James, and J. B. Jago, pp. 119-122, Australian Academy of Science, Canberra, 1983.

Kleinschmidt, G., and D. N. B. Skinner, Deformation styles in the basement rocks of north Victoria Land, Antarctica, Geol. Jahrb., Reihe B, 41, 155-199, 1981.

Kleinschmidt, G., and F. Tessensohn, Early Paleozoic westward directed subduction at the Pacific margin of Antarctica, this volume.

Kleinschmidt, G., N. W. Roland, and W. Schubert, The metamorphic basement complex in the Mountaineer Range, north Victoria Land, Antarctica, Geol. Jahrb., Reihe B, 60, 213-251, 1984.

Kreuzer, H., A. Höhndorf, H. Lenz, U. Vetter, F. Tessensohn, P. Müller, H. Jordan, W. Harre, and C. Besang, K/Ar and Rb/Sr dating of igneous rocks from north Victoria Land, Antarctica, Geol. Jahrb., Reihe B, 41, 267-273, 1981.

Laird, M. G., and J. D. Bradshaw, New data on the lower Paleozoic Bowers Supergroup, northern Victoria Land, in Antarctic Earth Science, edited by R. L. Oliver, P. R. James, and J. B. Jago, pp. 123-126, Australian Academy of Science, Canberra, 1983.

Laird, M. G., P. B. Andrews, and P. R. Kyle, Geology of northern Evans Névé, Victoria Land, Antarctica, N. Z. J. Geol. Geophys., 17, 587-601, 1974.

Nathan, S., K/Ar dates from the area between Priestley and Mariner glaciers, northern Victoria Land, Antarctica, N. Z. J. Geol. Geophys., 14, 504-511, 1971.

Odin, G. S., Interlaboratory standard for dating purposes, in Numerical Dating in Stratigraphy, edited by G. S. Odin, pp. 123-150, John Wiley, 1982.

Pitcher, W. S., Granite type and tectonic environment, in Mountain Building Processes, edited by K. S. Hsü, pp. 19-40, Academic, New York, 1982.

Plummer, C. C., R. S. Babcock, J. W. Sheraton, C. J. Adams, and R. L. Oliver, Geology of the Daniels Range, northern Victoria Land, Antarctica: A preliminary report, in Antarctic Earth Science, edited by R. L. Oliver, P. R. James, and J. B. Jago, pp. 113-117, Australian Academy of Science, Canberra, 1983.

Ravich, M. G., and A. J. Krylov, Absolute ages of rocks from East Antarctica, in Antarctic Geology, edited by R. J. Adie, pp. 579-589, North-Holland, Amsterdam, 1964.

Roland, N. W., G. M. Gibson, G. Kleinschmidt, and W. Schubert, Metamorphism and structural relations of the Lanterman metamorphics, north Victoria Land, Antarctica, Geol. Jahrb., Reihe B, 60, 319-361, 1984.

Skinner, D. N. B., The geology of Terra Nova Bay, in Antarctic Earth Science, edited by R. L. Oliver, P. R. James, and J. B. Jago, pp. 150-155, Australian Academy of Science, Canberra, 1983.

Steiger, R. H., and E. Jäger, Subcommission on geochronology: Convention on the use of decay constant in geo- and cosmochronology, Earth Planet. Sci. Lett., 36, 359-362, 1977.

Tessensohn, F., Geological and tectonic history of the Bowers Structural Zone, north Victoria Land, Antarctica, Geol. Jahrb., Reihe B, 60, 371-396, 1984.

Tessensohn, F., K. Duphorn, H. Jordan, G. Kleinschmidt, D. N. B. Skinner, U. Vetter, T. O. Wright, and D. Wyborn, Geological comparison of basement units in north Victoria Land, Antarctica, Geol. Jahrb., Reihe B, 41, 31-88, 1981.

Vetter, U., N. W. Roland, H. Kreuzer, A. Höhndorf, H. Lenz, and C. Besang, Geochemistry, petrography, and geochronology of the Cambro-Ordovician and Devonian-Carboniferous granitoids of northern Victoria Land, Antarctica, in Antarctic Earth Science, edited by R. L. Oliver, P. R. James, and J. B. Jago, pp. 140-155, Australian Academy of Science, Canberra, 1983.

Vetter, U., H. Lenz, H. Kreuzer, and C. Besang, Pre-Ross granites at the Pacific margin of the Robertson Bay Terrane, north Victoria Land, Antarctica, Geol. Jahrb., Reihe B, 60, 363-369, 1984.

Weaver, S. D., J. D. Bradshaw, and M. G. Laird, Geochemistry of Cambrian volcanics of the Bowers Supergroup and implications for the early Paleozoic tectonic evolution of northern Victoria Land, Antarctica, Earth Planet. Sci. Lett., 68, 128-140, 1984.

Webb, A. W., I. McDougall, and J. A. Cooper, Potassium-argon dates from the Vincennes Bay region and Oates Land, in Antarctic Geology, edited by R. J. Adie, pp. 597-600, North-Holland, Amsterdam, 1964.

Wright, T. O., Sedimentology of the Robertson Bay Group, north Victoria Land, Antarctica, Geol. Jahrb., Reihe B, 41, 127-138, 1981.

Wright, T. O., and R. H. Findlay, Relationship between the Robertson Bay Group and the Bowers Supergroup: New progress and complications from the Victory Mountains, north Victoria Land, Antarctica, Geol. Jahrb., Reihe B, 60, 105-116, 1984.

Wyborn, D., Granitoids of north Victoria Land, Antarctica--Field and petrographic observations, Geol. Jahrb., Reihe B, 41, 229-249, 1981.

Wyborn, D., Chemistry of Paleozoic granites of northern Victoria Land (abstract), in Antarctic Earth Science, edited by R. L. Oliver, P. R. James, and J. B. Jago, p. 144, Australian Academy of Science, Canberra, 1983.

A REVIEW OF THE PROBLEMS IMPORTANT FOR INTERPRETATION OF THE CAMBRO-ORDOVICIAN
PALEOGEOGRAPHY OF NORTHERN VICTORIA LAND (ANTARCTICA), TASMANIA, AND NEW ZEALAND

R. H. Findlay[1]

Department of Geology, University of Tasmania, Hobart, Tasmania, Australia 7001

Abstract. The Cambrian to Ordovician tectonics of northern Victoria Land may be explained in terms of a plate tectonic collision involving three terranes: the western foreland Wilson Terrane, the central island arc Bowers Terrane, and the eastern turbidite fan sequence of the Robertson Bay Terrane. The Cambro-Ordovician tectonics of Tasmania differ in that they were dominated by rifting, closure of rifts, and uplift. Likewise, although lithologies in the two regions are superficially comparable, the important units in either one of the regions cannot be identified in the other. New Zealand poses special problems. It contains Cambro-Ordovician sequences comparable to those in northern Victoria Land and Ordovician-Devonian sequences similar to those in Tasmania. New Zealand's Cambro-Ordovican tectonics resemble those of Tasmania, whereas New Zealand's Siluro-Devonian tectonics share close affinities with those of both Tasmania and northern Victoria Land. It is concluded that the three regions formed part of a tectonically linked system of Cambrian island arcs and rifted microcontinents which, in Late Cambrian and Early Ordovician times, collided with the then eastern margin of Gondwana (eastern Antarctica-South Australia). A subsequent Siluro-Devonian orogeny occurred along this newly accreted eastern margin of Gondwana. The newly reported west over east thrusting in northern Victoria Land could well have occurred during this event, rather than in Early Ordovican times.

Introduction

The 85 Ma tectonic reconstruction [cf. Grindley and Davey, 1982] (Figure 1) of the southwestern Pacific margin of Gondwana is well supported by the good fit between the shape of the continental shelves of Antarctica and Australia.

This Devonian-pre-Late Cretaceous fit is confirmed by (1) the Devono-Carboniferous volcanics and associated granitoids [cf. Grindley and Davey, 1982] of Marie Byrd Land, northern Victoria Land, northwest Nelson (New Zealand), Tasmania, and southeastern Australia (Figure 1); (2) the Carboniferous-Permian glacigene sediments in the Transantarctic Mountains (Figure 1), northern Victoria Land, Tasmania, and eastern Australia [Banks, 1962a; Crowell and Frakes, 1971, 1974; Barrett, 1981]; (3) the remarkably similar fluvial to shallow-marine sandstones of the Devono-Jurassic Beacon and Parmeener supergroups of the Transantarctic Mountains and Tasmania, respectively [Banks, 1962b; Clarke and Banks, 1974; Barrett, 1981]; and (4) the occurrence of the same dolerites throughout the Transantarctic Mountains (Ferrar Dolerite) and in Tasmania.

However, it has become increasingly evident that the 85 Ma reconstruction does not adequately explain the Cambro-Lower Ordovician paleogeography of the eastern Australo-Antarctic region of Gondwana [cf. Grindley and Davey, 1982; Brown et al., 1982]. This paper reviews the Cambrian and Ordovician events and sequences critical for understanding the lower Paleozoic paleogeography of northern Victoria Land, South Australia, and Tasmania, and also attempts to assess the relevance of the early Paleozoic of New Zealand to the regions. An interpretation is presented which considers the problems discussed.

Fits and Misfits

Tectonic Comparison

During Upper Cambrian and Lower Ordovician times, regional, more or less simultaneous tectonic events occurred in South Australia (Delamerian Orogeny), the Oates Coast, the western Ross Sea region of Antarctica (Ross Orogeny), Tasmania (Jukesian Movement), and New Zealand (Haupiri Disturbance) (Figure 1).

South Australia-Antarctica. The 85 Ma reconstruction (Figure 1) juxtaposes the Proterozoic basement of South Australia with the Proterozoic of East Antarctica. The two regions also display close similarities [Oliver, 1972; Ravich, 1982; Baillie, 1985] in their Cambro-Ordovician tectonics (Delamerian Orogeny of South Australia; Ross Orogeny of the Transantarctic Mountains).

Of importance are the Cambro-Ordovician Delamerian granites of South Australia and the Granite Harbour Intrusives of southern Victoria Land. These intrude Lower Cambrian shallow shelf sequences which are particularly rich in carbonates

---

[1] Now at Tasmanian Department of Mines, Rosny Park, Tasmania, Australia 7018.

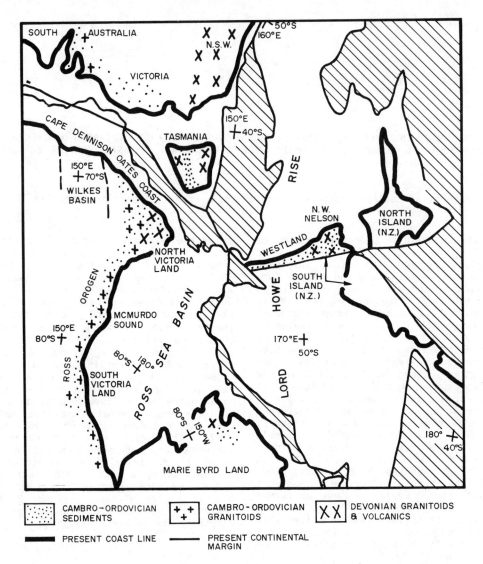

Fig. 1. Continental reconstruction immediately preceding Gondwana breakup. From Grindley and Davey [1982].

(Normanville Group of South Australia; Byrd Group of the Transantarctic Mountains). However, although the fit of the Proterozoic basement rocks and the continental shelves supports the 85 Ma reconstruction, in this fit there is a misalignment between the two belts of granite and their host sediments.

Recent geophysical studies [Steed, 1983] indicate that the Wilkes Subglacial Basin (Figure 1) is an important post-lower Paleozoic, pre-Cenozoic rift. Closure of the rift would rotate northern Victoria Land so that the Delamerian granites would line up with the Lower Ordovician Granite Harbour Intrusive Complex in the Wilson Hills and Morozumi Range of northern Victoria Land, and also align the flyschoid Berg Group [Soloviev, 1960] of the Oates Coast with similar rocks of the Kanmantoo group of South Australia. Although the Berg Group is thought to be Precambrian-Cambrian [Iltchenko, 1972], it has not been studied in detail, and future work could well lead to a reassesment of its age range.

Northern Victoria Land and Tasmania. Present reconstructions of Gondwana [Laird et al., 1977; Jago, 1980; Grindley and Davey, 1982] show either southern Tasmania juxtaposed with the northern coast of northern Victoria Land or southwest Tasmania abutting the northeastern coast of northern Victoria Land (models 1 and 2 in Figures 2c and 2d).

Tectonic developments in northern Victoria Land are listed in Figure 3. Thrust tectonics have been emphasized strongly in support for a model involving a Cambro-Ordovician plate-tectonic collision between the Robertson Bay Terrane and the island arc sequence of the Bowers Terrane, which is thought to have overlain a west dipping subduction zone [cf. Gibson and Wright, 1985]. The sup-

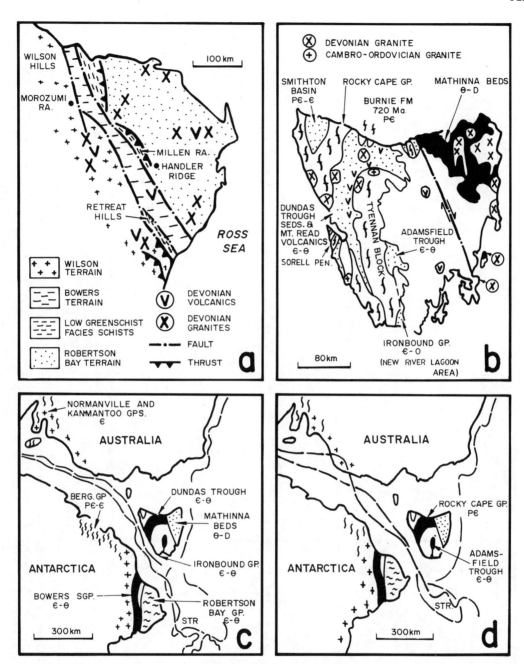

Fig. 2. (a) Locality and terrane map, northern Victoria Land; (b) summary geological map of Tasmania, with post-Devonian rocks omitted; (c) reconstruction of northern Victoria Land and Australia from Laird et al. [1977], model 1; (d) reconstruction of northern Victoria Land and Australia from Jago [1980], model 2.

posed plate suture is considered to be marked by the Millen Terrane [Mortimer, 1983; Gibson and Wright, 1985].

Gibson and Wright [1985] claim that thrusting was particularly important in closing to its present 6 km width the original basin between the Bowers Terrane island arc and the western, dominantly high-temperature/low-pressure, Wilson Terrane [see also Weaver et al., 1984].

There are at least three distinct phases of thrusting evident (Figure 3). Thrusting [Findlay and Field, 1983; Wright and Findlay, 1984] occurred within the Millen Terrane in concert with $_BF_1$, $_MF_1$, and $_RF_1$ (Figure 3) at 510 ± 15 Ma [Findlay, 1986]. The only known first-phase thrust occurs at Crosscut Peak in the Millen Range [cf. Crowder, 1968; Findlay and Field, 1982, 1983; Wright and Findlay, 1984; Bradshaw et al., 1985;

Event	Wilson Terrane (West)	Boundary	Bowers Terrane (Central)	Boundary	Robertson Bay Terrane (Eastern)
I	Time of end of sedimentation unknown; plausibly Late Cambrian.	Unknown	Cessation of deep to shallow marine sedimentation in Late Cambrian-Early Ordovician times [cf. Bradshaw et al. 1985; Bradshaw and Laird, 1983].	Unknown	Cessation of turbiditic sedimentation in Tremadoc [Burrett and Findlay 1984; Wright et al. 1984]. Note: age of basal Ordovician 500 ± 32 Ma [Palmer 1983].
II	Migmatisation, deformation, and massive intrusion of dominantly S-type Granite Intrusive granitoids [cf. Vetter et al. 1983; Tessensohn 1984]. Maximum cooling age 491 ± 6 Ma [Adams and Kreuzer 1984]. HP kyanite zone metamorphism in east, HT sillimanite/andalusite metamorphism in west [Grew et al. 1984] separated by Rennick-Aviator line.	No evidence for Leap year Fault	Open to tight folds ($_BF_1$) about NW trend, plunge at 0-70° in relatively uniform, steep axial plane cleavage yielding max. K/Ar slate age of 504 ± 4 Ma [Adams and Kreuzer 1984]. No preferred vergence except near west over east reverse faults in west of terrane [Tessensohn 1984].	"Crosscut Event". Millen terrane forms. Syn-M_1 metamorphism with cooling at 510 ± 8 Ma [Adams and Kreuzer 1984]. Isoclinal, possibly recumbent folds ($_MF_1$) with axial plane schistosity and thrusting.	Folding (F_1) at 510 ± 15 Ma. Noncylindrical upright folds plunge 0° - 60° in uniform NW trending, steep axial plane. Folds tighten progressively SW till interlimb angle = 25° [Findlay 1986]. Metamorphism very low grade [Kleinschmidt 1983] at 508 ± 12 Ma [Adams and Kreuzer 1984].
III	West over east thrusting juxtaposes high-grade rocks over parts of Bowers terrane [cf. Gibson et al. 1984].	? -- Lanterman Fault Zone -- ?	West over east reverse faults and minor folds of uncertain significance [cf. Gibson et al. 1984]. Age post-Early Ordovician to Devonian/Carboniferous [cf. Tessensohn 1984].	Upright open folds ($_MF_2$) about subhorizontal axes in a NW trending axial plane crenulation cleavage ($_MS_2$)	Shortening normal to fold axial planes, conjugate and reverse faults, ascribed to latter part of Ross Orogeny [Findlay 1986] but conceivably younger.
				? -- Leap Year Fault -- ?	
IV	Intrusion of Devonian Admiralty Granite.		Intrusion of Devonian Admiralty Granite and extrusion of associated volcanics. Granite intrudes some faults within Lanterman Fault Zone.	Intrusion of Devonian Admiralty Granite.	Intrusion of Devonian Admiralty Granite and extrusion of associated volcanics.

Fig. 3. Events (I, II, III, and IV) spanning Late Cambrian to Devonian times in northern Victoria Land. Terranes and their boundaries are arranged in their present geographical position from west (left) to east (right). The Ross Orogeny covers the radiometrically dated event (II) in all three terranes and is thought to include [cf. Tessensohn, 1984; Gibson and Wright, 1985] the west over east thrusting (event III) for which the age constraints are broader [see Tessensohn, 1984].

Findlay, 1986]. First-phase thrusting may be represented also by the low-grade schists in the southern part of the Bowers Terrane [Riddols and Hancox, 1968; Kleinschmidt et al., 1984; Gibson et al., 1984].

The most obvious thrusts (event III, Figure 3) dip moderately to steeply westward and are not known to be folded. They are younger than the regional folding in the Wilson and Bowers terranes but precede intrusion of the Devono-Carboniferous Admiralty Intrusives [Tessensohn, 1984]. A minimum age for the event III thrusts could well be the 420 ± 17 Ma K/Ar age derived from an actinolite-bearing shear plane in the eastern part of the Lanterman Range [Adams and Kruezer, 1984]. It is conceivable that this date could be evidence that event III formed some part of the much disputed Borchgrevink Orogeny [cf. Craddock, 1972; Tessensohn et al., 1981].

The youngest thrusts (event IV, Figure 3), or rather reverse faults [see Roland et al., 1984, Figure 1], postdate Jurassic dolerites of the Ferrar Group and are not discussed further.

In contrast to northern Victoria Land, the Cambro-Ordovician sediments of Tasmania occupy narrow belts between at least five Precambrian blocks (Smithton, Tyennan, Jubilee, Badger Head, and Forth blocks or "nuclei"). The most prominent of these belts is the Dundas Trough, wherein was deposited the Dundas Group, a possible [Laird et al., 1977] Tasmanian continuation of the Bowers Supergroup.

In western Tasmania, the major lower Paleozic tectonic events can be summarized [Corbett et al., 1977; Williams, 1978] as: (1) late Precambrian to possible Middle Cambrian separation of continental basement blocks and concomitant igneous activity; (2) introduction of mafics and ultramafics probably during early-Middle to Middle Cambrian times; (3) production of Middle to Late Cambrian olistostromes; (4) Early? Cambrian to possible late-Middle or early-Late Cambrian volcanism of Andean continental margin type, as exemplified principally by the Mount Read volcanics which flank the western edge of the Tyennan nucleus (Figure 2); (5) development of Late Cambrian-Early Ordovician local to regional unconformities during closure and uplift of the Cambrian troughs; and (6) shelf carbonate and siliciclastic sedimentation from Middle Ordovician to Silurian times.

Attempts have been made [cf. Solomon and Griffiths, 1972, 1974; Corbett et al., 1972; Crook and Powell, 1976; Crook, 1980; Green, 1983] to interpret this history in terms of collisional tectonics involving a subduction zone related to the Mount Read arc. However, the predominant opinion [cf. Corbett et al., 1977; Williams, 1978] disavows ocean-scale separation of the Precambrian blocks and subduction, within the Dundas Trough, of large amounts of oceanic crust below the Mount Read arc. Rather, the preferred view [Williams, 1978] favors continental rifting with minor production of oceanic crust, followed by closure of the rifts; indeed Powell [1984] interpreted the Dundas Trough as a relatively small intracontinental graben.

The major problem with the interpretation of Tasmanian tectonics is that there has been a major Middle Devonian event (Taberraberran Orogeny [cf. Williams, 1978]) which involved considerable folding and thrusting. Despite the masking effects of this event, it is indeed possible that thrust tectonics occurred during Late Cambrian-Early Ordovician times [Green, 1983; I. R. McLeod, unpublished data, 1955; N. Turner, personal communication, 1986; M. P. McClenaghan and R. Findlay, unpublished data, 1986; R. Findlay, unpublished data, 1984] during an important low-grade metamorphic event [Adams et al., 1985], and in this respect there may be similarities with northern Victoria Land. However, I reiterate the three important differences between the tectonic styles of these two regions: (1) in Tasmania, there is no obvious equivalent to the extensively migmatized Wilson Terrane; (2) in northern Victoria Land, there are no equivalents of the Precambrian blocks of Tasmania; and (3) the simple plate tectonic history proposed by Gibson and Wright [1985] for northern Victoria Land is very difficult to reconcile with the Cambro-Ordovician developments in Tasmania (see Williams [1978] and later discussion).

Lithostratigraphic Comparisons

Southern Victoria Land-South Australia. In both the Ross Orogen and the Delamerian Orogen (Delamerides of Powell [1984]), the Lower Cambrian sequences contain thick, Archaeocyath-bearing carbonates called the Shackleton Limestone [Laird et al., 1971] and the Normanville Group, respectively [cf. Cook, 1982; Jago et al., 1985]. Although these rocks indicate a Lower Cambrian epicontinental shelf flanking the South Australian and southern Victoria Land sectors of the Lower Cambrian Gondwana, there is no evidence that this carbonate shelf was continuous; that is, there are no carbonates of proven Lower Cambrian age known between southern Victoria Land and the Oates Coast (Figure 1). The Salmon Marble Formation [Blank et al., 1963; Findlay et al., 1984] in the McMurdo Sound region is the only possible formation for linking the two groups. However, the age of the Salmon Marble Formation is unknown; the tentative identification [Blank et al., 1963] of Archaeocyaths in these rocks is false [Findlay et al., 1984), and there is plausible evidence [Skinner, 1983] that the Salmon Marble could be Precambrian.

Mawson's report [Mawson, 1940] of thermally metamorphosed limestone erratics at Cape Denison, one of which (sample 316) may contain Archaeocyaths, could indicate that the Normanville Group is represented in Antarctica. However, the identification of Archaeocyaths has not been confirmed, and the erratics could have been derived from the Precambrian of the East Antarctic shield.

Both South Australia and the Oates Coast do, however, contain extensive flysch sequences, the Kanmantoo and Berg groups, respectively. The Kanmantoo Group overlies limestones of the Normanville Group and is assigned to the Early Cambrian [Daily and Milnes, 1973; Cook, 1982]. As the Berg Group is assigned on the basis of achritarchs [Iltchenko, 1972] to the late Precambrian-Cambrian, this unit could be the southern extension of the Kanmantoo Group.

Northern Victoria Land-Tasmania. Four important suites of rock must be considered when attempting stratigraphic correlations between northern Victoria Land and Tasmania: (1) the very extensive, flyschoid [Wright, 1981; Field and Findlay, 1983] Vendian to Lower Ordovician [Vetter et al., 1984; Burrett and Findlay, 1984] Robertson Bay Group of northern Victoria Land; (2) the extensive Precambrian units of western Tasmania; (3) the Eocambrian to Late Cambrian volcanic associations (Figure 4) of the west Tasmanian troughs and the Glasgow Formation volcanics of northern Victoria Land; and (4) the Millen Range schists of northern Victoria Land. If these units are not common to both regions, direct Cambrian links such as those advocated in models 1 and 2 (Figures 2c and 2d) are not plausible.

The Robertson Bay Group underlies approximately half of northern Victoria Land. The only Paleozoic Tasmanian flyschoid unit of comparable extent is formed by the Mathinna Beds of northeast Tasmania. However, although lithologically similar, the Mathinna Beds range in age from Early Ordovician [Banks and Smith, 1968] to Early Devonian [Banks, 1962a; Rickards and Banks, 1979]. In contrast, the Robertson Bay Group is no younger than approximately 490 Ma, when it was deformed in the Ross Orogeny [Adams and Kreuzer, 1984].

Early speculations that the Robertson Bay Group was of late Precambrian to Cambrian age [cf. Laird et al., 1977] led to its correlation with both the Precambrian of the Tyennan Block (model 1, Figure 2c) and with rocks in the Rocky Cape area (Rocky Cape Group and Burnie Formation) [Jago, 1980] (model 2, Figure 2d).

The clean quartzites and phyllites, common to the Rocky Cape Group and Tyennan Block, are dissimilar lithologically and sedimentologically from the quartz flysch of the Robertson Bay Group. Furthermore, these Tasmanian Precambrian rocks were metamorphosed and multiply deformed between 850 and 600 Ma [Råheim and Compston, 1977], probably before deposition of the Robertson Bay Group.

Although the Burnie Formation is a quartz flysch sequence similar to the Robertson Bay Group, it has a minimum age of about 720 Ma (Richards, cited by Solomon and Griffiths [1974]; see also Crook [1980]). Furthermore, the Burnie Formation was deformed and metamorphosed during the late Precambrian Penguin Orogeny [Williams, 1978], for which there is no evidence in northern Victoria Land and which probably preceded deposition of the Robertson Bay Group.

A Middle Cambrian to Ordovician proximal fan sequence (Ironbound Group; R. Findlay, unpublished data, 1984) occurs in the Ironbound Range-New River Lagoon region (see also Berry and Harley [1983]; Bischoff [1983]) of southern Tasmania. Although it is conceivable that this group could form a Tasmanian remnant of the Robertson Bay Group, paleocurrent directions and clast content indicate a northern, i.e., Tasmanian, source for the Ironbound Group. In contrast, the Robertson Bay Group was derived from the southeast [Wright, 1981; Field and Findlay, 1983]. Therefore, even if the two groups were to have shared the same Cambrian to Lower Ordovician depocenter [cf. Laird, 1981], they are unlikely to have been part of the same submarine fan system. The key regions for the resolution of this problem are the sparsely studied northwestern part of the Robertson Bay Terrane and the submarine South Tasman Rise.

Model 1 (see also Powell [1984] and Figures 1 and 2) links the Dundas Trough of Tasmania with the Bowers Terrane of northern Victoria Land. Both regions are notable for deep- to shallow-marine sedimentary sequences containing extensive volcanics and related intrusive rocks (Glasgow Formation volcanics of the Bowers Supergroup and the west Tasmanian volcanic-intrusive associations described in Appendix A).

The Glasgow Formation volcanics are Lower to Middle Cambrian pillow lavas, basaltic flows, and breccias of basalt and andesite, with subordinate dacite-rhyolite. The basic rocks are interpreted as primitive island-arc tholeiites and are high-Mg, low-Ti rocks [Weaver et al., 1984; Jordan et al., 1984]. The Eocambrian-Cambrian volcanics and intrusives of western Tasmania appear much more complex in character and no island-arc volcanics are known (Appendix A). It could be suggested that the voluminous dacite-rhyolite association is the west Tasmanian ensialic extension of the Bowers island arc, but this would appear preempted by their respective ages: the Glasgow Formation volcanics are Lower to lower Middle Cambrian [Cooper et al., 1983], whereas the Mount Read volcanics are late Middle to Upper Cambrian [Corbett, 1979]. At best, their overlap in time would seem slight.

The Millen Range schists form a narrow belt (Millen Terrane) of chlorite to biotite zone schists [Bradshaw et al., 1982; Wodzicki et al., 1982; Findlay and Field, 1983; Jordan et al., 1984; Findlay, 1986] trending northwest between the Robertson Bay and Bowers terranes. These schists were probably derived from mafic volcanic and flyschoid precursors, most likely of Middle Cambrian to Tremadoc age, which were deformed and metamorphosed at 510 ± 15 Ma [Findlay, 1986], during the Ross Orogeny. Although it has been implied that the Millen Terrane follows a plate suture between the Bowers and Robertson Bay terranes [Mortimer, 1983; Gibson and Wright, 1985], this opinion is not based on good evidence.

The Millen Terrane is an important tectono-metamorphic lineament extending across northern Victoria Land (Figure 1), yet is unknown in either southern (see model 1, Figure 2c) or southeastern Tasmania (model 2, Figure 2d). Burrett and Findlay [1984] speculated that the Millen Terrane might have a correlate in the Ironbound Range of southern Tasmania; this is incorrect (R. Findlay, unpublished data, 1984), as the only metamorphic rocks here are polydeformed Precambrian quartzites.

Given model 1 (Figure 2c), it would seem reasonable to expect correlates of the Millen Terrane in the Sorell Peninsula of western Tasmania. Yet here, the only metamorphics are low-grade Precambrian quartzites and phyllites of totally dissimilar character [see Baillie et al., 1985; R. Find-

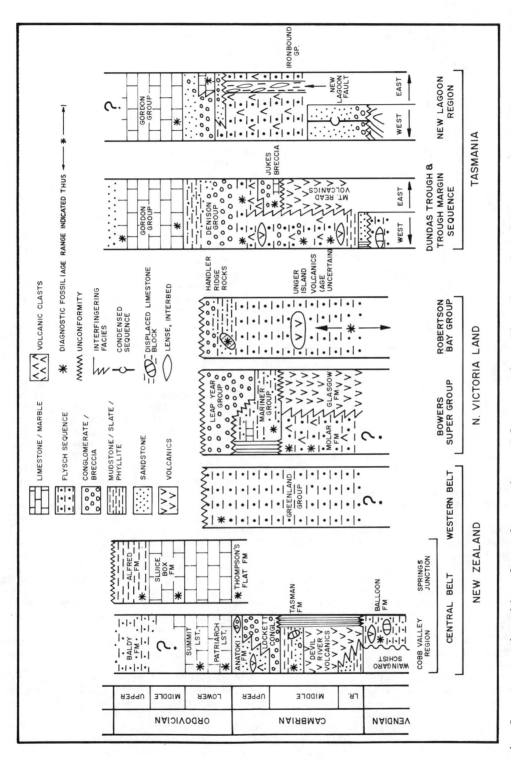

Fig. 4. Summary stratigraphic columns for northwest Nelson, northern Victoria Land, and Tasmania for Vendian–Lower Ordovician times.

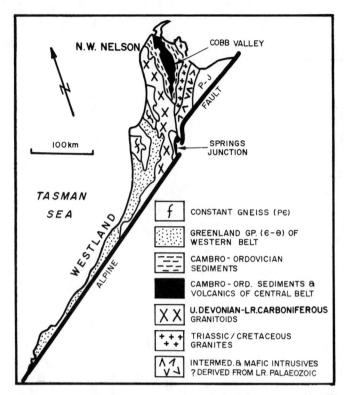

Fig. 5. Summary geological map of Westland and northwest Nelson, New Zealand, with post-Devonian rocks omitted.

lay, personal observation, 1986] to the Millen Terrane schists.

The Arthur Lineament of northwest Tasmania is the only Tasmanian schist lineament resembling the Millen Terrane. However, radiometric dating [see Williams, 1978] indicates this lineament to be some 200 million years older than the Millen Terrane.

These dissimilarities argue against Tasmania being contiguous with northern Victoria Land during Cambrian times, as suggested by Laird et al. [1977], Jago [1980], and Powell [1984]. Rather, I consider it more likely that the two regions were well separated. However, the following important features are common to both regions: (1) marine sedimentation and volcanism ended in Late Cambrian to Early Ordovician times; (2) Late Cambrian to Early Ordovician erosion is clearly recorded, both in Tasmania (Jukes and Haulage unconformities) and in northern Victoria Land (the Leap Year Group unconformity) (Figure 4); (3) radiometric evidence indicates a Late Cambrian-Early Ordovician metamorphic event in both regions; 4) in both regions shallow-marine to terrigenous siliciclastics dominate the Early Ordovician record; (5) in both regions, there are Devonian acidic/andesitic volcanics and contemporaneous granites; and (6) there is a remarkable similarity in their Carboniferous-Triassic sedimentary cover (Beacon and Parmeener supergroups). These common features suggest that even if the two regions were not contiguous in Cambrian times they were juxtaposed before Devonian times, possibly during the Ross Orogeny.

Where Does New Zealand Fit?

The lower Paleozoic rocks of New Zealand (Figure 4) are restricted to northwest Nelson and Westland (Figure 5). Here they form three belts which appear to mirror the distribution of the major rock units in northern Victoria Land (Appendix B, Figure 5, and Table 1).

According to accepted interpretatons [cf. Cooper, 1979], New Zealand has a Cambro-Ordovician volcanosedimentary history similar to that of northern Victoria Land and a Lower Ordovician-Devonian tectonic history similar in style to that of Tasmania. The sedimentary sequence of the eastern belt rocks (Appendix B) also broadly resembles the Ordovician-Devonian sequences in the west Tasmanian troughs.

All three regions contain Devonian to Carboniferous granites with attendant base metal mineralization; in northern Victoria Land, Grindley and Oliver [1983] speculated that this event was subduction related. The presence of Andean-type volcanics [Findlay and Jordan, 1984] may indicate here an Andean type of plate margin.

Other than the Haupiri Disturbance (Table 1), which appears to have involved neither major folding nor metamorphism, there is in New Zealand no accepted record of the Ross Orogeny. In this respect, the 495 Ma Rb/Sr event (Table 1) in the Greenland Group (Appendix B) is important. The Rb/Sr whole-rock isochron is defined by 15 well-aligned points from geographically well-separated slate samples. The cleavage in these samples was formed during one event [Laird and Shelley, 1975; Shelley, 1975a, b] and under low greenschist facies conditions [Shelley, 1975a, b; Adams, 1975; Nathan, 1978]. Therefore, the Rb/Sr isochron need not be regarded as indicating a near synsedimentary diagenetic age, as proposed by Adams [1975], especially so when the possible age range of the Greenland Group is considered. Therefore it is plausible that the Greenland event (Table 1) occurred at 495 Ma and could be a correlative of the Ross Orogeny; the Haupiri Disturbance would be the earliest manifestation of this event. Adams [1975] also reports four Greenland Group slate samples which form a poor Late Ordovician-Silurian Rb/Sr isochron. This age matches some of Adams' K/Ar whole-rock slate ages which he considers to represent the age of cleavage formation in the Greenland Group. However, it is plausible that these ages are an artifact of the Devonian intrusive event.

In northern Victoria Land, geological evidence confirms that the second phase of thrusting occurred at any time between late Early Ordovician and Early Devonian [cf. Tessensohn, 1984], thus allowing this event to be correlated with events IV, IVa, and/or V in northwest Nelson (Table 1). Similarly, this second-phase thrusting could also be related to the onset of the Tabberaberran Orogeny in Tasmania and the Siluro-Devonian deformations of the Lachlan Fold Belt in southeast Australia [cf. Cas, 1983].

TABLE 1. Tectono/Sedimentary Events in Northwest Nelson/Westland, New Zealand, Lower Paleozoic [see Cooper, 1979]

Western Belt		Central Belt Allochthonous Hypothesis [Grindley, 1961, 1971, 1978]		Central Belt Autochthonous Hypothesis [Cooper, 1979]
Middle to Late Devonian granites, local thermal metamorphism (equivalent to event VI)	VI	Middle to Late Devonian; intrusion of granites and thermal metamorphism	VI	Middle to Late Devonian granites
Folding of Devonian Reefton Group (equivalent to event V)	V	Pre-Late Devonian/Carboniferous upright F_2 folds on north/south trend, axial planes steep east; cleavage; folds Baton Formation	V	Fault block rotation and folding deforms Reefton Group and Baton Formation; possibly some F_2 folds
			IVa	Further uplift of central belt, folding in eastern belt in Early Devonian, some "F_2" folds
Upright folding about north/south and east/west axes with axial plane slaty cleavage. K/Ar whole-rock slate ages suggest Late Ordovician/Early Silurian [Adams, 1975], "Greenland Event" [Cooper, 1979]	IV	Probably Silurian; F_1 recumbent nappe folding; central belt thrust from south/southwest over eastern and western belts; L-S fabric	IV	Uplift of central belt along E-dipping reverse faults and thrusts in Late Ordovician/ Early Silurian; accompanied by Greenland Event; many central belt F_2 folds attributed to this phase
Rb/Sr homogenization at 495 ± 1 Ma [Adams, 1975]	III	Ending Late Cambrian/Early Ordovician sporadic vulcanism, sedimentation	III	As for allochthonous hypothesis
Sedimentation: Greenland Group formed of recycled quartzose flysch [Cooper, 1979] from sedimentary/metasedimentary sources in Australia/Antarctic sector of Gondwana [Laird et al., 1977]. Age: Cambrian to Early Ordovician (Lancfieldian) [Cooper, 1974]	II	End of Middle Cambrian; Haupiri Disturbance, uplift and erosion of Devil River arc, deposition of Lockett Conglomerate, and Anatoki Formation flysch wedges	II	As for allochthonous hypothesis
	I	Early to Middle Cambrian; Devil River volcanism	I	As for allochthonous hypothesis

Interpretation

During Late Cambrian-Early Ordovician times in northern Victoria Land, Tasmania, and New Zealand, volcanism and deep marine sedimentation were replaced by shallow marine to shelf siliciclastics and carbonates. This sedimentological change coincided with compressional tectonics in northern Victoria Land, and penecontemporaneous uplift (with less well understood deformation and low-grade metamorphism) in both Tasmania and New Zealand. These data (see also Figure 3, Table 1, and Appendix B) permit an interpretation for the three regions in terms of a tectonically linked system of island arcs and microcontinents in collision with the East Antarctic-Australian sector of the future Gondwana continent, whose Cambrian coastline followed approximately the carbonate belt formed by the Shackleton Limestone (southern Victoria Land) and the Normanville Group (South Australia).

Tectonic interpretations for northern Victoria Land vary. Weaver et al. [1984] discussed collisional models in which the Bowers volcanic arc overlay a west dipping subduction zone. The main problem they found was the present close proximity (6 km minimum) of the island arc sequence of the Bowers Terrane to the granitoids of the Wilson Terrane. They stated that either tectonic telescoping or transcurrent faulting must have juxtaposed the two terranes. As large-scale thrusting had not then been documented fully, Weaver et al. [1984] opted for juxtaposition of the two terranes by transcurrent faulting along the Leap Year and Lanterman faults.

The subsequent discovery [cf. Tessensohn, 1984] of pre-Middle Devonian reverse faults and thrusts allowed Gibson and Wright [1985] to adopt a collisional model. In that model, the Robertson Bay Group, partly overlying continental crust, collided with the Bowers arc which overlay a west dipping Benioff Zone. During collision, the Robertson Bay Group was overridden by the Bowers Terrane. In the last, Early Ordovician, phase of collision, tectonic telescoping on the pre-Middle Devonian faults juxtaposed the Bowers and Wilson terranes. Subsequent post-Jurassic faulting uplifted part of the underlying Robertson Bay Group to the west of the Bowers arc; these rocks are represented by the Morozumi Range flysch.

However, in the Gibson and Wright model, there is no appropriate suture between the Robertson Bay and Bowers terranes [Weaver et al., 1984; Gibson and Wright, 1985].

These problems would be overcome if the Bowers arc were well separated initially from the East Antarctic foreland by an ocean basin and an east dipping subduction zone which surfaced west of the arc. This would allow a simple interpretation for the Morozumi Range flysch. Rather than being allochthonous Robertson Bay Group, it would be an autochthonous or parautochthonous part of a Cambrian East Antarctic-South Australian submarine fan, of which the Berg Group (age unknown) and Kanmantoo Group would be parts also. In this model (Figure 6), the Robertson Bay Group would lie between the Bowers arc and an eastern to southeastern continental source (Iselin Plateau?). The Lower Ordovician Handler Ridge sequence would be the bounding unit between the Bowers Supergroup and Robertson Bay Group.

The collision of the Bowers arc with the South Australian-East Antarctic cratonic margin occurred during the "Crosscut Event" (Figure 3) at about 500-510 Ma. During this event, thrusting and production of schists (Millen Range schists; Retreat Hills schists; schists southwest of Marine Glacier) at the sole(s) of the thrust(s) occurred. Locally, oceanic mafics and ultramafics [cf. Roland et al., 1984] were incorporated into the thrusts' soles.

The later stages of the collision involved reversal of the subduction zone's polarity, as indicated by the melting of the rocks at the base of the Morozumi-Berg fan and formation of the 490-460 Ma migmatites and granites of the Wilson Terrane. Sillimanite zone metamorphism occurred in the western low-pressure part of the terrane. Kyanite zone metamorphism [Grew et al., 1984] affected the deeper easternmost rocks close to the plate suture (Rennick-Aviator Line [Grew et al., 1984] which lies between the Wilson and Bowers terranes) (Figure 7).

The post-Early Ordovician reverse faults and thrusts represent either the final phase of the Ross Orogeny or the result of renewed Siluro-Devonian plate convergence at an Andean-type convergent plate boundary [Grindley and Oliver, 1983; Findlay and Jordan, 1984]. There is unequivocal evidence for Siluro-Devonian compressional deformation in the then contiguous Australian part of the Siluro-Devonian Gondwana continent, as there is also in New Zealand.

It is evident that northern Victoria Land and Tasmania were juxtaposed before intrusion of their common Devonian-Carboniferous granitoids and contemporaneous volcanics, yet the two regions do not share a common sedimentary or tectonic history until the end of the Cambrian. The simplest interpretation is to regard the Tasmanian rifted microcontinent as lying well to the east of both its present position and of the Bowers arc, from which it was separated by an ocean basin which was consumed by an east dipping subduction zone off the coast of what is now western Tasmania. Given an easterly dip of 35° under a continental crust of 30 km thickness, the progressive evolution of this east dipping subduction zone could produce the Eocambrian to Cambrian rift-associated olivine-bearing quartz-normative volcanics, the pre-Middle to Middle Cambrian mafic-ultramafic association, and the Middle to Late Cambrian andesite-rhyolite association, as shown in Figure 8 (refer also to Appendix A).

In this model the Late Cambrian to Early Ordovician tectonism and metamorphism in Tasmania [Adams et al., 1985] would have occurred during the juxtaposition of Tasmania with northeastern northern Victoria Land, producing a fit similar to model 2 (Figure 2d).

Pre-Cretaceous Gondwana reconstructions [Griffiths, 1974; Weissel et al., 1977; Grindley and Davey, 1982] do not permit close proximity of New Zealand to either Tasmania or northern Victoria

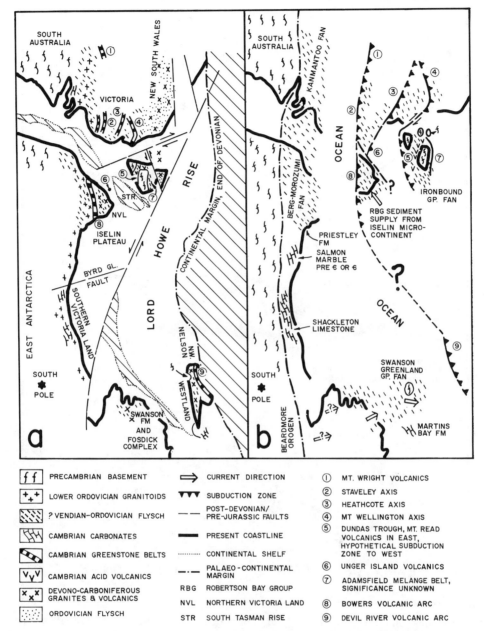

Fig. 6. Plate tectonic interpretation of the Australia-Antarctica-New Zealand sector of southwest Gondwana. (a) Pre-Carboniferous reconstruction derived from Grindley and Davey [1982] by (1) closing the Wilkes Basin; (2) closing the Murray Basin; (3) opening New South Wales Basin; (4) offsetting northwest Tasmania along the Tamar Graben; (5) closing the Tasman Basin by straightening the Lord Howe Rise and sliding it dextrally along a proposed post-Devonian/Carboniferous sinistral fault so that northwest Nelson falls into the ambit of the Australo-Antarctic Devono-Carboniferous granite province, and the Greenland Group aligns with the Swanson Formation (quartz flysch) of Marie Byrd Land. Note congruity of fault movements. (b) Plate tectonic scheme for southeast Gondwana in Cambrian times. The Victoria-Tasmanian-northern Victoria Land-northwest Nelson region consists of a tectonically linked complex of island arcs, localized deep to shallow marine deposits, and basement blocks overlying an east dipping subduction system. The Ross Orogeny occurred when this complex collided with the passive continental margin of the East Antarctic craton, in which is included Marie Byrd Land.

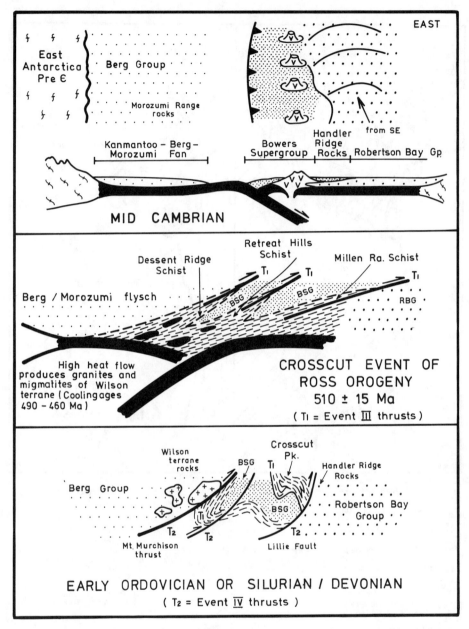

Fig. 7. Cartoons showing development of Ross Orogeny in northern Victoria Land (see text for explanation).

Land. Yet the reviewed geological data suggest that New Zealand could have been involved in the Cambro-Ordovician collisional event and most certainly was included in the Siluro-Devonian tectonism and granite intrusion along the Siluro-Devonian Gondwana margin.

There are two simple options for arranging northwest Nelson as part of the Cambro-Ordovician northern Victoria Land-Tasmanian collisional system. In the first option, the Devil River arc overlay a west dipping subduction zone [Shelley, 1975b] well separated by a backarc basin (Greenland Group depocenter) from Marie Byrd Land. Alternatively, the arc overlay an east dipping subduction zone and was brought into collision with a submarine fan (Greenland Group) off the coast of Marie Byrd Land, as in Figure 6. In either case, the Greenland Group would be a correlate of the Swanson Formation [see Adams, 1980; Bradshaw et al., 1983] of Marie Byrd Land.

Option 1 requires a transform fault between the Devil River arc and the Bowers arc. In option 1, the 495 Ma isochron from the Greenland Group could either be due to Cambro-Ordovician tectonism during the Ross Orogeny or indicate [Adams, 1975] age of diagenesis. Option 2 requires a Cambro-

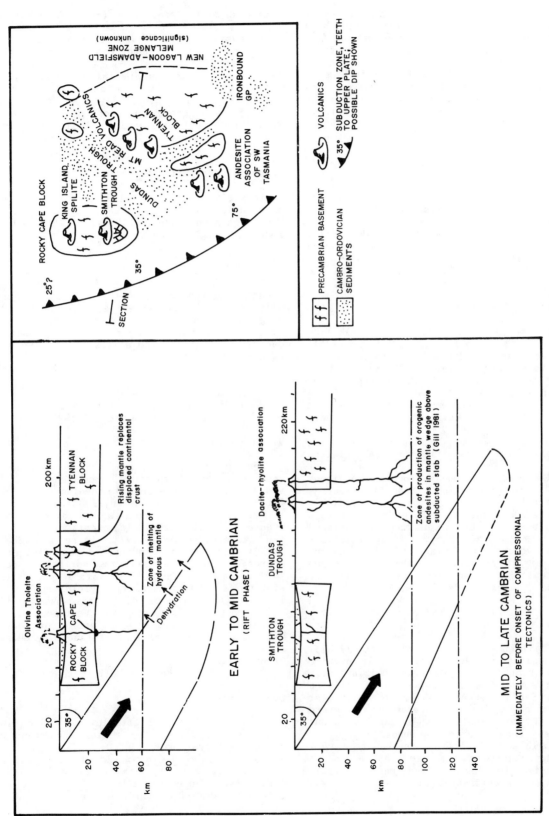

Fig. 8. Scale cartoons (left) and sketch paleogeographic map to illustrate relationships between tectonism and magma-genesis in the Cambrian troughs of western Tasmania.

Ordovician collision between the central and western belts, and in option 2 the 495 Ma isochron would represent a Ross orogenic deformation age. In either option, the Haupiri Disturbance forms part of the Ross Orogeny.

In either case, in New Zealand the Siluro-Devonian deformation (Tuhuan Orogeny) involves tectonic telescoping, and I suggest this includes the late west over east thrusting in northern Victoria Land (event III, Figure 3), as well as the deformation in the Lachlan Fold Belt of southeast Australia.

Appendix A:
Precambrian/Cambrian Volcanics and Related Intrusives in the West Tasmanian Troughs

R. Varne and J. D. Foden (unpublished data, 1986; see also Brown et al. [1980]) have grouped Precambrian/Cambrian volcanics and related intrusives in the west Tasmanian troughs into the following associations:

Alkaline Basalt Association

Late Precambrian; high-Al, mildly alkaline olivine basalt with high Ti, Zr, Nb [Brown, 1986]. May have been part of the alkalic phase at the onset of late Precambrian/Early Cambrian continental rifting (R. Varne and J. D. Foden, unpublished data, 1986; cf. Brown et al. [1980]). Brown [1986] prefers them as Precambrian lavas unrelated to the tholeiitic basalt association of the Dundas Trough.

Olivine Tholeiite or Tholeiitic Basalt Association

In the Smithton and Dundas troughs. Assigned [Brown, 1986] to Eocambrian Crimson Creek Formation and correlates. Low-K_2O tholeiites [Baillie and Crawford, 1984] and Heavy Rare Earth Elements (HREE) indicate its derivation from an enriched fertile mantle [Brown, 1986]. Field relations confirm the geochemical data for origin during rifting of sialic crust rather than at an ocean spreading center or in an intraoceanic island arc [Baillie and Crawford, 1984; Brown, 1986].

High-Mg Low-Ti Andesite Tholeiite (Ophiolite) Association

(Low-Ti ophiolite association of Varne [1978]). High-Mg andesites at 570-600 Ma locally overlie continental crust need not indicate an island arc; andesites precede low-Ti basalts by 50 million years; Rare Earth Element (REE) patterns suggest derivation from a depleted source by a second stage melt [Brown, 1986]. Low-Ti basalts, about 530 Ma, are youngest mafics in Dundas Trough sequence; possibly derived by a second-stage melt from high-Mg andesite magma, they are the product of progressive melting of the mantle diapir beneath the intracontinental rift system [Brown, 1986].

Dacite-Rhyolite Association (Mount Read Volcanics)

Extensive along the western flank of the Tyennam Block; Middle to Upper Cambrian; geochemically affine to: (1) orogenic "Andean-type" andesites; (2) volcanics in continental rift zones, e.g., rifted continental margin of southeast Africa, Baffin Bay, or geologically (see also Corbett [1979]) to volcanic terranes of the U.S. Great Basin; (3) Archaean greenstone belts [Brown et al. 1980; R. Varne and J. D. Foden, unpublished data, 1986].

Andesite Association of Southwest Tasmania

Possibly comagmatic with the dacite-rhyolite association (Brown et al. [1980]; see also McClenaghan and Corbett [1985]; R. Varne and J. D. Foden, unpublished data, 1986).

Ultramafic/Mafic Complexes

Pre-Middle to Middle Cambrian in all Tasmanian troughs and generally fault bounded. Cumulate ultramafics (orthopyroxenite; pyroxenite; harzburgite; lherzolite; dunite), gabbros, plagiogranites, and volcanics including high-Mg andesites, dacites, and quartz tholeiites indicative of low-pressure melting.

Appendix B:
Lower Paleozoic Units of Northwest Nelson and Westland from Cooper [1979]

Western Belt

Reefton Group. Devonian, shallow marine/beach/reef deposits of mudstone, sandstone, and limestone, 1450 m that overlies Greenland Group unconformably [Bradshaw and Hegan, 1983].

Greenland Group. ?Vendian to Lower Ordovician [Cooper, 1974] turbiditic quartz-rich submarine fan sequence [Laird, 1972; Laird and Shelley, 1975]. Limited data indicate a southeast or east source; paleoslope strikes northwest [Laird, 1972]. An Rb/Sr whole-rock slate isochron from 15 widely distributed localities yields age of 495 ± 11 Ma, supposedly of diagenesis [Adams, 1975]. This group occurs throughout Westland; supposed correlate of Aorangi Mine Formation of northwest Nelson, which is overlain conformably by mudstone-shale quartz sandstone sequences ranging to Upper Ordovician.

Separated from central belt by Anatoki Thrust.

Central Belt

Peel Formation and correlates. Late Ordovician dark phyllites and shales and minor sandstone are thrust over:

Mount Patriarch Group. Lower to ?Upper Ordovician, dominated by limestones and calcareous sediments. Three formations, the lowest (Mount Patriarch Formation) contains minor volcaniclastic sandstone. Uppermost unit (Baldy Formation) is 100 m of unfossiliferous sheared phyllite and siltstone.

Anatoki Formation. Possibly Late Cambrian dolomitic sandstone, thinly bedded siltstone, calcsiltstone, granule to cobble conglomerate lenses, with bands of basic volcanics. Volcanic detritus is common. This goes to 1300 m; deep marine in a subsiding basin flanked by a volcanic region. In part it overlies and in part it is laterally equivalent to:

Lockett Conglomerate. Middle to Upper Cambrian conglomerate of poorly sorted to normally/inversely graded pebbles, cobbles, and boulders of chert, sandstone, and limestone; assorted basic/intermediate volcanics/plutonics in a poorly sorted sandstone sequence. It has chromite grains and upper fan channel deposit [cf. Pound, 1985] or fanglomerate [Grindley, 1980].

Tasman Formation. Lower Middle Cambrian, laminated siliceous and calcareous siltstone and sandstone, chert conglomerate, and bands and lenses of limestone. Penecontemporaneous deformation and gravity sliding. To 1000 m; eastern provenance? Similar to possibly laterally equivalent Baloon Formation which contains diamictite and conglomerate also.

Devil River Volcanics. Lower Middle Cambrian, high-Mg, low-Ti andesitic island arc (A. R. Crawford, personal communication, 1984) volcanics of plausibly intraoceanic plate margin, with spilitized basic to intermediate volcanics flows, shallow sills, crystal and vitric tuffs, 1000-2000 m, with interbedded unfossiliferous limestones.

Separated from eastern belt by Devil River Thrust.

Eastern Belt

Baton Formation. Lochkovian/Pragian; dominantly mudstone; also muddy sandstone, limestone, and calcareous mudstone, with local angular unconformity with the Wangapeka Formation; the basal conglomerate has Haupiri Group clasts. 2600 m thick.

Hailes Quartzite. Probably Middle Silurian to Upper Devonian. Locally overlain by unfossiliferous phyllitic siltstone with sandstone and rare calcareous bands (Fowler Formation), and is at 450 m minimum.

Arthur Marble 2. Uppermost Ordovician fossils known. Black carbonaceous, pyritic limestone. Muddy and micaceous interbeds increase near the top. Possible lateral facies variant of the Wangapeka Formation, or excluded locally by disconformity, 300-1000 m thick.

Wangapeka Formation. Upper Ordovician, shale, fine sandstone, alternating quartz-rich sandstone and shale, local black limestone; 300-1000 m thick.

Arthur Marble 1. Early to Middle Ordovician, pale pure calcite marble, irregular siliceous nodules. Locally alternating pale quartzite and marble. Between 600-3000 m thick and gradational from:

Owen Formation. Unfossiliferous, probably pre-Middle Ordovician; phyllitic, calcareous, and dolomitic mudstone and sandstone.

Acknowledgments. This paper could not have been written without the financial support of the New Zealand Antarctic Division (1977-1981) and logistic support of GANOVEX III (1982-1983) during five Antarctic expeditions, and an ARGS sponsored research fellowship at the University of Tasmania (1982-1984). I thank my colleagues for much stimulating discussion: these include D. H. Green, C. F. Burrett, and R. F. Berry (University of Tasmania); M. G. Laird, B. D. Field, G. W. Grindley, and P. Oliver (N. Z. Geological Survey); J. D. Bradshaw (University of Canterbury); H. Jordan, G. Kleinschmidt, and F. Tessensohn (GANOVEX III), G. Gibson (Darling Downs Institute of Advanced Education); and T. O. Wright (National Science Foundation). The paper has benefited from the comments of two unknown reviewers and of P. W. Baillie, A. G. Brown, and K. D. Corbett (Tasmanian Department of Mines).

References

Adams, C. J. D., Discovery of Precambrian rocks in New Zealand: Age relations of the Greenland Group and Constant Gneiss, West Coast, South Island, Earth Planet. Sci. Lett., 28, 98-104, 1975.

Adams, C. J. D., Geochronological correlations of Precambrian and Paleozoic orogens in New Zealand, Marie Byrd Land (West Antarctica), northern Victoria Land (East Antarctica), and Tasmania, in Gondwana Five, edited by M. M. Creswell and P. Vella, pp. 191-197, A. A. Balkema, Rotterdam, 1980.

Adams, C. J. D., and H. Kreuzer, Potassium-argon age studies of slates and phyllites from the Bowers and Robertson Bay terranes, north Victoria Land, Antarctica, Geol. Jahrb., Reihe B, 60, 265-288, 1984.

Adams, C. J. D., L. P. Black, K. D. Corbett, and G. R. Green, Reconnaissance isotopic studies bearing on the tectonothermal history of the early Paleozoic and late Proterozoic sequences in western Tasmania, Aust. J. Earth Sci., 32, 7-36, 1985.

Baillie, P. W., A Paleozoic suture in eastern Gondwanaland, Tectonics, 4, 653-660, 1985.

Baillie, P. W., and A. J. Crawford, Excursion guide: Smithton Trough excursion, in Mineral Exploration and Tectonic Processes in Tasmania, edited by P. W. Baillie and P. L. F. Collins, pp. 59-64, Geological Society of Australia, (Tasmanian Division), Hobart, 1984.

Baillie, P. W., K. D. Corbett, and G. R. Green, Geological atlas, 1:50,000 Ser., sheet 57 (7913N) edited by Strahan, Explanatory Report of the Geological Survey of Tasmania, 76 pp., Government Printer, Hobart, Tasmania, 1985.

Banks, M. R., Silurian and Devonian system, J. Geol. Soc. Aust., 9, 177-178, 1962a.

Banks, M. R., Permian, Chapter V, in The Geology

of Tasmania, edited by A. H. Spry and M. R. Banks, pp. 189-215, Journal of the Geological Society of Australia, 9, 1962b.

Banks, M. R., and E. A. Smith, A graptolite from the Mathinna Beds, northeastern Tasmania, J. Geol. Soc. Tasmania, 26, 363-376, 1968.

Barrett, P. J., History of the Ross Sea region during the deposition of the Beacon Supergroup 400-180 million years ago, J. R. Soc. N. Z., 11, 447-458, 1981.

Berry, R. H., and S. Harley, Pre-Devonian stratigraphy and structure of the Prior Beach-Rocky Boat Inset-Osmiridium Beach coastal section, southern Tasmania, Pap. Proc. R. Soc. Tasmania, 117, 59-75, 1983.

Bischoff, K., The geology of the Rocky Boat Inlet--Surprise Bay area, Honours thesis, 213 pp., Dep. of Geol., University of Tasmania, Hobart, Australia, 1983.

Blank, H. R., R. A. Cooper, R. H. Wheeler, and I. A. Willis, Geology of the Koettlitz-Blue Glacier region, S. Victoria Land, J. R. Soc. N. Z., 2, 79-100, 1963.

Bradshaw, J. D., and M. G. Laird, The pre-Beacon geology of northern Victoria Land, a review, in Antarctic Earth Science, edited by R. L. Oliver, P. R. James, and J. B. Jago, pp. 98-101, Australian Academy of Science, Canberra, 1983.

Bradshaw, J. D., M. G. Laird, and A. Wodzicki, Structural style and tectonic history in northern Victoria Land, Antarctica, in Antarctic Geoscience, edited by C. Craddock, pp. 809-816, University of Wisconsin Press, Madison, 1982.

Bradshaw, J. D., P. B. Andrews, and B. D. Field, Swanson Formation and related rocks of Marie Byrd Land and a comparison with the Robertson Bay Group of northern Victoria Land, in Antarctic Earth Science, edited by R. L. Oliver, P. R. James, and J. B. Jago, pp. 274-279, Australian Academy of Science, Canberra, 1983.

Bradshaw, J. D., J. C. Begg, W. Buggisch, C. Brodie, F. Tessensohn, and T. O. Wright, New data on Paleozoic stratigraphy and structure in north Victoria Land, N. Z. Antarct. Rec., 6(3), 1-6, 1985.

Bradshaw, M. A., and B. D. Hegan, Stratigraphy and structure of the Devonian rocks of the Inangahua Outlier, Reefton, New Zealand, N. Z. J. Geol. Geophys., 26, 325-344, 1983.

Brown, A. V., The geology of the Dundas-Mt. Lindsay-Mt. Youngback region, Geol. Surv. Bull. Tasmania, 62, 221 pp., 1986.

Brown, A. V., M. J. Rubenach, and R. Varne, Geological environment, petrology and tectonic significance of the Tasmanian Cambrian ophiolitic and ultramafic complexes, in Ophiolites, Proceedings of the Internatinal Ophiolite Symposium, pp. 649-659, Cyprus Geological Survey Department, Nicosia, Cyprus, 1980.

Brown, A. V., et al., Late Proterozoic to Devonian sequences of southeastern Australia, Antarctica, and New Zealand, and their correlation, Spec. Publ. Geol. Soc. Aust., 9, 104 pp., 1982.

Burrett, C. F., and R. H. Findlay, Cambrian and Ordovician conodonts from the Robertson Bay Group, Antarctica, and their tectonic significance, Nature, 307, 723-725, 1984.

Cas, R., Paleogeographic and tectonic development of the Lachlan Fold Belt, southeastern Australia, Spec. Publ. Geol. Soc. Aust., 10, 104 pp., 1983.

Clarke, M. J., and M. R. Banks, The stratigraphy of the Lower Palaeozoic (Permo-Carboniferous) parts of the Parmeener Supergroup, Tasmania, in Gondwana Geology, edited by K. S. W. Campbell, pp. 453-467, Australian National University Press, Canberra, 1974.

Cook, P. J., The Cambrian paleogeography of Australia and opportunities for petroleum exploration, J. Aust. Pet. Expl. Assoc., 22, 42-64, 1982.

Cooper, R. A., Age of the Greenland and Waiuta groups, South Island, New Zealand, N. Z. J. Geol. Geophys., 17, 955-962, 1974.

Cooper, R. A., Lower Palaeozoic rocks of New Zealand, J. R. Soc. N. Z., 9, 29-44, 1979.

Cooper, R. A., A. J. Rowell, and P. Braddock, Age and correlation of the Cambrian-Ordovician Bowers Supergroup, northern Victoria Land, Antarctica, in Antarctic Earth Science, edited by R. L. Oliver, P. R. James, and J. B. Jago, pp. 128-131, Australian Academy of Science, Canberra, 1983.

Corbett, R. A., Stratigraphy, correlation, and evolution of the Mt. Read volcanics in the Queenstown, Jukes-Darwin and Mt. Sedgewick areas, Geol. Surv. Bull. Tasmania, 58, 75 pp., 1979.

Corbett, K. D., M. R. Banks, and J. B. Jago, Plate tectonics and the lower Paleozoic of Tasmania, Nature Phys. Sci., 240, 9-11, 1972.

Corbett, K. D., G. R. Green, and P. R. Williams, The geology of central western Tasmania, in Landscape and Man, A symposium, pp. 7-28, Royal Society of Tasmania, Hobart, Australia, 1977.

Craddock, C., Antarctic tectonics in Antarctic Geology and Geophysics, edited by R. J. Adie, pp. 449-455, Universitetsforlaget, Oslo, 1972.

Crook, K. A. W., Tectonic implications of some field relations of the Adelaidean Cooee Dolerite, Tasmania, J. Geol. Soc. Aust., 26, 353-361, 1980.

Crook, K. A. W., and M. A. Powell, The evolution of the southeastern part of the Tasman Geosyncline, in 25th International Geological Congress, Excursion Guide 17A, 122 pp., sponsored by Australian Academy of Science, Geological Society of Australia, and International Union of Geological Sciences, Sydney, 1976.

Crowder, D. E., Geology of part of N. Victoria Land, Antarctica, U.S. Geol. Surv. Prof. Pap., 600-D, D95-107, 1968.

Crowell, J. C., and L. A. Frakes, Late Paleozoic glaciation in Australia, J. Geol. Soc. Aust., 17, 115-155, 1971.

Crowell, J. C., and L. A. Frakes, Late Paleozoic Glaciation, in Gondwana Geology, edited by K. S. W. Campbell, pp. 313-331, Australian National University Press, Canberra, 1974.

Daily, B., and A. R. Milnes, Stratigraphy, structure and metamorphism in the Kanmantoo (Cambrian) in its type section east of Tunkilla Beach, South Australia, Trans. R. Soc. Aust., 97, 213-242, 1973.

Field, B. D., and R. H. Findlay, The sedimentology of the Robertson Bay Group, northern Victoria Land, in Antarctic Earth Science, edited by R. L. Oliver, P. R. James, and J. B. Jago, pp. 107-112, Australian Academy of Science, Canberra, 1983.

Findlay, R. H., Structural geology of the Robertson Bay and Millen terranes, northern Victoria Land, Antarctica, in Geological and Geophysical Investigations in Northern Victoria Land, Antarctica, Antarct. Res. Ser., edited by E. Stump, AGU, Washington, D. C., in press, 1986.

Findlay, R. H., and B. D. Field, Preliminary report on the structural geology of the Robertson Bay Group, north Victoria Land, Antarctica, N. Z. Antarct. Rec., 4(2), 15-19, 1982.

Findlay, R. H., and B. D. Field, Tectonic significance of deformations affecting the Robertson Bay Group and associated rocks, northern Victoria Land, Antarctica, in Antarctic Earth Science, edited by R. L. Oliver, P. R. James, and J. B. Jago, pp. 107-112, Australian Academy of Science, Canberra, 1983.

Findlay, R. H., and H. Jordan, The volcanic rocks of Mt. Black Prince and Lawrence Peaks, north Victoria Land, Antarctica, Geol. Jahrb., Reihe B, 60, 143-151, 1984.

Findlay, R. H., D. N. B. Skinner, and D. Craw, Lithostratigraphy and structure of the Koettlitz Group, McMurdo Sound, Antarctica, N. Z. J. Geol., 27, 513-536, 1984.

Gibson, G., F. Tessensohn, and A. J. Crawford, Bowers Supergroup rocks west of the Mariner Glacier and possible greenschist facies equivalents, Geol. Jahrb., Reihe B, 60, 289-318, 1984.

Gibson, G. M., and T. O. Wright, Importance of thrust faulting in the tectonic development of northern Victoria Land, Antarctica, Nature, 315, 480-483, 1985.

Gill, J., Orogenic Andesites and Plate Tectonics, 390 pp., Springer-Verlag, New York, 1981.

Green, G. R., The geological setting and formation of the Rosebery volcanic hosted massive sulphide orebody, Tasmania, Ph.D. thesis, University of Tasmania, Hobart, Australia, 1983.

Grew, E. S., G. Kleinschmidt, and W. Schubert, Contrasting metamorphic belts in north Victoria Land, Geol. Jahrb., Reihe B, 60, 253-265, 1984.

Griffiths, J. R., New Zealand and the southwest Pacific margin, in Gondwana Geology, edited by K. S. W. Campbell, pp. 619-637, Australian National University Press, Canberra, 1974.

Grindley, G. W., Geological map of New Zealand, scale 1:250,000, sheet 13 Golden Bay, 1st ed., Dep. of Sci. and Ind. Res., Wellington, 1961.

Grindley, G. W., Geological map of New Zealand, scale 1:63,360, sheet S8 Takaka, 1st ed., Dep. of Sci. and Ind. Res., Wellington, 1971.

Grindley, G. W., "West Nelson," in The Geology of New Zealand, edited by R. P. Suggate et al., pp. 80-93, Government Printer, Wellington, New Zealand, 1978.

Grindley, G. W., Geological map of New Zealand, scale 1:63,360, sheet 13 Cobb, 1st ed., Dep. of Sci. and Ind. Res., Wellington, 1980.

Grindley, G. W., and F. J. Davey, The reconstruction of New Zealand, Australia and Antarctica, in Antarctic Geoscience, edited by C. Craddock, pp. 15-29, University of Wisconsin Press, Madison, 1982.

Grindley, G. W., and P. J. Oliver, Post-Ross Orogeny cratonization of northern Victoria Land, in Antarctic Earth Science, edited by R. L. Oliver, P. R. James, and J. B. Jago, pp. 133-139, Australian Academy of Science, Canberra, 1983.

Iltchenko, L. N., Late Precambrian archritarchs of Antarctica, in Antarctic Geology and Geophysics, edited by R. J. Adie, pp. 599-602, Universitetsforlaget, Oslo, 1972.

Jago, J. B., Late Precambrian-early Paleozoic geological relationships between Tasmania and northern Victoria Land, in Gondwana Five, edited by M. M. Cresswell and P. Vella, pp. 199-204, A. A. Balkema, Rotterdam, 1980.

Jago, J. B., B. D. Dailly, C. C. von der Borch, A. Cernovski, and N. Saunders, First reported trilobites from the Lower Cambrian Fleurieu Group, Fleurieu Peninsula, South Australia, Trans. R. Soc. S. Aust., 108, 207-211, 1985.

Jordan, H., R. H. Findlay, M. Schmidt-Thomé, G. Mortimer, P. Muller, and A. Crawford, Geology of the northern Bowers Mountains, north Victoria Land, Geol. Jahrb., Reihe B, 60, 57-81, 1984.

Kleinschmidt, G., Trends in regional metamorphism and deformation in northern Victoria Land, Antarctica, in Antarctic Earth Science, edited by R. L. Oliver, P. R. James, and J. B. Jago, pp. 119-122, Australian Academy of Science, Canberra, 1983.

Kleinschmidt, G., N. Roland, and W. Schubert, The metamorphic basement complex in the Mountaineer Range, north Victoria Land, Antarctica, Geol. Jahrb., Reihe B, 60, 213-252, 1984.

Laird, M. G., Sedimentology of the Greenland Group in the Paparoa Range, West Coast, South Island, N. Z. J. Geol. Geophys., 15, 372-93, 1972.

Laird, M. G., Lower Paleozoic rocks of the Ross Sea and their significance in the Gondwana context, J. R. Soc. N. Z., 11, 425-438, 1981.

Laird, M. G., and D. Shelley, Sedimentation and early tectonic history of the Greenland Group, Reefton, New Zealand, N. Z. J. Geol. Geophys., 17, 839-854, 1975.

Laird, M. G., G. D. Mansergh, and J. M. A. Chappell, Geology of the central Nimrod Glacier area, Antarctica, N. Z. J. Geol. Geophys., 14, 427-468, 1971.

Laird, M. G., R. A. Cooper, and J. B. Jago, New data in the lower Paleozoic sequence of northern Victoria Land, Antarctica, and its significance for Australian-Antarctic relations in the Paleozoic, Nature, 265, 107-110, 1977.

Mawson, D., Sedimentary rocks, in Scientific Reports of the Australian Antarctic Expeditions 1911-1914, Ser. A, Geology, part 11, pp. 347-367, Government Printer, Sydney, 1940.

Mortimer, G., GANOVEX III, 1982/1983, N. Z. Antarct. Rec., 5(1), 54-60, 1983.

Nathan, S., Geological map of New Zealand, sheet 44 Greymouth, scale 1:63,360, 1st ed., Dep. of Sci. and Ind. Res., Wellington, 1978.

Oliver, R. L., Some aspects of Antarctic-Australian geological relationships, in *Antarctic Geology and Geophysics*, edited by R. J. Adie, pp. 859-864, Universitetsforlaget, Oslo, 1972.

Palmer, A. R., The decade of North American geology, 1983 geologic time scale, *Geology*, 11, 503-504, 1983.

Pound, K. S., The lower Paleozoic Lockett Conglomerate on northwest Nelson, New Zealand: A marine deposit, in *Abstracts of the Third Circum-Pacific Terrane Conference, Sydney*, edited by E. Leitch, pp. 190-192, Geological Society of Australia, Sydney, 1985.

Powell, C. Mc. A., Tectonic evolution of the Tasman Fold Belt, with emphasis on Paleozoic implications for Tasmania, in *Mineral Exploration and Tectonic Processes in Tasmania, Volume of Abstracts and Excursion Guide for Burnie Conference*, edited by P. W. Baillie and P. L. F. Collins, pp. 6-11, Geol. Soc. Aust., Hobart, 1984.

Råheim, A., and W. Compston, Correlations between metamorphic events and Rb/Sr ages in metasediments and eclogite from western Tasmania, *Lithos*, 10, 271-289, 1977.

Ravich, M. G., Comparisons of the folded complexes of the Adelaide (Australia) and Ross (Antarctica) Geosyncline areas, in *Antarctic Geoscience*, edited by C. Craddock, pp. 65-71, University of Wisconsin Press, Madison, 1982.

Rickards, R. B., and M. R. Banks, An Early Devonian monograptid from the Mathinna Beds, Tasmania, *Alcherigna*, 3, 307-311, 1979.

Riddols, B. W., and G. T. Hancox, The geology of the upper Mariner Glacier region, north Victoria Land, Antarctic, *N. Z. J. Geol. Geophys.*, 11, 881-889, 1968.

Roland, N. W., G. Gibson, G. Kleinschmidt, and W. Schubert, Metamorphism and structural relations of the Lanterman metamorphics, north Victoria Land, Antarctica, *Geol. Jahrb., Reihe B*, 60, 319-363, 1984.

Shelley, D., Temperature and metamorphism during cleavage and fold formation of the Greenland Group, north of Greymouth, *J. R. Soc. N. Z.*, 1, 65-75, 1975a.

Shelley, D., Metamorphic belt and volcanic arc migration in New Zealand, *Nature*, 258, 668-672, 1975b.

Skinner, D. N. B., The granites and two orogenies of southern Victoria Land, in *Antarctic Earth Science*, edited by R. L. Oliver, P. R. James, and J. B. Jago, pp. 160-163, Australian Academy of Science, Canberra, 1983.

Solomon, M., and J. R. Griffiths, Tectonic evolution of the Tasman Orogenic Zone, eastern Australia, *Nature*, 237, 3-6, 1972.

Solomon, M., and J. R. Griffiths, Aspects of the early history of the Tasman Orogenic Zone, in *The Tasman Geosyncline-A Symposium*, edited by A. K. Denmead, G. W. Tweedale, and A. F. Wilson, pp. 19-144, Geological Society of Australia, Brisbane, 1974.

Soloviev, D. S., The lower Paleozoic metamorphic slates of the Oates Coast, (In Russian), Sb. *Statei Geol. Antarkt.*, no. 2, 1960. (English translation available from G. Warren, N. Z. Geol. Surv., Wellington, 1960).

Steed, R. H. N., Structural interpretations of Wilkes Land, Antarctica, in *Antarctic Earth Science*, edited by R. L. Oliver, P. R. James, and J. B. Jago, pp. 567-573, Australian Academy of Science, Canberra, 1983.

Tessensohn, F., K. Duphorn, H. Jordan, G. Kleinschmidt, D. N. B. Skinner, U. Vetter, T. O. Wright, and D. Wyborn, Geological comparison of basement units in north Victoria Land, Antarctica, *Geol. Jahrb., Reihe B*, 41, pp. 31-88, 1981.

Tessensohn, F., Geological and tectonic history of the Bowers Structural Zone, north Victoria Land, Antarctica, *Geol. Jahrb., Reihe B*, 60, 371-396, 1984.

Varne, R., The Cambrian volcanic associations of Tasmania and their tectonic setting, in *Geology and Mineralization of N. W. Tasmania, Abstracts of Symposium*, edited by D. C. Green and P. R. Williams, unnumbered mimeo, Geological Society of Tasmania, Hobart, Australia, 1978.

Vetter, U., N. Roland, H. Kruezer, A. Höhndorf, H. Lenz, and C. Besang, Geochemistry, petrography, and geochronology of the Cambro-Ordovician and Devonian-Carboniferous granitoids of northern Victoria Land, Antarctica, in *Antarctic Earth Science*, edited by R. L. Oliver, P. R. James, and J. B. Jago, pp. 140-143, Australian Academy of Science, Canberra, 1983.

Vetter, U., H. Lenz, H. Kreuzer, and C. Besang, Pre-Ross granites at the Pacific margin of the Robertson Bay Terrane, *Geol. Jahrb., Reihe B*, 60, 363-370, 1984.

Weaver, S., J. D. Bradshaw, and M. G. Laird, Geochemistry of Cambrian volcanics of the Bowers Supergroup and implications for the early Paleozoic tectonic evolution of northern Victoria Land, Antarctica, *Earth Planet. Sci. Lett.*, 68, 128-140, 1984.

Weissel, J. K., D. E. Hayes, and E. M. Herron, Plate tectonic synthesis; the displacements between Australia, New Zealand and Antarctica since the Late Cretaceous, *Mar. Geol.*, 25, 231-277, 1977.

Williams, E., Tasman Fold Belt system in Tasmania, *Tectonophysics*, 48, 159-205, 1978.

Wodzicki, A., J. D. Bradshaw, and M. G. Laird, Petrology of the Wilson and Robertson Bay Group and the Bowers Supergroup, in *Antarctic Geoscience*, edited by C. Craddock, pp. 549-554, University of Wisconsin Press, Madison, 1982.

Wright, T. O., Sedimentology of the Robertson Bay Group, north Victoria Land, Antarctica, *Geol. Jahrb., Reihe B*, 41, 127-138, 1981.

Wright, T. O., and R. H. Findlay, Relationships between the Robertson Bay Group and the Bowers Supergroup: New progress and complications from the Victoria Mountains, northern Victoria Land, *Geol. Jahrb., Reihe B*, 41, 105-116, 1984.

Wright, T. O., R. J. Ross, Jr., and J. E. Repetski, Newly discovered youngest Cambrian or oldest Ordovician fossils from the Robertson Bay Terrane (formerly Precambrian), northern Victoria Land, Antarctica, *Geology*, 12, 301-305, 1984.

Copyright 1987 by the American Geophysical Union.

PALEOZOIC MAGMATISM AND ASSOCIATED TECTONIC PROBLEMS OF NORTHERN VICTORIA LAND, ANTARCTICA

S. G. Borg

Department of Earth and Space Sciences, University of California, Los Angeles, California 90024

E. Stump

Department of Geology, Arizona State University, Tempe, Arizona 85287

Abstract. The division of northern Victoria Land (NVL) into three north-northwest trending terranes is underscored by recent chemical and isotopic studies of the Paleozoic granitoids of the region. Early Paleozoic Granite Harbour Intrusives are found only in the (western) Wilson Terrane (WT), best interpreted as a continuation of the Ross Orogenic Belt, which developed along the margin of the East Antarctic Craton. Devonian Admiralty Intrusives, however, are found only within the (central) Bowers Terrane (BT) and the (eastern) Robertson Bay Terrane (RBT), and studies of chemical polarity indicate that these terranes are fragments of an allochthonous crustal block. In view of the key position occupied by NVL in reconstructions of Antarctica and Australia, recognition of the relationships outlined above raises important questions about the assembly and perhaps the breakup of this portion of Gondwana. Timing of juxtapositioning of the BT + RBT against the WT is problematic but is thought to have occurred after emplacement of the Devonian Admiralty Intrusives and must have occurred prior to deposition of the Beacon Supergroup in the Permian. Other geologic relations known at present do not provide definitive constraints on tectonic models, but there are unresolved points which may have special significance. For example, the tectonic setting of the Salamander Granite Complex, as well as the tectonic setting of middle Paleozoic volcanics exposed at Mount Black Prince, Lawrence Peaks, and Gallipoli Heights remain an important question. Clearly, the geology of NVL holds important clues to the tectonic history of this segment of Gondwana. Continued studies of the granitoids, coupled with new research on the middle Paleozoic volcanics, detailed structural studies with emphasis on timing and sense of movement on the terrane bounding faults, and more detailed provenance studies of all the sedimentary and metasedimentary rock units in NVL, would be helpful in evaluating and revising tectonic models. Furthermore, integration of the geology of NVL with that of Australia, New Zealand, and West Antarctica is of paramount importance.

Introduction

Recent investigations in northern Victoria Land (NVL), Antarctica, have resulted in a profound expansion of geological data as well as new interpretations of regional geologic relationships. As an expected consequence of these findings, a whole new set of questions has arisen concerning the geologic and tectonic development of the region. This paper is not intended to be a comprehensive review of NVL geology, nor is its purpose to describe yet another tectonic model for the development of NVL. Rather, its purpose is to assess what we know about some crucial aspects of the regional geology of NVL and, using new data obtained on the granites, to put constraints on tectonic models and to identify some problems of the tectonic development of the region which need to be addressed.

Background

That the pre-Permian metamorphic basement of northern Victoria Land is composed of three diverse fault-bounded lithologic groups has been known for some time [e.g., Gair et al., 1969]. These geologic relations are shown in Figure 1. From west to east in NVL, these groups are referred to as the Wilson Group (following the usage of Grindley and Warren [1964]), a diverse assemblage of amphibolite facies metasediments, the Bowers Supergroup (as recognized by Sturm and Carryer [1970] and as redefined by Laird et al. [1977]), most recently interpreted as a Cambrian volcanic arc assemblage [Weaver et al., 1984], and the Robertson Bay Group (following the usage of Harrington et al. [1964]), a monotonous graywacke-shale sequence interpreted as turbidites in a subsea fan complex. Lithologic characteristics of these units are reasonably well known, and recent summaries have been published by Kleinschmidt and Skinner [1981], Laird et al. [1977, 1982], Wright [1981], Field and Findlay [1983], and Tessensohn et al. [1981]. The complete extent of the Bowers Supergroup across NVL was first recognized by

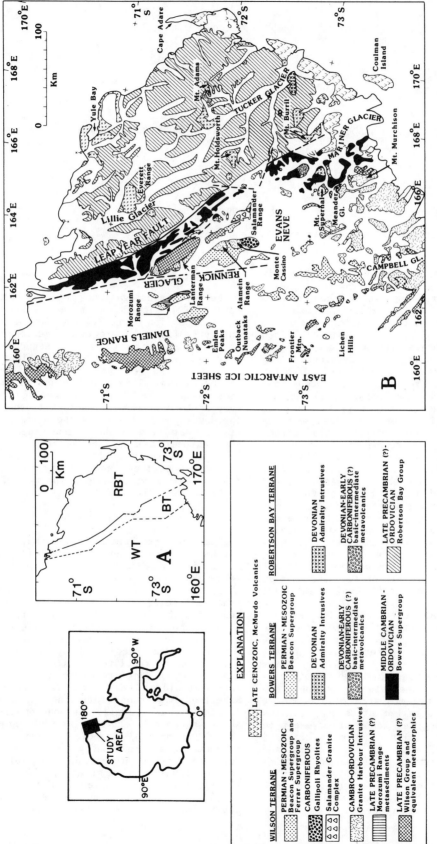

Fig. 1. Regional geology of northern Victoria Land. (a) Sketch map showing the relationship of the Wilson Terrane (WT), Bowers Terrane (BT), and Roberson Bay Terrane (RBT). (b) Geologic sketch map of northern Victoria Land. Compiled from maps by Gair et al. [1969], Stump et al. [1983], Borg et al. [1986a, b], Kleinschmidt et al. [1984], Tessensohn [1984], and Weaver et al. [1984].

Stump et al. [1983]. The Leap Year Fault had been recognized as the boundary between the Robertson Bay Group and the Bowers Supergroup, but its significance has been the object of considerable speculation [e.g., Wright and Findlay, 1984]. The nature of the boundary between the Wilson Group and the Bowers Supergroup was also somewhat speculative, but it has been identified as a fault in most places [e.g., Weaver et al., 1984].

In the past, geologic models for the development of NVL have been based on the concept of a relatively simple continental margin of Gondwana developing through successive orogenic cycles [e.g., Hamilton, 1967; Elliot, 1975]. Tessensohn et al. [1981] interpreted the geology from a different perspective and suggested that the diverse lithologic groups were merely the product of different crustal levels being exposed. Furthermore, they proposed that all the metasediments were deposited over the same Precambrian to Early Cambrian time interval, with sedimentation of the upper portion of the Bowers Supergroup occurring during Early Ordovician time. A key element of their interpretation, which ties the lithologic units together, is their assumption that turbidites in the Morozumi Range are correlative with the Robertson Bay Group. However, this assumption, as argued by Bradshaw and Laird [1983], was based on gross lithologic similarities and not on any firm evidence. More recently, arguments against correlation made by Borg [1984] and Borg et al. [1986a, b] have been based on sedimentological differences described by Wyborn [1983] and on recent fossil discoveries which define the maximum age of at least part of the Robertson Bay Group as Late Cambrian to Early Ordovician [Wright et al., 1984]. The latter point means that the Robertson Bay Group was being deposited at essentially the same time that the turbidites in the Morozumi Range were being intruded by a posttectonic granitoid pluton of the Granite Harbour Intrusives (emplacement and cooling about 515-480 Ma [Vetter et al., 1983]). Therefore, despite the arguments by Tessensohn et al. [1981], there appear to be no geologic relationships which require genetic links between these turbidite groups.

In a recent review paper, Bradshaw and Laird [1983] emphasized the spatial separation and apparent tectonic boundaries of the major basement units. They mention the proposal by Harrington [1980] that the Leap Year Fault may be an old plate boundary and point to recently developed models of orogenic belts as tectonic collages of diverse terranes [e.g., Davis et al., 1978] in suggesting that the tectonic development of NVL may have involved more than a succession of Wilson cycles on the margin of East Antarctica. Working from this point, Weaver et al. [1984] have proposed that the geology of NVL represents three different portions, or terranes, of a continuous early Paleozoic plate margin which have been brought together by complex movements on transform faults. Alternatively, Gibson and Wright [1985] argue for head-on collision of crustal blocks with substantial overthrusting, followed by block faulting, to explain the geology. Both papers infer that juxtapositioning of the terranes was accomplished prior to the Devonian, based on their interpretation that the Admiralty Intrusives are "terrane stitching" granitoids. Although highly speculative, these papers are applauded as the first attempts at actualistic plate tectonic models for the development of NVL.

The division of NVL into distinct geologic terranes is underscored by recent chemical and isotopic studies on the Paleozoic granitoids by Borg [1984] and Borg et al. [1984, 1986a, b]. Several points in these papers, which bear on the tectonic evolution of NVL, are summarized below. Granitoid plutons of the early Paleozoic Granite Harbour Intrusives (GHI) are only found intruding Wilson Group rocks, and this association defines the Wilson Terrane (WT) in the western part of NVL. To the north, in reconstructions of Australia and Antarctica, the WT may be correlated, approximately, with the Adelaide Geosyncline of South Australia [Craddock, 1969, 1972].

The central Bowers Terrane (BT) and the eastern Robertson Bay Terrane (RBT) experienced postfolding volcanism (Lawrence Peaks and Black Prince volcanics) and emplacement of granitoid plutons of the Admiralty Intrusives (AI) during middle Paleozoic time. Because the distribution of these rocks in NVL is very important to any discussion of regional geology, a review of several crucial relationships is in order. Grindley and Oliver [1983] postulated that the Black Prince volcanics (in the RBT), Lawrence Peaks volcanics (in the BT), and the Gallipoli Rhyolites (in the WT) were part of the same magmatic province. This statement, however, is pure speculation with no supporting evidence.

Many authors have inferred that the Admiralty Intrusives are present in the WT as well as the BT and RBT. This was based on three lines of evidence: (1) a K-Ar biotite age of 393 Ma from a granite in the Alamein Range; (2) the presumption that the Salamander Granite (formerly the Salamander Granodiorite of Laird et al. [1974]) is an AI and was emplaced into the WT; and (3) the presumption that the Supernal Granodiorite was emplaced into the WT. With regard to point 1, the published age is misleading because the dated sample was from an outcrop which is about 20 m below the projection of a major Ferrar sill and so may have lost Ar during the Jurassic. Also, samples of granite from the Alamein Range have subsequently yielded Rb-Sr mineral isochrons and a three-point whole-rock isochron of about 535 Ma, clearly a Granite Harbour age [Borg, 1984; E. Stump et al., unpublished data, 1983]. With regard to point 2, the Salamander Granite is petrographically and geochemically distinct from the AI [Borg et al., 1986a] and has yielded Rb-Sr mineral isochrons of 319 and 324 Ma, which are significantly younger than the 363 to 390 Ma Rb-Sr mineral isochrons known from the AI [Borg, 1984; E. Stump et al., unpublished data, 1983]. Thus, it clearly should not be correlated with the AI. Also, though our preferred interpretation is that it is within the WT, contacts with country rock are not known [Borg, 1984, Borg et al., 1986a], so it is not clear whether the Salamander Granite is a feature of the WT or the BT. With regard to point 3, the

Supernal Granodiorite is an AI and occurs near the WT-BT boundary. However, despite recent interpretive mapping [Gibson et al., 1984; Tessensohn, 1984], at its southwestern and eastern contacts it intrudes greenschist facies mafic volcanic and volcaniclastic rocks unlike anything known from the WT and similar to rocks described from the BT. On the basis of these field relations and the isotopic studies summarized below, we believe this granite to be a feature of the BT and so place the WT-BT boundary west of Mount Supernal. Finally, the point which we wish to make here is that these granitoids are not known from the WT and hence there is no evidence requiring the AI to be "terrane stitching" plutons as inferred by Weaver et al. [1984] and Gibson and Wright [1985].

Chemical and isotopic polarity of the granitoid groups is a feature which has important implications for tectonic models of the development of NVL. Figures 2a and 2b and Table 1 summarize data reported and discussed by Borg et al. [1986b] about the chemical and isotopic polarity of the AI and GHI in NVL. The polarity of the GHI indicates that these granites were emplaced along the margin of the East Antarctic Craton as a plutonic province associated with the Ross Orogeny. This is in keeping with previous tectonic models for the development of NVL and the Transantarctic Mountains to the south [e.g., Craddock, 1972; Elliot, 1975]. The Sr and Nd isotopic compositions of the GHI suggest that in NVL, only the continental "inboard" portion of the magmatic province is preserved. This interpretation finds additional support in the work of Grew et al. [1984], who describe metamorphism in the Wilson Group as possibly representing the inboard portion of a continental-margin paired metamorphic belt. The polarity of the AI is distinct and different from the polarity of the GHI and indicates greater contribution of old continental material from south-southwest to north-northeast, a direction away from the East Antarctic Craton. Furthermore, the Sr and Nd isotopic compositions indicate that the most "outboard" portion of the AI magmatic province is not exposed in the region. The chemical polarity of the AI is nearly opposite to the polarity which would be expected if the inference made by Weaver et al. [1984], that the AI developed over a westward dipping subduction zone, was correct. This leads to the interpretation that the BT and RBT are fragments of the margin of an allochthonous continental block or blocks. These data for the AI are hard evidence for allochthonous terranes in Victoria Land. Continuity of chemical and isotopic variations of the AI across the Leap Year Fault is evidence of lower crustal continuity between the BT and RBT at the time of generation of the AI. This is a very important point with respect to regional tectonics because it means that the Leap Year Fault is not a major crustal plate boundary, such as a transform fault, as suggested by Weaver et al. [1984]. Rather, the inference by Wright and Findlay [1984] that this fault represents a zone of thrusting between the BT and RBT and not a boundary between exotic terranes is more in accord with inferences from the granitoid studies mentioned above. The pronounced discontinuity of isotopic compositions between the GHI in the WT and the AI in the BT + RBT indicates the presence of a major lower crustal boundary, which has been interpreted by Borg and colleagues as the plate boundary along which the BT + RBT was juxtaposed to the WT, after emplacement of the Devonian AI.

New Tectonic Problems

In view of the key position occupied by NVL in reconstructions of Australia and Antarctica, recognition of the allochthonous nature of the BT + RBT raises important questions about the assembly of this part of Gondwana. Some of these questions are as follows: (1) Where were the BT and RBT formed, and what is their geologic history prior to assembly with the WT? (2) When and how were the BT and RBT juxtaposed to the WT? (3) Since the GHI in the WT appears to be the continental "inboard" part of a continental margin magmatic province, and since the AI in the BT + RBT appears to represent the "central" to "inboard" portions of a different continental margin magmatic province, what has happened to the "outboard" portions of the WT and the BT + RBT? (4) Are other recognizable fragments of the BT and RBT allochthon preserved in West Antarctica, Australia, or New Zealand? Although there are no unambiguous solutions to these questions at present, the known geologic relations, combined with the new information on the GHI and the AI, do provide some constraints and allow some new speculation on the tectonic evolution of the region.

Discussion

On the basis of interpretations from the granitoid studies, the tectonic evolution of NVL must be viewed in terms of two separate crustal entities, the WT and the BT + RBT, which had distinct geologic histories until at least the middle to late Paleozoic, after emplacement of the AI. This means that, although deposition of the Bowers Supergroup was approximately concurrent with and possibly related to that of the Robertson Bay Group, as suggested by Wright and Findlay [1984], these two units are not related in any way to the Wilson Group or to any portion of pre-late Paleozoic East Antarctica. Also, the BT + RBT experienced folding and metamorphism which, although apparently coincident with metamorphism in the WT (approximately Late Cambrian to Middle Ordovician [Adams et al., 1982; Adams and Kreuzer, 1984]), must be regarded as a separate event, unrelated to the Ross Orogeny. This folding and metamorphism probably records the suturing of the BT to the RBT at some location remote from the WT. In this same vein, the possibly late Precambrian or early Paleozoic granitoids described from the northern coast of the RBT by Vetter et al. [1984] cannot be correlated with the Ross Orogeny or to any events recorded in East Antarctica. Rather, these granitoids and their wallrocks would represent pieces of the allochthonous continental block along which the BT and RBT developed. The fact that recent work in NVL has identified possible ties between

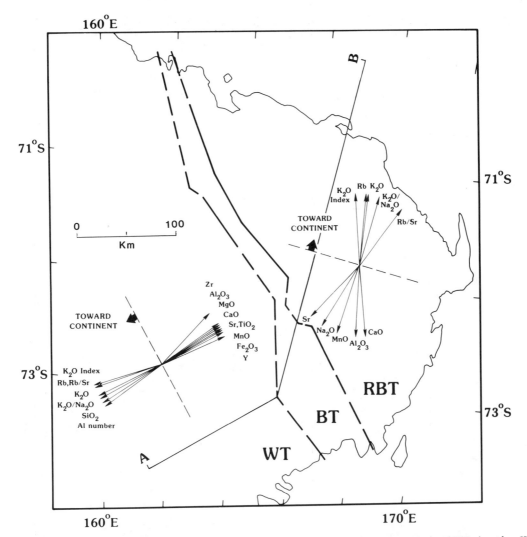

Fig. 2a. Summary of chemical polarity of the Granite Harbour Intrusives (GHI in the WT) and the Admiralty Intrusives (AI in the BT + RBT) in northern Victoria Land. Rose diagrams for the GHI and the AI showing the directions of maximum increase of various chemical parameters. The polarity of the GHI is not consistent with emplacement along the margin of the East Antarctic Craton. Rather, the polarity indicates emplacement along the margin of a continent, which, relative to present directions, lay to the north-northeast as depicted on the sketch. See text for further explanation. This information is the product of first-order trend-surface regression analysis of chemical data over the region. Only trends which were significant at a confidence level greater than 90% are shown. Magnitudes of trends are summarized in Table 1 and are similar to trends for the same parameters in other batholithic provinces around the world [cf. Baird et al., 1974; Reed et al., 1983]. K_2O index is defined by Bateman and Dodge [1970] as $(K_2O \times 1000)/(SiO_2 - 45)$. Al number is defined as the molar ratio $Al_2O_3/(K_2O + Na_2O + CaO)$. Modified from Borg [1984] and Borg et al. [1986b].

the Bowers Supergroup and the Robertson Bay Group [e.g., Wyborn, 1983; Wright and Findlay, 1984] and has forced a reappraisal of the nature of the contact between the Bowers Supergroup and the Wilson Group [e.g., Bradshaw and Laird, 1983; Wodzicki and Robert, 1986] fits well with a scenario of distinct pre-late Paleozoic geologic histories for the WT and the BT + RBT.

The problem as to where the BT and RBT were formed is almost pure speculation at this point. Based on the interpretations outlined above, the only choice is to exclude the Antarctic margin of Gondwana and, by default, form BT and RBT somewhere in the paleo-Pacific, perhaps to the northeast or east relative to present directions. Plate movements which juxtaposed the BT + RBT to

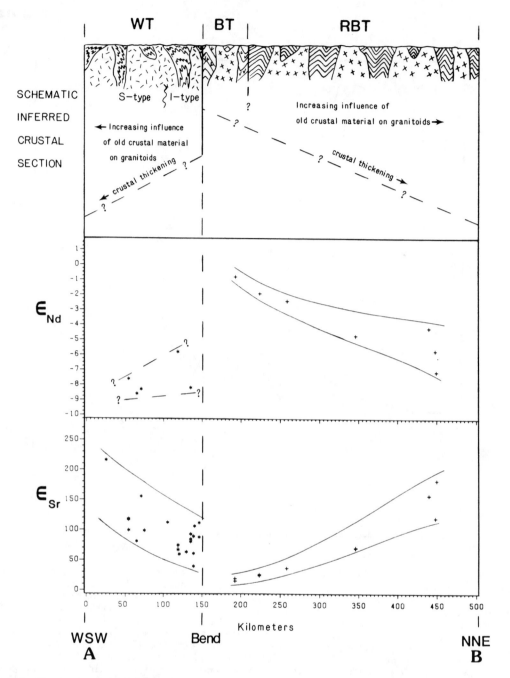

Fig. 2b. Summary of Sr and Nd isotopic polarity of the Granite Harbour Intrusives (GHI in the WT) and the Admiralty Intrusives (AI in the BT + RBT) in northern Victoria Land. Initial Σ_{Nd} and Σ_{Sr} of the AI and GHI projected onto section line A-B of Figure 2a. The legs of this cross section are perpendicular to the trends of the continental margins inferred in the WT and BT + RBT by major and trace element polarity of the GHI and the AI, respectively. Ratios are calculated assuming an age of 550 Ma for the GHI and 400 Ma for the AI following arguments of Borg [1984] and Borg et al. [1986a, b]. These data, along with the chemical polarity and the surface geology, allows interpretation of crustal structure. These data clearly show that the crust in the WT, characterized by the GHI, is distinct and different from the crust in the BT + RBT, characterized by the AI. The isotopic continuity among AI in the BT and RBT allows the interpretation that the lower crust in these terranes was a continuum during emplacement of the AI. The pattern of variation of these isotopic compositions is in accord with the chemical polarity, and together these data are strong evidence for an allochthonous origin of the BT and RBT. See text for further explanation. Modified from Borg [1984] and Borg et al. [1986b]. Notation follows Farmer and DePaolo [1983].

TABLE 1. Summary of First-Order Trend Surface Data for Granite Harbour Intrusives (GHI) and Admiralty Intrusives (AI)

Chemical Parameter	Percent of Variance Explained by the First-Order Model		Slope	
	GHI	AI	GHI	AI
SiO_2	30.25		5.66	
TiO_2	18.31		0.23	
Al_2O_3	15.17	20.17	1.27	0.38
Fe_2O_3	30.19		2.25	
MnO	24.47	19.29	0.03	0.01
MgO	26.68		1.53	
CaO	33.45	16.04	2.22	0.41
Na_2O		47.36		0.33
K_2O	33.03	41.43	1.64	0.54
K_2O/Na_2O	19.13	52.31	0.65	0.23
Al number	21.29		0.08	
K_2O index	20.12	44.39	30.83	15.40
Rb	26.53	44.51	87.2	37.0
Sr	15.77	26.63	100.4	54.0
Y	21.97		10.5	
Zr	7.97		69.8	
Rb/Sr	28.42	29.08	1.8	0.7

Data are summarized from Borg [1984] and Borg et al. [1986b]. The significance level of each trend reported is greater than 90%. Slopes are given in units per 100 km, oxides in percentage of weight, and trace elements in parts per million.

the WT are equally speculative. The apparent absence of ophiolites, which might be expected along a collision suture, is conveniently explained by an interpretation that at least the last movements were strike-slip along the WT-BT contact. This style of terrane assembly could also easily explain the observation that the chemical and isotopic data for both the GHI and the AI indicate that the most outboard portions of each of the plutonic provinces appear to be missing. However, these characteristics of the geology could also be explained by collision of crustal blocks accompanied by extreme overthrusting (WT over BT + RBT) and followed by normal faulting, along the lines of the model presented by Gibson and Wright [1985]. A problem with this overthrusting hypothesis is the apparent lack of evidence for the extreme crustal thicknesses which would have resulted in such a scenario. If the BT + RBT had been overridden to the degree suggested by Gibson and Wright [1985], one might expect to see evidence of substantially higher pressures in the metamorphics of the BT and RBT exposed today. The chemical and isotopic data available at present cannot rule out either of these models for the style of terrane assembly.

Constraints on the timing of terrane assembly are few. On the basis of chemical polarity of the AI and our assertion that Admiralty plutons do not intrude the WT, juxtaposing of the BT + RBT with the WT is inferred to have occurred after emplacement of the AI and may have occurred somewhat later. An important point in this regard is the tectonic setting of the Carboniferous (post-AI) magmatism, including the volcanic piles at Gallipoli Heights, Mount Black Prince, and Lawrence Peaks as well as the subvolcanic Salamander Granite Complex. The Gallipoli Rhyolites and probably the Salamander Granite Complex were emplaced in the WT during Carboniferous time. These are presently on the eastern margin of the WT, but no definitely related rocks are known from the adjacent BT or RBT farther east. Although it is possible that this magmatism is somehow related to the volcanics at Mount Black Prince and Lawrence Peaks in the RBT and BT, respectively, there is no evidence requiring a genetic relationship between these isolated volcanic piles. Thus, one may speculate that the BT + RBT had not been juxtaposed to the WT by this time. The only definite lower age limit which can be placed on terrane assembly, at present, is the age of the Beacon sediments (presumably Permian) which overlie the Bowers Supergroup in the Neall Massif and on the southwest side of the Leitch Massif [Tessensohn et al., 1981].

Recognition of the BT + RBT as an allochthonous crustal entity raises the possiblity of identifying related terranes in areas adjacent to NVL in Gondwana reconstructions. The Swanson Formation and associated Devonian granitoids in Marie Byrd Land are prime candidates which may be related to the RBT. It is also possible that continental fragments related to the BT + RBT are present in New Zealand and the Campbell Plateau as well as in southeastern Australia, east of the Adelaide Geosyncline.

Summary

Results of recent work on granitoids in NVL emphasize the distinction of geologic terranes and indicate that the BT + RBT is an allochthonous crustal entity which was emplaced against the WT (on the margin of East Antarctica) sometime after the Devonian and before deposition of the Beacon Supergroup. These interpretations raise important new questions and allow some new speculation about the tectonic development of the region. To confirm, modify, or reject these recently developed ideas of regional tectonics, several lines of research might be pursued, including (1) sedimentological studies of the metamorphic basement units, with emphasis on provenance of the sediments, (2) continued geochemical work on the granitoids until a complete regional data base is established and evaluated, (3) field and geochemical studies of the middle Paleozoic volcanics in order to determine their tectonic setting, (4) geophysical studies to investigate the possibility of discriminating between the terranes at depth and to identify and determine the nature of the lower crustal discontinuity inferred from the geochemical studies of the granitoids, (5) studies of deep crustal xenoliths from the Hallett-McMurdo Volcanic Province, which crosses all three terranes in southern NVL, and (6) studies of the nature, age, and movement history of the terrane bounding faults. Clearly, the geology of NVL holds important clues to the geologic evolution of this segment of Gondwana prior to breakup. Integration of the geologic relationships and tectonic models of NVL with those of southeastern Australia, Tasmania, New Zealand, and West Antarctic is of paramount importance for resolving the tectonic history of the region.

Acknowledgments. Funding for this work was provided by NSF grants DPP8019991 and DPP8216281.

References

Adams, C. J. D., and H. Kreuzer, Potassium-argon studies of slates and phyllites from the Bowers and Robertson Bay terranes, north Victoria Land, Antarctica, Geol. Jahrb., Reihe B, 60, 265-288, 1984.

Adams, C. J. D., J. E. Gabites, A. Wodzicki, M. G. Laird, and J .D. Bradshaw, Potassium-argon geochronology of the Precambrian-Cambrian Wilson and Robertson Bay groups and Bowers Supergroup, northern Victoria Land, Antarctica, in Antarctic Geoscience, edited by C. Craddock, pp. 543-548, University of Wisconsin Press, Madison, 1982.

Baird, A. K., K. W. Baird, and E. E. Welday, Chemical trends across Cretaceous batholithic rocks of southern California, Geology, 2, 493-496, 1974.

Bateman, P. C., and F. W. C. Dodge, Variations of major chemical constituents across the central Sierra Nevada batholith, Geol. Soc. Am. Bull., 81, 409-420, 1970.

Borg, S. G., Granitoids of northern Victoria Land, Antarctica, Ph.D. thesis, 355 pp., Department of Geology, Arizona State University, Tempe, 1984.

Borg, S. G., B. W. Chappell, M. T. McCulloch, E. Stump, D. Wyborn, and J. R. Holloway, Compositional polarity of granitoids with implications to regional geology, northern Victoria Land, Antarctica, Geol. Soc. Am. Abstr. Program, 16, 450, 1984.

Borg, S. G., E. Stump, and J. R. Holloway, Granitoids of northern Victoria Land, Antarctica: A reconnaissance study of field relations, petrography and geochemistry, in Geological Investigations in Northern Victoria Land, Antarctica, Antarct. Res. Ser., vol. 46, edited by E. Stump, AGU, Washington, D.C., in press, 1986a.

Borg, S. G., E. Stump, B. W. Chappell, M. T. McCulloch, D. Wyborn, R. L. Armstrong, and J. R. Holloway, Granitoids of northern Victoria Land, Antarctica: Implications of chemical and isotopic variations to regional crustal structure and tectonics, Am. J. Sci., in press, 1986b.

Bradshaw, J. D., and M. G. Laird, The pre-Beacon geology of northern Victoria Land: A review, in Antarctic Earth Science, edited by R. L. Oliver, P. R. James, and J. B. Jago, pp. 98-101, Australian Academy of Science, Canberra, 1983.

Craddock, C., Map of Gondwanaland, in Geologic Maps of Antarctica, edited by V. C. Bushnell and C. Craddock, Antarct. Map Folio Ser., 12, plate XXII, Am. Geogr. Soc., New York, 1969.

Craddock, C., Antarctic tectonics, in Antarctic Geology and Geophysics, edited by R. J. Adie, pp. 449-455, Universitetsforlaget, Oslo, 1972.

Davis, G. A., J. W. H. Monger, and B. C. Burchfiel, Mesozoic construction of the Cordilleran "collage," central British Columbia to central California, in Mesozoic Paleogeography of the Western United States, edited by D. G. Howell and K. A. McDougall, pp. 1-32, Society of Economic Paleontologists and Mineralogists, Pacific Section, Los Angeles, Calif., 1978.

Elliot, D. H., Tectonics of Antarctica: A review, Am. J. Sci., 275-A, 45-106, 1975.

Farmer, G. L., and D. J. DePaolo, Origin of Mesozoic and Tertiary granite in the western United States and implications for pre-Mesozoic crustal structure: 1, Nd and Sr isotopic studies in the geocline of the northern Great Basin, J. Geophys. Res., 88, 3379-3401, 1983.

Field, B. D., and R. H. Findlay, The sedimentology of the Robertson Bay Group, northern Victoria Land, in Antarctic Earth Science, edited by R. L. Oliver, P. R. James, and J. B. Jago, pp. 102-106, Australian Academy of Science, Canberra, 1983.

Gair, H. S., A. Sturm, S. J. Carryer, and G. W. Grindley, The geology of northern Victoria Land, in Geologic Maps of Antarctica, edited by V. C. Bushnell and C. Craddock, Antarct. Map Folio Ser., 12, plate XII, Am. Geogr. Soc., New York, 1969.

Gibson, G. M., and T. O. Wright, Importance of thrust faulting in the tectonic development of northern Victoria Land, Antarctica, Nature, 315, 480-483, 1985.

Gibson, G. M., F. Tessensohn, and A. Crawford, Bowers Supergroup rocks west of the Mariner Glacier and possible greenschist facies equivalents, Geol. Jahrb., Reihe B, 60, 289-318, 1984.

Grew, E. S., G. Kleinschmidt, and W. Schubert, Contrasting metamorphic belts in north Victoria Land, Antarctica, Geol. Jahrb., Reihe B, 60, 253-263, 1984.

Grindley, G. W., and P. J. Oliver, Post-Ross Orogeny cratonization of northern Victoria Land, Antarctic Earth Science, edited by R. L. Oliver, P. R. James, and J. B. Jago, pp. 133-139, Australian Academy of Science, Canberra, 1983.

Grindley, G. W., and G. Warren, Stratigraphic nomenclature and correlation in the western Ross Sea region, in Antarctic Geology, edited by R. J. Adie, pp. 314-333, North-Holland, Amsterdam, 1964.

Hamilton, W., Tectonic map of Antarctica, Tectonophysics, 4, 555-568, 1967.

Harrington, H. J., Reconstruction of the Australian sector of Gondwanaland from Cambrian to Early Permian (abstract) in Fifth Gondwana Symposium, p. 71, Wellington, New Zealand, 1980.

Harrington, H. J., B. L. Wood, I. C. McKellar, and G. J. Lenson, The geology of Cape Hallett-Tucker Glacier district, in Antarctic Geology, edited by R. J. Adie, pp. 220-228, North-Holland, Amsterdam, 1964.

Kleinschmidt, G., and D. N. B. Skinner, Deformation styles in the basement rocks of north Victoria Land, Antarctica, Geol. Jahrb., Reihe B, 41, 155-200, 1981.

Kleinschmidt, G., N. W. Roland, and W. Schubert, The metamorphic basement in the Mountaineer Range, north Victoria Land, Antarctica, Geol. Jahrb., Reibe B, 60, 213-251, 1984.

Laird, M. G., P. B. Andrews, and P. R. Kyle, Geology of northern Evans Neve, Victoria Land, Antarctica, N. Z. J. Geol. Geophys., 17, 587-601, 1974.

Laird, M. G., R. A. Cooper, and J. B. Jago, New data on the lower Palaeozoic sequence of northern Victoria Land, Antarctica, and its significance for Australian-Antarctic relations in the Palaeozic, Nature, 265, 107-110, 1977.

Laird, M. G., J. D. Bradshaw, and A. Wodzicki, Stratigraphy of the upper Precambrian and lower Paleozic Bowers Supergroup, northern Victoria Land, Antarctica, in Antarctic Geoscience, edited by C. Craddock, pp. 535-542, University of Wisconsin Press, Madison, 1982.

Reed, B. L., A. T. Miesch, and M. A. Lanphere, Plutonic rocks of Jurassic age in the Alaska-Aleutian Range batholith: Chemical variations and polarity, Geol. Soc. Am. Bull., 94, 1232-1240, 1983.

Stump, E., M. G. Laird, J. D. Bradshaw, J. R. Holloway, S. G. Borg, and K. E. Lapham, Bowers graben and associated tectonic features cross northern Victoria Land, Antarctica, Nature, 304, 334-335, 1983.

Sturm, A. G., and S. Carryer, Geology of the region between Matusevich and Tucker glaciers, northern Victoria Land, Antarctica, N. Z. J. Geol. Geophys., 13, 408-435, 1970.

Tessensohn, F., Geological and tectonic history of the Bowers Structural Zone, north Victoria Land, Antarctica, Geol. Jahrb., Reihe B, 60, 371-396, 1984.

Tessensohn, F., K. Duphorn, H. Jordan, G. Kleinschmidt, D. N. B. Skinner, U. Vetter, T. Vetter, T. O. Wright, and D. Wyborn, Geological comparison of basement units in north Victoria Land, Antarctica, Geol. Jahrb., Reihe B, 41, 31-88, 1981.

Vetter, U., N. W. Roland, H. Kreuzer, A. Höhndorf, H. Lenz, and C. Besang, Geochemistry, petrography, and geochronology of the Cambro-Ordovician and Devonian-Carboniferous granitoids of northern Victoria Land, Antarctica, in Antarctic Earth Science, edited by R. L. Oliver, P. R. James, and J. B. Jago, pp. 140-143, Australian Academy of Science, Canberra, 1983.

Vetter, U., H. Lenz, H. Kreuzer, and C. Besand, Pre-Ross granites at the Pacific margin of the Robertson Bay Terrane, north Victoria Land, Antarctica, Geol. Jahrb., Reihe B, 60, 363-370, 1984.

Weaver, S. D., J. D. Bradshaw, and M. G. Laird, Geochemistry of Cambrian volcanics of the Bowers Supergroup and implications for the early Paleozoic tectonic evolution of northern Victoria Land, Antarctica, Earth Planet. Sci. Lett., 68, 128-140, 1984.

Wodzicki, A., and R. Robert, Jr., Geology of the Bowers Supergroup, central Bowers Mountains, northern Victoria Land, in Geological Investigations in Northern Victoria Land, Antarctica, Antarct. Res. Ser., vol. 46, edited by E. Stump, AGU, Washington, D.C., in press, 1986.

Wright, T. O., Sedimentology of the Robertson Bay Group, north Victoria Land, Antarctica, Geol. Jahrb., Reihe B, 41, 127-138, 1981.

Wright, T. O., and R. H. Findlay, Relationships between the Robertson Bay Group and the Bowers Supergroup--New progress and complications from the Victory Mountains, north Victoria Land, Geol. Jahrb., Reihe B, 60, 105-116, 1984.

Wright, T. O., R. J. Ross, Jr., and J. E. Repetski, Newly discovered youngest Cambrian or oldest Ordovician fossils from the Robertson Bay terrane (formerly Precambrian), northern Victoria Land, Antarctica, Geology, 12, 301-305, 1984.

Wyborn, D., Chemical control on stratigraphic relations in northern Victoria Land and some possible relations with SE Australia (abstract), in Antarctic Earth Science, edited by R. L. Oliver, P. R. James, and J. B. Jago, pp. 145, Australian Academy of Science, Canberra, 1983.

CONSTRUCTION OF THE PACIFIC MARGIN OF GONDWANA DURING THE PANNOTIOS CYCLE

Edmund Stump

Department of Geology, Arizona State University, Tempe, Arizona 85287

Abstract. A cycle of sedimentation and tectonism affected the supercontinent of Gondwana from the middle or late Proterozoic until its culmination at approximately 500 Ma. This paper is a synthesis of the activity that occurred along the Pacific margin of the supercontinent in portions of present-day South America, Africa, Antarctica, and Australia. The cycle began at approximately 900 Ma with sedimentation in basins bounded and, in many cases, floored by older cratonic basement. Tectonism began in the late Precambrian in a zone from the Ribeira and Damara belts of South America and southern Africa through to the central Transantarctic Mountains. A narrow septum of cratonic rocks, unaffected by late Precambrian tectonism, is postulated to have extended from the Haag Nunataks in Antarctica through the Falkland Islands to the Rio de la Plata craton in South America. During Early and Middle Cambrian, sedimentation was renewed in regions affected by late Precambrian tectonism, and continued in the unaffected regions of South America, Antarctica, and Australia. Tectonism culminated in Cambro-Ordovician time, affecting both a more cratonal belt, where tectonism ceased, and a more marginal belt, where tectonism was renewed during the middle Paleozoic. The name Pannotios is proposed as a unifying term to encompass the late Precambrian-early Paleozoic cycle of activity that occurred throughout Gondwana.

Introduction

From the middle Proterozoic onward until its breakup in the Mesozoic, a series of geosynclines and orogenic belts were at play along the Pacific side of the Gondwana supercontinent. The rocks from these events are to be found in a broad, composite belt that stretches from eastern Australia and New Zealand through the Transantarctic Mountains and West Antarctica to southern Africa and South America. The record is one both of quasi-synchronous events occurring throughout extensive portions of the belt and of regional events not widely developed. The pivotal episode in the organization of the Pacific margin of Gondwana occurred approximately 500 million years ago, when intercratonal portions of the mobile belt consolidated cratonic blocks into the coherent supercontinent, and tectonic activity shifted toward the Pacific, where compressive interaction between ocean and continent continued up to and during the breakup of Gondwana.

The approximately 500 Ma activity marked the climax of an orogenic cycle common to the southern continents. In Africa it is the "Pan-African thermotectonic episode," originally defined by Kennedy [1964] on the basis on K-Ar dates of that age developed throughout widespread regions of the continent, in both mobile belts and older cratons. African belts of concern in this discussion include the Damara of Namibia (Figure 1) and, farther south along the coast, the Gariep of the Richtersveld [Kröner and Blignault, 1976]. In the Cape region of South Africa, equivalent rocks belong to the Malmesbury Group, which was deformed during Saldanian tectonism [Hartnady et al., 1974].

In South America the equivalent of the Pan-African is the Brazilian orogenic cycle which had widespread effects throughout the continent [Cordani et al., 1973]. Of local note for this paper are the Ribeira fold belt of southern Brazil, Uruguay, and eastern Argentina [Almeida et al., 1973] and the Pampean orogenic cycle of northwest Argentina [Aceñolaza and Miller, 1982] (Figure 1). In Antarctica, equivalent rocks belong to the Ross Supergroup, which was affected by the Ross Orogeny [Gunn and Warren, 1962]. On mainland Australia the Adelaide Supergroup and overlying Cambrian sequences were folded and intruded during the Delamerian Orogeny [Rutland et al., 1981]. In Tasmania the Jukesian Movement and in New Zealand the Haupiri Disturbance correspond in time, though not in magnitude, to the other orogenies.

For purposes of this discussion it is desirable to have a unifying term for this cycle of rocks and events common to the southern continents, identified by the legion of names listed above. Hurley [1974] introduced the term "Pangeaic Orogenic System" for all orogenic events occurring throughout the world in the 400 million years (650-250 Ma) prior to the breakup of Pangea. However, the megatectonics of Gondwana are distinct from those of Laurasia. The approximately 500 Ma episode marking the cessation of orogenic activity throughout much of Gondwana predated most or all of the tectonic activity of the Cordilleran, Appalachian, Caledonian, Hercynian, and Uralide systems. The term "Gondwanide" would be a logical

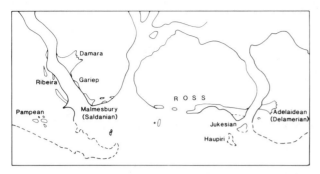

Fig. 1. Distribution of Pannotios belts discussed in this paper.

choice were it not already applied by du Toit [1937] to the Permo-Triassic folding along the Pacific margin of Gondwana. Alternatively, then, I would propose the term "Pannotios" from the Greek; pan meaning "all," and notios, "southern." The following discussion will highlight certain physical and temporal characteristics of the Pannotios belts along the Pacific margin of Gondwana. The widespread orogenesis and thermal reactivation throughout much of the remainder of South America, Africa, India, and the adjacent half of East Antarctica will not be considered.

Where appropriate, isotopic dates have been recalculated by using the decay constants of Steiger and Jager [1977].

The Pannotios Belts

Basement to the Pannotios belts. Older cratonic rocks appear to be closely associated with most of the Pannotios belts, both as bounding blocks and in some cases as inliers (Figure 2). The Ribeira belt in South America consists of several troughs bounded and separated by strips of cratonic rocks formed during the Trans-Amazonian cycle (ages: 2200-1800 Ma) [Almeida et al., 1973; Cordani et al., 1973]. The Damara province of Namibia is bounded on the north by the Congo craton and on the south by the Kalahari craton. In addition, numerous inliers of older basement are found both near the margins of the belt and throughout its central zone [Tankard et al., 1982; Martin, 1983]. Isotopic dates from these rocks fall generally into two groupings, approximately 2000 Ma and 1400-900 Ma, which are named the Eburnian and Kibaran cycles, and correspond to the Trans-Amazonian and Uruacuano cycles of South America [Porada, 1979; Almeida et al., 1973]. The Gariep belt is bounded to the east and separated from the Malmesbury to the south by the cratonic rocks of the Namaqua province (approximately 1100 Ma) whose predominant tectonism was Kibaran [Jackson, 1979].

In the Australian sector the Adelaide Geosyncline is bounded on the west by the Gawler craton with lower Proterozoic metamorphism overprinting Archean gneisses, and on the northeast by the Willyama inlier where the main metamorphism is dated at about 1700-1650 Ma [Plumb, 1979]. In addition, several inliers of lower Proterozoic basement (e.g., Haughton inlier, Mount Painter block) are found within the Adelaide Geosyncline itself [Rutland et al., 1981].

In Antarctica the East Antarctic craton crops out inboard of the Ross Orogen. Coastal exposures occur along the Adelie Coast toward Australia [James and Tingey, 1983] and in the Shackleton Range toward Africa, with a range of ages between 2700 and 1500 m.y. [Pankhurst et al., 1983]. In the central portion of the Transantarctic Mountains, crystalline rocks in the Miller and Geologists ranges crop out adjacent to folded sediments of the Ross Supergroup. The age of these rocks is equivocal. K-Ar dates cluster around 500 Ma, but range back to 1150 Ma [Adams et al., 1982]. Disturbed Rb-Sr systems suggest ages as old as 1900 Ma [Gunner and Faure, 1972].

In West Antarctica a small outcropping of crystalline rock at Haag Nunataks has concordant K-Ar dates on hornblende and biotite of about 1000 Ma [Clarkson and Brook, 1977]. Schopf [1969] suggested that the Ellsworth Mountains had shifted into their present position at some time during the Mesozoic from a position adjacent to the East Antarctic craton. Stump [1976] followed this suggestion, as did Clarkson and Brook [1977], who used the idea to explain the position of the Haag Nunataks. Paleomagnetic data of Watts and Bramall [1981] seemed to affirm the hypothesis.

However, new paleomagnetic data would suggest that the Ellsworth-Whitmore Mountains crustal block has not been rotated more than 20° and has been a part of West Antarctica since at least the Cambrian [Grunow et al., this volume]. It has always been difficult for me to rationalize the stratigraphy of the Ellsworth Mountains, conformable as it is from the Cambrian through the Permian, as having been deposited outboard of the Ross Orogen, showing no effects of the orogeny which left its mark throughout the Transantarctic Mountains. A more reasonable place for the sediments of the Ellsworth Mountains to have accumulated, I have argued [Stump, 1976, p. 121], would be in the notch between East Antarctica and Africa, adjacent to the craton, inboard of the effects of the Ross Orogeny and adjacent to rocks of the Gamtoos Valley in the eastern Cape of South Africa, the sole locality where pre-Cape (Malmesbury) rocks pass conformably up into the middle Paleozoic Cape Supergroup, unaffected by Saldanian folding [Frankel, 1936; Haughton et al., 1937].

If, however, the Ellsworth-Whitmore Mountains crustal block is essentially undisplaced, we must make models to accommodate it. The issue is very important with regard to the Ross Orogen, for to leave the Ellsworth Mountains and Haag Nunataks in West Antarctica at the time of the Ross Orogeny implies that a narrow unaffected region of continental crust existed between the oceanic crust of the Pacific and the evolving orogenic belt in the Transantarctic Mountains (Figure 3).

Further support for this suggestion exists in the Falkland Islands where gneissic basement (Cape Meredith Complex) is unconformably overlain by a sequence of folded sedimentary rocks, with affinities to the Cape and Karoo supergroups of South

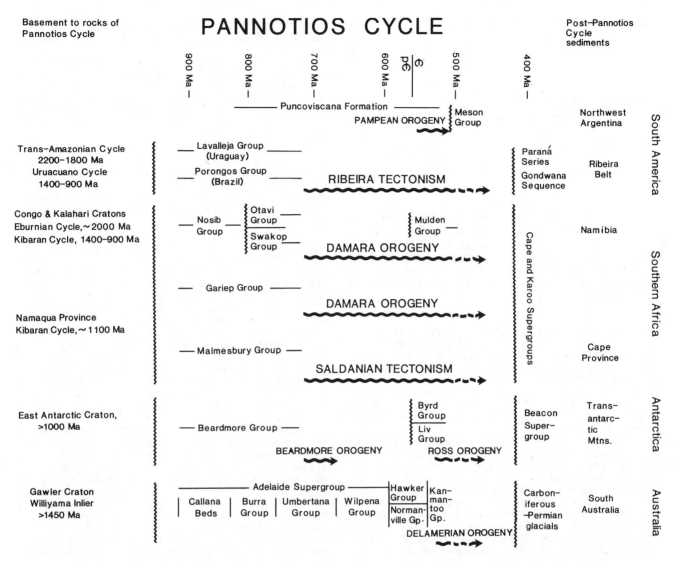

Fig. 2. Major components of the Pannotios cycle. Sawtooth lines indicate unconformities.

Africa [Baker, 1924; Greenway, 1972]. The Cape Meredith Complex apparently was not affected by Pannotios tectonism, for K-Ar dates on the complex are 977 ± 40 Ma and 953 ± 30 Ma [Rex and Tanner, 1982]. In a reconstructed Gondwana, if the Falklands remain in their present-day location within the South American plate, they must have been outboard of the Pannotios belt that connects from the Damara and Malmesbury of southern Africa through to the Ross Orogen of Antarctica (Figure 3). Unlike the Ellsworth Mountains, where deposition had begun by at least the Middle Cambrian, the sediments above the unconformity in the Falklands are fossil dated back to Lower Devonian and, by correlation with the Table Mountain Sandstone of South Africa, may be as old as Ordovician [Cocks et al., 1970].

One may speculate that this zone of rocks, unaffected by Pannotios tectonism, extended from West Antarctica to the Falkland Plateau and on through to cratonic rocks exposed in South America, passing west of the Ribeira and east of the Pampean belts. Published Precambrian dates from Patagonia are lacking, but most authors suggest a separation of the Pampean Ranges from the Ribeira and other belts of the Brazilian cycle [H. Miller, 1981; Aceñolaza and Miller, 1982; Coira et al., 1982]. Crystalline rocks of the Rio de la Plata craton are exposed in the vicinity of Buenos Aires, west and south of the Ribeira belt [Cordani et al., 1973; Dalla Salda, 1981]. Isotopic dates are approximately 2000 Ma, corresponding to the Trans-Amazonian cycle [Almeida et al., 1973].

Tasmania is an enigmatic bit of the puzzle. The lower Paleozoic sedimentation there has affinities with the marginal continental regime discussed below; however, basement to these sequences is a repeatedly deformed suite of predominantly sedi-

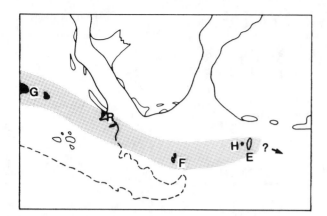

Fig. 3. Postulated region of Proterozoic cratonic rocks separating Pannotios belts in South America, Africa, and Antarctica from the Pacific. E is Ellsworth Mountains; F, Falkland Islands; G, Guapore craton; H, Haag Nunataks; and R, Rio de la Plata craton.

mentary rocks, metamorphosed no higher than greenschist facies. The first of the deformational and metamorphic events, the so-called Frenchman Orogeny, is isotopically dated at about 800 Ma [Råheim and Compston, 1977]. This was followed in the late Precambrian by the Penguin Orogeny. Granites on King Island, northwest of Tasmania, have isotopic dates ranging from 715 Ma to 750 Ma [McDougall and Leggo, 1965].

These dates are anomalous throughout the entire portion of Gondwana discussed in this paper. They are too young when compared with the basement sequences discussed in this section but are older than the apparent commencement of Pannotios tectonism as discussed below. Tasmania records an episode of orogeny unknown elsewhere on the Pacific side of Gondwana, making it difficult to place in regional models. Perhaps this crust was exotic to Gondwana in the late Precambrian during its early tectonic development.

Initiation of sedimentation. Although the beginnings of deposition of the Pannotios cycle are not well constrained in most areas, the cycle appears to have begun during a period spanning the middle to late Proterozoic. Aside from the dates on underlying basement that give a maximum age for the sedimentation, the best indications are from volcanic rocks erupted low in a section or from plutons emplaced early in the cycle.

In the Damara Supergroup the upper portion of the basal Nosib Group contains the Naauupoort Volcanics, which have been dated at 800 ± 20 Ma [Burger and Walraven, 1978]. The Gariep Group is assumed to be closely associated in time with a line of granitic and syenitic stocks (Richtersveld and Older Bremen igneous complexes) intruding the adjacent Namaqua basement, although no direct relationship is known. These intrusions have yielded a U-Pb zircon date of 920 ± 10 Ma, and a Rb-Sr whole-rock isochron of 911 ± 39 Ma [Allsopp et al., 1979]. A minimum age of sedimentation is set by a granitic stock that intruded the Stinkfontein Formation (basal Gariep Group) and that has been U-Pb dated at 780 ± 10 Ma [Allsopp et al., 1979].

Initiation of sedimentation in the Ribeira belt of South America is probably similar to that of the Damara [Porada, 1979; Bernasconi, 1983], but geochronology is lacking. The older rocks of the Pampean cycle are thought to be late Precambrian, but without constraint on their upper bound [Aceñolaza and Miller, 1982].

The only indication in Antarctica of the time of onset of sedimentation is in the Pensacola Mountains where felsic lavas are interbedded with the Patuxent Formation (basal Ross Supergroup) [Schmidt et al., 1965]. These have yielded a Rb-Sr whole-rock isochron of 1184 ± 74 Ma [Eastin, 1970].

In South Australia the age of the Wooltana Volcanics in the lower part of the Adelaide Supergroup is estimated to be about 800 Ma, although volcanics associated with Adelaidean on the Stuart Shelf have been dated as old as 1100 Ma and possibly 1300 Ma [Thomson, 1966, 1980; Rutland et al., 1981]. These later dates appear to contradict data from the Haughton inlier, an exposure of basement within the Adelaide belt that has been Rb-Sr isochron dated at 849 ± 31 Ma, the age being interpreted as the time of amphibolite metamorphism of the basement [Compston et al., 1966].

Pre-orogenic sedimentation. Once sedimentation had commenced it persisted for several hundred million years, ceasing when overtaken by orogeny at different times along the full extent of the Pannotios belts (Figure 2). Although each region has its distinct history of sedimentation and volcanism, some generalities can be stated that carry across continents. In both Namibia and South Australia, rift models have been proposed for the early histories of the Damara and Adelaide belts, in which sedimentation began in fault controlled basins or grabens, accompanied to a limited extent by rift-related volcanism [Martin, 1983; von der Borch, 1980]. In both cases, sedimentation occurred over attenuated continental crust, presently exposed on opposite sides of the belts and as inliers within them.

In Namibia the Damara forms a tripartite belt with northern and southern coastal segments and a central intracontinental branch that strikes northeasterly and disappears under the sands of the Kalahari [Martin, 1965; Clifford, 1967; Tankhard et al., 1982]. The entire width of the intracontinental branch is preserved; the coastal segments only partially remain, with their other side exposed along the facing coast in South America [Porada, 1979].

In the intracontinental segment, the initial sedimentation was in three parallel grabens [Martin and Porada, 1977]. The rocks are the Nosib Group, an assemblage of fluvial and shallow marine deposits. Two of the grabens contain rhyolites, and the northern one also contains a suite of alkaline and peralkaline volcanics called the Naauwpoort Formation [Frets, 1969; R. M. Miller, 1974]. Following Nosib deposition, sedimentation overstepped the graben boundaries and filled the belt. In the north, the calcareous and clean

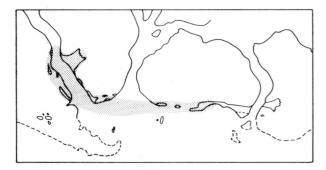

Fig. 4. Areas affected by Pannotios tectonism during the late Precambrian.

clastic shelf sediments of the Otavi Group were deposited on the gently subsiding margin of the Congo craton [Hedberg, 1979]. In the central and southern portion of the belt separated from the Otavi by a basement high, the largely flyschlike sediments of the Swakop Group (Khomas Subgroup) were being deposited in a deeply subsiding trough [Martin, 1965].

In Australia, the Adelaidean forms a sinuous belt which strikes inland from the coast. Sedimentation began when the evaporitic Callanna Beds were deposited in grabens or half grabens. The Wooltana Volcanics are associated with these initial sediments [Crawford, 1963]. These rocks are succeeded by the Burra, Umberatana, and Wilpena groups, which are largely deltaic and shallow water (probably lacustrine) deposits [Rutland et al., 1981]. At the top of the Wilpena Group, the Pound Quartzite, with its famous Ediacara fauna [Glassner and Wade, 1966; Wade, 1968], marks the first substantial marine incursion in the area [von der Borch, 1980]. Unlike the Damara, the Adelaidean did not develop a shelf and adjacent deepening flysch basin during the Precambrian. This did, however, come during the Cambrian, as discussed below.

The connection between southern Africa and Australia is the Transantarctic Mountains. Throughout the range, the late Precambrian portion of the Ross Supergroup is composed of extensive sequences of graywacke and shale, thought to have been deposited as turbidites [Grindley, 1981; Stump, 1982; Smit and Stump, 1986]. In the central Transantarctic Mountains these are named the Beardmore Group, and in the Pensacola Mountains, the Patuxent Formation.

Whether or not a rifted basin may be postulated for the onset of deposition in the Transantarctic Mountains is speculative, for the basal sequences are never seen. A spilite-keratophyre suite associated with the Patuxent Formation in the Pensacola Mountains [Schmidt et al., 1965; Eastin, 1970] may indicate magmatism in a rift setting. Also, whether the late Proterozoic basin in the Transantarctic Mountains was backed by the East Antarctic craton and faced the open Pacific, or was confined on both sides by continental crust, is uncertain. The septum of crystalline rocks postulated in the preceding section to extend from South America through the Falkland Islands to the Haag Nunataks may or may not have continued farther to the west across West Antarctica (Figure 3).

In the Pampean Ranges, the late Precambrian to early Cambrian Puncoviscana Formation is a turbidite sequence that appears to have accumulated in a passive margin setting open to the Pacific [Ježek et al., 1985; Ježek and Miller, this volume].

Late Precambrian orogenic activity. The orogenic activity that led to the terminal Pannotios event at approximately 500 Ma was timed differently throughout Gondwana. Portions of the Pannotios belts underwent tectonism during the late Precambrian, while in others tectonism did not commence until the Cambrian. The initial orogenic activity in the Damara caused ductile folding, intense metamorphism, and formation of granites in the core of the belt, the peak of which has been dated at approximately 660 Ma [Kröner et al., 1978].

In South America, orogenic activity in the Ribeira belt began in the late Precambrian, paralleling activity in the Damara belt [Porada, 1979]. Dates on syntectonic granites are as old as 670 Ma (discussed by Bernasconi [1983]). However, the Pampean Ranges of northwest Argentina appear not to have been affected by tectonism until the Paleozoic [Willner et al., this volume].

In Antarctica, in the Pensacola Mountains and central Transantarctic Mountains, deformation occurred during the late Precambrian Beardmore Orogeny [Grindley and McDougall, 1969]. This was primarily a folding event with accompanying metamorphism no higher than greenschist facies [Stump et al., 1986a]. In Australia, no late Precambrian orogeny is recorded in the Adelaidean, although a marked disconformity does occur at the Precambrian-Cambrian boundary [von der Borch, 1980].

In the McMurdo Sound area of Antarctica as well as in northern Victoria Land, the Beardmore Orogeny has not been recognized either by geological relationships or by isotopic dating. (However, see Vetter et al. [1983]). While this may simply signal the need for more work, that portion of the Transantarctic Mountains may not in fact have undergone late Precambrian orogeny, as is the situation in South Australia.

As outlined above, the initial stages of Pannotios tectonism appear to have begun in the late Precambrian in a belt which encompasses the adjacent continental margins of southern Africa and South America, as well as a major portion of the Transantarctic Mountains (Figure 4). In addition, the orogeny in the African-South American sector was much more intense, producing high-grade metamorphism and widespread plutonism, in contrast to the Antarctic sector in which deformation and low-grade metamorphism were characteristic.

When viewed in the broadest context, this tectonism appears to be related to the closure of cratonic blocks and to lessen in intensity in a direction from South America-Africa toward Antarctica, parallel to the narrowing septum of Proterozoic crystalline basement postulated above. The late Proterozoic tectonism of the Damara and Ri-

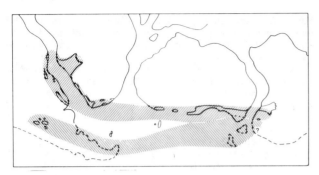

Fig. 5. Areas affected by Cambro-Ordovician Pannotios tectonism: interior belt welded to the craton and the marginal belt affected by later tectonism.

beira belts corresponds to initial activity in the remainder of the Pan-African and Brazilian belts now embedded in the continents of Africa and South America [Clifford, 1970; Cordani et al., 1973]. The initial Pannotios contractions in the late Precambrian were between cratonic blocks. In the early Paleozoic, orogenesis continued in the cratonal belts, but also involved interaction of the Gondwana margin with the oceanic crust of the Pacific.

Early to Middle Cambrian sedimentation. Following the late Precambrian tectonism, the Ribeira and Damara-Gariep belts remained positive areas above sea level. In the Damara, terrestrial molasse-type deposition of the Mulden Group occurred above an unconformity, probably during the Early Cambrian [Hedberg, 1979]. The seas, however, did return to the Transantarctic Mountains during the Early Cambrian, while marine deposition continued in the Pampean Ranges and the Adelaide belt.

In northwestern Argentina, deposition of the Puncoviscana Formation continued into the Cambrian [Aceñolaza and Miller, 1982]. In South Australia the disconformity above the late Precambrian Adelaide Supergroup is followed by Early Cambrian shelf sediments of the Normanville and Hawker groups to the west and the deep-water sediments of the Kanmantoo Group to the east [Daily and Milnes, 1973; Rutland et al., 1981].

In the central Transantarctic Mountains, the folded late Precambrian Beardmore Group is unconformably overlain by Early Cambrian limestones of the Byrd Group [Laird et al., 1971]. Outboard of this is a belt of volcanic rocks that are interbedded with limestones and clastics (Liv Group) [Stump, 1982]. The volcanic rocks are a bimodal suite, predominantly rhyolites and dacites but with subsidiary basalts [Stump, 1976]. Early Cambrian limestones are also found in the Argentina Range of Antarctica and the Weddell Sea area, but apparently volcanism and limestone deposition did not begin in the Pensacola Mountains until Middle Cambrian [Palmer and Gatehouse, 1972]. In the southern Victoria Land region of the Transantarctic Mountains most of the rocks are strongly metamorphosed and deformed. Lithologies cover the variety found in the central Transantarctic Mountains, but no fossils are known, so the age of the sedimentation is uncertain [Findlay et al., 1984]. In northern Victoria Land, the age of the Wilson Group is also uncertain. Lacking either limestones or volcanics, its wackestone lithology is similar to the late Precambrian sequences elsewhere in the Transantarctic Mountains, or perhaps the Cambrian Kanmantoo of South Australia or the Glenelg River beds of western Victoria [Stump et al., 1986b].

Von der Borch [1980], followed by Veevers [1984], has hypothesized that a rifting event occurred in the Adelaide Geosyncline around the Precambrian-Cambrian boundary, which removed an eastern portion of the belt, resulting in the slope and rise deposits of the Kanmantoo Group. The bimodal volcanic suite in the central Transantarctic Mountains may also signal an episode of rift tectonics in that region. While this postulate cannot be carried along the belt into Africa and South America, it does coincide with a proposed rifting event of global scale at this time [Bond et al., 1984].

Orogenic culmination. The culmination of the Pannotios cycle was Cambro-Ordovician; orogenesis affected all parts of the Pannotios belts and thermal reactivation occurred throughout large portions of the remaining cratonic portions of Gondwana (Figure 5). Complex deformation, high-grade metamorphism, and widespread plutonism characterize the event. In the Damara-Gariep-Malmesbury belt, tectonism that began in the late Precambrian continued its evolution. Mineral dates, Rb-Sr whole-rock isochrons, and U-Pb dates on metamorphic and plutonic rocks range from approximately 460 to 580 Ma [e.g., Cahen and Snelling, 1984]. The entire length of the Transantarctic Mountains was caught up in the Ross Orogeny, which overprinted effects of the Beardmore Orogeny. Numerous isotopic dates (K-Ar, Rb-Sr, U-Pb) range between 450 and 550 Ma [e.g., Grindley and McDougall, 1969; Faure and Jones, 1974; Faure et al., 1979; Gunner and Mattinson, 1975; Vetter et al., 1983]. The Adelaidean and Cambrian sediments of South Australia were tectonically affected for the first and only time during the Delamerian Orogeny. Granite emplacement during this event has been Rb-Sr whole-rock isochron dated between 477 and 504 Ma, and K-Ar and Rb-Sr mineral dates extend to about 450 Ma [White et al., 1967; Dasch et al., 1971; Milnes et al., 1977].

Pannotios tectonism consolidated these belts and affixed them in a broad swath to the supercontinent. Its signature is a marked unconformity unaffected by later orogenesis that spans southern Africa and adjacent South America, the Transantarctic Mountains, and South Australia. Sequences along the Pacific edge of Gondwana also were affected by the terminal Pannotios event, but in general not to the same extent as the more interior belt. In the more cratonal belt, sedimentation ceased, whereas in the more marginal belt, sedimentation and volcanism carried through the Middle and Late Cambrian and into the Ordovician, with limited interruptions. Until breakup, deposition and tectonism resulting from interaction with the Pacific plate continued along the

Gondwana margin, and one segment of the more cratonal belt (Cape Fold Belt, Pensacola Mountains, and Ellsworth Mountains) underwent further deformation around the Permo-Triassic boundary.

In northwest Argentina, the Pampean Orogeny caused multiple episodes of deformation and metamorphism during the Cambrian [Knüver and Miller, 1981; Aceñolaza and Miller, 1982; Willner et al., this volume]. Granites were generated during the Pampean Orogeny, but continued to be emplaced in the Pampean Ranges during the middle Paleozoic and perhaps as late as Permian [McBride et al., 1976; Rapela et al., 1982; Knüver and Miller, 1981]. In the exterior zone of the Pampean belt, an angular unconformity on the Puncoviscana Formation is overlain by the Upper Cambrian Meson Group, interpreted to be a molasse deposit [Aceñolaza and Miller, 1982]. This marks the onset of a new tectonic cycle in northwestern Argentina (the Famatinian) with activity from the Upper Cambrian through the Devonian [H. Miller, 1981]. The locus of deposition shifted westward from the Pampean Ranges to the Puna where a subduction-related volcanic arc is recognized (Faja Eruptiva), associated with mafic and ultramafic occurrences possibly of ophiolitic affinity [Coira et al., 1982; Allmendinger et al., 1983].

In northern Victoria Land, Antarctica, and southern Australia, rocks of the Ross and Delamerian orogens were consolidated during the terminal Pannotios event. However, outboard of these, sequences accumulated throughout the Middle and Late Cambrian, and later as well.

In northern Victoria Land, the Bowers Supergroup contains a volcanic arc sequence (Sledgers Group) overlain by shallow marine deposits (Mariner Group) that accumulated during the Middle and Late Cambrian. These rocks are unconformably overlain by clastic rocks of uncertain, but post-Late Cambrian age [Cooper et al., 1983]. Outboard of the Bowers Supergroup is the Robertson Bay Group, a turbidite sequence, at least in part, of uppermost Cambrian to lowermost Ordovician age [Burrett and Findlay, 1984; Wright et al., 1984]. Both the Bowers Supergroup and the Robertson Bay Group have K-Ar ages in the range 455 to 505 Ma, similar to Ross dates from the Wilson Group to the west; however, dates of the Bowers Supergroup also extend to 275 Ma [Adams and Kreuzer, 1984]. The contact between the Bowers Supergroup and the Wilson Group of the Ross Orogen is a complex fault. Ample models [Burmester et al., 1985; Findlay, this volume; Kleinschmidt and Tessensohn, this volume; Gibson and Wright, 1985; and Weaver et al., 1984] suggest that at least some of these deposits are subduction related, and that they accumulated in a position allochthonous to the Wilson Group in northern Victoria Land, being amalgamated into their present configuration by either subduction or strike-slip motion or some combination of the two. By current usage, the Wilson, Bowers, and Robertson Bay rocks are designated as terranes.

A similar configuration of terranes occurs in western Victoria, Australia [Stump et al., 1986b]. The westernmost Glenelg River beds were tectonized during the Delamerian Orogeny. East of this, the Stavely Group corresponds to the Sledgers Group, and the Grampians Group corresponds to the Leap Year Group in the Bowers Terrane [Harrington, 1979, 1980]. Farther east the Saint Arnaud beds [Brown et al., 1982] correspond to the Robertson Bay Group.

In Tasmania, sedimentation occurred during the Cambrian in several troughs, bounded by blocks or nuclei of deformed Precambrian rocks. The principal one of these was the Dundas Trough, in which sedimentary rocks interfinger with a suite of dominantly silicic volcanics (Mount Read volcanics) along its eastern flank [Corbett, 1981]. Ultramafic and mafic bodies also occur in the troughs. Sedimentation may have begun in the Vendian, but fossiliferous sediments are known with certainty from the Middle to Late Cambrian [Jago, 1979]. An upper Cambrian disturbance called the Jukesian Movement shed conglomerate (Owen Conglomerate) from the Precambrian nuclei over the trough sequences [Brown et al., 1982]. In some places the contact is conformable, in others unconformable.

In New Zealand, lower Paleozoic rocks are preserved in three elongate belts in northwest Nelson. The Western Sedimentary Belt contains a quartz-rich turbidite sequence (Greenland Group), fossil dated from Late Cambrian to Late Ordovician [Laird, 1972; Cooper, 1979]. The Central Sedimentary Belt contains the Haupiri Group, a dominantly volcanic sequence of Cambrian age. The basal Balloon Formation, which may be as old as Vendian, contains immature clastics and thick chert bodies [Grindley, 1980]. This is overlain by the Devil River Volcanics, primarily spilitized mafic to intermediate volcanics and volcaniclastics, in turn overlain by siltstones and sandstones of the Tasman Formation, fossil dated as late Middle Cambrian [Brown et al., 1982]. Conglomeratic rocks (Lockett Conglomerate), derived mainly from the Devil River Volcanics and Balloon Formation, mark a late Middle to early Late Cambrian uplift called the Haupiri Disturbance [Grindley, 1978]. As is the case in Tasmania, these conglomerates are conformable with underlying rocks in portions of the belt, unconformable in others.

The cycle of Pannotios tectonism began in the late Precambrian between the cratonal blocks that it would amalgamate to the Gondwana supercontinent; during the Cambrian it stepped out also to include rocks at the Pacific margin.

Post-Pannotios marginal development. Following Pannotios tectonism, the region marginal to Gondwana continued to receive sediments and volcanics through the Ordovician, Silurian, and Devonian, as recorded in deposits in southeastern Australia, Tasmania, New Zealand, and northwestern Argentina. This cycle culminated in the Devonian with deformation, metamorphism, and widespread emplacement of granitoid magmas. In Australia, this is called the Tabberabberan Orogeny; in New Zealand, the Tahuan Orogeny; and in Argentina, the Famatinian Orogeny. Devonian granites were also emplaced in the Bowers and Robertson Bay terranes of northern Victoria Land and in Marie Byrd Land, Antarctica.

This Devonian activity marked the final tectonism in southeastern Australia, Tasmania, and northern Victoria Land. In Australia, the cycle of sedimentation and tectonism again stepped out, this time to the New England Orogen during the upper Paleozoic, and continued in the Rangitata Orogen of New Zealand into the Mesozoic [Cawood, 1984]. Similarly, in northwest Argentina, following the Famatinian, the tectonic cycle stepped out to the west and south in Argentina and Chile, where it continued to develop in the upper Paleozoic and in the Mesozoic until breakup [Miller, 1981; Coira et al., 1982; Dalziel and Forsythe, 1985].

Whereas this pattern of successive cratonization, of stepping out of orogenic belts toward the Pacific margin, was characteristic of the period from the late Precambrian to the Mesozoic, one region unaffected by Devonian tectonism did undergo compression at the end of the Permian. This is the sector including the Cape Mountains of South Africa and the Pensacola and Ellsworth mountains of Antarctica. Variously called the Cape, Weddell, or Gondwanide Orogeny [Tankard et al., 1982; Ford, 1972; du Toit, 1937], this episode was primarily a folding event, probably too far removed from the active margin of the continent to have suffered extensive magmatism.

Summary

(1) The term Pannotios is proposed to encompass the tectonic cycle affecting Gondwana from the middle or late Proterozoic until its culmination approximately 500 Ma. (2) Pannotios belts are closely associated with cratonic basement, both as bounding blocks and as inliers. (3) It is postulated that a narrow septum of cratonic rocks connected from the Haag Nunataks in Antarctica through the Falkland Islands to the Rio de la Plata craton in South America, bounding the Pannotios belts that commenced orogenesis in the late Precambrian. (4) Tectonism began in the late Precambrian in a zone that is thought to have stretched from the Ribeira and Damara belts through to the central Transantarctic Mountains. (5) A rifting event possibly affected the region from Australia to the central Transantarctic Mountains in the Early Cambrian. (6) Pannotios tectonism culminated by Cambro-Ordovician time, affecting both a more cratonal belt where tectonism ceased and a more marginal belt where tectonism continued.

Acknowledgments. Research leading to this paper has been supported by NSF grants DPP7820624, DPP8109991, and DPP8216281.

References

Aceñolaza, F. G., and H. Miller, Early Paleozoic orogeny in southern South America, Precambrian Res., 17, 133-146, 1982.

Adams, C. J., and H. Kreuzer, Potassium-argon studies of slates and phyllites from the Bowers and Robertson Bay terranes, north Victoria Land, Antarctica, Geol. Jahrb., Reihe B, 60, 265-288, 1984.

Adams, C. J., J. E. Gabites, M. G. Laird, A. Wodzicki, and J. D. Bradshaw, Potassium-argon geochronology of the Precambrian-Cambrian Wilson and Robertson Bay groups and Bowers Supergroup, north Victoria Land, Antarctica, in Antarctic Geoscience, edited by C. Craddock, pp. 543-548, University of Wisconsin Press, Madison, 1982.

Allmendinger, R. W., V. A. Ramos, T. E. Jordan, M. Palma, and B. L. Isacks, Paleogeography and Andean structural geometry, northwest Argentina, Tectonics, 2, 1-16, 1983.

Allsopp, H. L., E. O. Kostlin, H. J. Welke, A. J. Burger, A. Kröner, and H. J. Blignault, Rb-Sr and U-Pb geochronology of late Precambrian-early Paleozoic igneous activity in the Richtersveld (South Africa) and southern South West Africa, Trans. Geol. Soc. S. Afr., 82, 185-204, 1979.

Almeida, F. F. M., G. Amaral, U. G. Cordani, and K. Kawashita, The Precambrian evolution of the South American cratonic margin south of the Amazon River, in The Ocean Basins and Margins, vol. 1, South Atlantic, edited by A. E. M. Nairn and F. G. Stehli, pp. 441-446, Plenum, New York, 1973.

Baker, H. A., Final Report on Geological Investigations in the Falkland Islands 1920-1922, pp. 1-38, Government Printer, Stanley, Falkland Is., 1924.

Bernasconi, A., Geological comparison of Precambrian and early Paleozoic terrains between the southern west coast of Africa and the southeast coast of South America, Precambrian Res., 23, 9-31, 1983.

Bond, G. C., P. Nickeson, and M. A. Kominz, Breakup of a supercontinent between 625 Ma and 555 Ma--New evidence and implications for continental histories, Earth Planet. Sci. Lett., 70, 325-345, 1984.

Brown, A. V., R. A. Cooper, K. D. Corbett, B. Daily, G. R. Green, G. W. Grindley, J. Jago, M. G. Laird, A. H. M. VandenBerg, G. Vidal, B. D. Webby, and H. E. Wilkinson, Late Proterozoic to Devonian sequences of southeastern Australia, Antarctica, and New Zealand and their correlation, Spec. Publ. Geol. Soc. Aust., 9, 103 pp., 1982.

Burger, A. J., and F. Walraven, Summary of age determinations carried out during the period April to March 1976, Ann. Geol. Surv. S. Afr., 11, 323-329, 1978.

Burmester, R., R. S. Babcock, and A. Wodzicki, The tectonic evolution of north Victoria Land, Antarctica (abstract), in Abstracts, Sixth Gondwana Symposium, Inst. Polar Stud. Misc. Publ., 231, p. 13, Ohio State University, Columbus, 1985.

Burrett, C. F., and R. H. Findlay, Cambrian and Ordovician conodont from the Robertson Bay Group, Antarctica, and their tectonic significance, Nature, 307, 723-726, 1984.

Cahen, L., and N. J. Snelling, The Geochronology and Evolution of Africa, 512 pp., Oxford University Press, New York, 1984.

Cawood, P. A., The development of the SW Pacific margin of Gondwana: Correlations between the Rangitata and New England orogens, Tectonics, 3, 539-553, 1984.

Clarkson, P. D., and M. Brook, Age and position of the Ellsworth Mountains crust fragment, Antarctica, Nature, 265, 615-616, 1977.

Clifford, T. N., The Damaran Episode in the upper Proterozoic-lower Palaeozoic structural history of southern Africa, Spec. Pap. Geol. Soc. Am., 92, 100 pp., 1967.

Clifford, T. N., The structural framework of Africa, in African Magmatism and Tectonics, edited by T. N. Clifford and I. G. Gass, pp. 1-26, Oliver and Boyd, Edinburgh, 1970.

Cocks, L. R. M., C. H. C. Brunton, A. J. Rowell, and I. C. Rust, The first lower Palaeozoic fauna proved from South Africa, J. Geol. Soc. London, 125, 583-601, 1970.

Coira, B., J. Davidson, C. Mpodozis, and V. Ramos, Tectonics and magmatic evolution of the Andes of northern Argentina and Chile, Earth Sci. Rev., 18, 303-332, 1982.

Compston, W., A. R. Crawford, and V. M. Bofinger, A radiometric estimate of the duration of sedimentation in the Adelaide geosyncline, South Australia, J. Geol. Soc. Aust., 13, 229-276, 1966.

Cooper, R. A., Lower Paleozoic rocks of New Zealand, J. R. Soc. N. Z., 9, 29-84, 1979.

Cooper, R. A., J. B. Jago, A. J. Rowell, and P. Braddock, Age and correlation of the Cambro-Ordovician Bowers Supergroup, northern Victoria Land, in Antarctic Earth Science, edited by R. L. Oliver, P. R. James, and J. B. Jago, pp. 128-131, Australian Academy of Science, Canberra, 1983.

Corbett, K. D., Stratigraphy and mineralization in the Mt. Read volcanics, western Tasmania, Econ. Geol., 76, 209-230, 1981.

Cordani, U. G., G. Amaral, and K. Kawashita, The Precambrian evolution of South America, Geol. Rundsch., 62, 309-317, 1973.

Crawford, A. J., The Wooltana volcanic belt, South Australia, Trans. R. Soc. South Aust., 87, 123-154, 1963.

Daily, B., and A. R. Milnes, Stratigraphy, structure and metamorphism of the Kanmantoo Group (Cambrian) in its type section east of Tungkilla Beach, South Australia, Trans. R. Soc. South Aust., 97, 213-215, 1973.

Dalla Salda, L., The Precambrian geology of El Cristo, southern Tandilia region, Argentina, Geol. Rundsch., 70, 1030-1042, 1981.

Dalziel, I. W. D., and R. D. Forsythe, Andean evolution and the terrane concept, in Tectonostratigraphic Terranes of the Circum-Pacific Regions, edited by D. G. Howell, pp. 565-581, Circum-Pacific Council for Energy and Mineral Resources, Houston, Texas, 1985.

Dasch, E. J., A. R. Milnes, and R. W. Nesbitt, Rubidium-strontium geochronology of the Encounter Bay granite and adjacent metasedimentary rocks, South Australia, J. Geol. Soc. Aust., 18, 259-266, 1971.

du Toit, A. L., Our Wandering Continents, 366 pp., Oliver and Boyd, Edinburgh, 1937.

Eastin, R., Geochronology of the basement rocks of the central Transantarctic Mountains, Antarctica, Ph.D. dissertation, 215 pp., Ohio State Univ., Columbus, 1970.

Faure, G., and L. M. Jones, Isotopic composition of strontium and geologic history of the basement rocks of Wright Valley, southern Victoria Land, Antarctica, N. Z. J. Geol. Geophys., 17, 611-627, 1974.

Faure, G., R. Eastin, P. T. Ray, D. McLelland, and C. H. Shultz, Geochronology of igneous and metamorphic rocks, central Transantarctic Mountains, in Fourth International Gondwana Symposium: Papers, edited by B. Laskar and C. S. Raja Rao, pp. 805-813, Hindustan Publishing, Delhi, 1979.

Findlay, R. H., A review of the problems important for interpretation of the Cambro-Ordovician paleogeography of northern Victoria Land (Antarctica), Tasmania, and New Zealand, this volume.

Findlay, R. H., D. N. B. Skinner, and D. Craw, Lithostratigraphy and structure of the Koettlitz Group, McMurdo Sound, Antarctica, N. Z. J. Geol. Geophys., 27, 513-536, 1984.

Ford, A. B., The Weddell orogeny-latest Permian to early Mesozoic deformation at the Weddell Sea margin of the Transantarctic Mountains, in Antarctic Geology and Geophysics, edited by R. J. Adie, pp. 419-423, Universitetsforlaget, Oslo, 1972.

Frankel, J. J., Geology of a portion of the Gamtoos valley, Trans. Geol. Soc. S. Afr., 39, 263-279, 1936.

Frets, D. C., Geology and structure of the Haub-Welwitschia area, South West Africa, Bull. Univ. of Cape Town Chamber Mines Precambrian Res. Unit, 5, 235 pp., 1969.

Gibson, G. M., and T. O. Wright, The importance of thrust faulting in the tectonic development of northern Victoria Land, Antarctica, Nature, 315, 480-483, 1985.

Glassner, M. F., and M. J. Wade, The late Precambrian fossils from Ediacara, South Australia, Paleontology, 9, 599-628, 1966.

Greenway, M. E., The geology of the Falkland Islands, Br. Antarct. Surv. Sci. Rep., 76, 42 pp., 1972.

Grindley, G. W., Tectonism of the early geosynclinal cycle: The Tahua Orogeny and the New Zealand Geanticline, in The Geology of New Zealand, vol. 1, edited by R. P. Suggate, G. R. Stevens, and M. T. Te Punge, pp. 117-135, Government Printer, Wellington, New Zealand, 1978.

Grindley, G. W., Geological Map of New Zealand, Cobb: sheet S13, scale 1:63360, Dep. of Sci. and Ind. Res., Wellington, New Zealand, 1980.

Grindley, G. W., Precambrian rocks of the Ross Sea region, J. R. Soc. N. Z., 11, 411-423, 1981.

Grindley, G. W., and I. McDougall, Age and correlation of the Nimrod Group and other Precambrian rock units in the central Transantarctic Mountains, Antarctica, N. Z. J. Geol. Geophys., 12, 391-411, 1969.

Grunow, A. M., I. W. D. Dalziel, and D. V. Kent, Ellsworth-Whitmore Mountains crustal block, western Antarctica: New paleomagnetic results and their tectonic significance, this volume.

Gunn, B. M., and G. Warren, Geology of Victoria Land between Mawson and Mulock glaciers, Antarctica, N. Z. Geol. Surv. Bull., 71, 157 pp., 1962.

Gunner, J., and G. Faure, Rubidium-strontium geochronology of the Nimrod Group, central Transantarctic Mountains, in Antarctic Geology and

Geophysics, edited by R. J. Adie, pp. 305-311, Universitetsforlaget, Oslo, 1972.

Gunner, J., and J. M. Mattinson, Rb-Sr and U-Pb isotopic ages of granites in the central Transantarctic Mountains, Geol. Mag., 112, 25-31, 1975.

Harrington, H. J., Tectonic reconstruction of Antarctica, eastern Australia, and New Zealand, J. Geol. Soc. Aust., 26, 276-277, 1979.

Harrington, H. J., Reconstruction of the Australian sector of Gondwanaland from Cambrian to early Tertiary, (abstract), Fifth Gondwana Symposium, Wellington, New Zealand, pp. 71-72, R. Soc. N. Z. and IUGS, 1980.

Hartnady, C. J., A. R. Newton, and J. N. Theron, The stratigraphy and structure of the Malmesbury Group in the southwestern Cape, Bull. Univ. Cape Town Chamber Mines Precambrian Res. Unit, 15, 193-213, 1974.

Haughton, S. H., H. F. Frommurze, and D. J. L. Visser, The geology of portions of the coastal belt near the Gamtoos valley, Cape Province: Expl. sheets 151 North & South (Gamtoos River), report, 55 pp., Geol. Surv. S. Afr., Gov. Printer, Pretoria, 1937.

Hedberg, R. M., Stratigraphy of the Ovamboland Basin, South West Africa, Bull. Univ. Cape Town Chamber Mines Precambrian Res. Unit, 24, 325 pp., 1979.

Hurley, P. M., Pangeaic Orogenic system, Geology, 2, 373-376, 1974.

Jackson, M. P. A., A major charnockite-granulite, province in southwestern Africa, Geology, 7, 22-26, 1979.

Jago, J. B., Tasmanian Cambrian biostratigraphy-- A preliminary report, J. Geol. Soc. Aust., 26, 223-230, 1979.

James, P. R., and R. J. Tingey, The Precambrian evolution of the East Antarctica metamorphic shield--A review, in Antarctic Earth Sciences, edited by R. L. Oliver, P. J. James, and J. B. Jago, pp. 5-10, Australian Academy of Science, Canberra, 1983.

Ježek, P., A. P. Willner, F. G. Aceñolaza, and H. Miller, The Puncoviscana trough--A large basin of late Precambrian to Early Cambrian age on the Pacific edge of the Brazilian shield, Geol. Rundsch., 74, 573-584, 1985.

Ježek, P., and H. Miller, Petrology and facies analysis of turbiditic sedimentary rocks of the Puncoviscana trough (upper Precambrian-Lower Cambrian) in the basement of the NW Argentine Andes, this volume.

Kennedy, W. Q., The structural differentiation of Africa in the Pan African (±500 m.y.) tectonic episode, Annu. Rep. Univ. Leeds Res. Inst. Afr. Geol. Dep. Earth Sci., 8, 48-49, 1964.

Kleinschmidt, G., and F. Tessensohn, Early Paleozoic westward directed subduction at the Pacific margin of Antarctica, this volume.

Knüver, M., and H. Miller, Ages of metamorphic and deformational events in the Sierra de Ancasti (Pampean Ranges, Argentina), Geol. Rundsch., 70, 1020-1029, 1981.

Kröner, A., and H. J. Blignault, Toward a definition of some tectonic and igneous provinces in western South Africa and southern South West Africa, Trans. Geol. Soc. S. Afr., 79, 232-238, 1976.

Kröner, A., M. Halpern, and R. B. Jacob, Rb-Sr geochronology in favor of polymetamorphism in Pan-African Damara belt of Namibia (South West Africa), Geol. Rundsch., 67, 688-705, 1978.

Laird, M. G., Sedimentology of the Greenland Group in the Paparoa Range, West Coast, South Island, N. Z. J. Geol. Geophys., 15, 372-393, 1972.

Laird, M. G., G. D. Mansergh, and J. M. A. Chappell, Geology of the central Nimrod Glacier area, Antarctica, N. Z. J. Geol. Geophys., 14, 427-468, 1971.

Martin, H., The Precambrian geology of South West Africa and Namaqualand, Bull. Univ. Cape Town Chamber Mines Precambrian Res. Unit, 1, 159 pp., 1965.

Martin, H., Overview of the geosynclinal, structural, and metamorphic development of the intracratonal branch of the Damara Orogen, in Intracontinental Fold Belts, edited by H. Martin and F. W. Eder, pp. 473-502, Springer-Verlag, New York, 1983.

Martin, H., and H. Porada, The intercratonic branch of the Damara orogen in South West Africa, 1, Discussion of geodynamic models, Precambrian Res., 5, 311-338, 1977.

McBride, S. L., J. C. Caelles, A. H. Clark, and E. Farrar, Paleozoic age provinces in the Andean basement, latitudes 25°-30°, Earth Planet. Sci. Lett., 29, 373-383, 1976.

McDougall, I., and P. J. Leggo, Isotopic age determinations on granitic rocks from Tasmania, J. Geol. Soc. Aust., 12, 295-332, 1965.

Miller, H., Pre-Andean orogenies of southern South America in the context of Gondwana, in Gondwana Five, edited by M. M. Cresswell and P. Vella, pp. 237-242, A. A. Balkema, Rotterdam, 1981.

Miller, R. M., The stratigraphic significance of the Naaupoort Formation of east central Damaraland, South West Africa, Trans. Geol. Soc. S. Afr., 77, 363-367, 1974.

Milnes, A. R., W. Compston, and B. Daly, Pre- to syntectonic emplacement of early Paleozoic granites in South Australia, J. Geol. Soc. Aust., 24, 87-106, 1977.

Palmer, A. R., and C. G. Gatehouse, Early and Middle Cambrian trilobites from Antarctica, U.S. Geol. Surv. Prof. Pap., 456-D, 37 pp., 1972.

Pankhurst, J. J., P. D. Marsh, and P. D. Clarkson, A geochronological investigation of the Shackleton Range, in Antarctic Earth Sciences, edited by R. L. Oliver, P. R. James, and J. B. Jago, pp. 176-182, Australian Academy of Science, Canberra, 1983.

Plumb, K. A., The tectonic evolution of Australia, Earth Sci. Rev., 14, 205-249, 1979.

Porada, H., The Damara-Ribeira Orogen of the Pan African-Brasiliano cycle in Namibia (Southwest Africa) and Brazil as interpreted in terms of continental collision, Tectonophysics, 57, 237-265, 1979.

Råheim, A., and W. Compston, Correlations between metamorphic events and Rb-Sr ages in metasediments from western Tasmania, Lithos, 10, 271-289, 1977.

Rapela, C. W., L. M. Heaman, and R. H. McNutt, Rb/Sr geochronology of granitoid rocks from the Pampean Ranges, Argentina, J. Geol., 90, 574-582, 1982.

Rex, D. C., and Tanner, P. W. G., Precambrian age of gneisses at Cape Meredith in the Falkland Islands, in Antarctic Geoscience, edited by C. Craddock, pp. 107-108, University of Wisconsin Press, Madison, 1982.

Rutland, R. W. R., A. J. Parker, G. M. Pitt, W. V. Preiss, and B. Murrell, The Precambrian of South Australia, in Precambrian of the Southern Hemisphere, edited by D. R. Hunter, pp. 309-360, Elsevier, New York, 1981.

Schmidt, D. L., P. L. Williams, W. H. Nelson, J. R. Ege, Upper Precambrian and Paleozoic stratigraphy and structure of the Neptune Range, Antarctica, U.S. Geol. Surv. Prof. Pap., 525-D, D112-D119, 1965.

Schopf, J. M., Ellsworth Mountains: Position in West Antarctica due to sea-floor spreading, Science, 164, 62-66, 1969.

Smit, J. H., and E. Stump, Sedimentology of the La Gorce Formation, La Gorce Mountains, Antarctica, J. Sediment. Petrol., in press, 1986.

Steiger, R. H., and E. Jager, Subcommission on geochronology: Convention on use of decay constants in geo- and cosmochronology, Earth Planet. Sci. Lett., 36, 359-362, 1977.

Stump, E., On the late Precambrian-early Paleozoic metavolcanic and metasedimentary rocks of the Queen Maud Mountains, Antarctica, and a comparison with rocks of similar age from southern Africa, Inst. of Polar Stud. Rep. 62, 212 pp., Ohio State Univ., Columbus, 1976.

Stump, E., The Ross Supergroup in the Queen Maud Mountains, in Antarctic Geoscience, edited by C. Craddock, pp. 565-569, University of Wisconsin Press, Madison, 1982.

Stump, E., J. H. Smit, and S. Self, Timing of events during the late Proterozoic Beardmore Orogeny, Antarctica: Geological evidence from the La Gorce Mountains, Geol. Soc. Am. Bull., 97, 953-965, 1986a.

Stump, E., A. J. R. White, and S. G. Borg, Reconstruction of Australia and Antarctica: Evidence from granites and recent mapping, Earth Planet. Sci. Lett., 79, 348-360, 1986b.

Tankard, A. J., M. P. Jackson, K. A. Erikson, D. K. Holiday, D. R. Hunter, and W. E. L. Minter, Crustal Evolution of Southern Africa, 523 pp., Springer-Verlag, New York, 1982.

Thomson, B. P., The lower boundary of the Adelaide System and older basement relationships in South Australia, J. Geol. Soc. Aust., 13, 203-228, 1966.

Thomson, B. P., Geological map of South Australia, scale 1:1,000,000, Geol. Surv. South Aust., Gov. Printer, Adelaide, 1980.

Veevers, J. J. (Ed.), Phanerozoic Earth History of Australia, 418 pp., Clarendon, Oxford, 1984.

Vetter, U., N. W. Roland, H. Dreuzer, A. Höhndorf, and C. Besand, Geochemistry, petrography, and geochronology of the Cambro-Ordovician and Devonian-Carboniferous granitoids of northern Victoria Land, Antarctica, in Antarctic Earth Sciences, edited by R. L. Oliver, P. R. James, and J. B. Jago, pp. 140-143, Australian Academy of Science, Canberra, 1983.

von der Borch, C. C., Evolution of late Proterozoic to early Paleozoic Adelaide foldbelt, Australia, comparisons with post-Permian rifts and passive margins, Tectonophysics, 70, 115-134, 1980.

Wade, M. J., Preservation of soft-bodied animals in Precambrian sandstones at Ediacara, South Australia, Lithaia, 1, 238-267, 1968.

Watts, D. R., and A. N. Bramall, Paleomagnetic evidence for a displaced terrain in western Antarctica, Nature, 293, 638-641, 1981.

Weaver, S. D., J. D. Bradshaw, and M. G. Laird, Geochemistry of Cambrian volcanics of the Bowers Supergroup and implications for the early Paleozoic tectonic evolution of northern Victoria Land, Antarctica, Earth Planet. Sci. Lett., 68, 128-140, 1984.

White, A. J. R., W. Compston, and A. W. Kleeman, The Palmer Granite--A study of a granite within a regional metamorphic environment, J. Petrol., 8, 29-50, 1967.

Willner, A. P., U. S. Lottner, and H. Miller, Early Paleozoic structural development of the NW Argentine basement of the Andes and its implications for geodynamic reconstructions, this volume.

Wright, T. O., R. J. Ross, Jr., and J. E. Repetski, Newly discovered youngest Cambrian or oldest Ordovician fossils from the Robertson Bay terrane (formerly Precambrian), northern Victoria Land, Antarctica, Geology, 12, 301-305, 1984.

EARLY PALEOZOIC WESTWARD DIRECTED SUBDUCTION AT THE PACIFIC MARGIN OF ANTARCTICA

G. Kleinschmidt

Geological Institute, University of Frankfurt, Frankfurt, Federal Republic of Germany

F. Tessensohn

Federal Institute for Geosciences and Natural Resources, Hannover, Federal Republic of Germany

Abstract. The tectonic history of the Antarctic continent during the Phanerozoic is best considered as part of the development of the circum-Pacific mobile belts. In early Paleozoic time, the continental margin at the Pacific side of Antarctica formed a 3000-km-long segment of the active Gondwana margin, which also includes Australia and parts of South Africa and South America. In the Antarctic segment, north Victoria Land forms one of the best exposed and investigated areas. This paper develops a plate tectonic reconstruction for this area based on several different lines of evidence; plate tectonic implications of paired metamorphic belts, granite types, submarine volcanism, a subduction complex, sedimentary basins, and thrust belts. It is concluded that a continuous westward directed subduction of oceanic crust under the East Antarctic craton has proceeded from the Early Cambrian to the Late Devonian. Three main episodes are documented by different criteria. Stage I is the early pre-Ross subduction phase; paired metamorphic belts formed the stage in the west and a volcanic island arc in the east; terminates with the accretion of the volcanic arc. Stage II occurs after the outward displacement of the subduction zone, when renewed subduction caused the magmatotectonic culmination of the Ross Orogeny in two steps: (1) Granite Harbour plutonism with S types in the west and I types in the east, associated low-pressure metamorphism and migmatitization being documented in the inner terranes, and (2) maximum of crustal shortening, folding, and thrusting in the outer sedimentary terranes. Stage III occurs when after a possible lull there is again subduction with development of andesitic volcanism and I-type plutonism of the Admiralty Intrusives.

Introduction

In the framework of plate tectonics, two types of mechanisms, collision and subduction, are widely recognized today as causes for orogenesis. Collision is typified in the Himalayan and Alpine mountain chains; subduction by many of the circum-Pacific mountain chains, especially the cordilleran-type belts of North and South America formed at the border between an older continental core and the vast oceanic realm of the Pacific [Fox, 1983].

Antarctica has experienced several orogenies, typified by the succession of mobile belts found between the craton and the Pacific Ocean. The West Antarctic Andean Orogen forms a section of the present circum-Pacific mobile belt. Several older fold belts, for example the Mesozoic Gondwanide Orogen and the Paleozoic Ross Orogen, are located between this youngest orogen and the East Antarctic craton. This paper concerns the Paleozoic Ross Orogen and addresses the question, Was the active Gondwana margin originally formed by accretion of an oceanic realm as postulated from other areas [Burchfiel, 1980; Cawood, 1982], or by collision with another continent after the complete closure of such an ocean?

Although much evidence has been lost by subsequent tectonics, erosion, and lack of present exposure, the knowledge of the Antarctic Ross Orogen has grown considerably in the last two decades, not only in the field of stratigraphy but also in sedimentology, tectonics, geochemistry, and petrology. Thus a plate tectonic analysis seems feasible as has been done in other orogens of a similar age [Dewey, 1969; Bird and Dewey, 1970; Scheibner, 1973; Leitch, 1975; Crook, 1980].

Geological and Tectonic Setting

North Victoria Land (NVL), situated at the Pacific end of the approximately 3000-km-long Ross Mobile Belt, forms an area particularly suitable for a plate tectonic analysis as it offers a good cross section through the mountain belt, contains a fossiliferous volcanosedimentary record, a large turbidite realm, an interesting suite of metamorphic rocks, and two well-dated generations of granites. In addition, there is enough exposure to study the main tectonic structures.

Early mapping efforts in the 1960s [Gair et al., 1969] have already shown that NVL can be divided into three different, NW-SE trending tectonic zones (Figure 1). During the last two

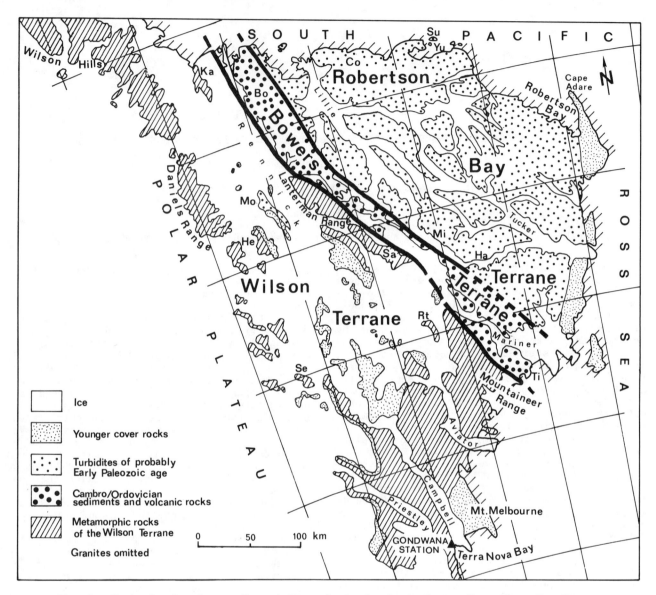

Fig. 1. Tectonic sketch map of north Victoria Land. Bo is Bowers Mountains; Co, Cooper Bluff; Ha, Handler Ridge; He, Helliwell Hills; Ka, Kavraiskyi Hills; Mi, Millen Range; Mo, Morozumi Range; Rt, Retreat Hills; Sa, Salamander Range; Se, Sequence Hills; Su, Surgeon Island; Ti, Tiger Gabbro; and Yu, Yule Bay.

decades the interpretation of the nature of these three zones has changed considerably, from horsts and grabens to exotic terranes.

The approximately 200-km-wide Robertson Bay Terrane (RBT) contains a very thick series of regularly folded turbiditic graywackes and mudstones of a distal character, the Robertson Bay Group [Harrington et al., 1964; Wright, 1981; Field and Findlay, 1983]. The lack of conglomeratic intercalation is a distinctive feature. The group was first considered to be late Precambrian in age, but now a Cambrian age is more probable because the recent discovery of fossiliferous exotic limestone blocks indicates a Tremadocian minimal age for the topmost beds [Burrett and Findlay, 1984; Wright et al., 1984].

The fault-bounded Bowers Terrane (BT) is a key area for the structural history of the region [Bradshaw and Laird, 1983; Tessensohn, 1984]. The base of the exposed succession is formed by the submarine Glasgow Volcanics and by the associated sandstones and conglomerates of the Middle Cambrian Molar Formation [Cooper et al., 1983]. The overlying Mariner Group is characterized by marine fossiliferous limestones and mudstones interbedded with coarse conglomerates and sandstones. Both sedimentary series show evidence for a slope environment. The uppermost member of the sequence, the Leap Year Group, is a thick series of molasse-like shallow marine to fluviatile quartzitic sandstones and quartz conglomerates. The rocks in the BT are folded and deformed like the rocks the Robertson Bay Group. A zone of stronger deformation parallels the fault on both sides of the BT.

The Wilson Terrane (WT), exposed in a 200-km-wide area in NVL, is named after the Wilson Hills where gneisses and migmatites were first described [Ravich et al., 1965; Sturm and Carryer, 1970]. The terrane is mainly composed of metamorphic rocks characterized by polyphase deformation and a wide variation in metamorphic grade. They are described in more detail in the section on the metamorphic belts.

Subduction-Related Geologic Features

There are several lines of evidence for subduction, including the presence of ophiolites, paired metamorphic belts, old volcanic arcs, thrust belts, and mélanges [Fox, 1983]. Additionally, the pattern of sedimentary basins [Karig, 1971; Dott and Shaver, 1974; Dickinson, 1974; Dickinson and Seely, 1979], the composition and provenance of sandstone suites [Crook, 1974; Schwab, 1975; Dickinson and Suczek, 1979], the geochemistry of volcanic rocks [Gilluly, 1971; Engel et al., 1974; Wood et al., 1979], and the petrological characteristics of granitic intrusions [Dickinson, 1970; Pitcher, 1984; Brown et al., 1984] provide useful information in plate tectonic reconstructions.

We will try to extend earlier analyses of NVL using the metamorphic pattern, the newly discovered thrust belts, and the character of the intrusive rocks as additional lines of evidence.

Paired Metamorphic Belts

Composition. While the rocks of the RBT and BT are generally of very low metamorphic grade and pass into low grade only at the western terrane margins [Wodzicki et al., 1982; Kleinschmidt, 1983; Gibson et al., 1984; Jordan et al., 1984], the WT shows a much more complex metamorphic pattern. Although it is composed mainly of medium- to high-grade metamorphic rocks, e.g., in the Daniels Range [Kleinschmidt, 1981], Lanterman Range [Roland et al., 1984], and Terra Nova Bay [Skinner, 1983], greenschist facies rocks are encountered as well, e.g. in the Morozumi Range [Engel, 1984], Retreat Hills [Riddols and Hancox, 1968], and Daniels Range [Plummer et al., 1983], and as overprint in the Lanterman Range [Roland et al., 1984].

These studies show that in rocks of suitable composition, no glaucophane has been found in very low and low-grade metamorphics, no omphacite, i.e., ecologite, has been encountered in medium-grade rocks, and sillimanite is present in all medium- to high-grade areas.

After all grades had been initially included in the low-pressure type of metamorphism [Kleinschmidt, 1981], pressure-dependent mineral assemblages were later recognized in the WT [Grew et al., 1984]. These show a distinct linear arrangement in the form of two parallel belts (Figure 2.)

The western belt extends principally along the edge of the Polar Plateau from the Pacific coast in the north to the coast of the Ross Sea in the south. This belt is characterized by the critical minerals and mineral assemblages andalusite, cordierite or cordierite plus muscovite, and the lack of kyanite and staurolite. This characteristic assemblage has been found in the Aviation Islands [McLeod and Gregory, 1967], Kavrayskiy Hills [Schubert et al., 1984], Daniels Range [Kleinschmidt, 1981; Plummer et al., 1983], Helliwell Hills [Grew and Sandiford, 1982], Mount Weihaupt [Grew et al., 1984], Sequence Hills [Gair 1967; Grew and Sandiford, 1982], and Terra Nova Bay with Cape Sastrugi, Snowy Point, and Gondwana Station [Skinner, 1983, Grew et al., 1984].

The P conditions were estimated in the range of 4.5-5 kbar [Schubert et al., 1984]. The same low-pressure character is indicated by the dramatic rise of metamorphic grade over distances of less than 10 km, as indicated by a rock sequence from phyllites to migmatites in the Daniels Range [Kleinschmidt, 1981] and Helliwell Hills [Grew and Sandiford, 1982].

The eastern belt, which includes the Lanterman Range and the Mountaineer Range, is composed of two separate areas at the eastern margin of the WT. Their pressure type of metamorphism is characterized by the general lack of andalusite and cordierite, and the occurrence of relics of kyanite and staurolite or staurolite plus talc [Grew and Sandiford, 1982, 1984; Grew et al., 1984; Roland et al., 1984; Kleinschmidt et al., 1984]. The relict paragenesis staurolite plus talc plus corundum plus chlorite in the Lanterman Range is referred to as a first stage of metamorphism with $T = 650°-700°C$ and $P = 7.5-9.5$ kbar [Grew and Sandiford, 1984]. The occurrence of (deformed) kyanite and staurolite at Dessent Ridge indicates also a medium-pressure on the order of 6-7 kbar, [Grew et al., 1984].

Along the eastern boundary of the Lanterman and Salamander ranges a train of mafic to mostly ultramafic lenses occurs at a length of more than 60 km. They consist mainly of serpentinites and actinolite-bearing rocks, and sometimes contain talc, chlorite, and magnesite. Where the relation with the country rocks is exposed, the ultramafic pods appear to be aligned in the direction of the regional schistosity. Talc and talc-actinolite-bearing rocks are especially associated with cataclasis or mylonitization in the ultrabasites and the adjacent country rocks.

The most peculiar variety occurs east-southeast of Mount Moody near the base of Index Spur. The main constituents are 20-30% (vol) garnet, $(Pyr_{53}Alm_{32}Gross_{14}Spess_1)$, 10-15% (vol) olivine $(Fa_{20}Fo_{80})$, 30-60% (vol) hornblende plus clinopyroxene plus orthopyroxene in varying proportions (hbl > cpx > opx), and about 5% (vol) diablastesis around garnets. The orthopyroxene is near the enstatite-bronzite boundary $(En_{88}Fs_{12})$. Parallelism of the elongate minerals gives the rock a distinct metamorphic fabric. Grew et al. [1984] called it a "garnet-olvine pyroxenite," but garnet peridotite may be correct as well.

Garnet peridotites are interpreted as products

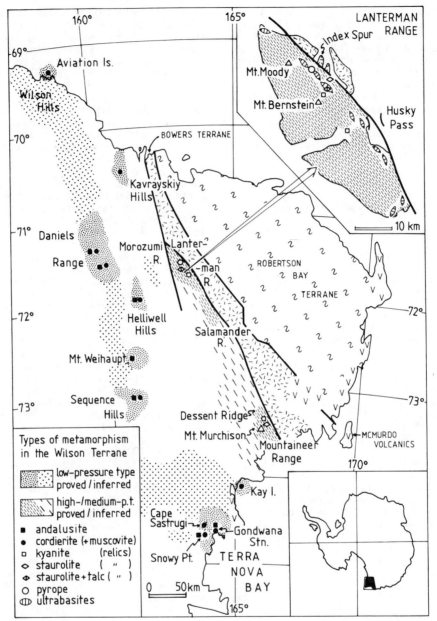

Fig. 2. Paired metamorphic belts in the basement rocks of the Wilson Terrane (WT), north Victoria Land (NVL).

of high-pressure conditions of 10 kbar or more according to O'Hara and Mercy [1963]. The composition of the rock is also close to "pyrolite" the formation of which requires high pressures within all temperature ranges [Schreyer, 1966]. For the temperature range derived from the surrounding Lanterman Metamorphics this would mean approximately 5-10 kbar. According to Boyd and England [1959] "the presence of pyrope-rich garnets appears to reveal the influence of high-lithostatic pressure during their formation." For temperatures between 300° and 700°C, the stability of pure pyrope would start at about 10-15 kbar; substitution by almandine would somewhat lower the pressure [Boyd and England, 1959].

Grew et al. [1984] have derived P conditions of 9-12 kbar or more for this rock type. It is possible that this type of high-pressure metamorphism was in fact the primary metamorphism for the entire Lanterman Range. This would fit the pressure calculation derived from the staurolite-talc assemblage, and would not contradict the occurrence of the relict kyanite in a somewhat higher-tectonic level. If, however, only medium-pressure

Fig. 3. Multiple deformation in gneiss of the Terra Nova Bay region, Gondwana Station, Gerlache Inlet. Pencil for scale.

type is derived for the metamorphics of the Lanterman Range [Grew et al., 1984], the ultrabasites have to be considered as bodies not formed in place.

After the work of Grew et al. [1984], it thus appears (Figure 2) that within the WT a western belt of low-pressure metamorphism extends from the Aviation Islands in the north to the Terra Nova Bay in the south, whereas an eastern belt with minerals and rock types indicative of medium-pressure, in part, even high-pressure, metamorphism is documented in the Lanterman, Salamander, and Mountaineer ranges.

The eastern of these nearly north-south trending parallel belts is 50 km wide and 300 km long. The western belt has a width of more than 100 km and a proven length of 600 km. The belts resemble the "paired metamorphic belts" of Miyashiro [1973] and would indicate formation above a westward directed subduction zone; the western belt as roots of a volcanic front formed on continental crust, the eastern one more directly related to the trench and the subduction zone. This model is, of course, based on the assumption that it is justified to replace the high-pressure type metamorphism of Miyashiro [1973] by our medium-pressure type. The scarcity of high-pressure belts however, seems to be a general problem for mobile belts of pre-Mesozoic age. Thickening of the earth's crust or a major change in subduction mechanism have been put forward as possible explanations for this phenomenon [Miyashiro et al., 1982; Ernst, 1972].

Age relations. At least three intense phases of deformation (Figure 3) have been reported from both belts [Kleinschmidt and Skinner, 1981; Plummer et al., 1983; Roland et al., 1984; Kleinschmidt et al., 1984]. They are outlasted by amphibolite facies metamorphism in these areas. One additional deformation related to thrusting postdates this metamorphism in the Lanterman and Mountaineer ranges.

The overall pattern of radiometric ages in NVL is discussed by Kreuzer et al. [this volume]. For our considerations the ages in the WT are of particular importance.

"Rb-Sr whole-rock and U-Pb monazite ages from the Wilson Gneisses in the northern Daniels Range [Adams, 1986] indicate that widespread high-grade regional metamorphism and associated synmetamorphic plutonism occurred about 490 ± 33 Ma [Kreuzer et al. 1981; Vetter et al., 1983; Kreuzer et al., this volume]. Mineral (and total rock) K-Ar age data from the southern Daniels Range reflect both these events, and the ages 460-480 Ma date various stages in the cooling of the complex" [Adams and Kreuzer, 1984]. These ages document the main event of the Ross Orogeny.

Some older dates, however, have been recorded as well: metasediments, Lanterman Range, Rb/Sr isochron, 610 ± 10, (H. Lenz, personal communication, 1985); metasediments, Salamander Range, Rb/Sr isochron, 550 ± 26, (H. Lenz, personal communication, 1985); orthogneiss, Wilson Hills, K/Ar whole-rock, 550 recalculated, (Ravich and Krylov [1965], in the work of Gair et al., [1969]); Rennick Schist, Daniels Range, Rb/Sr isochron, 532 ± 14, [Adams, 1986]; granites, Terra Nova Bay, Rb/Sr isochron, 535 ± 26, [Vetter et al., 1984]. The list shows clearly a concentration of dates from 550 to 530 Ma. More dates within this range have recently been obtained by

E. Stump (personal communication, 1985). It thus appears that parts of the WT show radiometric indications of a distinct pre-Ross event. The observed polarity of pressure in the two belts is most likely connected with this event.

During the subsequent Ross event, which was the main and last deformational episode in the whole WT, the metamorphic character changed in the former eastern belt as documented by the decomposition of the staurolite-talc paragenesis and the transition kyanite to sillimanite. This was due to either a decrease of pressure [Grew and Sandiford, 1984; Grew et al., 1984] or an increase in temperature [Kleinschmidt et al., 1984].

In the western belt no change in the pressure character of metamorphism can be observed. Thus it appears that during the main Ross event the whole area was affected by the low-pressure/high-temperature type of metamorphism.

The Volcanic Record

The Glasgow Volcanics of the BT were geochemically interpreted and used as a tool for a first plate tectonic reconstruction by Weaver et al. [1984]. Their results are summarized as follows. The Glasgow Volcanics [Laird et al., 1982], a thick, mainly submarine suite of pillow lavas, lava flows, and volcanic breccias, occur over the whole length of the BT. The volcanic rocks, tholeiites, and andesites with subordinate dacites and rhyolites interdigitate with the clastic sediments of the Molar Formation. By using the relative abundance of immobile elements (Ti, Zr, Y, Nb, and Cr), the Glasgow Volcanics are geochemically classified as primitive island arc tholeiites originating most probably from an intraoceanic setting (or possibly a back arc environment). The suite shows marked similarities to the Oman ophiolite. Because of the narrow width of the BT, the Glasgow Volcanics and the accompanying sediments are taken to represent only the submerged margin of an island arc which has been formed in connection with a westward dipping subduction zone. The later welding of this arc to the continental crust of the WT is explained by strike-slip movements.

Other investigators (P. Müller, personal communication, 1985) [Wodzicki and Robert, 1986], find geochemical evidence for a mid-oceanic ridge basalt origin of some of the volcanics.

Subduction Complex

Weaver et al. [1984] conclude that there are strong indications for loss of crust along the faulted contact of WT and BT, but that there are no fragments of a subduction complex along this contact. However, we feel that a subduction complex is present.

Highly deformed "conglomerates" occur exclusively along the boundary of the WT and the BT. Two elongated outcrop zones are known: the classic exposure along the northeastern border of the Lanterman Range [Crowder, 1968; Laird et al., 1982; Wodzicki et al., 1982; Kleinschmidt and Skinner, 1981; Grew and Sandiford, 1982; Laird and Bradshaw, 1983; Gibson, 1984; Wodzicki and Robert, 1986] and a second zone in the Mountaineer Range [Stump et al., 1983; Gibson et al., 1984; Kleinschmidt et al., 1984; Tessensohn, 1984]. In both areas the rocks occur on either side of the terrane boundary: in the metamorphics of the WT (Lanterman Conglomerate and Dessent Conglomerate) and in the greenschist facies rocks of the BT (Husky Conglomerate and Black Spider Conglomerate).

We are somewhat reluctant to include the psammitic Lanterman Conglomerate in a possible subduction complex, but think that the other three complexes conform with the definition of a tectonic mélange as given by Hsü [1968]. In particular the Husky Conglomerate, consisting of a bulk of amphibolitic and metavolcanic clasts mixed with other components in a fine-grained and schistose mafic matrix [Gibson, 1984, figures 3, 7 and 10], is similar to other examples of mélanges described in the literature (Hsü, 1974; Karig and Moore, 1975; Kay, 1976; Moore and Wheeler, 1978]. The "conglomerate" is a breccia in parts. Larger angular blocks and tectonic lenses of the adjacent (tectonically overlying) Lanterman Conglomerate are incorporated [Gibson, 1984]. Geochemically the amphibolitic clasts show affinities to boninites of the Mariana trench [Weaver et al., 1984].

The Black Spider Conglomerate contains less numerous volcanogenic components. But like the Husky Conglomerate, it is made up of completely deformed parts alternating with parts still in the original sedimentary context.

The metaconglomerate in the Dessent Ridge again is fairly basic in character. It contains mainly calc-silicate components in a para-amphibolitic matrix. Whereas the formation of the mélange bodies is related to subduction, it is supposed that the main deformation of the clasts is the result of later tectonism (see below).

Some basic volcanic rocks of oceanic lithosphere character are incorporated in the described mélange bodies, especially in the Husky Conglomerate. Others occur as small mafic and gabbroic bodies and ultramafic cumulates associated with lavas and pillow lavas in the greenschist belt as described by Gibson et al. [1984, p. 305].

More significant, however, are two bodies of layered gabbro which occur at the southern end of the BT. One of them, the "Tiger gabbro" with an areal extent of 35 km^2, was discovered by Vetter during the season 1982/1983. The samples are currently under investigation. The other gabbro of unknown size was found by F. Tessensohn in the 1984/1985 season (Figure 4). It is located in WT at the mouth of the Wylde Glacier. The rocks consist of alternating cumulate layers of clinopyroxene, basic plagioclase, orthopyroxene, and accessory olivine. The "Tiger gabbro" is tilted about 60° to the NE and is cut by numerous shear zones. It has intruded a clastic sedimentary series in supposed Glasgow Volcanics which it has contact metamorphosed (S. Engel, unpublished thesis, 1984). A K/Ar age of 523 Ma ± 10 was obtained from a pegmatitic hornblende [Kreuzer et al., this volume].

Fig. 4. Layered ultramafic intrusion, Wylde Glacier, Mountaineer Range. Height of cliff is approximately 150 m, lower part obscured by fog on sea ice.

According to other described examples these gabbros may well represent the deeper parts of an ophiolite suite. They can be interpreted as fossil magma chambers formed beneath an axis of spreading. A geochemical classification is in progress.

Thrust Belts: Structural Evidence for Subduction

One of the most puzzling features noted during previous research in NVL was the apparent lack of vergence in the folded sedimentary strata. The folds are generally regular and upright.

Fig. 5. Thrust plane in the Mountaineer Range, Wylde Glacier, north of Cape King. Upper light-colored part is granite, lower darker sequences are metasediments. Height of exposed section is approximately 80 m.

Only recently, new observations along both boundaries of the BT [Wright, 1982; Findlay and Field, 1983; Gibson, 1984; Gibson et al., 1984; Jordan et al., 1984; Kleinschmidt et al., 1984; Wright and Findlay, 1984] have confirmed that thrusting localized in narrow schist zones has a strong vergence toward the NE [Gibson and Wright, 1985].

Associated with the western boundary fault is a narrow belt of greenschist rocks, which among other lithologies contains bodies of highly deformed conglomerates [Gibson, 1984].

Thrusting at this fault can most clearly be observed at its southern end in the Mountaineer Range, where high-grade metamorphic rocks are thrust over unmetamorphosed sediments. A whole series of thrust planes is developed in an upside-down tectonic pile with the following units from bottom to top: unmetamorphosed sediments, greenschists, kyanite-bearing metasediments, gneisses and migmatites, and granites. Thrust planes are exposed within the kyanite-bearing metasediments [Kleinschmidt et al., 1984] and between migmatites and granites (Figure 5). Mylonites and deformed conglomerates are encountered in the greenschist unit above the BT sediments.

A similar contact is exposed in the Lanterman Range (Figure 2), where medium-grade metamorphic rocks are thrust up against fossiliferous Cambrian sediments of the BT. The contact zone is characterized by highly deformed "conglomerates." In both areas the conglomerates are cut by innumerable shear zones [Gibson, 1984]. We suppose that thrust movements along these zones are responsible for the stretching component of the clast deformation. Greenschist metamorphism with retrograde effects on the Wilson side is associated with, and has outlasted, the thrust-related deformation. Gibson and Wright [1985] explain that this later metamorphism occurs because the Wilson unit is still hot when thrust over the Bowers rocks.

Both thrusting and greenschist metamorphism are clearly later than the subduction process, which causes the medium- to high-grade metamorphism of the WT.

The late thrust event has presumably cut out a major part of the former subduction complex and of the "leading edge" of the continental WT, thus causing considerable telescoping.

The eastern boundary fault is steep along most of its length, and accompanied by a narrow schist zone (Millen-Schist [Findlay and Field, 1982]). Although these schists seem to consist mainly of deformed Robertson Bay turbidites, some minor volcanic rocks (of the BT?) have also been reported [Jordan et al., 1984]. No deformed conglomerates have been found at this fault. A horizontal, folded thrust plane is well developed only in the Millen Range [Wright and Findlay, 1984, figure 5]. The occurrence of pillow lavas and other volcanic rocks in the upper unit suggests that rocks of the BT are carried several kilometers northeastward over folded Robertson Bay turbidites. The latter are only schistose close to the thrust plane, whereas the volcanics in the upper unit are strongly schistose and metamorphosed to greenschist facies.

These new observations indicate that the structural deformation, while generally moderate and regular within the BT and the RBT, has been very intense along the terrane boundaries, where it was partly associated with greenschist metamorphism.

Recently, a new tectonic interpretation of NVL [Gibson and Wright, 1985] has been presented which emphasizes the role of thrust tectonics and implies severe structural telescoping of entire terranes.

Sedimentary Basins

Each of the three terranes discussed contains sedimentary sequences which can be analyzed in terms of their plate tectonic position. Although there is some biostratigraphic age control for the sediments, it is difficult to relate them to the radiometrically dated igneous, metamorphic, and structural events. At present, there is a difference of up to 60 m.y. in the radiometric time scales for the Cambrian [Harland et al., 1982; Odin et al., 1982; Odin and Gale, 1985]. Our tentative interpretation of the spatial and temporal relations is presented in Figure 6.

A high proportion of the recognizable metasedimentary sequences in the WT consist of low-grade turbidites which are best exposed in the Morozumi Range and Daniels Range [Kleinschmidt and Skinner, 1981]. Apart from the apparent lack of trace fossils, the thick turbidite sequence of the Morozumi Range is comparable in its sedimentological features to the Robertson Bay Group. But a deformational age of 530 Ma from the Daniels Range [Adams, 1986] and a Tremadocian sedimentation age (505 to 495 Ma on all time scales) on the Handler Ridge in the RBT seem to preclude a correlation.

The Robertson Bay turbidites today cover an area about 200 km wide across the strike which indicates a minimum prefolding distance of 270 km. This implies a fairly extensive depositional area, especially as only the distal parts of the sedimentary fan are preserved. Lithologically the graywackes consist of quartz, lithic fragments of metamorphic and igneous rocks, and subordinate volcanic components. In a provenance analysis, Wright [1980, 1981] deduces sedimentary and metamorphic source areas with a consistent minor input of a mafic volcanic source. Farther south in the outcrop area, Field and Findlay [1983] found a mixture of sedimentary and igneous fragments, but no high-grade metamorphic and basic igneous fragments. Both authors stress the high content of angular, partly polycrystalline quartz.

Wright [1985] has analyzed the graywacke composition in terms of plate tectonic settings using the diagrams of Dickinson and Suczek [1979]. He deduced a continent-continent collisional origin for the graywackes.

The sedimentary sequence in the BT poses some problems for a plate tectonic interpretation. It spans a stratigraphic range from at least the early Middle Cambrian to the Ordovician, and should therefore include some evidence of the major tectonic events deduced for this time interval, i.e., a major break of sedimentation or an unconformity.

The lowermost sedimentary suite (Molar Formation) contains graywackes which plot in the arc orogen field [Wright, 1985]. The suite is associated with the submarine Glasgow Volcanics. On the one hand, this volcanism has the character of a primitive island arc [Weaver et al., 1984]; on the other, there is definitely continental material mainly of a granitic nature reworked in the conglomerates of the encompassing sediments [Laird et al., 1982]. These two facts are difficult to reconcile.

Possible explanations can either challenge the validity of the derived primitive island arc character of the Glasgow Volcanics, or assume a time interval between arc formation and continent-derived sedimentation.

To derive the Robertson Bay turbidites as a deepwater facies from a continental source in the west across a buried or dissected Glasgow arc does not pose unsurmountable problems.

Granitic Intrusions

In recent years, a series of investigations have dealt with the granitic rocks of NVL [Kreuzer et al., 1981, this volume; Wyborn, 1981, 1983; Babcock et al., 1983; Plummer et al., 1983; Vetter et al., 1983, 1984; Borg et al., 1984; Vetter et al., 1984; Borg and Stump, this volume]. The granitoids occur in three different age groups. Some singular early intrusions are followed by the main bulk of the Cambro-Ordovician Granite Harbour Intrusives (GHI) and the Devonian-Carboniferous Admiralty Intrusives (AI).

Only a few scattered granitoids with ages greater than 500 Ma are known. They include the "Terra Nova Granite" in the WT, dated at 535 Ma with an initial Sr^{87}/Sr^{86} ratio of 0.709, and two examples from the RBT: the granitoids of Cooper Bluff and Surgeon Island [Vetter et al., 1984; Kreuzer et al., this volume]. In both cases, Rb/Sr whole-rock isochrone ages differ from younger Rb/Sr mineral determinations. There are some doubts about the Cooper Bluff Granite as it has an apparent age of 525 ± 15 Ma, which is

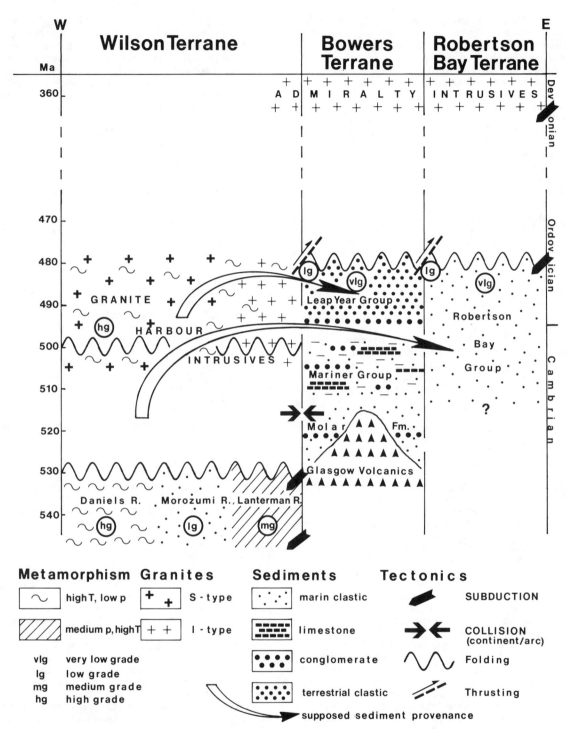

Fig. 6. Time relations of sedimentary and tectonic events in the tectonic terranes of north Victoria Land.

earlier than the deformation ages of the intruded and contact metamorphosed sediments. Conversely, the Surgeon Island granite (599 ± 21 Ma) shows an internally consistent age. It is strongly sheared and this later event is dated by muscovite ages of 480 Ma. Its occurrence in the middle of supposed oceanic terrane lacks a convincing explanation at the present time. It may be a continental sliver caught up in the subduction process like the island of Sumba in Indonesia situated on the oceanward side of the volcanic arc [Hamilton, 1979].

The first major granitic event was the formation of the GHI with ages clustering around 480 Ma. Petrographically, they are mainly 2-mica granites and migmatites; but granodiorite, tonalite, and diorite also occur. The intrusives in the west of the WT are predominantly hornblende-free peraluminous S-type granitoids with initial strontium ratios of 0.712 to 0.716, derived from anatectic melting of pelitic and psammitic sediments [Wyborn, 1981, 1983; Babcock et al., 1983; Plummer et al., 1983; Vetter et al., 1983; Borg et al., 1984]. Coeval hornblende-bearing I-type granitoids with initial strontium ratios of 0.707 to 0.709 occur only subordinately in the west, but are more widespread in the east, particularly in the Lanterman Range [Wyborn, 1981; Borg et al., 1984; Roland et al., 1984].

The parallel alignment of an inner S-type and an outer I-type belt can best be interpreted as a typical active margin setting [Pitcher, 1984]. Thus the distribution of the granitoids supports the model already derived from other evidence. (Compare also Borg and Stump, [this volume].) The GHI can be regarded as "stitching granites," transforming the entire WT to continental crust.

The Devonian-Carboniferous AI form the second major pulse of plutonism in NVL. These granitoids which, with a few exceptions, show a remarkably narrow age range around 360 Ma [Kreuzer et al., 1981; this volume] are petrographically somewhat ambiguous.

The compositional range spans granites and granodiorites [Wyborn, 1981; Vetter et al., 1983] and in addition, tonalites and monzogranites [Borg et al., 1984]. They show typical I-type characteristics [Wyborn, 1981] except for high initial strontium ratios of 0.710 to 0.718 [Vetter et al., 1983]. These may indicate crustal contamination. Borg and Stump [this volume] have presented very convincing polarity trends in the granites, but as the Supernal Granite cuts across the western boundary fault, we cannot accept their assumption that the AI are earlier than the collision of RBT/BT with the WT.

Late Devonian andesitic and partly rhyolitic volcanism occurring in all three terranes is most probably related to the AI plutonism. The products of subaerial activity are still preserved in isolated piles resting unconformably on the older deformed sequences (Mount Black Prince in RBT, Lawrence Peaks and Mariner Plateau in BT, and Mount Anakiwa and Gallipoli Hights in WT).

The tectonic environment of the AI is not easy to deduce, although most characteristics plot in the field of I-type granites. But within this field they show features of both Cordilleran and Caledonian I types according to the classification of Pitcher [1984].

The compositional range tonalite-diorite-monzogranite as given by Borg et al. [1984], the content of hornblende plus biotite together with magnetite and sphene, the restitic xenoliths, and a Cu/Mo mineralization would classify the Admiralties as andinotype marginal arc (Cordilleran I type).

Conversely, a granodiorite-granite compositional range as given by Vetter et al. [1983], the pattern of isolated multiple plutons, the association with basalt-andesite lava "plateaux," the short duration of the plutonism, and the general absence of associated tectonism are arguments for a postcollisional granitic phase (Caledonian I type).

Interpretation: Previous Models

Several plate tectonic interpretations have been presented during the last 2 years. Weaver et al. [1984] made the first attempt based mainly on volcanic geochemistry and the pattern of sedimentary basins. Bradshaw et al. [1985] argue on the same lines of evidence. They explain the present pattern of terranes in NVL by strike-slip movements. Objections against this model are (1) structural evidence for strike-slip movements is not obvious in the field, whereas northeastward directed thrusting is quite spectacular in some segments along the terrane boundaries; (2) it is difficult to explain how coarse continent-derived clastic sediments on the Glasgow island arc could have passed the deep turbidite basin in its postulated position between continent and arc; and (3) the mechanism for the postulated tectonic transport of the RBT, from a backarc position at the inside of the arc to the present position outside of it, involves a complicated and not very plausible passing maneuver.

Wodzicki and Robert [1986] present additional evidence derived from the geochemistry of the volcanics, and the sequence of sedimentary events. Although based on only a very narrow section, their model is quite elaborate. The geology of the central part of NVL is explained by a combination of subduction and strike-slip motion. It is concluded that a collision between the WT and BT was the ultimate cause for the Ross Orogeny. This is consistent with our ideas. However, an assumed northeastern continental source area for the Molar sediments at the base of the Bowers sequence is the main argument for an eastward directed subduction. The absence of a subduction complex [Weaver et al., 1984] is explained by postsubduction strike-slip motion.

Gibson and Wright [1985] emphasized the role of thrust tectonics with significant loss of intervening crust by telescoping. A collision of Antarctica with a westward drifting eastern continental block is deduced as cause for this tectonism. This model has the advantage of providing a different eastern source for the Robertson Bay turbidites, but it does not provide a source for the coarse continental conglomerates on the oceanic Glasgow island arc. Regional metamorphism and Granite Harbour intrusions are related to the accretion of this island arc; whereas the later thrusting and crustal thickening is attributed to continent-continent collision at the boundary between RB and BT. This late (Ordovician) phase is called the Ross Orogeny, although it is not associated with strong regional metamorphism which could be expected by this kind of collision. In addition, a continent-continent collision would cause considerable crustal thickening followed by uplift and erosion. Yet the erosinal level reached in the low-grade RBT is much too shallow to account for such a cause. The greatest problem in this model is the fact that it does not explain where the assumed colliding continent east of the RBT has later disappeared.

Wright [1985] has analyzed the clastic sedimentary series of two terranes and deduces for the RBT a continent-continent collisional environment, for the BT a derivation from an undissected arc.

Crawford et al. [1984] have recently presented a model for the Australian Lachlan Fold Belt, similar in age and geological setting to NVL. In this model a Late Proterozoic splitting off of small slivers of continental crust is postulated. Small ocean basins are formed between these outer microcontinents and the continent, which show features of backarc and island arc environments similar to the present western Pacific. The closing of these oceans is supposed to have caused first island arc continent collisions and finally microcontinent-continent collisions. This Australian model may be useful to explain the collisional features discussed by Wright [1985] and Gibson and Wright [1985].

Borg and Stump [this volume] base their interpretation of the three terranes of NVL mainly on the distribution and chemical and isotopic polarity of the granitoids. Their data show a clear difference in isotopic and chemical polarity between the two granite generations. It is claimed that the GHI are restricted to the WT and the AI to RBT and BT, which leads them to make the western boundary fault the main tectonic suture line of NVL. BT and RBT are supposed to be fragments of one allochthonous continental block. The juxtaposition is supposed to have occurred after the implacement of the AI which therefore cannot be "stitching granites." Although the polarity arguments are quite convincing, the basic assumption that the western boundary fault is also the dividing line between the granite generations is unacceptable to us; as with Mount Supernal, at least one Admiralty granite cuts across the western boundary fault, and indicates that the implacement of the AI was later than the main tectonism. A close relationship of RBT and BT is derived from the polarity of the AI; both terranes are interpreted as marginal fragments of an allochthonous continental block in the east. No explanation, however, is provided for the primitive island arc volcanism of the Glasgow Volcanics in the BT, which contradicts such a model.

Conclusions

Evidence for relics of a former passive margin or for a continent-continent collision is not conclusive. In our opinion, rather, the geological record indicates a subduction-related accretional process which took place over a considerable period of time. During the entire period, from at least 550 Ma to about 350 Ma, the sense of movement remained the same with a Benioff zone dipping westward under the Precambrian continental crust of East Antarctica. The subduction process must have taken place semicontinuously in several episodes. Different phases of this process are documented by varying criteria in the rock record.

The following three major steps can be recognized (Figure 7): (I) an early subduction phase which terminates with the accretion of a volcanic island arc; (II) after the outward displacement of the subduction zone, renewed subduction causing the magmatotectonic culmination of the Ordovician Ross Orogeny in two distinct pulses: (a) Granitic Harbour plutonism and associated low-pressure metamorphism documented mainly in the inner terrane and (b) maximum of crustal shortening: thrusting and folding in the outer sedimentary terranes; and (III) after a possible lull, again subduction with development of andesitic volcanism and Admiralty plutonism in the Devonian.

Stage I (Figure 7) is documented in the WT by a set of paired metamorphic belts formed around 550 Ma. Evidence for the inner low-pressure belt is found mainly in the form of radiometric ages near 530 Ma in low-grade metasedimentary areas. Granitic intrusions of this age are rare. The outer medium- to high-pressure belt is associated with ultrabasic lenses and serpentinites. This belt indicates the proximity of a subduction zone which, in addition, is documented by relics of a subduction complex, ophiolitic rocks, and mélanges along the boundary with the BT. The turbidites of the Morozumi Range are interpreted as forearc sediments deposited between arc (low-pressure belt) and trench (medium-pressure belt).

Contemporaneously, the oceanic island arc of the Glasgow Volcanics, with its primitive geochemical character, was transported toward the continental WT. Both terranes collided between 530 and 500 Ma. The island arc must have formed earlier in an intraoceanic region. If the Cambrian sedimentary envelope of the Glasgow Volcanics (Molar Formation, Mariner Group) is approximately time equivalent with the metamorphic radiometrically dated belts, then it can only have formed on the arc before the collision. If it is later, it can be regarded as postcollisional and derived from the continent.

Stage II(a) (Figure 7) documents the main event of the Ross Orogeny immediately after the collision with the island arc. The subduction zone has moved to the outside, east of the RBT. Widespread granitization (GHI) and low-pressure metamorphism affected the roots of the entire WT.

Most of the GHI show S-type characteristics in the western area (Daniels Range, Morozumi Range), and I-type characteristics in the east (Lanterman Range). This indicates a typical active margin setting; the I types derived from an underplating of oceanic crust under a continental lip, and the S types by melting of older continental crust caused by crustal thickening farther inland. The granite distribution therefore matches the earlier pattern of paired metamorphic belts.

Uplift and erosion of the WT followed. This is indicated by the maximum of cooling ages around 480 Ma (Ross Orogeny), and by the onset of the molasse-type sedimentation of the Leap Year Group.

According to the Tremadocian Handler Ridge fauna, the Robertson Bay Group is a time equivalent of parts of the Bowers sequence. It is also best explained as a postcollisional sequence fed by the crystalline rocks of the WT across the buried or dissected Glasgow island arc. This

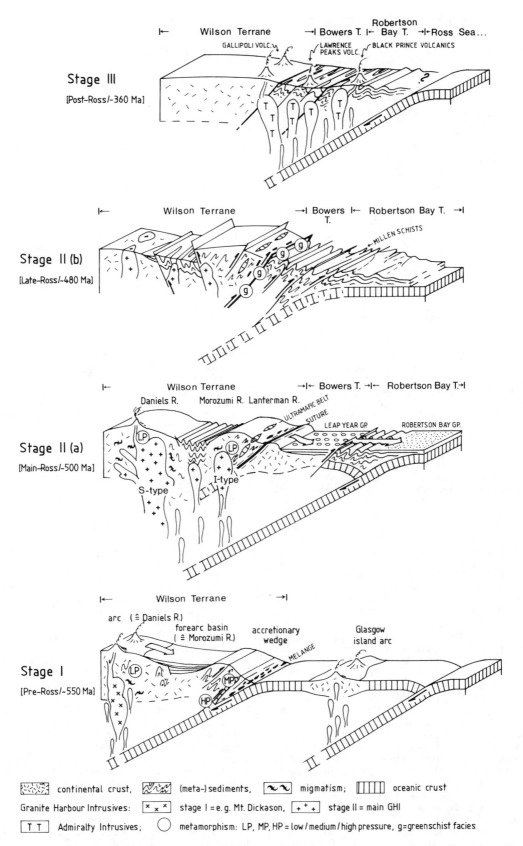

Fig. 7. Subduction stages at the Pacific margin of Gondwana based on evidence from north Victoria Land, Antarctica. For further explanation see conclusions in text.

would account for the high amount of quartz in the graywackes as compared to the volcanic-derived graywackes of the arc.

Stage II(b) (Figure 7) is a distinct second phase which followed closely after the main event of the Ross Orogeny. It contains the maximum of crustal shortening and is documented mainly by structural evidence; by the main deformation (folding and very low-grade metamorphism) within the two sedimentary terranes, BT and RBT, and by thrusting at the terrane boundaries.

The fold event is dated in the RBT by a series of K/Ar determinations on phyllites ranging in age from 505 to 455 Ma [Adams and Kreuzer, 1984]. The maximum age lies around 480 Ma. In the BT a similar fold event affected the whole sedimentary sequence including the Leap Year Group. But here, the radiometric evidence is greatly obscured by a suite of younger dates reaching into the Carboniferous.

The thrusting event which took place at both boundary faults of the BT is not directly dated. Along the western boundary fault, the site of the former subduction zone, it was accompanied by greenschist facies metamorphism. This event is supposed to be responsible for the loss of crust along this line and the present juxtaposition of the two terranes.

In the Mountaineer Range, where migmatites and granites are thrust northeastward over the medium-pressure belt, additional thrust planes can be observed some distance back in the crystalline WT.

Stage III (Figure 7) occurred at the Devonian-Carboniferous boundary, after a long tectonic lull. The evidence for this stage is provided by the Admiralty granitoids and associated volcanics. No corresponding structural event has been found in NVL, although it may be present in other corresponding Gondwana fragments such as Tasmania.

Contrary to the opinion of Borg and Stump [this volume], we are convinced that both granitoids and volcanics occur in all three terranes, and in some cases cut across the boundary faults. In the WT both are restricted to the eastern margin.

It is suggested that the igneous rocks represent the arc of a third stage which has its fore-arc basin and trench in the present Ross Sea, or possibly in Tasmania where deformation and metamorphism of the Devonian Tabberabberan Orogeny [Adams, 1981] could be related tectonic phenomena.

We are fairly confident from the evidence discussed that the assumption of a uniformly directed, long-term but episodic process of subduction-accretion can well explain the development of the active Gondwana margin in this part of Antarctica. However, we do not consider all aspects of our model as a final solution for the complicated geological pattern. Weak points in particular are not very conclusive explanations for the sliver of continental crust in the RBT (at Surgeon Island) and the inconclusive discussion of the spatial and temporal relations between the sedimentary and tectonic processes leading to speculations at the present time.

Future research should be directed to improve these parameters, to extend the area of investigation to more western outcrops, and to carry out more comparative geochemical work between the terranes of NVL and other related Gondwana fragments such as Tasmania, the Tasman Rise, New Zealand, and Marie Byrd Land.

Acknowledgments. This analysis is largely based on our own fieldwork in NVL since 1979 (four seasons). Intensive discussions with colleagues in the field during recent expeditions in 1982/1983 and 1984/1985, and during a workshop in 1984, have played an important role in the development of the concept. We would especially like to thank Thomas Wright, George Gibson, Edward Grew, Ulrich Vetter, John Bradshaw, and Malcolm Laird for stimulating discussions and Hans Kreuzer, Christopher Adams, Axel Höhndorf, and Heinz Lenz for providing the radiometric background. The Deutsche Forschungsgemeinschaft has funded parts of this research program, logistic support was provided by the German Federal Institute for Geosciences and Natural Resources (BGR), and organized by Jurgen Kothe.

References

Adams, C. J., Geochronological correlations of Precambrian and Paleozoic orogens in New Zealand, Marie Byrd Land (West Antarctica), northern Victoria Land (East Antarctica), and Tasmania, in Gondwana Five, edited by M. M. Cresswell and P. Vella, pp. 191-197, A. A. Balkema, Rotterdam, 1981.

Adams, C. J., Age and ancestry of the metamorphic rocks of the Wilson Group of the Daniels Range, USARP Mountains, Antarctica, in Geological Investigations in Northern Victoria Land, Antarctic Res. Ser., edited by E. Stump, AGU, Washington, D.C., in press, 1986.

Adams, C. J., and H. Kreuzer, Potassium-argon age studies of slates and phyllites from the Bowers and Robertson Bay terranes, northern Victoria Land, Antarctica, Geol. Jahrb., Reihe B, 60, 265-288, 1984.

Babcock, R. S., C. C. Plummer, J. S. Sherator, C. J. Adams, and R. L. Oliver, Geology of the Daniels Range intrusive complex, northern Victoria Land, Antarctica (abstract), in Antarctic Earth Sciences, edited by R. L. Oliver, P. R. James, and J. B. Jago, p. 118, Australian Academy of Science, Canberra, 1983.

Bird, J. M., and J. F. Dewey, Lithosphere plate-continental margin tectonics and the evolution of the Appalachian Orogen, Geol. Soc. Am. Bull., 81, 1031-1060, 1970.

Borg, S. G., and E. Stump, Paleozoic magmatism and associated tectonic problems of northern Victoria Land, Antarctica, this volume.

Borg, S. G., B. W. Chappell, M. T. McCulloch, E. Stump, D. Wyborn, and J. R. Holloway, Compositional polarity of granitoids with implications to regional geology, northern Victoria Land, Antarctica, Geol. Soc. Am. Abstr. Programs, 16(6), 1984.

Boyd, F. R., and J. L. England, Pyrope., Year Book Carnegie Inst. Washington, 58, pp. 83-87, 1959.

Bradshaw, J. D., and M. G. Laird, Pre-Beacon geology of northern Victoria Land: A review, in

Antarctic Earth Sciences, edited by R. L. Oliver, P. R. James, and J. B. Jago, pp. 98-101, Australian Academy of Science, Canberra, 1983.

Bradshaw, J. D., S. D. Weaver, and M. G. Laird, Suspect terranes and Cambrian tectonics in north Victoria Land, Antarctica, in Tectonostratigraphic Terranes of the Circum-Pacific Region, edited by D. G. Howell, pp. 467-479, Circum-Pacific Council for Energy and Mineral Resources, Earth Sci. Ser. 1, Houston, 1985.

Brown, G. C., R. S. Thorpe, and P. C. Webb, The geochemical characteristics of granitoids in contrasting arcs and comments on magma sources, J. Geol. Soc. London, 141(3), 413-426, 1984.

Burchfiel, B. C., Plate tectonics and the continents: A review, in Continental Tectonics-Studies in Geophysics, pp. 15-25, National Academy of Sciences, Washington, 1980.

Burrett, C. F., and R. H. Findlay, Cambrian and Ordovician conodonts from the Robertson Bay Group, Antarctica and their tectonic significance, Nature, 307(5953), 723-725, 1984.

Cawood, P. A., Structural relations in the subduction complex of the Paleozoic New England fold belt, eastern Australia, J. Geol., 90, 381-392, 1982.

Cooper, R. A., J. B. Jago, A. J. Rowell, and P. Braddock, Age and correlation of the Cambrian-Ordovician Bowers Supergroup, northern Victoria Land, in Antarctic Earth Science, edited by R. L. Oliver, P. R. James, and J. B. Jago, pp. 128-131, Australian Academy of Science, Canberra, 1983.

Crawford, A. J., W. E. Cameron, and R. R. Keays, The association boninite low-Ti andesite-tholeiite in the Heathcote greenstone belt, Victoria; ensimatic setting for the early Lachlan Fold Belt, Aust. J. Earth Sci., 31, 161-175, 1984.

Crook, K. A. W., Lithogenesis and geotectonics: the significance of compositional variation in flysch arenites (graywackes), in Modern and Ancient Geosynclinal Sedimentation, Spec. Publ. 19, edited by R. H. Dott and R. H. Shaver, pp. 304-310, Society of Economic Paleontologists and Mineralogists, Tulsa, Okla., 1974.

Crook, K. A. W., Fore-arc evolution in the Tasman Geosyncline: The origin of the southeast Australian continental crust, J. Geol. Soc. Aust., 27, 235-340, 1980.

Crowder, D. F., Geology of a part of north Victoria Land, Antarctica, U.S. Geol. Surv. Prof. Pap., 600-D, 95-107, Washington, 1968.

Dewey, J. F., Evolution of the Appalachian/Caledonian Orogen, Nature, 22, 124-129, 1969.

Dickinson, W. R., Relation of andesites, granites, and derivative sandstones to arch-trench tectonics, Rev. Geophys. Space Phys., 8, 813-860, 1970.

Dickinson, W. R., Plate tectonics and sedimentation, in Tectonics and Sedimentation, Spec. Publ. 22, edited by W. R. Dickinson, pp. 1-27, Society of Economic Paleontologists and Mineralogists, Tulsa, Okla., 1974.

Dickinson, W. R., and D. R. Seely, Structure and stratigraphy of fore-arc regions, Am. Assoc. Pet. Geol. Bull., 63, 2-31, 1979.

Dickinson, W. R., and C. A. Suczek, Plate tectonics and sandstone compositions, Am. Assoc. Pet. Geol. Bull., 63, 2164-2182, Tulsa, Okla., 1979.

Dott, R. H., Jr., and R. H. Shaver, Modern and ancient geosynclinal sedimentation, Soc. Econ. Paleontol. Mineral. Spec. Publ., 19, 345-357, 1974.

Engel, A. E. J., S. P. Itson, C. G. Engle, D. M. Stickney, and E. J. Cray, Jr., Crustal evolution and global tectonics: A petrogenic view, Geol. Soc. Am. Bull., 85, 843-858, 1974.

Engel, S., Petrogenesis of contact schists in the Morozumi Range, north Victoria Land, Antarctica, Geol. Jahrb., Reihe B, 60, 167-185, 1984.

Ernst, W. G., Occurrence and mineralogic evolution of blueschist belts with time, Am. J. Sci., 272, 657-668, 1972.

Field, B. D., and R. H. Findlay, The sedimentology of the Robertson Bay Group, northern Victoria Land, in Antarctic Earth Science, edited by R. L. Oliver, P. R. James, and J. B. Jago, pp. 102-106, Australian Academy of Science, Canberra, 1983.

Findlay, R. H., and B. D. Field, Preliminary report on the structural geology of the Robertson Bay Group, North Victoria Land, Antarctica, N. Z. Antarct. Rec., 4(2), 15-19, 1982.

Findlay, R. H., and B. Field, Tectonic significance of deformations affecting the Robertson Bay Group and associated rocks, northern Victoria Land, Antarctica, in Antarctic Earth Science, edited by R. L. Oliver, P. R. James, and J. B. Jago, pp. 107-112, Australian Academy of Science, Canberra, 1983.

Fox, K. F., Jr., Mélanges and their bearing on late Mesozoic and Tertiary subduction and interplate translation at the west edge of the North American plate, U.S. Geol. Surv. Prof. Pap., 1198, 1-40, Washington, 1983.

Gair, H. S., The geology from the upper Rennick Glacier to the coast, northern Victoria Land, Antarctica, N. Z. J. Geol. Geophys., 10, 309-344, 1967.

Gair, H. S., A. Sturm, S. J. Carryer, and G. W. Grindley, The geology of northern Victoria Land, Antarct. Map Folio Ser., folio 12, plate XII, Am. Geogr. Soc., New York, 1969.

Gibson, G., Deformed conglomerates in the eastern Lanterman Range, northern Victoria Land, Antarctica, Geol. Jahrb., Reihe B., 60, 117-141, 1984.

Gibson, G., and T. O. Wright, Importance of thrust faulting in the tectonic development of northern Victoria Land, Antarctica, Nature, 315(6019), 480-483, London, 1985.

Gibson, G., F. Tessensohn, and A. R. Crawford, Bowers Supergroup rocks west of the Mariner Glacier and possible greenschist facies equivalents, Geol. Jahrb., Reihe B, 60, 289-318, 1984.

Gilluly, J., Plate tectonics and magmatic evolution, Geol. Soc. Am. Bull., 83, 2382-2396, 1971.

Grew, E. S., and M. Sandiford, Field studies of the Wilson and Rennick Groups, Rennick Glacier area, northern Victoria Land, Antarct. J. U. S., 17(5), 7-8, 1982.

Grew, E. S., and M. Sandiford, A staurolite-talc assemblage in tourmaline-phlogopite-chlorite schist from northern Victoria Land, Antarctica, and its petrogenetic significance, Contrib. Mineral. Petrol., 87, 337-350, 1984.

Grew, E. S., G. Kleinschmidt, and W. Schubert, Contrasting metamorphic belts in north Victoria Land, Antarctica, Geol. Jahrb., Reihe B, 60, 253-263, 1984.

Hamilton, W., Tectonics of the Indonesian region, U.S. Geol. Surv. Prof. Pap., 1078, 1-345, Washington, 1979.

Harland, W. B., A. V. Cox, P. G. Llewellyn, C. A. G. Pickton, A. G. Smith, and R. Walters, A Geologic Time Scale, 131 pp., Cambridge University Press, New York, 1982.

Harrington, H. J., B. L. Wood, I. C. McKellar, and G. L. Lensen, The geology of Cape Hallett-Tucker Glacier district, in Antarctic Geology, edited by R. J. Adie, pp. 200-228, North Holland, Amsterdam, 1964.

Hsü, K. J., Principles of mélanges and their bearing on the Franciscan-Knoxville paradox, Geol. Soc. Am. Bull., 79, 1063-1074, 1968.

Hsü, K. J., Mélanges and their distinction from olistostromes, in Modern and Ancient Geosynclinal Sedimentation, Spec. Publ. 19, edited by R. H. Dott, Jr., and R. D. Shaver, pp. 321-333, Society of Economic Paleontologists and Mineralogists, Tulsa, Okal., 1974.

Jordan, H., R. H. Findlay, G. Mortimer, M. Schmidt-Thome, A. Crawford, and P. Müller, Geology of the northern Bowers Mountains, north Victoria Land, Antarctica, Geol. Jahrb., Reihe B, 60, 57-81, 1984.

Karig, D. E., Origin and development of marginal basins in the western Pacific, J. Geophys. Res., 76, 2542-2561, 1971.

Karig, D. E., and G. F. Moore, Tectonically controlled sedimentation in marginal basins, Earth Planet. Sci. Lett., 26, 233-238, 1975.

Kay, M., Dunnage mélange and the subduction of the Protoacadic Ocean, northeast Newfoundland, Spec. Pap. Geol. Soc. Am., 175, 1-49, 1976.

Kleinschmidt, G., Regional metamorphism in the Robertson Bay Group area and in the southern Daniels Range, north Victoria Land, Antarctica-A preliminary comparison, Geol. Jahrb., Reihe B, 41, 201-228, 1981.

Kleinschmidt, G., Trends in regional metamorphism and deformation in northern Victoria Land, Antarctica, in Antarctic Earth Science, edited by R. L. Oliver, P. R. James, and J. B. Jago, pp. 119-122, Australian Academy of Science, Canberra, 1983.

Kleinschmidt, G., and D. N. B. Skinner, Deformation styles in the basement rocks of North Victoria Land, Antarctica, Geol. Jahrb., Reihe B, 41, 155-199, 1981.

Kleinschmidt, G., N. W. Roland, and W. Schubert, The metamorphic basement complex in the Mountaineer Range, north Victoria Land, Antarctica, Geol. Jahrb., Reihe B, 60, 213-251, 1984.

Kreuzer, H., A. Höhndorf, H. Lenz, U. Vetter, F. Tessensohn, P. Müller, H. Jordan, W. Harre, and C. Besang, K/Ar and Rb/Sr dating of igneous rocks from north Victoria Land, Antarctica, short note, Geol. Jahrb., Reihe B, 41, 267-273, 1981.

Kreuzer, H., A. Höhndorf, H. Lenz, P. Müller, and U. Vetter, Radiometric ages of pre-Mesozoic rocks from northern Victoria Land, Antarctica, this volume.

Laird, M. G., and J. D. Bradshaw, New data on the Lower Paleozoic Bowers Supergroup, northern Victoria Land, in Antarctic Earth Science, edited by R. L. Oliver, P. R. James, and J. B. Jago, pp. 123-126, Australian Academy of Science, Canberra, 1983.

Laird, M. G., J. D. Bradshaw, and A. Wodzicki, Stratigraphy of the late Precambrian and early Paleozoic Bowers Supergroup, northern Victoria Land, Antarctica, in Antarctic Geoscience, edited by C. Craddock, pp. 535-542, University of Wisconsin Press, Madison, 1982.

Leitch, E. C., A plate tectonic interpretation of the Paleozoic history of the New England Fold Belt, Geol. Soc. Am. Bull., 86, 141-144, 1975.

McLeod, I. R., and C. M. Gregory, Geological investigations along the Antarctic coast between longitudes 108°E and 166°E, Rep. Bur. Miner. Resour., Geol. Geophys. Aust., 78, 1-53, 1967.

Miyashiro, A., Metamorphism and Metamorphic Belts, 492 pp., Allen and Unwin, London, 1973.

Miyashiro, A., K. Aki, and A. M. C. Sengör, Orogeny, 242 pp., John Wiley, New York, 1982.

Moore, G. F., and R. L. Wheeler, Structural fabric of a mélange, Kodiak, Alaska, Am. J. Sci., 278, 739-765, 1978.

Odin, G. S., and N. H. Gale, Calibration of the Phanerozoic time scale, Bull. Liaison Informations, International Geological Correlation Programme Proj. 196, 4, 1985.

Odin, G. S., D. Curry, N. H. Gale, and W. J. Kennedy, The Phanerozoic time scale in 1981, in Numerical dating in stratigraphy, vol. 2, edited by G. S. Odin, pp. 957-960, John Wiley, New York, 1982.

O'Hara, M. J., and E. L. P. Mercy, Petrology and petrogenesis of some garnetiferous peridotites, Trans. R. Soc. Edinburgh, 65, 264-310, 1963.

Pitcher, W. S., Granite type and tectonic environment, in Mountain Building Processes, edited by K. J. Hsü, pp. 19-40, Academic, New York, 1984.

Plummer, C. C., R. S. Babcock, J. W. Sheraton, C. J. D. Adams, and R. L. Oliver, Geology of the Daniels Range, Antarctica: A preliminary report, in Antarctic Earth Science, edited by R. L. Oliver, P. R. James, and J. B. Jago, pp. 113-117, Australian Academy of Science, Canberra, 1983.

Ravich, M. G., L. V. Klimov, and D. S. Solov'ev, The Precambrian of East Antarctica (in Russian), (English translation, Israel Program for Scientific Translation, Jerusalem, 1968.) National Science Foundation, Washington, D. C., 1965.

Riddols, B. W., and G. T. Hancox, The geology of the upper Mariner Glacier region, north Victoria Land, Antarctica, N. Z. J. Geol. Geophys., 11 (4), 881-899, 1968.

Roland, N. W., G. Gibson, G. Kleinschmidt, and W. Schubert, Metamorphism and structural relations of the Lanterman Metamorphics, Geol. Jahrb., Reihe B, 60, 319-361, 1984.

Scheibner, E., A plate tectonic model of the Paleozoic tectonic history of New South Wales, J. Geol. Soc. Aust., 20, 1-283, 1973.

Schreyer, W., Zur mineralogischen Konstitution des Erdmantels, Naturwiss. Rundsch., 19, 184-189, 1966.

Schubert, W., M. Olesch, and K. Schmidt, Paragneiss-Orthogneiss relationships in the Kavrayskiy Hills, north Victoria Land, Antarctica, Geol. Jahrb., Reihe B, 60, 187-211, 1984.

Schwab, F. L., Framework mineralogy and chemical composition of continental margin-type sandstone, Geology, 3, 487-490, 1975.

Skinner, D. N. B., The geology of Terra Nova Bay, in Antarctic Earth Science, edited by R. L. Oliver, P. R. James, and J. B. Jago, pp. 150-155, Australian Academy of Science, Canberra, 1983.

Stump, E., M. G. Laird, J. D. Bradshaw, J. R. Holloway, S. G. Borg, and K. E. Lapham, Bowers graben and associated tectonic features cross northern Victoria Land, Antarctica, Nature, 304 (5924), 334-336, 1983.

Sturm, A., and S. J. Carryer, Geology of the region between the Matusevich and Tucker glaciers, northern Victoria Land, Antarctica, N. Z. J. Geol. Geophys., 13(2), 408-435, 1970.

Tessensohn, F., Significance of the Bowers structural zone in north Victoria Land, Antarctica, Geol. Jahrb., Reihe B, 60, 371-396, 1984.

Vetter, U., N. W. Roland, H. Kreuzer, A. Höhndorf, H. Lenz, and C. Besang, Geochemistry, petrography, and geochronology of the Cambro-Ordovician and Devonian-Carboniferous granitoids of northern Victoria Land, Antarctica, in Antarctic Earth Science, edited by R. L. Oliver, P. R. James, and J. B. Jago, pp. 140-143, Australian Academy of Science, Canberra, 1983.

Vetter, U., H. Lenz, H. Kreuzer, and C. Besang, Pre-Ross granites at the Pacific margin of the Robertson Bay Terrane, north Victoria Land, Antarctica , Geol. Jahrb., Reihe B, 60, 363-369, 1984.

Weaver, S. D., J. D. Bradshaw, and M. G. Laird, Geochemistry of Cambrian volcanics of the Bowers Supergroup and implications for the Early Paleozoic tectonic evolution of northern Victoria Land, Antarctica, Earth Planet. Sci. Lett., 68, 128-140, 1984.

Wodzicki, A., and R. Robert, Jr., Geology of the Bowers Supergroup, central Bowers Mountains, northern Victoria Land, in Geological Investigations in Northern Victoria Land, Antarctic Res. Ser., edited by E. Stump, AGU, Washington, D.C., in press, 1986.

Wodzicki, A., J. D. Bradshaw, and M. G. Laird, Petrology of the Wilson and Robertson Bay groups and Bowers Supergroup, northern Victoria Land, Antarctica, in Antarctic Geoscience, edited by C. Craddock, pp. 549-554, University of Wisconsin Press, Madison, 1982.

Wood, D. A., J. L. Joron, and M. Treuil, A reappraisal of the use of trace elements to classify and discriminate between magma series erupted in different tectonic settings, Earth Planet. Sci. Lett., 45, 326-336, 1979.

Wright, T. O., Sedimentology of the Robertson Bay Group, northern Victoria Land, Antarctica, Antarct. J. U. S., 15(5), 6-9, 1980.

Wright, T. O., Sedimentology of the Robertson Bay Group, north Victoria Land, Antarctica, Geol. Jahrb., Reihe B, 41, 127-138, 1981.

Wright, T. O., Structural study of the Leap Year Fault, northern Victoria Land, Antarct. J. U. S., 17(5), 11-13, 1982.

Wright, T. O., Late Precambrian and early Paleozoic tectonism and associated sedimentation in northern Victoria Land, Antarctica, Geol. Soc. Am. Bull., 96, 1332-1339, 1985.

Wright, T. O., and R. H. Findlay, Relationships between the Robertson Bay Group and the Bowers Supergroup-New progress and complications from the Victory Mountains, northern Victoria Land, Geol. Jahrb., Reihe B, 60, 105-116, 1984.

Wright, T. O., R. Ross, Jr., and J. Repetski, Newly discovered youngest Cambrian or oldest Ordovician fossils from the Robertson Bay Terrane (formerly Precambrian), northern Victoria Land, Antarctica, Geology, 12, 301-305, 1984.

Wyborn, D., Granitoids of north Victoria Land, Antarctica-Field and petrographic observations, Geol. Jahrb., Reihe B, 41, 229-249, 1981.

Wyborn, D., Chemistry of Palaeozoic granites of northern Victoria Land (abstract), in Antarctic Earth Science, edited by R. L. Oliver, P. R. James, and J. B. Jago, p. 144, Australian Academy of Science, Canberra, 1983.

Copyright 1987 by the American Geophysical Union.

JOINT U.K.-U.S. WEST ANTARCTIC TECTONICS PROJECT: AN INTRODUCTION

Ian W. D. Dalziel[1]

Lamont-Doherty Geological Observatory of Columbia University, Palisades, New York 10964

Robert J. Pankhurst

British Antarctic Survey, High Cross, Cambridge CB3 0ET, United Kingdom

Abstract. The relationship of West Antarctica to the Precambrian craton of East Antarctica is the longest standing major tectonic problem in Antarctic geology. It has a bearing on Gondwanaland reconstruction but has even more important geotectonic, paleoenvironmental, and paleobiogeographic implications. A joint United Kingdom-United States project was initiated in 1980-1981 to study certain aspects of the overall problem. Major field programs were undertaken in the 1983-1984 and 1984-1985 Antarctic seasons. Additional efforts will be made in 1986-1987 and 1987-1988. This contribution briefly introduces several accompanying papers describing the preliminary results of geological, geophysical, and geochemical aspects of the project as well as a preliminary tectonic synthesis based on the work completed at the time of the Sixth Gondwana Symposium.

Introduction

Antarctica has long been known as the "key" to Gondwanaland. Alex du Toit, in his classic book Our Wandering Continents [du Toit, 1937], incorporated in his basic reconstruction of the supercontinent the results of discoveries by scientists of the Antarctic expeditions led by Bruce, Scott, and Shackleton. In so doing, he accepted the distinction between the Precambrian craton of East Antarctica and the portion of the circum-Pacific mobile belt known as West Antarctica. The latter portion of the continent he portrayed as a tectonic zone broadly comparable to the volcanic arcs fringing the eastern margin of the Asian continent.

The tectonic history of West Antarctica and its relationship to the East Antarctic craton remains critical to a modern assessment of Antarctic and Gondwana geology. Two major problems are the unacceptable "overlap" of the Antarctic Peninsula with the Falkland Plateau of South America in some modern reconstructions of the supercontinent [e.g., Norton and Sclater, 1979] and the widely accepted need to restore the Paleozoic Gondwana sedimentary sequence of the Ellsworth Mountains to a position close to the Gondwanaland Precambrian craton and an orientation in which its trend can be reconciled with that of the Cape Fold Belt-Transantarctic Mountains [Schopf, 1969; Watts and Bramall, 1981; Dalziel and Elliot, 1982]. A solution to these and related tectonic problems is vital, not only to solving the refit of Gondwanaland but also to understanding the geological processes involved in its breakup, the subsequent development of circulation in the "Southern Ocean," climate change in the late Mesozoic and Cenozoic (including the onset of glaciation), and faunal dispersion in the southern hemisphere. Current analyses of the circum-Pacific mobile belt in terms of potentially displaced "terranes" [Howell, 1985] have emphasized the long recognized possibility that parts of West Antarctica may be exotic to that continent, although published paleomagnetic results make it unlikely that any major portions are far-traveled microcontinents (for review see Dalziel and Grunow [1985]).

With these problems in mind, a joint project was initiated by the British Antarctic Survey (BAS) and the United States Antarctic Research Program, Division of Polar Programs (DPP), to elucidate the tectonic history of West Antarctica and its relation to East Antarctica.

Field Program

The project was initially confined to that part of West Antarctica closest to South America and the Weddell Sea and concentrated on the tectonic history and relationships of the Antarctic Peninsula, Ellsworth-Whitmore Mountains crustal block, and Thurston Island-Eights Coast crustal block [see Dalziel et al., this volume, Figure 1].

In addition to basic field geology, structural geology, and petrology-geochemistry, the project has involved extensive collecting for paleomagnetic studies and for isotope geochemistry-geochronology. Airborne ice radio-echosounding and magnetic surveys were also included, extending and locally intensifying the coverage of the recent Scott Polar Research Institute-National Science

---

[1]Now at Institute for Geophysics, University of Texas at Austin, Austin, Texas 78751.

Foundation-Technical University of Denmark program [Drewry, 1983]. Some of the initial results of the new airborne program obtained in 1980-1981 have already been published [Doake et al., 1983]. The geophysical flights have been flown by teams of BAS geophysicists along lines planned jointly with U.S. scientists. Following the initial airborne geophysical program in 1980-1981, the joint U.K.-U.S. project has involved a combined geological and geophysical program in the Ellsworth Mountains-Whitmore Mountains-Thiel Mountains region (1983-1984) [Dalziel and Pankhurst, 1984] and a geological season in the Thurston Island-Jones Mountains area (1984-1985) [Dalziel and Pankhurst, 1985]. Future fieldwork is planned for the area between the Antarctic Peninsula, Ellsworth Mountains, Thurston Island, and Marie Byrd Land (airborne geophysics), and for the Pensacola Mountains (geology, 1987-1988).

Publication

Preliminary results were outlined at the Sixth Gondwana Symposium, and the following papers in this volume expand on these oral reports. Garrett et al. [this volume] describe airborne geophysical data bearing on the region between the Antarctic Peninsula and the Ellsworth Mountains. Storey and Dalziel [this volume] outline the structural geology of the Ellsworth-Thiel Mountains ridge. The geochemisty of granitic and mafic plutons is presented in two papers by Vennum and Storey [this volume, a, b]. Millar and Pankhurst [this volume] present the first Rb-Sr whole-rock geochronological study of the area, and the paleomagnetic results are analyzed by Grunow et al. [this volume]. Finally, a synthesis of all the new results [Dalziel et al., this volume] focuses on the role of the Ellsworth-Whitmore Mountains crustal block in the tectonic evolution of West Antarctica.

Acknowledgments. This project has benefited from the support of the Directors of the British Antarctic Survey (R. M. Laws) and of the Division of Polar Programs, National Science Foundation (E. Todd and P. Wilkness). Logistic efforts by the staff of both the BAS and DPP have been outstanding; special appreciation is due John Hall (BAS) and David Breshnahan and Erick Chiang (DPP). The fieldwork could not have been undertaken without the support of the Air Unit of BAS and the Antarctic Development Squadron (VXE6) of the United States Navy. Funding for U.S. science participation was made available through a National Science Foundation grant to I. W. D. Dalziel (DPP 82-13798). The participants are grateful to Campbell Craddock for providing advice and copies of unpublished sketch maps of several groups of nunataks.

References

Dalziel, I. W. D., and D. H. Elliot, West Antarctica: Problem child of Gondwanaland, Tectonics, 1, 3-19, 1982.
Dalziel, I. W. D., and A. M. Grunow, The Pacific margin of Antarctica, in Tectonostratigraphic Terranes of the Circum-Pacific Region, Earth Sci. Ser., vol. 1, edited by D. G. Howell, pp. 555-564, Circum-Pacific Council for Energy and Mineral Resources, Houston, Texas, 1985.
Dalziel, I. W. D., and R. J. Pankhurst, West Antarctica: Its tectonics and its relationship to East Antarctica, Antarct. J. U. S., 19, 35-36, 1984.
Dalziel, I. W. D., and R. J. Pankhurst, Tectonics of West Antarctica and its relation to East Antarctica: USARP-BAS geology/geophysics project (1984-1985), Antarct. J. U. S., 20(5), in press, 1986.
Dalziel, I. W. D., S. W. Garrett, A. M. Grunow, R. J. Pankhurst, B. C. Storey, and W. R. Vennum, The Ellsworth-Whitmore Mountains crustal block: Its role in the tectonic evolution of West Antarctica, this volume.
Doake, C. S. M., R. D. Crabtree, and I. W. D. Dalziel, Subglacial morphology between Ellsworth Mountains and Antarctic Peninsula: New data and tectonic significance, in Antarctic Earth Science, edited by R. L. Oliver, P. R. James, and J. B. Jago, pp. 270-273, Australian Academy of Sciences, Canberra, 1983.
Drewry, D. J. (Ed.), Antarctica: Glaciological and Geophysical Folio, Scott Polar Research Institute, Cambridge, England, 1983.
du Toit, A. L., Our Wandering Continents, 366 pp., Oliver and Boyd, Edinburgh, 1937.
Garrett, W. W., L. D. B. Herrod, and D. R. Mantripp, Crustal structure of the area around Haag Nunataks, West Antarctica: New aeromagnetic and bedrock elevation data, this volume.
Grunow, A. M., I. W. Dalziel, and D. V. Kent, Ellsworth-Whitmore Mountains crustal block, western Antarctica: New paleomagnetic results and their tectonic significance, this volume.
Howell, D. G., Tectonostratigraphic Terranes of the Circum-Pacific Region, Earth Sci. Ser., vol. 1, 581 pp., Circum-Pacific Council for Energy and Mineral Resources, Houston, Tex., 1985.
Millar, I. L., and R. J. Pankhurst, Rb-Sr geochronology of the region between the Antarctic Peninsula and the Transantarctic Mountains: Haag Nunataks and Mesozoic granitoids, this volume.
Norton, I. O., and J. G. Sclater, A model for the evolution of the Indian Ocean and the break-up of Gondwanaland, J. Geophys. Res., 84, 6803-6830, 1979.
Schopf, J. M., Ellsworth Mountains: Position in West Antarctica due to sea floor spreading, Science, 164, 63-66, 1969.
Storey, B. C., and I. W. D. Dalziel, Outline of the structural and tectonic history of the Ellsworth Mountains-Thiel Mountains ridge, West Antarctica, this volume.
Vennum, W. R., and B. C. Storey, Correlation of gabbroic and diabasic rocks from the Ellsworth Mountains, Hart Hills, and Thiel Mountains, West Antarctica, this volume, a.
Vennum, W. R., and B. C. Storey, Petrology, geochemistry, and tectonic setting of granitic rocks from the Ellsworth-Whitmore Mountains crustal block and Thiel Mountains, West Antarctica, this volume, b.
Watts, D. R., and A. M. Bramall, Paleomagnetic evidence for a displaced terrain in western Antarctica, Nature, 293, 638-641, 1981.

Copyright 1987 by the American Geophysical Union.

CRUSTAL STRUCTURE OF THE AREA AROUND HAAG NUNATAKS, WEST ANTARCTICA: NEW AEROMAGNETIC AND BEDROCK ELEVATION DATA

S. W. Garrett, L. D. B. Herrod, and D. R. Mantripp

British Antarctic Survey, Natural Environment Research Council
High Cross, Cambridge CB3 0ET, United Kingdom

Abstract. Five thousand line kilometers of new airborne geophysical data are presented from the area between the Antarctic Peninsula and Ellsworth Mountains. The surveys involved total-field magnetic measurements complemented by simultaneous determinations of ice thickness by radio-echo sounding. The southernmost end of the Antarctic Peninsula shows isolated aeromagnetic anomalies of over 400-nT amplitude which correspond to discrete plutonic bodies. The central area of Fowler Peninsula shows a continuous zone of high-amplitude (600 nT) short-wavelength (5 km) anomalies attributed to metamorphic rocks which outcrop beneath the ice cover. This interpretation is supported by measurements of the magnetic properties of Precambrian rocks exposed at Haag Nunataks. Anomalies of up to 400-nT amplitude and 50-km wavelength suggest a magnetic basement between 5 and 15 km below Evans Ice Stream, Carlson Inlet, Fletcher Promontory, and Rutford Ice Stream. A more subdued magnetic signature to the west of Rutford Ice Stream suggests that the presumed crystalline basement beneath the Paleozoic sediments of the Ellsworth-Whitmore Mountains crustal block may either lie at great depth (>20 km) or have a significantly lower magnetization than the rocks exposed at Haag Nunataks. A correlation of steep bedrock scarps, magnetic anomalies, and depth to magnetic sources indicates major faults delimiting horst and graben structures formed during northeast to southwest extension. The Ellsworth Mountains uplift is attributed to a thermal response of the lithosphere on the flank of this rift system.

Introduction

Airborne geophysical data were gathered during the 1983-1984 field season as part of the joint British Antarctic Survey (BAS)/U.S. Antarctic Research Program (USARP) investigation of the tectonic evolution of West Antarctica. Airborne surveys were dependent upon fuel depots established by USARP LC-130 aircraft at sites in the Sweeney Mountains (75°09.4'S, 70°37.7'W), Siple Station (75°56.4'S, 84°15.0'W), the Ellsworth Mountains (79°05.8'S, 85°58.5'W), and Martin Hills (81°54.4'S, 87°48.9'W). A BAS DHC-6C Twin Otter aircraft was used for the survey. A total coverage exceeding 17,000 line kilometers was achieved.

We present 5,000 line kilometers of data gathered between the Antarctic Peninsula and Ellsworth Mountains (Figure 1). This area contains the boundaries between the Ellsworth-Whitmore Mountains, Haag Nunataks, and Antarctic Peninsula crustal blocks [Dalziel and Elliot, 1982]. An additional 12,000 line kilometers of data were gathered to the south of the Ellsworth Mountains, a preliminary discussion of which is included in the paper by Dalziel et al. [this volume]. Both data sets are at present being more fully interpreted by two of the authors (L.D.B.H. and S.W.G.), M. P. Maslanyj, BAS, and D. Damaske, Bundesanstalt für Geowissenschaften und Rohstoffe (BGR), Hannover.

Data Acquisition and Processing

At the start of survey operations, a Geometrics G-866 base station magnetometer was deployed in the survey area to monitor diurnal variations. It was found that the ambient magnetic field frequently became disturbed between 2400 and 1200 UT owing to auroral activity, variations of over 200-nT amplitude being common [Maslanyj and Damaske, 1986]. Survey flying was therefore restricted to 1200-2000 UT; during this time, short-period (i.e., less than 1 hour) variations were generally less than 20 nT in amplitude, and long-period variations seldom exceeded 50 nT.

A flight-line spacing of between 25 and 50 km was achieved (Figure 2), most lines running perpendicular to predominant topographic and structural trends. Ice thickness information from a 60-MHz radio-echo sounder was recorded on 35-mm film. An optimum terrain clearance of 150 m allowed ice-bottom reflections to be obtained from ice up to 1.5 km thick. Magnetic measurements were made with a Geometrics G-803 total-field magnetometer connected to a wing-tip sensor. Digital magnetic and navigation data were recorded at 1-s intervals along flight track, equivalent to a reading every 55 m at a ground speed of 200 km per hour.

Navigation was provided by a Decca Tactical Air Navigation System (TANS) linked to a Doppler radar

110 WEST ANTARCTICA, NEW AEROMAGNETIC AND BEDROCK DATA

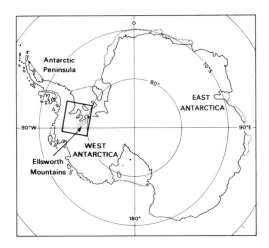

Fig. 1. Location Map. The area of Figures 2 and 3 is outlined.

unit, giving both ground velocities and an in-flight latitude and longitude. The use of a magnetic compass heading reference for the TANS required the manual input of magnetic declination to an accuracy of $0.3°$. The navigation error was measured by overflying a fuel depot, usually at the start and end of each flight. The positions of these depots had been previously fixed to within 0.5-km accuracy by using a Magnavox MX2102 satellite navigator. End-of-flight position errors up to 10 km were encountered, which is within 1% of the distance flown. As most of the errors were in a consistent direction [Herrod and Garrett, 1986] it was assumed that they had accumulated linearly throughout the flight. Correction vectors were applied accordingly during data processing, and the adjusted tracks are considered accurate to within 2 km.

The magnetic data were filtered to remove high-frequency noise due to interference from aircraft electrical systems. The 1980 International Geomagnetic Reference Field [Barraclough, 1981] was used to remove the global component of earth's magnetic field, and the residual magnetic anomaly profiles are shown in Figure 2. The data do not warrant contouring as the line spacing is too large in comparison with anomaly wavelengths. Diurnal variations were not applied because data acquisition was confined to magnetically quiet periods with less than 20-nT variation on any line.

The combination of barometric heighting and radar altimeter data allows ice surface elevations to be deduced. Bedrock elevations may then be obtained where ice thicknesses are known. Atmospheric pressure differences of 20 mbar might occur between the two ends of a 200-km survey line,

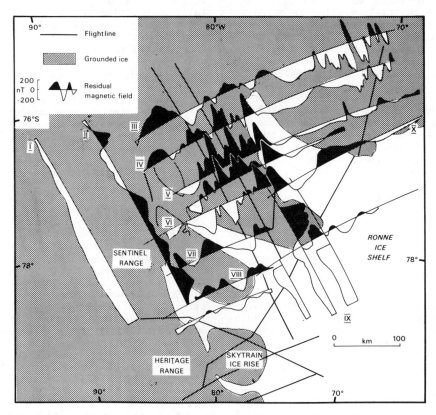

Fig. 2. Flight lines and residual magnetic anomalies. Grounded areas are stippled. Numbered profiles are referred to in the text.

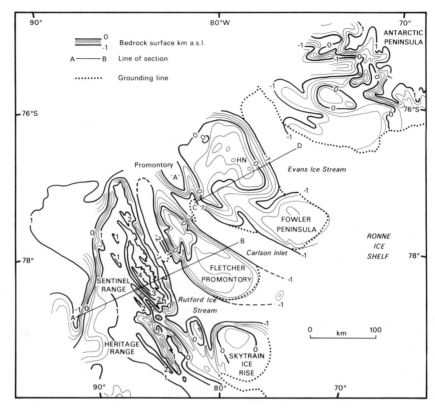

Fig. 3. Bedrock topography, contoured at 250-m intervals. HN is Haag Nunataks. The Ellsworth Mountains include the Sentinel Range and the Heritage Range. A cross section is shown in Figure 4.

equating to a potential discrepancy of the order of 200 m between the true and barometric altitudes at sea level. Surface elevations were, however, obtained to 25-m accuracy over the floating ice streams by assuming that the measured thickness of ice was in hydrostatic equilibrium. The adjustment of surface elevations of grounded areas to this datum reduced the mismatches at line intersections to within 25 m. The maximum error in the calculated bedrock elevation is estimated to be 100 m. The adjusted data have been used with the existing bedrock data [Behrendt, 1964; Swithinbank, 1977; Jankowski and Drewry, 1981; Doake et al., 1983] and Landsat images in compiling a bedrock elevation map (Figure 3).

Bedrock Morphology and Magnetic Anomalies

The bedrock elevation map (Figure 3) serves to clarify several topographic features recognized from geophysical surveys which formed the first part of the BAS-USARP project [Doake et al., 1983]. In this paper the area of grounded ice to the northeast of Evans Ice Stream will be considered to be part of the Antarctic Peninsula. Bedrock elevations fall abruptly from over 1,000 m above sea level at the southernmost nunataks of the Antarctic Peninsula to less than -1,500 m to the southwest [Behrendt, 1964]. Linear bedrock depressions bounded by steep escarpments are filled by Evans Ice Stream, Carlson Inlet, and Rutford Ice Stream. Between the ice streams are blocks showing bedrock elevations between -500 m and +500 m. Our data show that bedrock beneath Skytrain Ice Rise rises above sea level, and that this elevated block is bounded to the north and southwest by steep topographic scarps (Figure 3). The remaining elevated blocks show varying morphology. Bedrock beneath central Fowler Peninsula shows relief of up to 500 m over distances of a few kilometers, whereas southeastern Fowler Peninsula, Fletcher Promontory, and promontory A show smoother bedrock surfaces (Figure 4). Rutford Ice Stream is flanked by the linear massif of the Ellsworth Mountains, where bedrock elevations reach 1.5 km (Heritage Range) and 4 km (Sentinel Range) above sea level.

The aeromagnetic profiles indicate several distinct regions of contrasting magnetic character (Figure 2) which are best described with reference to the bedrock topography (Figure 3). The flattest magnetic signature is observed to the west of the Ellsworth Mountains, where small anomalies of less than 50-nT amplitude and 60-km wavelength are superimposed on an otherwise smooth background (Figure 2, profile I). To the east of the Ellsworth Mountains the field becomes significantly more disturbed. Profile II along Rutford Ice Stream shows anomalies of up to 400-nT amplitude and 60-km wavelength, complemented by smaller

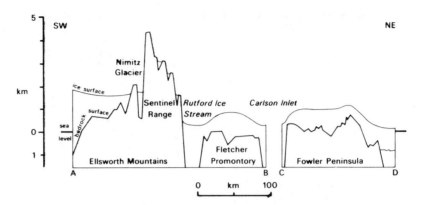

Fig. 4. Topographic cross section in the vicinity of the Ellsworth Mountains. Lines of section A-B and C-D are shown in Figure 2.

anomalies of less than 30-nT amplitude and 10-km wavelength. This pattern extends over Fletcher Promontory, although amplitudes of the smaller wavelength components (profiles VI, VII, and VIII) decrease. The central axis of the same topographic block is coincident with a negative magnetic anomaly, and its northern grounding line trends parallel to an associated positive anomaly. Carlson Inlet is marked by a negative magnetic anomaly of -300-nT amplitude and 50-km wavelength with a shallow western limb and a steep eastern limb (profiles VI and VII). A positive magnetic anomaly of 200-nT amplitude overlies promontory A (profile IV). The southwestern edge of Fowler Peninsula coincides with significant changes in magnetic signature (profiles IV, V, VI, and VII), for to the east of this line the amplitude of anomalies exceeds 600 nT, and wavelengths may be as short as 5 km. Several anomalies may be correlated between adjacent profiles (e.g., profiles IV and V). Profile IX shows a reduced amplitude of the higher-frequency component of the residual magnetic field associated with the southeastern corner of Fowler Peninsula. The northeastern grounding line of the Fowler Peninsula is marked by a sharp negative magnetic anomaly (profiles IV, V, and VI). This gives way to a continuous and well-defined positive magnetic anomaly over the western half of Evans Ice Stream, which has an amplitude of 500 nT, a wavelength of about 40 km, and a continuous length along strike of more than 250 km (profiles IV, V, VI, and X). Results from the 1982-1984 BAS Ronne Ice Shelf project [Herrod and Garrett, 1986] show that this anomaly terminates southeast of the Fowler Peninsula. It's northwesterly extent is as yet undefined, although it appears to wane toward the edge of the survey area.

The northeastern edge of Evans Ice Stream and that part of the Antarctic Peninsula with bedrock elevations below sea level show anomalies of up to 200-nT amplitude and 40-km wavelength (profiles III, IV, V, and X). This quiet pattern is occasionally punctuated by sharp magnetic anomalies of up to 400-nT amplitude corresponding to isolated subglacial topographic peaks. The area of the Antarctic Peninsula where bedrock elevation exceeds +1 km (Figure 3) displays a series of short-wavelength, high-amplitude anomalies which increase in amplitude to the northwest and appear to be superimposed on a regional magnetic low (profiles III, IV, and V).

Interpretation of the Magnetic Anomalies

Werner-based deconvolution. Werner-based deconvolution of magnetic profiles is an effective method of calculating depths to sources by digital filtering of magnetic anomalies [Hartman et al., 1971]. The method assumes that the magnetic sources conform to a dike or fault of arbitrary dip and magnetization striking perpendicular to the magnetic profile. The solutions represent the maximum depth to the upper surface of the causative magnetic bodies but are not diagnostic of the shape or magnetization. Modeling studies [e.g., Jankowski, 1981] suggest that the results may be accurate only to within a few kilometers. All quoted source elevations are relative to sea level.

Large-amplitude, short-wavelength anomalies over the Fowler Peninsula and Antarctic Peninsula suggest sources between 0- and -3-km elevation. Source elevations between -3 km and -8 km are implied over Fletcher Promontory and the southwestern half of Evans Ice Stream. Long-wavelength anomalies over Carlson Inlet, southeastern Fowler Peninsula, and the adjacent part of the Ronne Ice Shelf suggest source elevations between -7 and -14 km. The clearest relationship between magnetic anomalies and outcrop geology might be expected in the parts of the Antarctic Peninsula and central Fowler Peninsula where bedrock lies above sea level and magnetic sources are shallow.

Two-dimensional modeling. We have produced straightforward crustal models to explain the observed magnetic anomalies using a two-dimensional interpretation computer program [Lee, 1979]. It has been assumed that profiles trend perpendicular to strike, and that body magnetization is parallel to earth's present field. The initial models (Figure 5) are presented for qualitative comparison with the observed magnetic signature on profiles III and V (Figure 2).

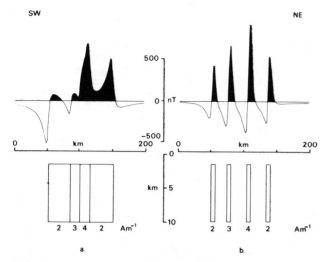

Fig. 5. Simple modeled anomalies at the latitude of the survey area: (a) continuous magnetic basement; compare with profile V, Figure 2, and (b) discrete magnetic bodies; compare with profile III, Figure 2.

A broad block of composite magnetization gives a series of connected positive anomalies (Figure 5a). Coincident with the edges of the body are a negative anomaly to the southwest and a positive anomaly to the northeast. This pattern corresponds to that seen (Figure 2, profile V) over the part of Fowler Peninsula where bedrock lies above sea level (Figure 3). The edges of such a block of magnetic material may be marked by the negative anomaly over Carlson Inlet and the positive anomaly over Evans Ice Stream (Figure 2).

We suggest that the shallow continuous magnetic basement beneath Fowler Peninsula is composed of metamorphic rocks similar to those exposed at Haag Nunataks [Clarkson and Brook, 1977; Storey and Dalziel, this volume]. Three samples of granitic orthogneiss from this locality showed volume susceptibilities in the range $11-20 \times 10^{-3}$ SI (amperes per metre) which give an intensity of magnetization of 1 A m^{-1} in an ambient field of 50,000 nT. Six samples of amphibolitic gneiss had susceptibilities $0.2-1.0 \times 10^{-3}$ SI, and three samples from a recrystallized intermediate dike showed susceptibilities of up to 50×10^{-3} SI. These compositional variations within the basement could be partly responsible for the magnetic anomalies. Remanent intensities of 0.1 (A m^{-1}) were measured by A. M. Grunow (personal communication, 1985) on rocks from Haag Nunataks, but as the direction of magnetization was random and unstable it is difficult to estimate the importance of this contribution to the total source magnetizations.

The part of profile IV (Figure 2) crossing promontory A (Figure 3) and Fowler Peninsula was modeled by using a nonlinear-optimization technique in a computer program written by G. K. Westbrook and supplied by the Department of Geological Sciences, University of Durham, England. The upper surface of the magnetic body was constrained at the bedrock surface beneath central Fowler Peninsula but allowed to vary beneath promontory A and Evans Ice Stream. The magnetic body was divided into vertical sections, and the magnetization within each was allowed to vary within 40° of the direction of the earth's present field. An acceptable fit between observed and calculated magnetic anomalies (Figure 6) was obtained with intensities of magnetization between 1.1 and 4.0 A m^{-1}. These values fall at the upper limit of the volume susceptibilities of granitic gneiss exposed at Haag Nunataks. The computed magnetization values in the model would be reduced if the magnetic body extended to a depth greater than 8 km. Alternatively, remanence could be influential in increasing source magnetizations.

Some of the short-wavelength magnetic anomalies observed over central Fowler Peninsula coincide with short-wavelength irregularities in the bedrock surface (Figure 4). However, the interpretation showed that the majority of these anomalies over central Fowler Peninsula must be caused by compositional variations within the basement which may themselves also have been responsible for the bedrock morphology.

A striking feature of the model (Figure 6) is that the basement surface is vertically displaced by several kilometers between Fowler Peninsula and promontory A with magnetic basement elevations of about -6 km beneath the promontory. Magnetic anomalies of similar amplitude and wavelength (Figure 2) suggest that the magnetic metamorphic basement may occur at comparable depths beneath Rutford Ice Stream, Fletcher Promontory, Carlson Inlet, and southeastern Fowler Peninsula. The linear magnetic anomaly along southwestern Evans Ice Stream (Figure 2) appears (Figure 6) to be due to a fault marking the northeasterly limit of sub-

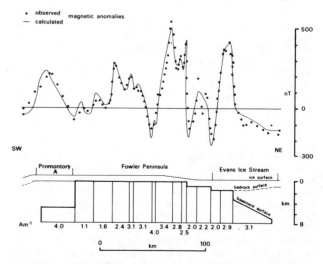

Fig. 6. Detailed interpretation of magnetic anomalies over Fowler Peninsula (profile IV, Figure 2). Body magnetizations vary between 1.1 and 4.0 A m^{-1}. Body magnetization vectors have an inclination of -30° to -76°. The dip of the earth's field in the plane of the profile is -71°.

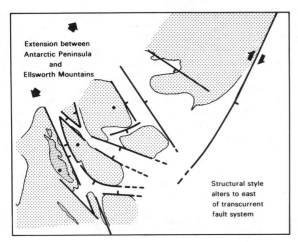

Fig. 7. Proposed development of crustal extension. Faults are indicated by bedrock topography and magnetic anomalies, and a sense of throw is determined by changes in magnetic anomaly wavelength and the depth of magnetic source. Topographically elevated blocks are stippled; the highest point of a block is indicated by a large dot.

glacial outcrop of the basement. This fault lies beneath central Evans Ice Stream and displaces the basement by over 2 km. Long-wavelength magnetic anomalies observed in the southernmost Antarctic Peninsula [Behrendt, 1964] (Figure 2) suggest that magnetic basement may also be present at several kilometers depth to the north of this fault (Figure 6).

A simple model (Figure 5b) consisting of four discrete bodies each of 10-km width gives a series of short-wavelength positive anomalies superimposed on an apparent long-wavelength negative anomaly. This pattern corresponds to that observed (e.g., Figure 2, profile II) over the part of the Antarctic Peninsula where bedrock elevation exceeds sea level. We attribute this magnetic pattern to discrete plutonic bodies intruded into nonmagnetic sedimentary rocks. This interpretation concurs with the exposed geology [Rowley et al., 1983].

Discussion

The sinuous topographic scarp trending northwest-southeast at 76°S (Figure 3) may mark the southern limit of the Antarctic Peninsula crustal block. It is marked by an abrupt northward decrease in Bouguer anomalies [Behrendt, 1964] and by an increase in magnetic activity (Figure 2). The magnetic anomalies are associated with igneous bodies of the Mesozoic-Cenozoic magmatic arc which may be traced northward along the length of the Antarctic Peninsula together with the elevated bedrock plateau and associated Bouguer anomaly minimum [Renner et al., 1985].

The small-amplitude magnetic anomalies observed to the southwest of the Ellsworth Mountains (Figure 2) [Jankowski and Drewry, 1981] would be produced if the crystalline basement underlying the thick Paleozoic sediments [Craddock, 1969] was either at great depth (>20 km) and not disturbed by major faulting or was less magnetic than the metamorphic rocks exposed at Haag Nunataks. Whatever the cause, there is a strong magnetic discontinuity at the southwestern edge of Rutford Ice Stream where the flat magnetic signature to the southwest yields to the active magnetic signature to the northeast. This may define the northeasterly limit of the Ellsworth-Whitmore Mountains crustal block [Dalziel et al., this volume].

The bedrock topography (Figure 3) suggests a pattern of horst and graben structures formed by northeast to southwest extension between the Antarctic Peninsula and Ellsworth-Whitmore Mountains crustal blocks [Doake et al., 1983]. Faults bounding the horsts are indicated by the steep topographic scarps and by the magnetic fabric (Figure 6). A simple geometrical fit of elevated blocks (Figure 7) suggests a direction of extension of 30°. This extension may have resulted from the northward motion of the Antarctic Peninsula along a dextral transcurrent fault during Gondwana breakup (Figure 7) [Barker and Griffiths, 1977]. The truncation of the magnetic and topographic expression of the rift system to the southeast of Fowler Peninsula is indirect evidence for the existence of such a fault. Closure of the rift system during Gondwana reconstruction places the Ellsworth-Whitmore Mountains block closer to the Antarctic Peninsula, yet the blocks still maintain their present relative orientations. Such a restoration is compatible with new paleomagnetic data [Grunow et al., this volume] and helps to reduce problems of overlap between the Antarctic Peninsula and South America encountered in many Mesozoic reconstructions of the supercontinent [e.g., Norton and Sclater, 1979].

Satellite magnetic anomalies reveal a regional magnetic low over this area [Ritzwoller and Bentley, 1983] which may reflect an elevated Curie isotherm associated with high heat flow in a region of thin, extended lithosphere. We suggest that the Ellsworth Mountains uplift (Figure 4) formed in response to lateral heat conduction [Turcotte, 1983; Karner and Weissel, 1984] or small-scale mantle convection [Keen, 1985] on the flank of the rift system between the Antarctic Peninsula and Ellsworth-Whitmore Mountains blocks. The timing of extension is uncertain, but the dramatic topography and exposures of Cenozoic alkali basalts within the region [O'Neill and Thomson, 1986] suggest that much of the movement may have occurred during the Cenozoic.

Summary and Conclusions

The interpretation of airborne geophysical data gathered during the 1983-1984 season is still in progress and further surveys are planned. Nevertheless, several significant features of the crustal structure of this area are already apparent. High-frequency high-amplitude magnetic anomalies and irregular bedrock morphology beneath central Fowler Peninsula indicates that the Precambrian metamorphic rocks exposed at Haag Nunataks may occur beneath the ice cover throughout this topographically elevated block. Longer wavelength anomalies with amplitudes in excess of 300 nT suggest the continued presence of this basement at several kilometers depth between Rutford Ice Stream and Evans Ice Stream. The geophysical signature of this area contrasts with the featureless magnetic field to the west of the Ellsworth Mountains and with the high-amplitude anomalies corresponding to discrete plutonic bodies within the Antarctic Peninsula magmatic arc. Extension between the Antarctic Peninsula and Ellsworth-Whitmore Mountains crustal blocks has resulted in horst and graben structures showing large displacements of the magnetic basement. This is illustrated by models suggesting basement elevations of more than 0 km in central Fowler Peninsula, about −5 km beneath promontory A and about −10 km beneath the graben occupied by Carlson Inlet. Much of this rifting may have occurred during the Cenozoic. Closure of this rift system during reconstruction of Gondwana reduces the overlap between South America and the Antarctic Peninsula, which is a common feature of Mesozoic assemblies of the supercontinent.

Acknowledgments. Thanks are given to D. Damaske, M. P. Maslanyj, and H. E. Thompson for their assistance in data collection during the 1983-1984 field season. We are also grateful to the air crew of R. Hasler and A. B. Carter. This work was partially supported by NSF grant DPP 82-13798 to I. W. D. Dalziel.

References

Barker, P. F., and D. H. Griffiths, Toward a more certain reconstruction of Gondwanaland, Philos. Trans. R. Soc. London, Ser. B, 279, 143-159, 1977.

Barraclough, D. R., The 1980 International Geomagnetic Reference Field, Nature, 294, 14-15, 1981.

Behrendt, J. C., The crustal geology of Ellsworth Land and southern Antarctic Peninsula from gravity and magnetic anomalies, J. Geophys. Res., 69, 2047-2063, 1964.

Clarkson, P. D., and M. Brook, Age and position of the Ellsworth Mountains crustal fragment, Antarctica, Nature, 265, 615-616, 1977.

Craddock, C., Geology of the Ellsworth Mountains, Antarct. Map Folio Ser., edited by V. C. Bushnell and C. Craddock, folio 12, plate IV, Am. Geogr. Soc., New York, 1969.

Dalziel, I. W. D., and D. H. Elliot, West Antarctica: Problem child of Gondwanaland, Tectonics, 1, 3-19, 1982.

Dalziel, I. W. D., S. W. Garrett, A. M. Grunow, R. J. Pankhurst, B. C. Storey, and W. R. Vennum, The Ellsworth-Whitmore Mountains crustal block: Its role in the tectonic evolution of West Antarctica, this volume.

Doake, C. S. M., R. D. Crabtree, and I. W. D. Dalziel, Subglacial morphology between Ellsworth Mountains and Antarctic Peninsula: New data and tectonic significance, in Antarctic Earth Science, edited by R. L. Oliver, P. R. James, and J. B. Jago, pp. 270-273, Australian Academy of Science, Canberra, 1983.

Grunow, A. M., I. W. D. Dalziel, and D. V. Kent, Ellsworth-Whitmore Mountains crustal block, western Antarctica: New paleomagnetic results and their tectonic significance, this volume.

Hartman, R. R., D. J. Tesky, and J. L. Friedberg, A system of rapid digital aeromagnetic interpretation, Geophysics, 36, 891-918, 1971.

Herrod, L. D. B., and S. W. Garrett, Geophysical fieldwork on the Ronne Ice Shelf, Antarctica, First Break, 4, 9-14, 1986.

Jankowski, E. J., Airborne geophysical investigations of subglacial structure of West Antarctica, Ph.D. thesis, 293 pp., University of Cambridge, Cambridge, 1981.

Jankowski, E. J., and D. J. Drewry, The structure of West Antarctica from geophysical studies, Nature, 291, 17-21, 1981.

Karner, G. D., and J. K. Weissel, Thermally induced uplift and lithospheric flexural readjustment of the eastern Australian Highlands (abstract), Geol. Soc. Aust. Abstr., 12, 293-294, 1984.

Keen, C. E., The dynamics of rifting: Deformation of the lithosphere by active and passive driving forces, Geophys. J. R. Astron. Soc., 80, 95-120, 1985.

Lee, M. K., Two dimensional gravity and magnetic interpretation (version 2), program GAM2D, Computer Program Rep. Ser., 32, 30 pp., Applied Geophysics Unit, British Geological Survey, Nottingham, 1979.

Maslanyj, M. P., and D. Damaske, Lessons regarding aeromagnetic surveying during magnetic disturbances in polar regions, Br. Antarct. Surv. Bull., in press, 1986.

Norton, I. O., and J. G. Sclater, A model for the evolution of the Indian Ocean and the break-up of Gondwanaland, J. Geophys. Res., 84, 6803-6830, 1979.

O'Neill, J. M., and J. W. Thomson, Tertiary mafic volcanic and volcaniclastic rocks of the English Coast, Antarctica, Antarctic J. U. S., in press, 1986.

Renner, R. G. B., L. J. S. Sturgeon, and S. W. Garrett, Reconnaissance gravity and aeromagnetic surveys of the Antarctic Peninsula, Br. Antarct. Surv. Sci. Rep., 110, 50 pp., 1985.

Ritzwoller, M. H., and C. R. Bentley, Magnetic anomalies over Antarctica measured from MAGSAT, in Antarctic Earth Science, edited by R. L. Oliver, P. R. James, and J. B. Jago, pp. 504-507, Australian Academy of Science, Canberra, 1983.

Rowley, P. D., W. R. Vennum, K. S. Kellog, R. S. Laudon, P. E. Carrara, J. M. Boyles, and M. R. A. Thomson, Geology and plate tectonic setting of the Orville Coast and eastern Ellsworth Land, Antarctica, in Antarctic Earth Science, edited by R. L. Oliver, P. R. James, and J. B. Jago, pp. 245-250, Australian Academy of Science, Canberra, 1983.

Storey, B. C., and I. W. D. Dalziel, Outline of the structural and tectonic history of the Ellsworth Mountains-Thiel Mountains ridge, West Antarctica, this volume.

Swithinbank, C. W. M., Glaciological research in the Antarctic Peninsula, Philos. Trans. R. Soc. London, Ser. B, 279, 161-183, 1977.

Turcotte, D. L., Mechanisms of crustal deformation, J. Geol. Soc. London, 140, 701-724, 1983.

Copyright 1987 by the American Geophysical Union.

OUTLINE OF THE STRUCTURAL AND TECTONIC HISTORY OF THE ELLSWORTH MOUNTAINS-
THIEL MOUNTAINS RIDGE, WEST ANTARCTICA

B. C. Storey

British Antarctic Survey, Natural Environment Research Council, High Cross
Cambridge, CB3 OET, United Kingdom

I. W. D. Dalziel[1]

Lamont-Doherty Geological Observatory of Columbia University, Palisades, New York 10964

Abstract. The Ellsworth Mountains-Thiel Mountains ridge and adjoining areas are divided into three tectonic provinces: (1) Haag Nunataks, (2) Thiel Mountains, and (3) Ellsworth-Whitmore Mountains crustal block. Haag Nunataks are part of a Precambrian tectonic province the overall extent of which is not clearly known. The Thiel Mountains are part of a distinctive Transantarctic Mountains province that is separated by a major tectonic break from deformed sedimentary rocks of the Ellsworth-Whitmore Mountains crustal block. The crustal block is divided, on the basis of a detailed structural analysis, into two domains: the Ellsworth and Marginal domains. The sedimentary rocks throughout the Ellsworth domain are correlated with parts of the Paleozoic succession forming the Ellsworth Mountains themselves. These rocks were all deformed by a single phase of northwest-southeast trending structures, whereas in the Marginal domain the fold history is more complex and structures trend northeast-southwest. The tectonic significance of the Marginal domain is discussed but is not clearly understood. Mount Woollard has a unique lithological association with the Ellsworth-Whitmore Mountains crustal block; it consists of paragneiss and pegmatite of possible Middle Jurassic age and has a structural trend parallel to the Ellsworth domain structures.

Introduction

The Ellsworth Mountains-Thiel Mountains ridge is the name given [Craddock, 1983] to the prominent sub-ice topographic high at the head of the Weddell Sea extending across West (Lesser) Antarctica between the Antarctic Peninsula and the Transantarctic Mountains (Figure 1). Between the Ellsworth Mountains in the north and the Thiel Mountains in the south, there are several scattered groups of nunataks, namely the Whitmore Mountains, the Hart, Martin, Nash, Pirrit, and Stewart Hills, mounts Johns, Moore, and Woollard, and Pagano Nunatak (Figure 2). The Thiel Mountains are composed of a basement of upper Precambrian to lower Paleozoic porphyries, volcaniclastic sedimentary rocks, and granites overlain by sedimentary strata of the Beacon Supergroup intruded by mafic igneous rocks of the Ferrar Supergroup [Schmidt and Ford, 1969]. The isolated Haag Nunataks consist of gneiss and minor intrusives yielding Precambrian radiometric ages [Clarkson and Brook, 1977]. These rocks may form the Precambrian basement to the deformed Paleozoic sedimentary succession of the spectacular Ellsworth Mountains (successions are greater than 10 km thick [Craddock et al., 1964; Webers and Sporli, 1983]). This succession has affinities with the Beacon Supergroup of the Transantarctic Mountains. This led Schopf [1969] to suggest that the Ellsworth Mountains are allochthonous and originated within the present Weddell Sea embayment between the Transantarctic Mountains and southern Africa. The scattered nunataks emerging through the ice along the Ellsworth Mountains-Thiel Mountains ridge have seldom been seen, let alone studied geologically. Reconnaissance work, undertaken in the 1960s, revealed that they are composed mainly of low-grade metasedimentary strata intruded by several Mesozoic granitic plutons and locally by gabbro of unknown age [Thiel, 1961; Craddock, 1972, 1983; Webers et al., 1982, 1983]. Most of the country rocks were taken by these authors to be correlative with the Ellsworth Mountains succession. Some more highly deformed strata in the Stewart Hills were believed to represent an older continental basement. The boundary between East and West Antarctica has been drawn in the narrow gap between the Hart and Stewart Hills [Jankowski et al., 1983] and on the north side of the Hart Hills [Craddock, 1983]. The entire elevated area between the Thiel Mountains and the base of the Antarctic Peninsula has been incorporated within the "Ellsworth-Whitmore Mountains crustal block" of West Antarctica by Dalziel and Elliot [1982], Watts and Bramall [1981], and Grunow et al. [this volume] in

---

[1]Now at Institute for Geophysics, The University of Texas at Austin, Austin, Texas 78751.

117

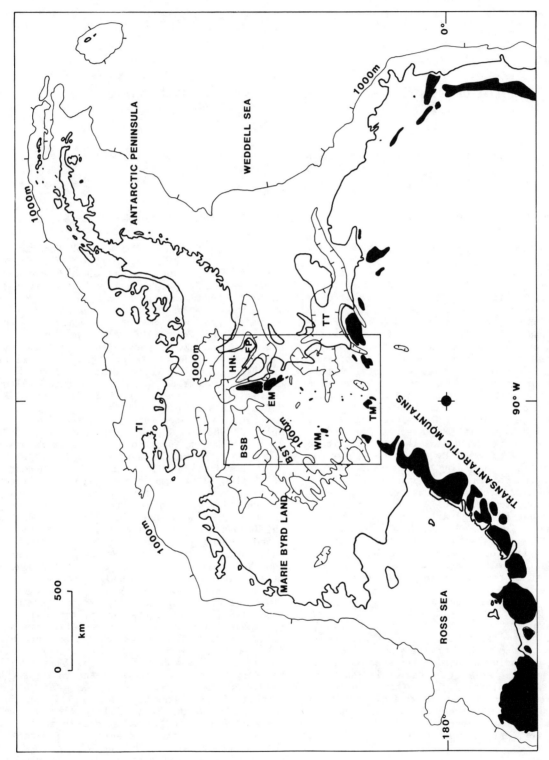

Fig. 1. Location map of study area within West Antarctica, with 1000-m bathymetric contour. TI is Thurston Island; EM, Ellsworth Mountains; TT, Thiel trough; WM, Whitmore Mountains; TM, Thiel Mountains; BSB, Byrd Subglacial Basin; BST, Bentley Subglacial Trench; HN, Haag Nunataks; and FP, Fowler Peninsula.

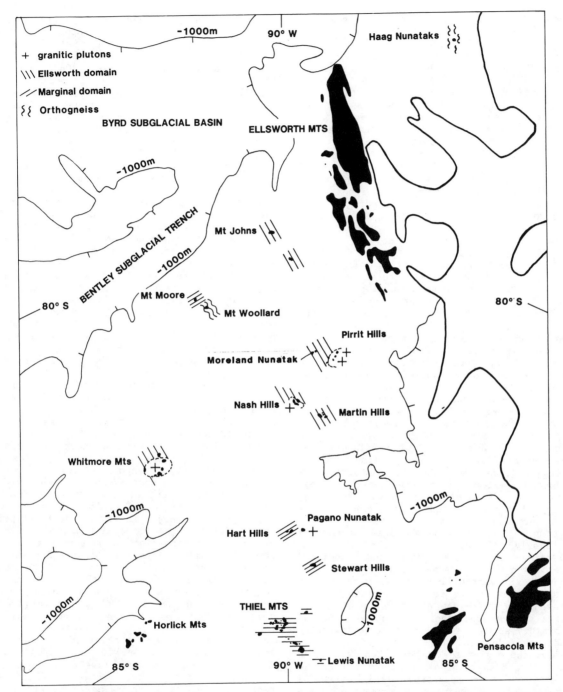

Fig. 2. Geological sketch map of the Ellsworth-Thiel Mountains ridge showing the main tectonic provinces and structural domains.

considering possible displacement of the Ellsworth Mountains within Gondwanaland.

As part of the Joint U.K.-U.S. West Antarctic Tectonics Project during the 1983-1984 field season, we studied the geology of all the groups of nunataks in the Ellsworth Mountains-Thiel Mountains ridge including Haag Nunataks. No detailed work was undertaken in the Ellsworth Mountains themselves as they were the subject of intensive study during the 1979-1980 season [Splettstoesser and Webers, 1980]. We have divided the area into three tectonic provinces: (1) Thiel Mountains, (2) Haag Nunataks, and (3) Ellsworth-Whitmore Mountains crustal block. As previously mentioned, the

Thiel Mountains form part of the Transantarctic Mountains. We note a clear distinction between the virtually undeformed upper Precambrian basement, lower Paleozoic granite rocks, Beacon Supergroup, and Ferrar diabase there and the strongly deformed metasedimentary country rock of the Hart and Stewart Hills, which are part of the Ellsworth-Whitmore Mountains block. The Haag Nunataks had not been previously studied by geologists. The samples originally dated as Precambrian by Clarkson and Brook [1977] were collected in the course of glaciological work by C. W. M. Swithinbank. They are unique in this region with regard to their structural style and geometry and their metamorphic and igneous history. A Precambrian age for the gneiss has been confirmed by Millar and Pankhurst [this volume] using the Rb-Sr method.

On the basis of a structural analysis, we assign the nunataks between the Ellsworth Mountains and the Thiel Mountains to domains as follows: (1) Ellsworth domain (i.e., the structural domain of the Ellsworth Mountains)-Mount Johns, the Martin, Nash, and Pirrit Hills, Whitmore Mountains, and Pagano Nunatak; (2) Marginal domain-Mount Moore and the Stewart and Hart Hills; and (3) Mount Woollard. The rocks forming the nunataks of the Ellsworth domain consist predominantly of shallow marine sedimentary strata intruded by granitic plutons. The latter are exposed in the Nash and Pirrit Hills, the Whitmore Mountains, and Pagano Nunatak. At Pagano Nunatak, the country rock is not exposed and may in fact consist of rocks of the adjoining Marginal domain. Conversely, none of the granitic rocks are exposed in the Ellsworth Mountains. The country rocks of the Ellsworth domain are deformed by geometrically simple northwest trending folds with subhorizontal hinge lines and a weak to moderately developed axial planar spaced or slaty cleavage. Together with the strata of the Ellsworth Mountains, they form a structurally homogeneous domain.

The country rocks of the Marginal domain are more intensely deformed than those of the Ellsworth domain. The dominant folds and associated cleavage of the sandstone-shale sequences at Mount Moore and the Stewart Hills trend at right angles to the structures in the Ellsworth domain. While the metasedimentary strata of the Hart Hills are geometrically complex, their polyphase deformational history appears to relate them to the rocks of the Marginal domain.

Finally, the migmatites of Mount Woollard could be part of the Ellsworth domain structurally. We treat them here as a separate domain because of their distinctive high-grade metamorphism. Although the age of these rocks is uncertain, it is unlikely that they are as old as Proterozoic or that they form a basement to the Paleozoic sedimentary rocks [Millar and Pankhurst, this volume]. Their results suggest that some of the pegmatites may be of Middle Jurassic age.

Thiel Mountains

Within the Thiel Mountains (Figure 2), which form part of the Transantarctic Mountains, late Precambrian quartz monzonite porphyries are interbanded with a sequence of contemporaneous volcaniclastic sedimentary rocks up to 100 m thick and intruded by a discordant Cambrian-Ordovician alkali-granite [Ford and Aaron, 1962; Ford, 1964]. The porphyries are locally discordant with the flat-lying rocks and were most likely emplaced as sills. The sedimentary rocks are an interbedded sequence of thin- to very thick-bedded conglomerates, gray to green sandstones and shales with conspicuous carbonate horizons, pyroclastic breccias and lapilli, and crystal-lithic tuff units. They contain a variety of sedimentary structures including tabular and trough cross bedding, ripple cross laminations, mud cracks, normal and reverse grading, rip-up clasts, load clasts, and small slump structures. The sandstones are predominantly lithic and subarkosic arenites derived from reworked pyroclastic porphyry deposits. Quartz is the main detritus, but porphyritic lithic fragments and feldspar crystal fragments make up a considerable fraction of some samples. Some sandstones contain biotite and muscovite detrital fragments and conspicuous heavy mineral horizons of garnet, tourmaline, apatite, zircon, and opaque phases. The pyroclastic breccias and tuffs are petrographically similar to the interbanded porphyry. This provides good evidence that sedimentation was contemporaneous with the late Precambrian magmatic activity. Stromatolites have been recorded from the sequence [Craddock, 1985], and some carbonate horizons contain organic remains.

Although a complete facies analysis is not presented here, the sedimentary rocks may broadly be considered to be part of a shallow water marine sequence that was deposited close to a shoreline along a volcanically active margin. The interbanding of contemporaneous silicic volcanic and volcaniclastic sedimentary rocks is characteristic of the central part of the Transantarctic Mountains during the late Precambrian and Early Cambrian [Stump, 1983]. The Thiel Mountains porphyry and sedimentary rocks may be lateral equivalents of the Wyatt Formation, a uniformly massive porphyry generally believed to be late Precambrian in age, and the Ackerman Formation, a sequence of alternating clastic and volcanic rocks which conformably overlies the Wyatt Formation in the La Gorce Mountains [Stump, 1983]. Stump [1983] has suggested that the Ackerman Formation represents a return to marine conditions in the central Transantarctic Mountains in the latest Precambrian following the folding, uplift, and erosion of the Beardmore Group. Although the Ackerman Formation contains an incipient cleavage, the sedimentary rocks in the Thiel Mountains are flat lying and show no cleavage development; some large-scale open folds are present within the porphyry that is also cut by numerous microshear zones.

Lewis Nunatak

To the south of the Thiel Mountains, Lewis Nunatak is part of a massive dolerite sill of the Jurassic Ferrar Supergroup [Vennum and Storey, this volume; Millar and Pankhurst, this volume]. However, at the base of the sill, thermally

Fig. 3. Orthogneiss and concordant mafic enclaves and aplogranite sheet at Haag Nunataks.

altered flat-lying sedimentary rocks have recently been exposed above the snow and scree line. The rocks are interbedded black laminated shales and very thick poorly sorted greenish-gray sandstones which contain a small percentage of rounded granite gneiss, granitoid, and carbonate boulders up to 1 m in diameter. The sandstones are matrix-supported quartz and feldspar-rich wackes with small proportions of lithic clasts. The clasts also include folded mica-schists, composite quartz-feldspar, and intraformational shale and sandstone. Opaque phases, garnet, biotite, chlorite, muscovite, and zircon are common detrital phases. Some of the quartz crystal fragments are rounded and embayed and may be derived from the Thiel Mountains porphyry. As only a small sequence of hornfelsed sedimentary rocks are exposed beneath the sill, it is difficult to draw comparisons with neighboring areas. However, the petrography, flat-lying attitude, lithology, and association with mafic sills suggest that these sedimentary rocks could be part of the Devonian-Permian Beacon Supergroup characteristic of the Transantarctic Mountains. Furthermore, the presence of large boulders within the massive sandstones suggests that they may be of glacial origin comparable to part of the Beacon Supergroup.

Haag Nunataks

Haag Nunataks (Figure 2) are a group of small nunataks along the central spine of the Fowler Peninsula between the Ellsworth Mountains and the Antarctic Peninsula (Figure 1). They form part of a Precambrian (= 1000 Ma) basement crustal block [Clarkson and Brook, 1977; Millar and Pankhurst, this volume]. The main lithology is a granodiorite-orthogneiss with large concordant rafts and inclusions of biotite-hornblende mafic and ultramafic rocks (Figure 3). A strong fabric defined by aligned mafic minerals and felsic segregations dips gently to the east and west (Figure 4). A quartz mineral lineation plunges gently toward 110° and 290°. Second-phase shear zones and folds deform the S1 fabric. Extensive aplite and pegmatite sheets intrude the deformed gneisses and are in turn cut by gray microgranitoid sills. A small intrusion of intermediate composition, up to 10 m wide, cuts all the above lithologies and has a conspicuous chilled margin.

122 ELLSWORTH-THIEL MOUNTAINS RIDGE

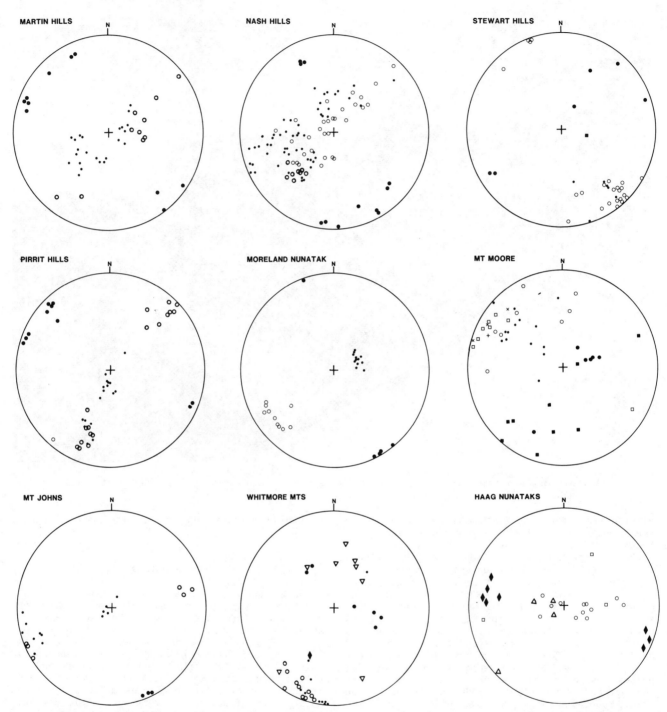

Fig. 4. Stereograms of structural data (Lambert equal-area projection, lower hemisphere plot).

Ellsworth-Whitmore Mountains Crustal Block

Ellsworth Domain

Within the central and northern part of the Thiel Mountains-Ellsworth Mountains ridge the Martin, Pirrit, and Nash Hills, Whitmore Mountains, and Mount Johns and associated nunataks are part of a structural domain within which the structural history and orientation of the main fabrics are similar to those of the Ellsworth Mountains (Figures 2 and 4). In the Nash and Pirrit Hills and

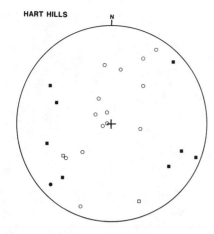

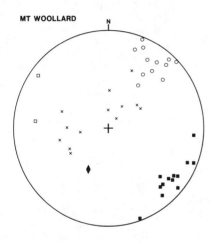

- • pole to bedding
- ○ pole to S1 cleavage
- ● hinge of F1 fold and/or S0/S1 intersection
- + pole to axial surface of F1 fold
- ▫ pole to S2 cleavage
- ▪ hinge of F2 fold
- × pole to axial surface of F2 fold
- ▽ pole to gneissic fabric in granite
- ♦ mineral lineation
- △ pole to granitic dykes and sills

Fig. 4. (continued)

Whitmore Mountains, deformed sedimentary rocks are preserved mainly as remnants of a contact aureole of large Early to Middle Jurassic granitic intrusions [Millar and Pankhurst, this volume], whereas in the Martin Hills, although the sedimentary rocks are thermally altered, only small porphyritic felsite plugs intrude the deformed strata.

The sedimentary rocks within the Martin, Nash, and Pirrit Hills are interbedded brown and green calcareous and noncalcareous sandstone, carbonate flat-pebble and mud-flake conglomerates, black shales, and gray limestones, some of which contain deformed oolites and pisoliths characteristic of a shallow marine and intertidal sequence; three large carbonate-clastic sequences are present in the Nash Hills (Figure 5). Sedimentary structures include large- and small-scale trough and tabular cross beds, ripple cross laminations, and load structures. At Moreland Nunatak, 20 km west of the Pirrit Hills, at Mount Johns, and at a series of small unnamed and previously undocumented nunataks (79°58'S, 89°37'W) between Mount Johns and the Ellsworth Mountains, the sedimentary rocks are interbedded parallel and cross-bedded red and green sandstones, shales, and mud-flake conglomerates. Ripple cross laminations and trough cross bedding are well preserved, and mud-flake conglomerates infill deep channels. Sedimentary rocks within the Whitmore Mountains are for the most part poorly preserved hornfelsed rafts of phyllite, calcareous sandstone, and quartzite. However, in the nunataks on the northern side of the pluton there are interbedded green sandstone-shale units with well-developed graded bedding and a range of sole structures characteristic of turbidity currents.

Although there is some facies variation in this domain, the sedimentary rocks contain the same clast population and were most likely derived from the same source. The sandstones predominantly contain detrital quartz and feldspar crystal fragments with some composite grains derived from a granitic or granite-gneiss terrane. Deformed muscovite and biotite crystals are common, and apatite, zircon, tourmaline, garnet, and opaque phases are common accessories. Some felsite volcanic fragments have also been identified.

The structural history within the domain is relatively simple, and the data are summarized in Figure 4. The sedimentary rocks are folded by one main fold phase, and a prominent slaty and pressure-solution cleavage is moderately to well developed in some lithologies. The fold hinges and cleavage consistently trend within the northwest to southeast quadrant, although there is some variation (see Figure 4). Folds and bedding/cleavage intersection lineations plunge gently toward both the northwest and southeast, although in the Whitmore Mountains the intersection lineation plunges moderately to steeply toward the east-southeast.

Within the Nash Hills the structural history is more complex, the bedding is downward facing on the cleavage; the cleavage is not related to the main phase macroscopic folds and can clearly be seen to truncate the fold limbs. The tectonic significance of this is not understood, but it is possible that some of the folds may be slump structures. A second crenulation cleavage and a series of veins infilled by quartz and magnetite postdate the hornfels minerals in the Nash Hills and may be related to granite emplacement.

In the Martin Hills a large-scale dislocation zone dipping gently toward the southeast occurs on

Fig. 5. The contact zone of a posttectonic Middle Jurassic granite (left) and deformed sedimentary rocks (right) of the Nash Hills. The sedimentary rocks show conspicuous pale-colored carbonate horizons.

the eastern side. Large-scale recumbent folds are developed above and below the dislocation plane, and bedding has been severely disrupted. Faulting and fold closures are common, as are low-angle normal and reverse faults on the upper and lower fold limbs, respectively. Fold vergence directions indicate that the upper zone has moved southeastward relative to the lower zone.

In the Whitmore Mountains, in contrast to the Nash and Pirrit Hills, the margins of the granite are deformed, and feldspar porphyroblasts in the aureole form augen within the sheared margin. The mafic phases are also aligned, and parallel shear zones occur within the pluton, indicating synemplacement or postemplacement fabric. In the Nash and Pirrit Hills the metamorphic aureole minerals overgrow the S1 fabric. The metasedimentary rocks close to the contact contain a medium-grade hornfels assemblage: biotite + muscovite ± cordierite ± andalusite ± clinopyroxene ± staurolite ± opaque phases. Most of the hornfels porphyroblasts are replaced by muscovite-biotite-chlorite-opaque aggregates.

Marginal Domain

In marked contrast to the Thiel Mountains, which lie less than 100 km to the south, the Stewart and Hart Hills are highly deformed metasedimentary rocks. Together with Mount Moore (Figure 2) they form a domain extending from the northwestern side of the Ellsworth-Whitmore Mountains block that is structurally more complex than the Ellsworth domain and within which the main folds and cleavage planes trend toward the northeast-southwest, normal to the trend of the Ellsworth domain (Figure 4). Fourteen kilometers northeast of the Hart Hills, a Middle Jurassic granite forms Pagano Nunatak [Webers et al., 1983; Millar and Pankhurst, this volume]. This occurrence indicates that part of this domain may be intruded by the same Middle Jurassic magmatic suite as the Ellsworth domain.

The Stewart Hills mainly consist of green and black phyllites with interbedded massive and graded green sandstone beds and interlaminated sandstone-shale units. The sandstones are subarkosic and quartz arenites with subrounded to rounded detrital fragments derived from a similar provenance to the sedimentary rocks of the Ellsworth domain. The sedimentary rocks are folded into tight asymmetric upright northeast trending, gently plunging mesoscopic folds. A slaty cleavage dipping steeply toward the northwest and southeast and defined by a muscovite and chlorite fabric is well developed in the fine-grained lithologies; a spaced anastomosing pressure-solution cleavage is present in the sandstones. A second crenulation cleavage and associated minor folds deform the main fabric.

The Hart Hills are formed of interbedded massive green and gray subarkosic arenites and greenish-gray and black quartz-mica phyllites with some carbonate-rich horizons and sheared intrusive quartz-gabbro sills [Webers et al., 1983]. The structure is more complex and less clearly understood than that of the Stewart Hills. It is dominated by a flat-lying muscovite-chlorite slaty cleavage, axial planar to reclined subisoclinal folds plunging at low to moderate angles. Later crenulation cleavages within the southwest and southeast quadrants and associated small-scale chevron folds deform the main phase structures. The structural inhomogeneity may be due in part to the large quartz-gabbro sills. Craddock [1983] inferred that the sills were emplaced before the deformation, but petrofabric and field relations clearly demonstrate that the aureole minerals (?andalusite), now replaced by calcite and opaque phases, postdate the S1 cleavage and are deformed by the D2 crenulations.

The deformed metasedimentary rocks at Mount Moore are thick interbedded green and greenish-brown calcareous and noncalcareous sandstones and gray to green slates. They show considerable tectonic disruption of the strata and are folded by tight main-phase folds which plunge steeply toward the northeast, with northeast-southwest trending axial planar slaty cleavage. A second-phase crenulation cleavage is well developed, and associated minor folds plunge moderately to steeply toward the south-southwest.

Mount Woollard

Mount Woollard, situated less than 25 km southeast of Mount Moore (Figure 2), forms the most distinctive lithological association in the Ellsworth-Whitmore Mountains crustal block. In marked contrast to the deformed metasedimentary rocks of the Ellsworth and Marginal domains, Mount Woollard consists of deformed garnet-andesine-biotite paragneiss, orthopyroxene-clinopyroxene-biotite-hornblende amphibolite, and massive garnet-feldspar-quartz pegmatites. There is a complex history of pegmatite injection, and pegmatites form up to 60% of the total exposure. A metamorphic layering folded by tight F1 folds in the paragneiss is defined by variation in the modal proportions of biotite and amphibole and by thin concordant pegmatites. Garnet-bearing pegmatites (MP1), up to 3 m wide, are deformed by a prominent second fold phase (Figure 6) which plunges gently toward the southeast and has a variably inclined axial planar fabric. A further injection of massive pegmatites (MP2) cuts the F2 folds, and these are in turn cut by occasional micro-granite dikes or sills. Although the paragneiss and pegmatite complex at Mount Woollard are unique in this crustal block, the structural trend of the first- and second-phase structures is similar to the main structural trend within the Ellsworth domain but markedly different from that in Mount Moore, the neighboring nunatak.

Preliminary isotopic work [Millar and Pankhurst, this volume] suggests that the pegmatites and microgranites are derived from the Ellsworth Mountains succession, and that Mount Woollard is not part of a crystalline basement to these sedimentary rocks. The complex may be a deep migmatitic zone associated with the Middle Jurassic granitic suite. However, as the pegmatites are syntectonic, deformation within the Ellsworth domain may have overlapped with the time of formation of the granitic suite. The occurrence of these rocks of medium to high metamorphic grade within an area of low-grade sedimentary rocks may be due to uplift of a horst block along the edge of the Bentley Subglacial Trench, one of a series of large extensional features within West Antarctica (for a review, see Dalziel et al. [1986]). The uplift and erosion of this block may have exposed rocks of higher metamorphic grade within the Ellsworth domain.

Tectonic Synthesis

On the basis of a regional structural analysis the Ellsworth-Thiel Mountains ridge has been divided into three tectonic provinces. Haag Nunataks are the oldest known rocks in the area and are part of a Precambrian tectonic province of West Antarctica that occurs between the base of the Antarctic Peninsula and the Ellsworth Mountains [Garrett et al., this volume]; the overall extent of this province is not presently known. The Thiel Mountains are part of a distinct Transantarctic Mountains province which lay close to a volcanically active margin during the late Precambrian and was overlain by Paleozoic platform sediments; it has been subjected to little or no tectonic activity prior to uplift in the Mesozoic or Cenozoic. In marked contrast to this province, the Stewart Hills, less than 75 km to the north, are on the southern edge of a large area of deformed sedimentary rocks of the Ellsworth-Whitmore Mountains crustal block. A major tectonic boundary separates the deformed sedimentary rocks of the block from the undeformed rocks of the Thiel Mountains. This may be viewed as the boundary between East and West Antarctica that separates the stable cratonic area of East Antarctica from the Phanerozoic mobile belts of West Antarctica. Although the age of the deformed sedimentary rocks of the Ellsworth-Whitmore Mountains block is not well constrained, they are considered to be lithologic correlatives of part of the Middle to Late Cambrian Heritage Group or the carbonates of the Minaret Formation, the basal part of the thick

Fig. 6. A large pegmatite folded by an F2 fold at Mount Woollard.

Ellsworth Mountains succession [Webers and Sporli, 1983], and may indicate widespread shallow-marine conditions during the early Paleozoic. Mount Johns has previously been correlated with the Devonian Crashsite quartzite [Craddock, 1983], and our observations would support this.

Two structural domains, the Ellsworth and Marginal domains, have been recognized within the Ellsworth-Whitmore Mountains crustal block. Within the Ellsworth domain the sedimentary rocks were deformed by a single phase of northwest-southeast trending structures, whereas in the Marginal domain the fold history is more complex and trends northeast-southwest. It is possible that these structural domains may represent two different orogenic episodes; Craddock [1983] previously correlated the Stewart Hills, included here in the Marginal domain, with the Patuxent Formation of the Pensacola Mountains and considered them to be part of an early Paleozoic Ross orogenic belt. If that is correct, the lack of superimposition of the early Mesozoic Ellsworth domain folds on the marginal domain trends and the geographical location of both Mount Moore and the Stewart Hills on the edge of the crustal block are curious; Mount Moore occurs on the edge of the Bentley Subglacial Trench and the Stewart Hills close to the major tectonic boundary highlighted above. It is possible that these represent structural trends which may have been rotated during movement of the Ellsworth-Whitmore Mountains block or by strike-slip motion during formation of the bordering extensional rift systems. Alternatively, the variation in structural trend may represent a primary variation within the early Mesozoic Ellsworth Orogen. Abrupt changes in structural style and trend are relatively common within other early Mesozoic fold belts that have been correlated with the Ellsworth fold belt; for example, in the Falkland Islands an east-west trend and a northeast-southwest trend both occur [Greenway, 1972], and in the Cape Fold Belt a north-south trend is formed in a western province and an east-west trend in a southern province [Söhnge and Hälbich, 1983]. These variations may be associated with zones of higher strain or may be controlled by preexisting basement structures and/or the structural control of basin development.

The data presented in this paper emphasize the geological differences between East and West Antarctica and contrast the different geological histories of two distinctive crustal blocks in West Antarctica. It is hoped that further work will help to resolve the tectonic significance of the structural trends within one of these blocks - the Ellsworth-Whitmore Mountains crustal block.

Acknowledgments. This work is part of the joint BAS-USARP program investigating the tectonic history of West Antarctica. We are very grateful to our colleagues (A. M. Grunow, R. J. Pankhurst, and W. Vennum) on this project for much useful discussion and to the BAS air unit, VXE6 of the U.S. Navy, and our field assistants for their support during the field program. Support for the U.S. side of the project was supplied by the Division of Polar Programs, National Science Foundation, through grant DPP 82-13798 to I.W.D.D. We are grateful to Campbell Craddock for supplying us with unpublished sketch maps of the major nunataks.

References

Clarkson, P. D., and M. Brook, Age and position of the Ellsworth Mountains crustal fragment, Antarctica, Nature, 265, 615-616, 1977.

Craddock C. (compiler), Geologic map of Antarctica, 1:500,000, American Geographical Society, New York, 1972.

Craddock, C., The East Antarctica-West Antarctica boundary between the ice shelves: A review, in Antarctic Earth Science, edited by R. L. Oliver, P. R. James, and J. B. Jago, pp. 367- 371, Australian Academy of Science, Canberra, 1983.

Craddock, C., Proterozoic stromatolites from the Thiel Mountains, Antarctica, in Abstracts, Sixth Gondwana Symposium, IPS Misc. Publ. 231, p. 24, Ohio State University, Columbus, 1985.

Craddock, C., J. J. Anderson, and G. F. Webers, Geologic outline of the Ellsworth Mountains, in Antarctic Geology, edited by R. J. Adie, pp. 155-170, North-Holland, Amsterdam, 1964.

Dalziel, I. W. D., and D. H. Elliot, West Antarctica: Problem child of Gondwanaland, Tectonics, 1, 3-19, 1982.

Dalziel, I. W. D., B. C. Storey, S. W. Garrett, A. M. Grunow, L. D. B. Herrod, and R. J. Pankhurst, Extensional tectonics and the fragmentation of Gondwanaland, in Extensional Tectonics, Spec. Publ., edited by J. F. Dewey, M. P. Coward, and P. Hancock, in press, Geological Society of London, 1986.

Ford, A. B., Cordierite-bearing, hypersthene-quartz-monzonite porphyry in the Thiel Mountains and its regional importance, in Antarctic Geology, edited by R. J. Adie, pp. 429-441, North-Holland, Amsterdam, 1964.

Ford, A. B., and J. M. Aaron, Bedrock geology of the Thiel Mountains, Antarctica, Science, 137, 751-752, 1962.

Garrett, S. W., L. D. B. Herrod, and D. R. Mantripp, Crustal structure of the area around Haag Nunataks, West Antarctica: New aeromagnetic and bedrock elevation data, this volume.

Greenway, M. E., The geology of the Falkland Islands, Sci. Rep. Br. Antarct. Surv., 76, 42 pp., 1972.

Grunow, A. M., I. W. D. Dalziel, and D. V. Kent, Ellsworth-Whitmore Mountains crustal block, western Antarctica: New paleomagnetic results and their tectonic significance, this volume.

Jankowski, E. J., D. J. Drewry, and J. C. Behrendt, Magnetic studies of upper crustal structure in West Antarctica and the boundary with East Antarctica, in Antarctic Earth Science, edited by R. L. Oliver, P. R. James, and J. B. Jago, pp. 197-203, Australian Academy of Science, Canberra, 1983.

Millar, I., and R. J. Pankhurst, Rb-Sr geochronology of the region between the Antarctic Peninsula and the Transantarctic Mountains: Haag Nunataks and Mesozoic granitoids, this volume.

Schmidt, D. L., and A. B. Ford, Geology of the

Pensacola and Thiel Mountains, in *Geologic Maps of Antarctica*, *Antarct. Map Folio Ser.*, Folio 12, plate 5, edited by V. C. Bushnell and C. Craddock, American Geological Society, New York, 1969.

Schopf, J. M., Ellsworth Mountains: Position in West Antarctica due to sea-floor spreading, *Science*, *164*, 63-66, 1969.

Söhnge, A. P. G., and I. W. Hälbich, Geodynamics of the Cape Fold Belt, *Spec. Publ. Geol. Soc. S. Afr.*, *12*, 184 pp., 1983

Splettstoesser, J., and G. F. Webers, Geological investigations and logistics in the Ellsworth Mountains, 1979-1980, *Antarct. J. U. S.*, *15*, 36-39, 1980.

Stump, E., Type locality of the Ackerman Formation, La Gorce Mountains, Antarctica, in *Antarctic Earth Science*, edited by R. L. Oliver, P. R. James, and J. B. Jago, pp. 170-174, Australian Academy of Science, Canberra, 1983.

Thiel, E. C., Antarctica, one continent or two?, *Polar Rec.*, *10*, 335-348, 1961.

Vennum, W. R., and B. C. Storey, Correlation of gabbroic and diabasic rocks from the Ellsworth Mountains, Hart Hills, and Thiel Mountains, West Antarctica, this volume.

Watts, D. R., and A. M. Bramall, Paleomagnetic evidence for a displaced terrain in western Antarctica, *Nature*, *293*, 638-642, 1981.

Webers, G. F., and K. B. Sporli, Paleontological and stratigraphic investigations in the Ellsworth Mountains, West Antarctica, in *Antarctic Earth Science*, edited by R. L. Oliver, P. R. James, and J. B. Jago, pp. 261-265, Australian Academy of Science, Canberra, 1983.

Webers, G. F., C. Craddock, M. A. Rogers, and J. J. Anderson, Geology of the Whitmore Mountains, in *Antarctic Geoscience*, edited by C. Craddock, pp. 841-847, University of Wisconsin Press, Madison, 1982.

Webers, G. F., C. Craddock, M. A. Rogers, and J. J. Anderson, Geology of Pagano Nunatak and the Hart Hills, in *Antarctic Earth Science*, edited by R. L. Oliver, P. R. James, and J. B. Jago, pp. 251-255, Australian Academy of Science, Canberra, 1983.

Copyright 1987 by the American Geophysical Union.

CORRELATION OF GABBROIC AND DIABASIC ROCKS FROM THE ELLSWORTH MOUNTAINS, HART HILLS, AND THIEL MOUNTAINS, WEST ANTARCTICA

Walter R. Vennum

Department of Geology, Sonoma State University, Rohnert Park, California 94928

Bryan C. Storey

British Antarctic Survey, High Cross, Cambridge CB3 0ET, England

Abstract. Gabbroic stocks and/or diabasic sills crop out in the southern Heritage Range of the Ellsworth Moutains, in the Hart Hills 400 km southwest of the Ellsworth Mountains, and in the Thiel Mountains 200 km south of the Hart Hills. In the Ellsworth Mountains and Hart Hills these mafic igneous rocks have undergone pumpellyite-actinolite and/or greenschist facies regional metamorphism. Relic clinopyroxene compositions, alteration-resistant trace elements, rare earth element data, and the nature of intruded sedimentary host rocks suggest that (1) gabbroic stocks and diabasic dikes and sills in the southern Ellsworth Mountains were emplaced in a continental area undergoing extensional tectonism and are geochemically dissimilar to both the Ferrar Supergroup of the Transantarctic Mountains and the Hart Hills sill, (2) an unmetamorphosed diabase sill at Lewis Nunatak is correlative with the Ferrar Supergroup, and (3) the Hart Hills sill does not appear to have been emplaced in a withinplate setting, but no further resolution of its tectonic environment is possible. Although geochemical data from the Hart Hills and Lewis Nunatak sills are very similar, the Hart Hills body is not considered to be metamorphosed Ferrar Supergroup. The nearest possible correlatives of the Ellsworth Mountains mafic intrusive rocks are an extensive suite of basaltic and diabasic sills and dikes intrusive into the late Precambrian Patuxent Formation in the Neptune Range of the Pensacola Mountains.

Introduction

During the past 10 to 15 years it has become apparent that West Antarctica consists of at least four discrete continental fragments [Dalziel and Elliot, 1982]. Any attempted Gondwanaland reconstruction must take into consideration the relationship of these microplates (the Antarctic Peninsula, the Eights Coast-Thurston Island area, the Ellsworth Mountains, and Marie Byrd Land), both to each other and to the Precambrian shield of East Antarctica. One of the most critical areas to examine in this respect is the Ellsworth-Whitmore Mountains crustal block (EWM), a series of widely scattered nunataks, hills, and small mountain ranges which extend 500 km southward from the Ellsworth Mountains to the Thiel Mountains of the Transantarctic range (the edge of the East Antarctic craton). These outcrops are the only known extensive rock exposures which transect the West-East Antarctic boundary.

Hjelle et al. [1982], von Gizycki [1983], and Vennum et al. [1986] all independently concluded that abundant gabbroic and diabasic stocks, sills, and dikes exposed in the southern Heritage Range of the Ellsworth Mountains were (1) emplaced in a continental area undergoing extensional tectonism, (2) are probably no younger than Late Cambrian, and (3) are geochemically not equivalent to the Ferrar Supergroup, an extensive suite of diabasic sills and basaltic lava flows which are exposed throughout the length of the Transantarctic Mountains. One of the objectives of the 1983-1984 U.S.-British Antarctic Survey investigation of the EWM was a petrologic and geochemical study of gabbroic and diabasic rocks which crop out in the Hart Hills and at Lewis Nunatak in the Thiel Mountains. In this paper we attempt to use alteration-resistant trace elements and relic clinopyroxene compositions to draw geochemical comparisons between gabbroic and diabasic rocks exposed in these areas and to predict the tectonic setting in which they were emplaced. All important localities mentioned in the text are shown in Figure 1.

Methods

Whole-rock analyses were performed on a Phillips PW1400 X ray fluorescence spectrometer at Bedford College, University of London. Major elements were determined on lithium tetraborate/lithium carbonate fusion beads, and trace elements were determined on pressed powder disks. All calibrations were effected using international and laboratory standards. Mass absorption corrections for trace elements were made by monitoring the Ag- and W-tube lines or by using mass absorption

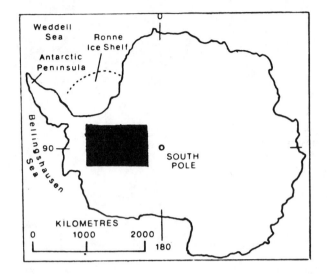

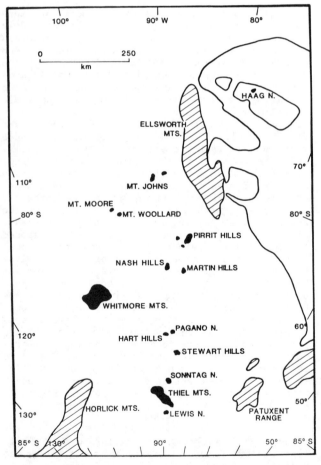

Fig. 1. Map of Antarctica showing location of all important localities mentioned in text.

coefficients determined from major element analyses. Ferrous iron was determined by titration, and H_2O by loss on ignition. Rare earth element data were obtained on a Phillips 65-channel inductively coupled plasma emission spectrometer at Kings College, University of London. These data were normalized to the average of 10 ordinary chondrites using the standards obtained by Nakamura [1974].

Pyroxene compositions were determined at the University of California, Davis, with an energy dispersive ARL-EMX microprobe with the following operating conditions: excitation potential of 15 kV, specimen current of 300 nA, and a beam diameter of less than 5 microns. Corrections were made for background, atomic number effects, absorption, characteristic fluorescence, and instrumental drift. A diopside 85:jadeite 15 glass was used as a standard for determining Al, Ca, Mg, Na, and Si, and a naturally occurring cossyrite which had previously been analyzed by wet chemical methods was used as a standard for determining Fe, Mn, and Ti.

Field Relations, Mineralogy, and Petrography

Introduction. Gabbroic and diabasic rocks crop out at three localities in the study area. Two diabasic sills (100 m and 300 m thick), three gabbroic stocks (outcrop areas each less than 0.5 km^2), a spessartite lamprophyre plug (200 m diameter), and numerous mafic aphanitic to porphyritic dikes, all of which have undergone pumpellyite-actinolite or greenschist facies metamorphism, are exposed in the southern Heritage Range of the Ellsworth Mountains [Vennum et al., 1986]. A 107-m-thick quartz gabbro sill, which has also been metamorphosed to the greenschist facies, is exposed in the Hart Hills [Webers et al., 1983], and an unmetamorphosed diabase sill (60 m of exposed thickness) crops out at Lewis Nunatak (85°40'S, 88°04'W) in the Thiel Mountains. Ice-covered escarpments in the southern Thiel Mountains are, at least in part, probably also underlain by diabasic sills [Schmidt and Ford, 1969], but these outcrops were not visited.

Ellsworth Mountains. All gabbroic and diabasic rocks of the Ellsworth Mountains are intrusive into metasedimentary and metavolcanic rocks of the Heritage Group, a 7200-m-thick sequence of phyllite, argillite, black shale, quartzite, conglomerate, mafic to felsic lava flows, and minor marble of Middle to Late Cambrian age. The Heritage Group is the basal unit in a >13,000-m-thick section of Cambrian to Permian metasedimentary rocks which underlie the Ellsworth Mountains. Lack of intrusive rocks, especially dikes, in units overlying the Heritage Group and the close correspondence in chemical composition between the intrusive rocks and basaltic lava flows which compose 10 to 15% of the 1000-m-thick Liberty Hills Formation of the Heritage Group, strongly suggest that the intrusive igneous rocks of the southern Heritage Range are cogenetic with the volcanic rocks and were emplaced no later than Late Cambrian time [Vennum et al., 1986].

The mineralogy and petrography of all Ellsworth Mountains mafic intrusive rocks are similar. Hand samples are fine-grained (<1 mm) to medium-grained (1-5 mm) greenish-black rocks with subophitic to equigranular textures, scattered clinopyroxene phenocrysts, and sparse plagioclase phenocrysts.

TABLE 1. Representative Electron Microprobe Analyses and Structural Formulae of Clinopyroxenes From the Ellsworth Mountains, Hart Hills, and Lewis Nunatak, West Antarcitca

	V11a	V11c	V11f	V7f	V7c	79V5c	79V5n	79V14g	79V14j	79V6h
Major Element Oxides, in Weight Percent										
SiO_2	53.40	52.34	52.58	51.22	50.74	51.91	53.08	51.48	50.91	50.79
TiO_2	0.33	0.45	0.31	0.53	0.73	0.51	0.58	1.15	1.10	1.25
Al_2O_3	1.96	2.19	1.82	3.33	1.56	2.95	2.33	2.76	4.14	2.77
FeO^*	6.51	6.51	9.63	12.22	19.11	7.19	7.15	9.20	7.51	12.18
MgO	18.76	16.51	16.39	15.17	11.93	16.63	17.67	15.21	14.61	12.66
MnO	0.65	0.54	0.38	nd	0.46	0.43	0.43	0.47	0.28	0.34
CaO	18.65	21.21	18.41	17.90	16.36	19.22	18.99	19.81	21.67	19.81
Na_2O	0.17	nd	nd	nd	nd	0.28	0.17	0.21	0.24	0.36
Total	100.43	99.74	99.52	100.36	100.90	99.12	100.40	100.30	100.47	100.16
Structural Formulae Based on Six Oxygens										
Si	1.943	1.933	1.955	1.907	1.939	1.925	1.938	1.909	1.879	1.913
Al^{IV}	0.057	0.067	0.045	0.093	0.061	0.075	0.062	0.091	0.121	0.087
Al^{VI}	0.027	0.028	0.034	0.053	0.009	0.054	0.038	0.030	0.059	0.036
Ti	0.009	0.013	0.009	0.015	0.021	0.014	0.016	0.032	0.030	0.035
Fe	0.198	0.201	0.300	0.381	0.611	0.223	0.218	0.286	0.232	0.384
Mn	0.020	0.017	0.012		0.015	0.013	0.013	0.015	0.009	0.011
Mg	1.017	0.909	0.908	0.842	0.680	0.919	0.962	0.841	0.804	0.711
Ca	0.727	0.839	0.733	0.714	0.670	0.764	0.743	0.787	0.857	0.800
Na	0.012					0.020	0.012	0.015	0.017	0.027
X + Y	2.010	2.007	1.996	2.005	2.006	2.007	2.002	2.006	2.008	2.004
Atomic Percentages										
Ca	37.1	42.7	37.5	36.9	33.9	39.8	38.4	40.8	45.0	42.0
Mg	51.8	46.2	46.5	43.4	34.4	47.9	49.7	43.6	42.3	37.3
Fe + Mn	11.1	11.1	16.0	19.7	31.7	12.3	11.9	15.6	12.7	20.7
$\frac{100\ Mg}{Mg+Fe+Mn}$	82.3	80.7	74.4	68.8	52.1	79.6	80.6	73.6	76.9	64.3

Here nd means not detected above background. Samples are V11a and V11f, interior of Hart Hills sill; V11c, chilled base Hart Hills sill; V7f, chilled base Lewis Nunatak sill; V7c, interior of Lewis Nunatak sill; 79V5c, mafic dike High Nunatak, Ellsworth Mountains; 79V5n, chilled margin gabbro stock, High Nunatak; 79V14g, chilled top Hyde Glacier sill, Ellsworth Mountains; 79V14j, interior of Hyde Glacier sill; 79V6h, interior of Wilson Nunatak sill, Ellsworth Mountains.
*Total iron as FeO.

Both sills become coarser grained and more feldspathic upward, but pegmatitic segregations occur only in the upper part of the thicker Wilson Nunataks sill (80°01'S, 80°44'W). The groundmass of all rocks is composed of variable mixtures of actinolite, albite, chlorite, clinozoite, leucoxene, pumpellyite, sphene, calcite, hematite, pyrite, and apatite. Quartz occurs only as interstitial grains in the uppermost parts of the sills. Relic cores remain in some clinopyroxene phenocrysts, but most of these grains are at least partially replaced by actinolite and lesser chlorite. Reddish-brown pleochroic magnesian hastingsite occurs in the upper part of the Wilson Nunatak sill, in the pegmatitic segregations associated with the Wilson Nunatak sill, and in most dikes which crosscut both the Wilson Nunatak sill and the gabbro stock at High Nunatak (80°04'S, 82°36'W).

Hart Hills. The oldest rocks in the Hart Hills are polydeformed clastic metasedimentary rocks which are largely quartz-mica phyllites and subarkosic sandstones. These metasedimentary rocks, named the Hart Hills formation by Webers et al. [1983], are intruded by a mass of sheared locally schistose quartz gabbro after a first main phase of deformation but prior to a second later period of deformation. The gabbro-metasedimentary rock contact, where exposed along the western side of the Hart Hills, is sharp, chilled, and locally discordant. Contact relations suggest, however, that the Hart Hills gabbro is a sheetlike body from which numerous dikes and small apophyses extend outward into its host rocks. Hand samples are grayish-green and vary from fine grained to medium grained. Most contain numerous small (2-3 mm) altered black pyroxene phenocrysts set in a much finer-grained groundmass. Strongly retrograded andalusite (?) porphyroblasts are locally developed in the phyllitic rocks within a few meters of the contact. The age of all these rocks is unknown.

Alteration of the Hart Hills gabbro as viewed microscopically is intense. The groundmass appears to have originally been an intergrowth of feldspar, pyroxene, opaque minerals, and minor quartz, but this is now largely a mixture of unidentified cryptocrystalline alteration products

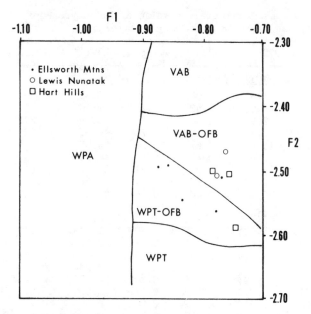

Fig. 2. Plot of discriminant functions F_1 against F_2 for clinopyroxene analyses from the Ellsworth Mountains, Hart Hills, and Lewis Nunatak. $F_1 = (0.0026\ Al_2O_3 + 0.0087\ MgO) - (0.012\ SiO_2 + 0.0807\ TiO_2 + 0.0012\ FeO* + 0.0026\ MnO + 0.0128\ CaO + 0.0419\ Na_2O)$. $F_2 = (0.0085\ CaO + 0.016\ Na_2O) - (0.0469\ SiO_2 + 0.0818\ TiO2 + 0.0212\ Al_2O_3 + 0.0041\ FeO* + 0.1435\ MnO + 0.0025\ MgO)$. Fields of various basalt types are from Nisbet and Pearce [1977]. OFB is ocean floor basalt; VAB, volcanic arc basalt; WPT, within-plate tholeiite; and WPA is within-plate alkalic basalt.

and lesser chlorite, sericite, sphene, actinolite, leucoxene, epidote and/or clinozoisite, quartz, and opaque minerals. Some quartz appears to represent deformed amygdules and is thus secondary. Many clinopyroxene grains, although extensively altered to actinolite and chlorite, still retain relic cores.

<u>Lewis Nunatak</u>. The uppermost 60 m of Lewis Nunatak in the Thiel Mountains is composed of flat-lying massive diabase which intrudes a horizontally bedded sedimentary sequence. The base of the diabase is in contact with a 2-m-thick bed of hornfelsed black slate which is underlain by an unknown thickness of greenish-gray diamictite containing clasts of both sedimentary and granitic rocks up to 1 m long. Black aphantitic rock of the lower chilled margin grades quickly upward into medium-grained grayish-black diabase, all of which is strongly stained with iron oxides.

The chill zone contains numerous pyroxene plagioclase varioles in which length to width ratios of prismatic crystals reach 10:1 or 15:1. Radiate intergrowths of plagioclase and pyroxene similar to those described by MacKenzie et al. [1982] and quenched plagioclase hopper crystals are also present. The plagioclase is fresh, but pyroxene and most of the groundmass is altered to a golden brown cryptocrystalline mica. Above the chilled margin, textures are subophitic to intergranular, and the alteration is less intense. Pyroxenes are altered only marginally, and the groundmass is composed of feldspar, pyroxene, quartz, opaque minerals, granophyric intergrowths of quartz and felsdspar, and minor amounts of apatite, amphibole, opaque minerals, biotite, and cryptocrystalline alteration products.

Clinopyroxene Compositions

Numerous studies [e.g., Hervé et al., 1983; Snoke, 1979; Barron, 1976] have shown that composition of relic clinopyroxene crystals often provides valuable information concerning the original magmatic composition and hence the tectonic setting of altered mafic igneous rocks. Relic clinopyroxene grains are not common in mafic igneous rocks from the Ellsworth Mountains and Hart Hills; those in coarser-grained rocks are often partially altered to actinolite, chlorite, and other hydrous minerals, while those in fine-grained rocks (i.e., chilled margins) are often completely pseudomorphed. Consequently, our samples from these two localities represent the least altered grains available from a representative suite of rocks. All analyses listed in Table 1 are augites in the classification of Poldervaart and Hess [1951].

We have plotted our clinopyroxene analyses in the F_1-F_2 discrimination diagram (Figure 2) devised by Nisbet and Pearce [1977]. Except for one sample, Hart Hills and Lewis Nunatak pyroxenes fall in the overlapping fields of ocean floor basalt (OFB) and volcanic arc basalt (VAB), where VAB is basalt erupted above subduction zones in island arcs or at active continental margins. Although these results are ambiguous, they do rule out a within-plate magma genesis. Except for one sample, the Ellsworth Mountains pyroxenes plot in the OFB-WPT field (WPT, within-plate tholeiite) and are clearly separated from the Hart Hills and Lewis Nunatak pyroxenes. Similar results were also obtained on other discrimination diagrams devised by Nisbet and Pearce [1977] but not shown here (i.e., SiO_2-TiO_2, TiO_2-MnO-Na_2O and MgO/FeO^*-TiO_2). All pyroxene analyses are characterized by low Al_2O_3 and TiO_2 and consequently plot in the subalkaline field of LeBas's [1962] SiO_2-Al_2O_3 diagram, also not shown. This is a geochemical characteristic of pyroxenes from tholeiitic rocks from both VAB and OFB magmas, but not from within-plate alkalic basalts. This situation is apparently normal, as Nisbet and Pearce [1977] state that while pyroxene geochemistry does provide constraints, it often does not give totally unambiguous results, especially when attempting to discriminate VAB and OFB magmas.

Geochemistry and Discrimination Diagrams

<u>Introduction</u>. It is well known that basaltic rocks are prone to extensive chemical modification when subjected to either hydrothermal alteration or low-grade metamorphism [Bass et al., 1973; Pearce, 1975]. In 1970, Cann [1970] showed that amounts of Nb, Ti, Y, and Zr in ocean floor basalt fall within a fairly restricted range, and that

TABLE 2. Chemical Data of Gabbroic and Diabasic Rocks From the Ellsworth Mountains, Hart Hills, Thiel Mountains, and Neptune Range, West Antarctica

	V11a	V11c	V11f	V7f	V7c	79V5n	79V7f	79V14g	A	B	C	D	E
Major Element Oxides, in Weight Percent													
SiO_2	51.33	52.17	53.90	54.67	55.51	47.67	46.27	43.96	55.65	53.75	50.40	47.08	52.93
TiO_2	0.83	0.81	0.97	1.21	1.33	0.69	1.99	1.11	1.03	0.70	0.44	1.40	2.01
Al_2O_3	15.04	13.34	14.39	13.09	13.35	13.34	13.75	17.10	13.95	14.33	15.51	14.85	13.68
Fe_2O_3	2.36	2.94	3.23	2.72	2.98	2.25	3.04	2.57	3.24	2.33	0.99	1.68	1.43
FeO	6.39	6.55	7.91	10.27	9.75	7.21	13.28	8.91	7.38	7.61	7.83	8.57	10.01
MnO	0.17	0.17	0.18	0.20	0.19	0.13	0.25	0.15	0.17	0.18	0.17	0.18	0.19
MgO	7.52	8.03	5.63	4.24	3.83	10.06	6.13	9.84	4.50	6.64	10.60	7.34	4.88
CaO	10.41	10.71	6.23	8.43	8.34	10.76	7.94	6.02	8.51	10.60	10.87	9.74	5.94
Na_2O	2.05	1.51	3.21	2.27	2.25	1.72	3.92	2.52	2.50	1.83	1.42	1.33	4.35
K_2O	0.35	0.81	1.30	1.27	1.33	2.40	0.50	1.36	1.45	0.81	0.37	3.48	0.51
P_2O_5	0.09	0.08	0.11	0.13	0.17	0.06	0.18	0.04	0.23	0.18	0.08	0.24	0.37
H_2O	3.10	3.24	2.91	1.03	0.72	3.05	1.96	5.81	1.67	1.32	1.55	3.44	2.98
Total	99.64	100.36	99.97	99.53	99.03	99.34	99.21	99.39	100.28	100.18	100.23	99.33	99.28
Trace Elements, in parts per million													
Ba	262	139	298	331	327	---	---	185	376	232	157	1170	250
Cr	419	405	78	18	10	523	89	152	59	142	352	400	46
Nb	6	6	10	10	8	---	13	16	---	---	---	50	50
Ni	104	88	66	32	34	138	83	126	53	85	249	120	13
Rb	13	33	63	47	50	95	8	30	50	30	12	---	---
Sr	403	108	181	128	132	268	244	414	138	126	100	360	113
Th	5	2	7	4	5	11	---	---	---	---	---	---	---
V	220	227	251	298	293	---	---	---	---	---	---	280	280
Y	24	26	35	39	36	34	22	31	---	---	---	67	68
Zr	87	99	142	167	151	158	147	132	157	83	53	142	190
Rare Earth Elements, in parts per million													
La	---	12.36	17.85	19.31	---	17.02	12.75	8.96					
Ce	---	24.88	37.02	39.43	---	33.40	30.06	20.09					
Pr	---	3.32	4.96	5.07	---	4.15	4.50	2.89					
Nd	---	13.52	20.09	20.40	---	16.82	21.05	12.58					
Sm	---	3.40	4.73	4.80	---	3.66	5.67	3.35					
Eu	---	0.90	1.10	1.22	---	0.99	2.14	1.44					
Gd	---	3.55	4.90	5.07	---	3.35	6.25	3.71					
Dy	---	3.88	5.25	5.48	---	3.18	6.13	3.90					
Ho	---	0.81	1.10	1.14	---	0.65	1.23	0.83					
Er	---	2.45	3.33	3.58	---	1.99	3.58	2.40					
Yb	---	2.25	3.10	3.31	---	1.68	2.98	2.12					
Lu	---	0.40	0.51	0.56	---	0.30	0.49	0.37					
$\Sigma 14REE$	---	72.6	105.2	110.7	---	88.8	98.3	63.6					

Dashes mean element not reported. Sample 79V7f is chilled base Wilson Nunatak sill, Ellsworth Mountains. Other sample locations are given in Table 1. A, average of four analyses, Ferrar Supergroup pigeonite tholeiite sill margins. B, average of five analyses, Ferrar Supergroup hypersthene tholeiite sill margins. C, one analysis, Ferrar Supergroup oliving tholeiite sill margin. A, B, and C are from Gunn [1966]. D, chilled top Neptune Range sill. E, interior Neptune Range basalt flow. D and E are from D.L. Schmidt (U.S. Geological Survey, written communication, 1986).

abundances of these elements are strongly resistant to modification by secondary processes. A number of other petrologists [Pearce and Cann, 1973; Pearce and Norry, 1979; Pearce et al., 1975; Floyd and Winchester, 1975] have since devised a variety of geochemical discrimination diagrams utilizing these four elements as well as Cr, P, and Sr to determine the original chemistry and tectonic setting of altered basaltic rocks. Pearce and Cann [1973] use the Y/Nb ratio to indicate whether a specific rock is alkalic or tholeiitic; Y/Nb ≤ 1 is indicative of alkaline chemistry, and Y/Nb ≥ 2 is indicative of tholeiitic chemistry in their scheme.

Ellsworth Mountains. Chemical analyses of representative Ellsworth Mountains samples are given in Table 2. All are olivine normative. All analyzed Ellsworth Mountains mafic rocks have Y/Nb ratios that fall in the transitional basalt chemistry range (Y/Nb ≥ 1, but ≤ 2). When these anal-

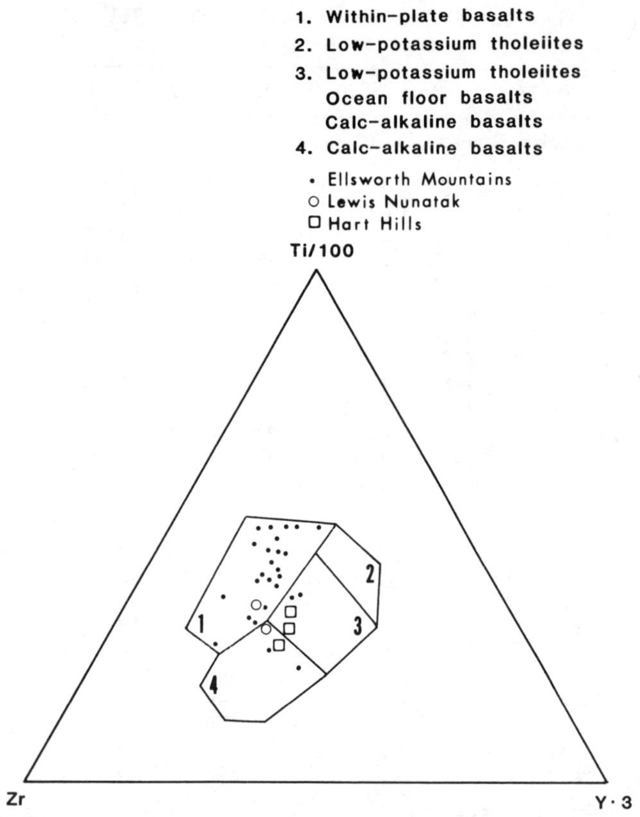

1. Within-plate basalts
2. Low-potassium tholeiites
3. Low-potassium tholeiites
   Ocean floor basalts
   Calc-alkaline basalts
4. Calc-alkaline basalts

- Ellsworth Mountains
○ Lewis Nunatak
□ Hart Hills

Fig. 3. Variation in Ti, Zr, and Y for chemically analyzed samples from the Ellsworth Mountains, Hart Hills, and Lewis Nunatak. Fields of various basalt types are from Pearce and Cann [1973].

yses are plotted in a Ti-Zr-Y discrimination diagram (Figure 3), 22 of 26 points plot in or along the edge of the "within-plate basalt" field, a category that includes both ocean island and continental basalt [Pearce and Cann, 1973]. On a Ti-Y/Nb plot (Figure 4), all analyzed samples fall in the continental tholeiite field but in areas overlapped by the fields of either oceanic tholeiites or continental alkalic basalt. Twenty-five of 27 points plotted on a Zr/Y-Sr diagram (not shown) fall in or just outside the "within-plate field" [Pearce and Norry, 1979].

The scarcity of olivine, the composition of relic clinopyroxene cores, and the relatively large ratio of plagioclase to pyroxene in the Ellsworth Mountains rocks are suggestive of tholeiitic chemistry. The lack of orthopyroxene and the presence of hastingsitic amphibole are characteristic of alkalic basalts. Thus study of alteration-resistant trace element abundances and relic primary mineralogy suggests that the mafic igneous rocks of the southern Heritage Range are continental basalts whose original chemical compositions were mildly tholeiitic or transitional between tholeiitic and alkalic basalt. Their intrusion into a thick sequence of carbonate and clastic sedimentary rocks is not compatible with an oceanic island setting. Mafic igneous rocks of the southern Heritage Range are part of a bimodal volcanic and hypabyssal suite characterized by large amounts of mafic rocks, less abundant felsic rocks, and sparse rocks of intermediate composition [Vennum et al., 1986]. Basaltic or strongly bimodal basaltic-rhyolitic suites with scarce andesite are characteristic of continental areas subjected to plate separation and extension. We therefore conclude that the igneous rocks of the southern Heritage Range were emplaced in a continental area undergoing extensional tectonism.

Hart Hills. Three chemical analyses of the Hart Hills gabbro are listed in Table 2; all are diopside-hypersthene-quartz normative, and all have Y/Nb ratios between 3.5 and 4.3. These geochemical features are suggestive of tholeiitic chemistry, as are the compositions of clinopyroxenes from these rocks. Weaver et al. [1972] have shown that the ratio Zr/Nb is useful for discriminating between tholeiitic and alkalic basalt, values less than 8 being typical of alkaline chemistry. This parameter varies between 14.2 and 16.5 for the three Hart Hills analyses. On the TiO_2-Y/Nb diagram (Figure 4) these analyses plot in or near the overlapping fields of continental tholeiitic basalt (CTB) and oceanic tholeiitic basalt (OTB) but are clearly separated from the Ellsworth Mountains rocks. Plots of these analyses in several other discrimination diagrams (TiO_2-Zr, P_2O_5-Zr, Ti-Zr/P_2O_5, and Nb/Y-Zr/P_2O_5, not shown) also fall in the overlapping CTB-OTB fields. However, on graphs of P_2O_5-TiO_2 (Figure 5) [Hawkins, 1980], Ti/Cr-Ni (Figure 6) [Beccaluva et al., 1979], Ti/100-Zr/3Y (Figure 3), and Ti/100-Sr/2 (not shown) these three analyses plot as island arc tholeiite, island arc tholeiite and ocean floor basalt, low potassium tholeiite and calc-/alkaline basalt, and low potassium tholeiite (basaltic magmas erupted on oceanic crust at converging plate margins typically close to deep ocean trenches), respectively.

Lewis Nunatak. Chemical analyses of the chilled margin and coarser interior of the unaltered Lewis Nunatak sill are listed in Table 2. Both analyses are diopside-hypersthene-quartz normative, and both have Y/Nb ratios equal to or slightly less than 4. These geochemical features, as well as the clinopyroxene compositions, are suggestive of tholeiitic chemistry.

Rare earth elements. Chondrite-normalized rare earth element (REE) patterns for six samples (chill zones of the High Nunatak stock, two Ellsworth Mountains sills, the Hart Hills and Lewis Nunatak sills, and the coarser interior of the Hart Hills sill) are shown in Figure 7. Abundances of REE are listed in Table 2. Σ14REE was derived by estimating the amounts of Tb and Tm from the chondrite-normalized diagram and adding these figures to the reported total of Σ12REE. All patterns show moderate light REE (LREE) enrichment and are moderately fractionated (Ce_N/Yb_N = 2.4-3.0 for all samples except the High Nunatak stock, which has a value of 5.1). The averaged pattern for the two Ellsworth Mountains chilled sill margins has a small positive europium anomaly (Eu/Eu* = 1.15); all other patterns have

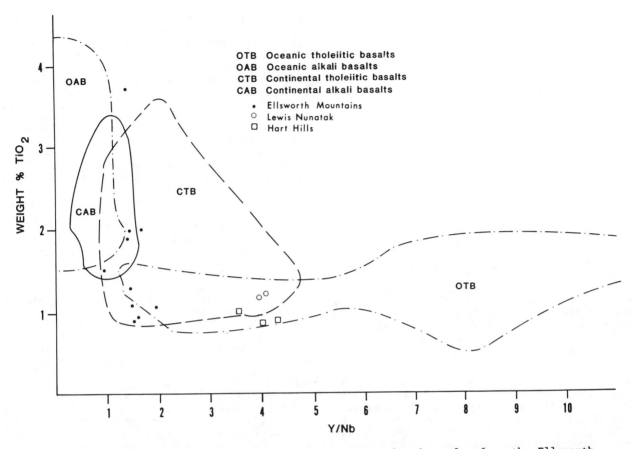

Fig. 4. Variation in TiO_2 and Y/Nb for chemically analyzed samples from the Ellsworth Mountains, Hart Hills, and Lewis Nunatak. Fields of various basalt are types from Floyd and Winchester [1975].

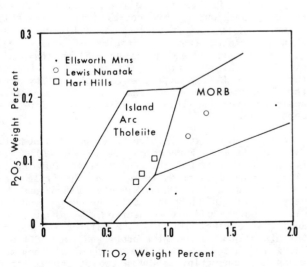

Fig. 5. P_2O_5-TiO_2 variation diagram. Fields of various basalt types are from Hawkins [1980]. MORB is mid-ocean ridge basalt.

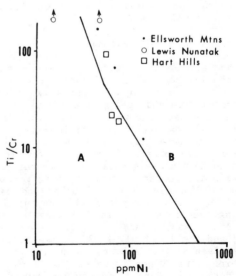

Fig. 6. Ti/Cr-Ni variation diagram. Fields of various basalt types are from Beccaluva et al. [1979]. Field A is island arc tholeiite, and field B is ocean floor tholeiite. The Ti/Cr ratio for Lewis Nunatak samples is greater than 650.

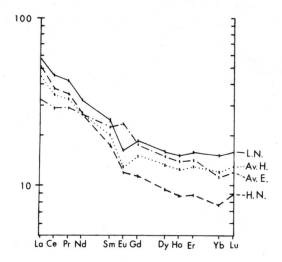

Fig. 7. Chondrite-normalized rare earth element patterns for chemically analyzed samples from the Ellsworth Mountains, Hart Hills, and Lewis Nunatak. Av. E is average of two Ellsworth Mountains sill margins; Av. H, average of two Hart Hills gabbros; H.N., chilled margin High Nunatak gabbro stock, Ellsworth Mountains; and L.N., Lewis Nunatak sill margin.

negative europium anomalies (Eu/Eu* = 0.7-0.8). $\Sigma 14REE$ ranges from 63 to 110. None of the patterns show the strong depletion of LREE typical of tholeiitic N-type ocean ridge basalts; none are as strongly fractionated as most E-type basalts from anamolous ridge sections or most oceanic islands, and none are as flat as the patterns from typical transitional ridge segments [Saunders, 1984]. They are, however, typical of patterns produced by continental intrusive and island arc tholeiites [Cullers and Graf, 1984].

Correlations

It has already been suggested that the gabbroic and diabasic rocks of the Ellsworth Mountains are part of a suite of igneous rocks emplaced in a continental area undergoing extensional tectonism, and that they are chemically mildly tholeiitic or transitional between tholeiitic and alkalic basalt. Vennum et al. [1986] have also shown on the basis of comparative geochemistry that the Ellsworth Mountains igneous rocks are not correlative with the Ferrar Supergroup, a very extensive series of diabasic sills and dikes and basaltic lava flows of Jurassic age which crop out for 3000 km along the Transantarctic Mountains. A similar conclusion was reached earlier by von Gizycki [1983] and Hjelle et al. [1982]. See Table 2 for a comparison of the Ellsworth Mountains rocks with the three major magma types of the Ferrar Supergroup defined by Gunn [1966] in the McMurdo Sound area.

The closest possible correlatives of the Ellsworth Mountains igneous rocks are in the Pensacola Mountains. Schmidt et al. [1978], Schmidt and Ford [1969], and Frischbutter and Vogler [1985] have mapped an extensive suite of basaltic and diabasic sills and dikes of uncertain, but probable late Precambrian age that are intrusive into the late Precambrian Patuxent Formation in the Neptune Range of the Pensacola Mountains. Plots of selected major and trace element ratios [Frischbutter and Vogler, 1985] and chemical analyses (D. L. Schmidt, U.S. Geological Survey, written communication, 1986) of these mafic rocks (Table 2) are similar to those from the southern Heritage Range.

The diabase sill at Lewis Nunatak is presumed to be part of the Ferrar Supergroup because (1) it intrudes flat-lying undeformed clastic sedimentary rocks, a setting comparable to Ferrar Supergroup rocks throughout the Transantarctic Mountains, (2) it is geochemically similar to Gunn's [1966] pigeonite tholeiite magma type (Table 2), (3), its REE pattern is similar to those reported from other Ferrar Supergroup rocks [Kyle, 1977; Kyle et al., 1983], and (4) its Rb, Sr, and $^{87}Sr/^{86}Sr$ ratios are comparable to values of these parameters recorded from Ferrar Supergroup rocks elsewhere in the Transantarctic Mountains [Millar and Pankhurst, this volume].

Analysis of clinopyroxene compositions (Figure 2) and of the various discrimination diagrams mentioned earlier do not yield an unambiguous answer to the tectonic setting of the Hart Hills sill. This rock body appears to have originally been tholeiitic in composition and not to have been emplaced in a within-plate setting, but any further distinction is not possible using our data. Clinopyroxene compositions and alteration-resistant trace element ratios for the Hart Hills sill differ considerably from similar types of data from the Ellsworth Mountains mafic rocks (Figures 2-5), suggesting that these two groups of rocks are not genetically related. The close similarity of some of the geochemical data from the Hart Hills and Lewis Nunatak sills (Figures 2-4, and 7) could invite the speculation that the Hart Hills sill is an outcrop of metamorphosed Ferrar Supergroup. This correlation is unlikely because (1) no post-Jurassic metamorphic event has been documented from this sector of Antarctica, (2) Storey and Dalziel [this volume] on the basis of structural analysis regard the Hart Hills as part of the EWM, a separate tectonic entity from the Transantarctic Mountains, and (3) only five analyses from two rock bodies are available for comparison.

Acknowledgments. This research was financed by National Science Foundation grant DPP82-13798 to I.W.D. Dalziel of Columbia University. Our logistical problems involved in reaching the field area were solved by U.S. Navy Squadron VXE-6 and two very capable British Antarctic Survey Twin Otter pilots, Gary Studd and Ed Mehrten. We thank Peter Schiffman of the University of California, Davis, for his help with the microprobe analyses, Barbara Young for her help with the geochemical computations and manuscript, and Ron Leu of Sonoma State University for making the thin sections. The figures were drafted by David Fowler of Sonoma State University.

References

Barron, B. J., Recognition of the original volcanic suite in altered mafic rocks at Sofala, New South Wales, Am. J. Sci., 276, 604-636, 1976.

Bass, M. N., R. Moberly, J. M. Rhodes, C. Shih, and S. E. Church, Volcanic rocks cored in the central Pacific, leg 17, deep sea drilling project, in Initial Reports of the Deep Sea Drilling Project, vol. 17, edited by E. S. Winterer et al., pp. 429-503, U.S. Government Printing Office, Washington, D. C., 1973.

Beccaluva, L., D. Ohnenstetter, and M. Ohnenstetter, Geochemical discrimination between ocean floor and island arc tholeiites, Ofioliti, 4(1), 67-72, 1979.

Cann, J. R., Rb, Sr, Y, Zr, and Nb in some ocean floor basaltic rocks, Earth Planet. Sci. Lett., 10, 7-11, 1970.

Cullers, R. L., and J. L. Graf, Rare earth elements in igneous rocks of the continental crust: Predominately basic and ultrabasic rocks, in Rare Earth Element Geochemistry, edited by P. Henderson, pp. 237-274, Elsevier, New York, 1984.

Dalziel, I. W. D., and D. H. Elliott, West Antarctica: Problem child of Gondwanaland, Tectonics, 1, 3-19, 1982.

Floyd, P. A., and J. A. Winchester, Magma type and tectonic setting discrimination using immobile elements, Earth Planet. Sci. Lett., 27, 211-218, 1975.

Frishbutter, A., and P. Vogler, Contributions to the geochemistry of magmatic rocks in the upper Precambrian-lower Paleozoic profile of the Neptune Range, Transantarctic Mountains, Antarctica, Z. Geol. Wiss., 13, 345-357, 1985.

Gunn, B. M., Modal and element variation in Antarctic tholeiites, Geochim. Cosmochim. Acta, 30, 881-920, 1966.

Hawkins, J. W., Petrology of back-arc basins and island arcs: Their possible role in the origin of ophiolites, in Proceedings of the International Ophiolite Symposium, edited by A. Panayiotou, pp. 244-254, Min. of Agric. Nat. Resour., Geol. Surv. Dep., Cyprus, 1980.

Hervé, F., E. Godoy, and J. Davidson, Blueschist relic clinopyroxenes of Smith Island (South Shetland Islands): Their composition, origin, and some tectonic implications, in Antarctic Earth Science, edited by R. L. Oliver, P. R. James, and J. B. Jago, pp. 363-366, Australian Academy of Science, Canberra, 1983.

Hjelle, A., Y. Ohta, and T. S. Winsnes, Geology and petrology of the southern Heritage Range, Ellsworth Mountains, in Antarctic Geoscience, edited by C. Craddock, pp. 599-608, University of Wisconsin Press, Madison, 1982.

Kyle, P. R., Petrogenesis of Ferrar Group rocks, Antarct. J. U. S., 12, 107-110, 1977.

Kyle, P. R., R. J. Pankhurst, and R. R. Bowman, Isotopic and chemical variations in Kirkpatrick Basalt Group rocks from southern Victoria Land, in Antarctic Earth Science, edited by R. L. Oliver, P. R. James, and J. B. Jago, pp. 234-237, Australian Academy of Science, Canberra, 1983.

LeBas, M.J., The role of aluminum in igneous clinopyroxenes with relation to their parentage, Am. J. Sci., 260, 267-288, 1962.

MacKenzie, W. S., C. H. Donaldson, and C. Guilford, Atlas of Igneous Rocks and Their Textures, 148 pp., Halsted, New York, 1982.

Millar, I. L., and R. J. Pankhurst, Rb/Sr geochronology of the region between the Antarctic Peninsula and the Transantarctic Mountains: Haag Nunataks and Mesozoic granitoids, this volume.

Nakamura, N., Determination of REE, Ba, Fe, Mg, Na, and K in carbonaceous and ordinary chondrites, Geochim. Cosmochim. Acta, 38, 757-775, 1974.

Nisbet, E. G., and Pearce, J. A., Clinopyroxene compositions in mafic lavas from different tectonic settings, Contrib. Mineral. Petrol., 63, 149-160, 1977.

Pearce, J. A., Basalt geochemistry used to investigate past tectonic environments on Cyprus, Tectonophysics, 25, 41-67, 1975.

Pearce, J. A., and J. R. Cann, Tectonic setting of basic volcanic rocks determined using trace element analysis, Earth Planet. Sci. Lett., 19, 290-300, 1973.

Pearce, J. A. and M. J. Norry, Petrogenetic implications of Ti, Zr, Y, and Nb variations in volcanic rocks, Contrib. Mineral. Petrol., 69, 33-47, 1979.

Pearce, T. H., B. E. Gorman, and T. C. Birkett, The TiO_2-K_2O-P_2O_5 diagram: A method of discriminating between oceanic and nonoceanic basalts, Earth Planet. Sci. Lett., 24, 419-426, 1975.

Poldervaart, A., and H. H. Hess, Pyroxenes in the crystallization of basaltic magmas, J. Geol., 59, 472-489, 1951.

Saunders, A. D., The rare earth element characteristics of igneous rocks from the ocean basins, in Rare Earth Element Geochemistry, edited by P. Henderson, pp. 205-236, Elsevier, Amsterdam, 1984.

Schmidt, D. L., and A. B. Ford, Geology of the Pensacola and Thiel Mountains, scale 1:1,000,000, Geologic Maps of Antarctica, edited by V. C. Bushnell and C. Craddock, Antarctic Map Folio Series, Folio 12, plate 5, American Geographical Society, New York, 1969.

Schmidt, D. L., P. L. Williams, and W. H. Nelson, Geologic map of the Schmidt Hills Quadrangle and part of the Gamborta Peak Quadrangle, Pensacola Mountains, Antarctica, scale 1:250,000 U.S. Geol. Surv. Antarc. Geol. Map A-8, 1978.

Snoke, A. W., Relic pyroxenes from the Preston Peak ophiolite, Klamath Mountains, California, Am. Mineral., 64, 865-873, 1979.

Storey, B. C., and I. W. D. Dalziel, Outline of the structural and tectonic history of the Ellsworth Mountains-Thiel Mountains ridge, West Antarctica, this volume.

Vennum, W. R., P. von Gizycki, V. V. Samsanov, A. G. Markovich, and R. J. Pankhurst, Igneous petrology and geochemistry of the southern Heritage Range, Ellsworth Mountains, Antarctica, in Geology of the Ellsworth Mountains, edited by G. F. Webers, J. Splettstoesser, and C. Craddock, Geol. Soc. Am. Mem., in press, 1986.

von Gizycki, P., Die magmatischen gesteine der Ellsworth Mountains, West-Antarktis, Neues

Jahrb. Geol. Palaeontol. Abh., 167(1), 65-88, 1983.

Weaver, S. D., J. S. C. Sceal, and I. L. Gibson, Trace element data relevant to the origin of trachytic and pantelleritic lavas in the East African rift system, Contrib. Mineral. Petrol., 36, 181-194, 1972.

Webers, G. F., C. Craddock, M. A. Rogers, and J. J. Anderson, Geology of Pagano Nunatak and the Hart Hills, in Antarctic Earth Science, edited by R. L. Oliver, P. R. James, and J. B. Jago, pp. 251-255, Australian Academy of Science, Canberra, 1983.

Copyright 1987 by the American Geophysical Union.

PETROLOGY, GEOCHEMISTRY, AND TECTONIC SETTING OF GRANITIC ROCKS FROM THE ELLSWORTH-WHITMORE MOUNTAINS CRUSTAL BLOCK AND THIEL MOUNTAINS, WEST ANTARCTICA

Walter R. Vennum

Department of Geology, Sonoma State University, Rohnert Park, California 94928

Bryan C. Storey

British Antarctic Survey, High Cross, Cambridge CB3 OET, England

Abstract. The Ellsworth-Whitmore Mountains crustal block (EWM) is a belt of small mountain ranges, hills, and nunataks which trend northward 500 km from the Thiel Mountains of the Transantarctic range to the Ellsworth Mountains. Granitic rocks compose most of the Pirrit Hills, Nash Hills, and Whitmore Mountains, and all of Pagano Nunatak. Mount Woollard is a migmatized complex of biotite schist, amphibolite, pegmatite, and massive biotite granite. A rhyodacite stock crops out in the Martin Hills. The Mount Seelig hornblende-bearing biotite granite of the Whitmore Mountains and the Mount Woollard biotite granite are metaluminous diopside-normative granitoids. All other granitic rocks of the EWM are mildly peraluminous, corundum-normative biotite (locally muscovite-bearing) leucogranites. Previously published radiometric ages range from 163 to 190 Ma. Petrography, geochemistry, and isotopic data suggest that these rocks are largely highly differentiated leucocratic S-type granites formed by anatexis of either metasedimentary rocks or more deeply seated, more mafic, plagioclase-rich granitoids. In these aspects they strongly resemble granites developed in intracontinental terranes such as the Hercynian belt of Europe and in continental collision-type settings such as the Bhutan and Nepalese Himalaya and the Seward Peninsula of western Alaska. We ascribe their origin to posttectonic emplacement in a neutral or extensional (rifted) within-plate setting following deformation of the EWM. S-type Cambro-Ordovician granitic rocks and Precambrian quartz monzonite porphyries of the Thiel Mountains are geochemically similar to the EWM granites, but are correlative with the Granite Harbour Intrusive Series of the Transantarctic Mountains and the Wyatt Formation of the nearby La Gorce Mountains, respectively.

Introduction

During the past 10 to 15 years it has become apparent that West Antarctica consists of at least four discrete continental fragments [Dalziel and Elliot, 1982]. Any attempted Gondwanaland reconstruction must take into consideration the relationship of these microplates (the Antarctic Peninsula, the Eights Coast-Thurston Island area, the Ellsworth Mountains, and Marie Byrd Land) both to each other and to the Precambrian shield of East Antarctica. One of the most critical areas to examine in this respect is the Ellsworth-Whitmore Mountains crustal block (EWM), a series of widely scattered nunataks, hills, and small mountain ranges which extend 500 km southward from the Ellsworth Mountains to the Thiel Mountains of the Transantarctic range (the edge of the East Antarctic craton). These outcrops are the only known extensive rock exposures which transect the West-East Antarctica boundary.

Although all known rock outcrops in the EWM have previously been examined in reconnaissance style [Craddock, 1983], detailed geochemical and isotopic data were not previously available for the abundant granitic rocks exposed in this region. Dalziel and Elliot [1982] and Longshaw and Griffiths [1981] both considered these granitic rocks to represent the Pacific margin of Gondwanaland. One of the main objectives of the 1983-1984 U.S.-British Antarctic Survey investigation of the EWM was a petrologic and geochemical study of the granitic rocks discussed above. In this paper we attempt to use the results of this project to deduce the tectonic setting of granitic rocks from both the EWM and the Thiel Mountains. A companion paper [Vennum and Storey, this volume] similarly treats gabbroic and diabasic rocks exposed in the southern Heritage Range of the Ellsworth Mountains, in the Hart Hills, and at Lewis Nunatak in the Thiel Mountains. All important localities mentioned in the text are shown in Figure 1.

Methods

Whole-rock analyses were performed on a Phillips PW1400 X ray fluorescence spectrometer at Bedford College, University of London. Major elements were determined on lithium tetraborate/lithium carbonate fusion beads, and trace elements were determined on pressed powder disks. All cal-

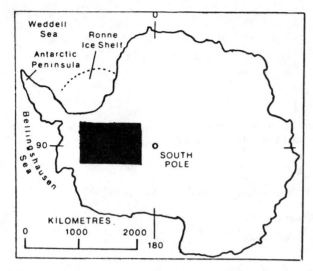

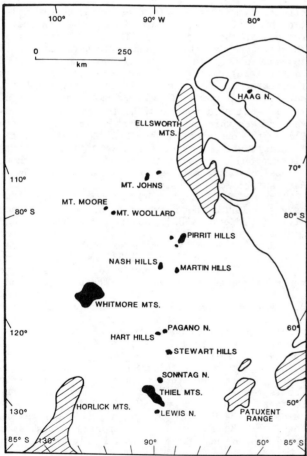

Fig. 1. Map of West Antarctica showing location of important geographic features mentioned in text.

ibrations were effected using international and laboratory standards. Mass absorption corrections for trace elements were made by monitoring the Ag- and W-tube lines or by using mass absorption coef- ficients determined from major element analyses. Ferrous iron was determined by titration, and H_2O by loss on ignition. Rare earth element (REE) data were obtained on a Phillips 65-channel inductively coupled plasma emission spectrometer at Kings College, University of London. These data were normalized to the average of 10 ordinary chondrites using the standards obtained by Nakamura [1974]. Granitic rock names were derived from Streckeisen's [1976] classification.

Geologic Setting

Igneous rocks crop out at seven locations in the EWM. From north to south these are Mount Woollard; the Pirrit, Nash, and Martin Hills; the Whitmore Mountains; Pagano Nunatak; and the Hart Hills. The Pirrit and Nash Hills, the Whitmore Mountains, and Pagano Nunatak are composed largely of felsic plutonic rocks which, except at Pagano Nunatak, are locally capped by metasedimentary roof pendants. Radiometric ages from plutonic rocks exposed at these four localities range from 163 to 190 Ma [Webers et al., 1982, 1983; Millar and Pankhurst, this volume]. Mount Woollard consists of a migmatized complex of biotite schist, amphibolite, pegmatite, and massive granite. Preliminary radiometric dates are inconclusive, but do not suggest that these rocks are an isolated exposure of Precambrian crystalline basement [Millar and Pankhurst, this volume].

Deformed metasedimentary rocks that are largely clastic, but which include some limestone, underlie most of the Martin Hills. This metasedimentary sequence is intruded by an as yet undated porphyritic rhyodacite stock. Deformed clastic metasedimentary rocks of the Hart Hills are intruded by a quartz gabbro sill; no radiometric ages are available. Sedimentary rocks which crop out at Mount Johns and metasedimentary rocks exposed at Mount Moore and in the Stewart Hills are discussed by Storey and Dalziel [this volume]. A K-Ar whole-rock(?) radiometric date of 508 Ma has been obtained from deformed clastic metasedimentary rocks of the Stewart Hills [Craddock et al., 1982]. Craddock et al. [1982] and Craddock [1983] have written short summaries of the geology of the EWM. More detailed accounts of the geology of the Whitmore Mountains [Webers et al., 1982] and of the Hart Hills and Pagano Nunatak [Webers et al., 1983] are available.

A 100-m-thick flat-lying or gently dipping sequence of volcaniclastic metasedimentary rocks crop out in the southeastern part of the Thiel Mountains. These sedimentary rocks are interbedded with and are locally cross cut by quartz monzonite porphyry which underlies most of the Thiel Mountains. Storey and Dalziel [this volume] suggest, on the basis of petrographic similarity, that the sedimentary rocks and the quartz monzonite porphyry formed contemporaneously, and that the volcanic rocks are both intrusive and extrusive. On the basis of extensive petrographic analysis, Ford and Sumsion [1971] concluded that the quartz monzonite porphyry represents a thick sequence of crystal-rich silicic tuff whose textures have been modified by recrystallization during cooling or by thermal metamorphism imparted by younger granitic intrusions. Ford et al. [1963]

obtained late Precambrian to earliest Paleozoic zircon lead alpha ages from the quartz monzonite porphyry. Granodiorite plutons which intrude the quartz monzonite porphyry yield Ordovician and Cambrian ages. Schmidt and Ford [1969] published a geological map of the Thiel Mountains, and Ford and Aaron [1962] and Ford [1964] described the bedrock geology.

Field Relations of Granitic Rocks

Pirrit Hills. The major constituent of the Pirrit Hills is coarse (5-10 mm) or very coarse grained (greater than 1 cm) pink to pinkish-white leucocratic granite that is locally porphyritic and contains 3- to 4-cm-long potassium feldspar crystals. Biotite and muscovite are present only in accessory amounts. This unit is, however, texturally highly variable and often grades into porphyritic or nonporphyritic medium-grained (1-5 mm) or even alaskitic phases over a distance of just a few meters. Outcrop areas of these finer-grained phases are usually irregularly shaped and have areal extents of one hundred to several hundred square meters. The alaskitic phases contain segregations of tourmaline-beryl-muscovite pegmatite up to 5 m in diameter which are made conspicuous by pale green beryl crystals up to 2 cm long. All phases of the Pirrit Hills intrusion are cut by numerous, often garnet-bearing, aplite dikes up to 10 m thick, some of which have biotite-rich margins. At Bradley Nunatak, strongly hydrothermally altered coarse-grained granite is cut by numerous veins of quartz-muscovite greisen. Inclusions are virtually nonexistent and were found only at Harter Nunatak.

Pagano Nunatak. Virtually all exposed bedrock is a light gray massive medium- to coarse-grained biotite granite named the Pagano granite by Webers et al. [1983]. This unit locally develops a seriate texture, but is generally (greater than 90%) strongly porphyritic and contains as much as 30 to 40% white potassium feldspar phenocrysts as long as 4 cm. The Pagano granite is cut by sparse muscovite-bearing aplite dikes up to 2 m wide, some of which contain minor biotite and/or tourmaline. Massive quartz veins up to 1 m wide which often have vuggy cores postdate the aplite dikes. Inclusions are uncommon in the Pagano granite.

Nash Hills. Plutonic rocks of the Nash Hills are uniformly massive, homogeneous, medium- to coarse-grained biotite granite. The unit is, however, often locally porphyritic and may contain from 1 to 10% potassium feldspar phenocrysts at any given locality. These phenocrysts average 2.5 to 4 cm in length. In the east central Nash Hills, a 3- to 5-m-thick aplitic border zone forms a chilled margin against overlying metasedimentary rocks. Aplite dikes both predate and postdate hydrothermal quartz veins. Inclusions are uncommon in plutonic rocks of the Nash Hills.

Whitmore Mountains. A grayish-white coarse- to very coarse-grained porphyritic biotite granite, named the Mount Seelig granite by Webers et al. [1982], composes about 95% of the bedrock exposures in the Whitmore Mountains. This unit is characterized by 20 to 25% white potassium feldspar phenocrysts that average 3 to 4 cm in length, but reach a maximum size of 7.5 cm. Biotite generally occurs in clots that occasionally define a weak to moderately well developed foliation. Trace amounts of hornblende are locally present.

A light gray fine-grained (less than 1 mm) equigranular muscovite biotite granite named the Linck Nunataks granite by Webers et al. [1982] crops out in the Linck Nunataks group and at Mount Chapman. This unit occupies all of West Linck Nunataks, but at East Linck Nunataks it crosscuts and contains inclusions of the Mount Seelig granite. At Mount Chapman the Linck Nunataks granite occurs as sheetlike masses as much as 25 m thick and 200 m long which crosscut and contain inclusions of the Mount Seelig granite. No metasedimentary inclusions were found in the Mount Seelig or Linck Nunataks granites.

A wide variety of late-stage dikes crosscut the two intrusive units of the Whitmore Mountains. Two stages of aplite dikes, some of which are garnet bearing, crosscut the Linck Nunataks granite and each other in the central Linck Nunataks group. At Mount Chapman both intrusive units are cut by aplite, pegmatite, and composite pegmatite-aplite dikes. At Mount Seelig the Seelig granite is cut by a swarm of aplite dikes.

Mount Woollard. The west and northwest slopes of Mount Woollard are underlain by a strongly deformed complex of migmatized micaceous paragneiss, amphibiotite, and pegmatite. At least three phases of pegmatite are present, all of which are cut by later microgranite dikes and sills. These microgranite intrusions are petrographically similar to fine-grained massive homogeneous biotite granite which underlies the more easterly part of the Mount Woollard massif. These rocks are described in more detail by Storey and Dalziel [this volume].

Martin Hills. On the southeast flank of the Martin Hills, sedimentary rocks are intruded by a porphyritic rhyodacite stock which has an outcrop area of less than 0.25 km^2. This body has the shape of a vertical plug from which many dikes, up to 20 m thick, branch laterally and vertically. A 30- to 40-cm-thick aphanitic grayish-green chill zone containing sparse 2 mm crystals of β quartz grades rapidly into a coarsely porphyritic rock with 10 to 15% β quartz phenocrysts (5 mm), 2 to 5% white potassium feldspar phenocrysts (1 cm), and less than 1% sausseritized plagioclase crystals (3-5 mm).

Thiel Mountains. The most widespread rock type in the Thiel Mountains is fine to medium grained, dark gray to almost black quartz monzonite porphyry. Feldspar phenocrysts are prominent, especially on weathered surfaces, but the only mafic minerals discernible in hand sample are chlorite and widely scattered grains of violet glassy cordierite. This rock is remarkably massive and uniform in appearance throughout the Thiel Mountains, but Ford [1964] has noted locally developed layering and interpreted it as the result of magmatic flowage or gravitational settling of crystals accompanied by magmatic flowage.

At several localities, mainly in the northern and southern parts of the range, the quartz monzonite porphyry is intruded by massive homogeneous granodiorite. The granodiorite is white to gray, coarse grained, and porphyritic with white potassium feldspar phenocrysts that average 1.5 to 2.0

TABLE 1. Representative Chemical Data for Granitic Rocks From the Ellsworth-Whitmore Mountains Crustal Block and Thiel Mountains, West Antarctica

	V42 MW-g	V39d PH-g	V39c PH-ap	V39b PH-al	V52b PH-gi	V28-38[a] N-g	V32c N-ap	V24 M	V22a WM-Sg	V20-43[a] WM-Lg	V21f WM-ap	V12g P-g	V12b P-ap	V4-10[a] T-gr	V17c T-al	V6-8[a] T-q	Av.27 T-q
						Major Element Oxides, in Weight Percent											
SiO_2	66.76	74.36	74.45	76.65	75.85	71.45	74.98	75.11	72.41	71.56	75.49	71.62	76.43	71.12	75.27	69.78	69.8
TiO_2	0.45	0.17	0.04	0.07	0.19	0.40	0.08	0.13	0.32	0.29	0.04	0.35	0.10	0.45	0.11	0.63	0.69
Al_2O_3	14.01	13.38	14.11	12.85	14.34	13.64	14.39	12.41	13.97	14.15	13.61	13.80	12.90	14.07	12.81	14.13	14.5
Fe_2O_3	0.71	0.35	0.16	0.27	0.61	0.47	0.17	0.56	0.87	0.71	0.24	1.13	0.19	0.67	0.42	0.95	1.1
FeO	1.95	0.86	0.49	0.69	1.31	2.18	0.60	0.55	1.24	1.19	0.45	0.99	0.32	2.30	0.77	3.10	3.0
MnO	0.02	0.07	0.18	0.11	0.05	0.06	0.06	0.04	0.05	0.05	0.13	0.05	0.03	0.06	0.04	0.04	0.02
MgO	2.01	0.13	0.01	0.03	0.43	0.70	0.05	0.45	0.35	0.51	0.03	0.55	0.03	0.81	0.17	1.05	1.2
CaO	1.59	0.88	0.40	0.50	0.70	1.44	0.35	2.50	1.83	1.09	0.53	1.26	0.47	1.63	0.33	1.95	2.2
Na_2O	3.33	3.43	4.67	4.13	0.21	2.92	4.11	1.06	3.35	3.25	4.45	3.09	3.62	2.77	2.57	2.69	2.5
K_2O	7.64	5.65	4.10	4.21	4.81	5.05	4.03	4.47	4.93	5.51	4.46	5.14	5.13	4.73	6.66	4.01	3.8
P_2O_5	0.34	0.04	0.03	0.03	0.14	0.13	0.11	0.03	0.07	0.20	0.05	0.05	0.07	0.13	0.13	0.17	0.18[b]
H_2O	1.60	1.17	1.31	0.81	1.71	1.57	0.61	2.97	0.79	1.24	0.51	1.34	0.32	0.95	0.67	1.58	1.1[b]
Total	100.41	100.49	99.95	100.35	100.35	100.01	99.54	100.28	100.20	99.73	99.99	99.60	99.65	99.69	99.95	100.11	100.1
						Trace Elements, in Parts Per Million											
Ba	606	121	18	25	187	448	38	324	397	436	47	376	33	430	148	495	
Cr	2	nd	nd	nd	12	10	nd	nd	nd	13	nd	3	nd	13	nd	11	
Nb	12	59	54	54	18	25	124	62	27	23	55	26	36	12	8	16	
Ni	nd	1	2	2	5	2	2	1	nd	4	nd	1	14	6	nd	5	
Sr	117	60	1	3	7	130	11	148	178	90	12	114	14	101	40	144	
Rb	251	503	978	950	655	361	1183	338	259	307	279	307	375	247	281	167	
Th	27	50	26	31	202	35	22	82	47	21	6	20	10	17	9	19	
V	98	3	nd	nd	12	33	nd	2	16	35	nd	21	nd	35	3	35	
Y	42	123	119	130	17	64	21	106	32	26	26	42	36	49	29	57	
Zr	184	142	85	71	152	227	28	133	177	165	59	180	62	184	65	333	
						Rare Earth Elements, in Parts Per Million											
La	44.78	41.49	14.27	8.10	26.01	51.13	2.71	51.15	52.25	30.66	5.67	35.76	6.82	31.79	7.54	47.12	
Ce	97.16	88.87	47.70	21.29	59.71	108.00	8.76	112.08	90.91	63.63	7.96	73.49	15.42	65.89	17.09	98.67	
Pr	9.76	10.81	6.19	3.08	6.32	12.73	1.40	13.31	9.39	7.83	1.19	8.68	1.90	7.95	2.10	11.65	
Nd	43.27	40.37	23.30	12.78	24.27	49.15	5.56	50.84	31.58	29.71	4.68	33.56	7.18	31.10	7.92	46.95	
Sm	9.10	11.21	11.07	6.97	4.78	10.41	2.62	13.14	5.60	6.89	1.46	7.52	2.56	7.00	2.59	9.88	
Eu	0.58	0.64	0.03	0.09	0.28	1.05	0.01	0.55	1.10	0.73	0.12	0.93	0.09	1.17	0.41	1.59	
Gd	8.83	11.48	10.45	8.64	3.66	8.90	1.75	12.66	4.54	5.93	1.55	6.74	2.56	6.60	2.49	9.10	
Dy	5.16	14.79	18.24	13.89	3.35	8.35	2.18	14.09	4.45	4.92	2.89	6.26	4.27	6.93	3.70	8.03	
Ho	0.95	3.10	3.60	2.88	0.61	1.63	0.33	2.79	0.90	0.84	0.61	1.18	0.87	1.37	0.72	1.59	
Er	2.36	9.76	12.98	9.75	1.82	4.79	1.14	8.42	2.77	2.21	2.10	3.41	3.09	4.07	2.32	4.51	
Yb	1.86	9.58	23.73	12.76	1.69	4.48	2.19	8.14	2.54	1.67	2.75	3.12	4.48	3.58	2.61	3.80	
Lu	0.23	1.46	3.67	2.04	0.26	0.71	0.38	1.25	0.42	0.28	0.49	0.52	0.76	0.57	0.41	0.61	
Σ14REE	226.3	247.4	179.9	105.5	134.6	263.3	29.6	291.7	207.5	156.4	32.1	182.7	51.0	169.6	50.7	245.5	
Ce_N/Yb_N	13.3	2.4	0.5	0.4	9.0	6.3	1.0	3.5	9.1	9.8	0.7	6.0	0.9	4.8	1.7	6.6	
La_N/Sm_N	3.0	2.3	0.8	0.7	3.3	3.0	0.6	2.4	5.7	2.7	2.4	2.9	1.6	2.8	1.8	2.9	
Gd_N/Yb_N	3.8	1.0	0.3	0.5	1.7	1.6	0.6	1.2	1.4	2.9	0.4	1.7	0.5	1.5	0.8	1.9	
Eu/Eu^a_N	0.2	0.2	0.01	0.04	0.2	0.3	0.02	0.1	0.7	0.3	0.2	0.4	0.1	0.5	0.5	0.5	

Av.27, average of 27 analyses [Ford and Himmelberg, 1976]. MW, Mount Woollard; PH, Pirrit Hills; N, Nash Hills; M, Martin Hills; WM, Whitmore Mountains; P, Pagano Nunatak; T, Thiel Mountains; g, granite; ap, aplite; al, alaskite; gi, greisen; Sg, Seelig granite; Lg, Linck Nunatak granite; gr, granodiorite; q, quartz monzonite porphyry; nd, not detected.
^aH$_2$O$^+$.
^bAverage of two analyses.

cm long. Clots of slightly chloritized biotite are the only varietal minerals present except for sporadic grains of muscovite. Dikes of coarse-grained granodiorite porphyry, pegmatite (locally tourmaline bearing), aplite, and muscovite biotite alaskite crosscut both the granodiorites and the quartz monzonite porphyry.

Petrography

Granites and granodiorites. Plagioclase is usually weakly or moderately sausseritized, has compositions in the oligoclase-andesine range, and is zoned through a range of only 10 to 20% of the anorthite molecule. Potassium feldspar is generally strongly perthitic; grid twinning is not commonly developed. Biotite (Z = Y, dark brown to reddish brown; X, pale yellow) is the only mafic mineral present in the Pirrit Hills, Pagano Nunatak, Nash Hills, Linck Nunataks, and Mount Woollard granites and the Thiel Mountains granodiorites. It is usually slightly to moderately chloritized and often contains numerous pleochroic haloes. Hornblende, accompanied by biotite, occurs only in the Mount Seelig granite. Primary muscovite (using the criteria of Miller et al. [1981] is a minor constituent of the Pagano Nunatak, Pirrit Hills, and Linck Nunataks granites and the Thiel Mountains granodiorites. Zinnwaldite (a colorless Li-bearing mica) occurs in some of the Pirrit Hills aplites. Color indices of all plutonic units except the Mount Seelig granite are less than 5. Electron microprobe analyses of micas and amphiboles are available from the senior author on request. Apatite, sparse zircon, and small uncommon opaque grains are the only conspicuous accessory minerals.

The groundmass of the Martin Hills stock is a very fine grained (less than 1 mm) equigranular quartz-feldspar intergrowth that is extensively altered to sericite and minor amounts of clinozoisite, carbonate, and clay minerals. Plagioclase phenocrysts are totally replaced by sericite. Potassium feldspar phenocrysts show minor alteration to sericite and clinozoisite, but β quartz crystals are unaltered and show only minor effects of resorption. Texture and composition suggest that the groundmass represents divitrified glass.

Quartz monzonite porphyry. Phenocrysts are largely angular plagioclase with strongly developed oscillatory zoning and anhedral, often resorbed, quartz along with lesser amounts of potassium feldspar, hypersthene (strongly pleochroic in shades of red and green and often rimmed with biotite and/or chlorite), and cordierite (usually partially altered to chlorite). Staining techniques indicate that the very fine grained groundmass is an equigranular mixture of quartz and potassium feldspar. Ford and Himmelberg [1976] published analyses of cordierite and hypersthene phenocrysts.

Geochemistry

A representative set of chemical data for granitic rocks forming the EWM and Thiel Mountains is given in Table 1. Additional analyses not listed

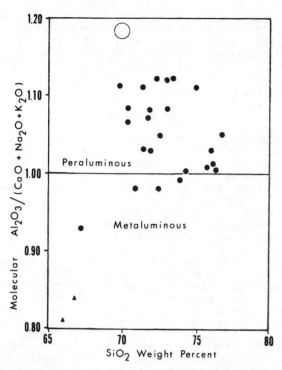

Fig. 2. Plot of percentage of SiO_2 versus molecular $Al_2O_3/(CaO+Na_2O+K_2O)$ for igneous rocks of the EWM and Thiel Mountains. Open circle represents average of 27 analyses of Thiel Mountains quartz monzonite porphyry [Ford and Himmelberg, 1976]. Solid triangles represent analyses from Mount Woollard.

in this table can be obtained from the senior author by request. The Mount Seelig hornblende-bearing biotite granite of the Whitmore Mountains and the Mount Woollard biotite granite are metaluminous diopside-normative granitoids. All other plutonic rocks discussed in this paper are mildly peraluminous corundum-normative biotite (locally muscovite bearing) granitoids. This peraluminous character is well displayed in Figure 2, a plot of molecular $Al_2O_3/(CaO+Na_2O+K_2O)$ versus SiO_2. All peraluminous rocks are leucogranites with color indices of 5 or less in which biotite predominates over muscovite. Except in the Whitmore Mountains, only one plutonic unit is exposed at each locality, and these units show remarkably uniform compositions (i.e., little or no fractionation). Within these individual plutonic units, SiO_2 varies by only 2 to 3% (it changes by only 5% in the two units of the Whitmore Mountains), and this is accompanied by only very slight (a few tenths of one percent) decreases in FeO, Fe_2O_3, MgO, CaO, TiO_2, and sometimes P_2O_5, slight increases or no change in Na_2O, and relatively constant K_2O. When compared to their host rocks, aplite dikes, however, show significant decreases in FeO, Fe_2O_3, MgO, CaO, and TiO_2, significant increases in Na_2O, but a drop in K_2O (except in the Whitmore Mountains aplites, where K_2O remains relatively constant,

and in the Thiel Mountains aplites, where it increases).

Although there are individual exceptions, this suite of rocks generally possesses the low iron, calcium, magnesium, and titanium and moderately high potassium and silica (70-76%) contents typical of many highly differentiated leucocratic peraluminous granites [Armstrong and Boudette, 1984]. They are not, however, especially rich in aluminum. Except for the Mount Woollard granite, which is only very slightly alkaline, all analyzed rocks are subalkalic in the classification of Irvine and Barager [1971]. Some, however, are too iron rich to be considered calc-alkaline. On a plot (not shown here) of SiO_2 versus (FeO* + MgO), where FeO* is total iron, granites from the Thiel Mountains, the Nash Hills, Pagano Nunatak, and Mount Woollard and the Linck Nunataks granite of the Whitmore Mountains plot in the calc-alkaline field of Miyashiro [1974], but the quartz monzonite porphyry of the Thiel Mountains, the Mount Seelig granite of the Whitmore Mountains, and the Pirrit Hills granites plot in the tholeiitic field.

There is little fractionation of trace elements (we have plotted Ba, Sr, Rb, Y, and Zr against SiO_2; Rb and Sr against Ba; and Rb against Sr) within any of the individual plutons and little significant difference in trace element content between any of the individual plutons. An exception to this is the very low Ba, Sr, and Sr/Ba ratio and very high Rb, Y, Rb/Sr, and Rb/Ba ratios of the Pirrit Hills granites relative to the other plutons. Aplite dikes from all plutons are strongly depleted in Ba, Sr, and Zr relative to their host rocks.

Abundances of REE are listed in Tables 1 and 2. Total REE contents were determined by estimating the amount of Tb and Tm from chondrite-normalized diagrams (see below) and adding these amounts to the reported total of $\Sigma 12REE$. Chondrite-normalized REE patterns from the Pirrit Hills (Figure 3) show moderate REE enrichment and approximately equal amounts of both light REE (LREE) and heavy REE (HREE), virtually no fractionation of either LREE or HREE, and strongly developed negative europium anomalies (Eu/Eu* ranges 0.01-0.02). There is a great deal of similarity between the patterns for the remaining plutonic centers of the EWM and the Thiel Mountains (Figures 4-7). All show moderate to strong LREE enrichment (100-250 times chondrite values) and moderate LREE fractionation (La_N/Sm_N averages 2.3-3.2). Heavy REE are less fractionated (Gd_N/Yb_N averages 1.3- 1.9). Europium anomalies are moderately negative (Eu/Eu* averages 0.3-0.6), and $\Sigma 14REE$ ranges from 128 to 292. There is, however, a progressive increase of overall REE fractionation between the various plutonic centers of the EWM as measured by the Ce_N/Yb_N ratio. Average values for this parameter are Pirrit Hills, 2.4; Martin Hills, 3.5; Pagano Nunatak, 6.0; Nash Hills, 6.3; the Mount Seelig granite, 9.1; the Linck Nunataks granite, 9.3; and Mount Woollard, 13.3. REE patterns from aplite and pegmatite dikes are virtually identical in shape to those of their host rocks, but in all cases $\Sigma 14REE$ is lower by a factor of 3-6 times,

TABLE 2. Rare Earth Element Content of Pegmatite Dikes and Inclusions From Granitic Rocks of the Ellsworth-Whitmore Mountains Crustal Block and Thiel Mountains, West Antarctica

	V21g	V39e	V1c	V2f	V4b	V12f	V38c	V48d
La	4.60	3.38	42.38	54.98	28.91	38.03	62.30	84.19
Ce	9.98	9.65	84.51	109.13	60.35	75.81	132.86	219.58
Pr	1.14	1.63	10.05	12.58	7.45	9.00	15.62	29.09
Nd	4.68	5.69	39.14	48.56	29.47	35.25	61.65	128.11
Sm	2.24	2.84	7.81	9.34	6.11	7.07	12.85	46.38
Eu	0.05	0.02	1.73	1.49	0.64	1.21	0.98	0.66
Gd	2.79	3.45	6.87	7.27	5.22	6.14	10.88	46.97
Dy	4.45	6.04	6.59	5.33	4.57	5.61	9.66	56.27
Ho	0.77	1.21	1.32	1.08	0.87	1.11	1.86	11.77
Er	2.21	4.06	3.87	3.49	2.56	3.37	5.38	37.65
Yb	2.39	6.28	3.48	3.44	2.04	2.87	5.05	41.22
Lu	0.35	0.97	0.56	0.59	0.33	0.48	0.81	5.78
$\Sigma 14Ree$	37.1	47.3	209.9	258.8	149.6	187.3	322.3	721.2
Ce_N/Yb_N	1.1	0.4	6.2	8.1	7.5	6.7	6.7	1.3
La_N/Sm_N	1.3	0.7	3.3	3.6	2.9	3.3	3.0	1.1
Gd_N/Yb_N	0.9	0.4	1.6	1.7	2.0	1.7	1.7	0.9
Eu/Eu^*	0.06	0.02	0.7	0.5	0.3	0.3	0.2	0.04

V21g, garnet-tourmaline pegmatite, Mount Chapman, Whitmore Mountains; V39e, biotite-beryl pegmatite, Mount Tidd, Pirrit Hills; V1c, biotite-plagioclase-hypersthene-cordierite granofels from quartz monzonite porphyry host, Hamilton Cliff, Thiel Mountains; V2f, biotite-green spinel-hypersthene-plagioclase-cordierite-granofels from quartz monzonite porphyry host, Hamilton Cliff; V4b, biotite-feldspar-sericite-quartz schist from granodiorite host, Reed Ridge, Thiel Mountains; V12f, garnet-muscovite-feldspar-biotite-quartz granofels, Pagano Nunatak; V38c, feldspar-muscovite-biotite-quartz granofels, Nash Hills; V48d, feldspar-biotite-quartz granofels, Harter Nunatak, Pirrit Hills.

and europium anomalies are much more strongly negative (0.01-0.2).

Source of Granitic Magmas

Igneous rocks of the EWM and from the Thiel Mountains thus show many characteristics of S-type granitoids [White and Chappell, 1983]: high but narrow range of SiO_2 (70-76% SiO_2); predominance of leucogranites, peraluminous mineralogy, molecular $Al_2O_3/(CaO+Na_2O+K_2O) > 1.05$; (xenoliths predominantly metasedimentary); scarcity of composite plutons; normative corundum; high K_2O/Na_2O and low CaO and Sr; and greisen-type alteration. It should be pointed out, however, that these rocks are only mildly peraluminous, and although primary muscovite is present, this mineral is not accompanied (except in the Thiel Mountains cordierite-bearing quartz monzonite porphyries) by any of the other peraluminous minerals commonly found in "two-mica" granites (garnet, cordierite, and the Al_2SiO_5 polymorphs).

The development of peraluminous granite has been attributed to (1) anatexis of peraluminous source rocks, especially pelitic metasediments; (2) fractional crystallization of metaluminous magma; or (3) reaction between late-stage magmas or subsolidus rocks and either hydrothermal fluids or host rocks [Clark, 1981]. We present here two models for the formation of the EWM peraluminous

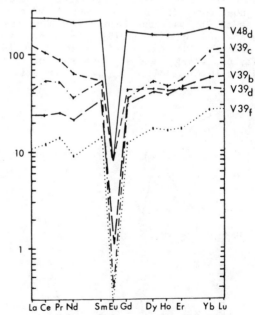

Fig. 3. REE distribution in igneous rocks of the Pirrit Hills. V39b is fine-grained granite; V39c, aplite; V39d, coarse-grained granite; V39f, biotite beryl pegmatite; and V48d, metasedimentary inclusion.

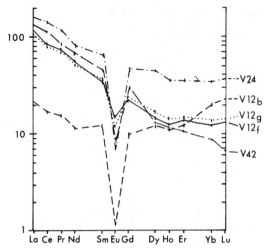

Fig. 4. REE distribution in igneous rocks of the Martin Hills, Mount Woollard, and Pagano Nunatak. V24 is chilled margin Martin Hills stock; V42, Mount Woollard granite; V12g, Pagano granite; V12b, Pagano aplite; and V12f, metasedimentary inclusion, Pagano Nunatak.

granites, both of which require substantial melting of continental crust. Xenoliths are uncommon to rare in our suite of rocks (none were found in the Whitmore Mountains), but those which are present all appear to be metasedimentary or granulitic.

REE abundances of selected inclusions (Table 2) are, in all cases, almost identical to those of their host rocks (Figures 3-5, and 7) and are very similar to REE patterns from shales described by Wildeman and Haskin [1973] and from average post-Archaean Australian sedimentary rocks [Nance and Taylor, 1976]. REE are less subject to fractionation in the sedimentary cycle than most other elements so the distribution of REE in sedimentary rocks is considered to be similar to their distribution in the continental crust [Nance and Taylor, 1976]. This suggests that the suite of plutonic rocks described in this paper is derived from the partial melting of continental crust; more particularly, it substantiates the idea proposed earlier [Ford, 1964] that the quartz monzonite porphyry of the Thiel Mountains was derived as a partial melt of charnockitic rocks of the East Antarctic shield, rocks which are not presently exposed in the Thiel Mountains.

Except for a more pronounced negative europium anomaly and slightly higher ΣREE, the REE distribution pattern of the Pirrit Hills granite is very similar to that of plagiogranite and other differentiates of mafic magma [Cullers and Graf, 1984]. This geochemical characteristic is consistent with the relatively low (0.708) initial $^{87}Sr/^{86}Sr$ ratio of these rocks [Millar and Pankhurst, this volume]. Cullers and Graf [1984] have suggested that large negative europium anomalies such as those shown by the EWM granites require at least a two-stage melting or crystallizaton history. A more deeply seated, more mafic plagioclase-rich batholith could produce an anatectic melt with a large negative europium anomaly such as the Pirrit Hills granite. If this is the case, then the increased REE fractionation of the various EWM plutonic centers mentioned above and the higher initial $^{87}Sr/^{86}Sr$ ratios [Millar and Pankhurst, this volume] shown by several of these plutons (Nash Hills, 0.712; Pagano Nunatak, 0.716; and Linck Nunataks granite, 0.722) suggest that at least some of the EWM granites are Pirrit Hills type differentiates contaminated with increasingly larger amounts of continental basement.

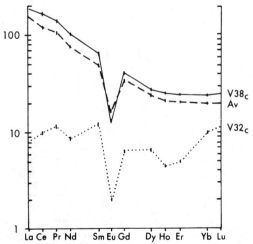

Fig. 5. REE distribution in igneous rocks of the Nash Hills. Av is average of two granites; V32c, aplite; and V38c, metasedimentary inclusion.

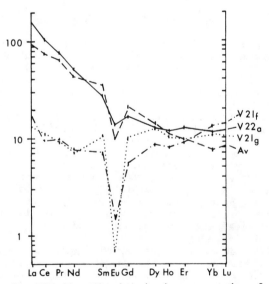

Fig. 6. REE distribution in igneous rocks of the Whitmore Mountains. Av is average of two Linck Nunataks granites; V21f, aplite; V21g, garnet tourmaline pegmatite; and V22a, Mount Seelig granite.

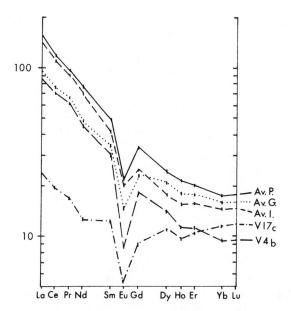

Fig. 7. REE distribution in igneous rocks of the Thiel Mountains. Av.G is average of two granodiorites; Av.I., average of two inclusions from quartz monzonite porphyry; Av.P., average of two quartz monzonite porphyries; V4b, metasedimentary inclusion from granodiorite; and V17c, alaskite dike.

Tectonic Setting

EWM granites represent a consanguinous suite of plutonic rocks that show great uniformity in chemical composition and a small spread in ages over a large distance. Although there is some between-pluton variation in initial $^{87}Sr/^{86}Sr$ ratio, petrographically, geochemically, and isotopically the EWM granites strongly resemble suites of peraluminous leucogranites developed in intracontinental terranes such as the Hercynian belt of Europe (summary by Pearce et al. [1984]) and in continental collision-type settings such as the Bhutan and Nepalese Himalaya [Dietrich and Gansser, 1981; Le Fort, 1981] and the Seward Peninsula of western Alaska [Hudson and Arth, 1983].

A collisional environment, however, is not compatible with the postulated Middle Jurassic tectonic setting of the EWM. The classical thrust- and nappe-type structures and medium-grade metamorphic host rocks typically found in association with collision zone granites are not present. In addition, all EWM granitic rocks were emplaced posttectonically following the compressional event which deformed their host rocks. Y/Nb ratios of the EWM granites when plotted (Figure 8) in the discrimination diagram of Pearce et al. [1984] also suggest a within-plate setting.

Contemporaneous magmatic activity in the Transantarctic Mountains (the Ferrar Supergroup), South Africa (the Karoo Dolerites), and southern South America (silicic volcanic rocks of the Tobifera Series and S-type granites of the Cordillera Darwin) suggests a broad pattern of regional extensional tectonism in the southern hemisphere during the time of emplacement of the EWM granites. We therefore attribute the origin of the EWM granites to the melting of continental crust during a Middle Jurassic thermal event followed by their emplacement in a neutral or extensional environment. Pitcher [1982] has proposed a similar model whereby highly siliceous S-type magmas may form in an extensional (rifted) environment which develops following the relaxation of a compressional event, and Wickham and Oxburgh [1985] and Kontak et al. [1985] have documented this mode of granite emplacement in the eastern Pyrenees and the eastern Cordillera of Peru, respectively.

Our proposed model is supported by aeromagnetic data [Garrett et al., this volume] that suggest a more extensive sub-ice east-west development of granitic rocks more or less perpendicular to the long axis of the EWM and by the large-scale sub-ice rifting reported in West Antarctica by Masolov et al. [1981]. The aeromagnetic data can, however, be interpreted in two ways: (1) granitic magmas were intruded as much larger masses than those presently exposed in the study area, and these masses were later disrupted by extensional tectonic activity, or (2) granitic magmas were emplaced in east-west-oriented fractures during extensional tectonic activity. The former explanation is preferred, as contemporaneous extrusive rocks are completely lacking, there are no characteristics of peralkaline magmatism present in the EWM granites, and we know of no other S-type leucogranite suites where plutonic rocks occur in linear belts oriented at high

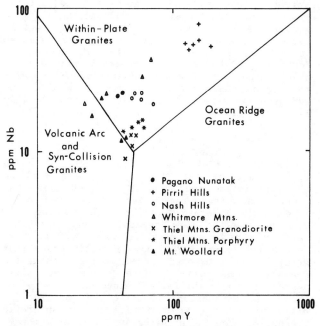

Fig. 8. Plot of Nb versus Y for igneous rocks of the EWM and Thiel Mountains. Fields of various granite types from Pearce et al. [1984].

angles to the major axis of an elongated outcrop belt.

Strong analogies exist between the settings of the posttectonic Hercynian Cornubian batholith of southwest England [Exley and Stone, 1982], the tin granites of the Seward Peninsula of Alaska [Hudson and Arth, 1983], and the EWM granites. The Cornubian batholith consists of several major plutons which delineate a southwest-northeast elongated belt. Gravity data suggest that these plutons are all cupolas rising from a buried continuous body of granite which is locally cut by major troughs. Tin granites of the Seward Peninsula are much smaller in volume and were emplaced at higher crustal levels than slightly older, much more extensive granites and granodiorites. Hudson and Arth [1983] concluded that the tin granites were derived by fractional crystallization of the more deeply seated granites and granodiorites and were then emplaced as highly evolved epizonal stocks or cupolas.

Except for the Mount Seelig granite of the Whitmore Mountains, all EWM granites are highly evolved rocks when geochemically compared to average continental granite. In addition to the above mentioned comparisons, their trace element and REE contents also approach those of the peraluminous pegmatitic granites of southeastern Manitoba [Goad and Černéy, 1981]. More specifically, values of Rb, Rb/Sr, and K/Ba are noticeably higher, and those of Ba, Sr, and K/Rb are noticeably lower than those of average continental granite [Taylor, 1964] or low-Ca granite [Turekian and Wedepohl, 1961]. Granitic averages for the above elements and ratios are 150-170, 1.7, 50, 600-850, 100-285, and 225.

The above mentioned features suggest that the EWM granites represent highly evolved cupolas derived by fractional crystallization of more deeply seated, nonexposed, less evolved granitic rocks. The EWM granites thus appear very similar in setting to the tin granites of the Seward Peninsula, although our study area is not quite as deeply eroded. The Mount Woollard massif appears to be a deep-seated migmatite zone related to the EWM granitic magmatism which has been uplifted by later rifting.

Correlations

During the 1984-1985 austral summer a detailed field study of the granitic rocks exposed on Thurston Island and along the Eights Coast was undertaken. These rocks have not yet been examined in the laboratory, and at this time no conclusions can be made regarding their possible relationship to the granitic rocks of the EWM. Vennum and Rowley [1986] have recently shown that granitic rocks exposed in the southern Antarctic Peninsula (the Lassiter Coast Intrusive Suite) represent a suite of Early Cretaceous I-type granitoids formed during subduction of the Pacific plate beneath the western edge of the Antarctic Peninsula. The EWM granites appear to be geochemically similar to granitic rocks of the S-type Darwin granite suite which is exposed in the Cordillera Darwin of southern Chile (I.W.D. Dalziel, University of Texas, personal communication, 1985), but data are again insufficient to make meaningful comparisons.

Granitic rocks of the Thiel Mountains, although geochemically similar to those of the EWM, have previously been radiometrically dated as Cambrian-Ordovician and are thus not consanguineous. The Thiel Mountains granodiorites are most likely part of the Granite Harbour Intrusive Series, which is exposed in the basement complex throughout the length of the Transantarctic Mountains [Vetter et al., 1983]. We consider the Thiel Mountains quartz monzonite porphyries to be correlative with the upper Precambrian Wyatt Formation of the nearby La Gorce Mountains of the Transantarctic range [Stump, 1983].

Acknowledgments. This research was financed by National Science Foundation grant DPP82-13798 to I.W.D. Dalziel, then at Columbia University. Our logistical problems involved in reaching the field area were solved by U.S. Navy Squadron VXE-6 and two very capable British Antarctic Survey Twin Otter pilots, Gary Studd and Ed Mehrten. We thank Peter Schiffman of the University of California, Davis, for his help with the microprobe analyses, Barbara Young for her help with the geochemical computations and manuscript, and Ron Leu of Sonoma State University for making the thin sections. The figures were drafted by David Fowler of Sonoma State University.

References

Armstrong, F. C., and E. L. Boudette, Two-mica-granites: Part A, Their occurrence and petrography, Open File Rep. U.S. Geol. Surv., 84-173, 1984.

Clarke, D. E., Peraluminous granites, an introduction, Can. Mineral., 19, 1-2, 1981.

Craddock, C., The East Antarctic-West Antarctic boundary between the ice shelves: A review, in Antarctic Earth Science, edited by R. L. Oliver, P. R. James, and J. B. Jago, pp. 94-97, Australian Academy of Science, Canberra, 1983.

Craddock, C., G. F. Webers, and J. J. Anderson, Geology of the Ellsworth Mountains-Thiel Mountains Ridge (abstract), in Antarctic Geoscience, edited by C. Craddock, p. 849, University of Wisconsin Press, Madison, 1982.

Cullers, R. L., and J. L. Graf, Rare earth elements in igneous rocks of the continental crust: Intermediate and silicic rocks--ore petrogenesis, in Rare Earth Element Geochemistry, edited by P. Henderson, pp. 275-316, Elsevier, New York, 1984.

Dalziel, I. W. D., and D. H. Elliot, West Antarctica: Problem child of Gondwanaland, Tectonics, 1, 3-19, 1982.

Dietrich, V., and A. Gansser, The leucogranites of the Bhutan Himalaya, Schweiz. Mineral. Petrogr. Mitt., 61, 177-201, 1981.

Exley, C. S., and M. Stone, Geological setting of the Hercynian granites, in Igneous Rocks of the British Isles, edited by D. S. Sutherland, pp. 287-292, John Wiley, New York, 1982.

Ford, A. B., Cordierite-bearing hypersthene-quartz monzonite-porphyry in the Thiel Mountains and its regional importance, in Antarctic Geology, pp. 429-441, North-Holland, Amsterdam, 1964.

Ford, A. B., and J. M. Aaron, Bedrock geology of the Thiel Mountains, Antarctica, Science, 137(3532), 751-752, 1962.

Ford, A. B., and G. R. Himmelberg, Cordierite and orthopyroxene megacrysts in late Precambrian volcanic rocks of the Thiel Mountains, Antarctica, Antarct. J. U. S., 11, 260-262, 1976.

Ford, A. B., and R. S. Sumsion, Late Precambrian silicic pyroclastic volcanism in the Thiel Mountains, Antarctica, Antarct. J. U. S., 6, 185-186, 1971.

Ford A. B., H. A. Hubbard, and T. W. Stern, Lead-alpha ages of zircon in quartz monzonite porphyry, Thiel Mountains, Antarctica, A preliminary report, U.S. Geol. Surv. Prof. Paper, 450-E, 105-107, 1963.

Garrett, S. W., L. D. B. Herrod, and D. R. Mantripp, Crustal structure of the area around Haag Nunataks, West Antarctica: New aeromagnetic and bedrock elevation data, this volume.

Goad, B. E., and P. Cerný, Peraluminous pegmatitic granites and their pegmatitic aureoles in the Winnipeg River district, southeastern Manitoba, Can. Mineral., 19, 177-194, 1981.

Hudson, T., and J. G. Arth, Tin granites of Seward Peninsula, Alaska, Geol. Soc. Am. Bull., 94, 768-790, 1983.

Irvine, T. N., and W. R. A. Barager, A guide to the chemical classification of the common volcanic rock types, Can. J. Earth Sci., 8, 523-548, 1971.

Kontak, D. J., A. H. Clark, E. Farrar, and D. F. Strong, The rift-associated Permo-Triassic magmatism of the Eastern Cordillera: A precursor to the Andean Orogeny, in Magmatism at a Plate Margin, edited by W. S. Pitcher, M. P. Atherton, E. J. Cobbing, and R. D. Beckinsale, pp. 36-44, John Wiley, New York, 1985.

Le Fort, P., Manaslu leucogranite: A collision signature of the Himalaya: A model for its genesis and emplacement, J. Geophys. Res., 86, 10,545-10,568, 1981.

Longshaw, S., and D. H. Griffiths, A paleomagnetic study of some rocks from the Antarctic Peninsula and its implications for Gondwanaland reconstructions, Geol. Soc. London Newsl., 10(5), 8-9, 1981.

Masolov, V. N., R. G. Kurinin, and G. E. Grikurov, Crustal structures and tectonic significance of Antarctic rift zones (from geophysical evidence), in Gondwana Five, edited by M. M. Cresswell and P. Vella, pp. 303-310, A. A. Balkema, Rotterdam, 1981.

Millar, I. L., and R. J. Pankhurst, Rb/Sr geochronology of the region between the Antarctic Peninsula and the Transantarctic Mountains: Haag Nunataks and Mesozoic granitoids, this volume.

Miller, C. F., E. F. Stoddard, L. F. Bradfish, and W. A. Dollase, Composition of plutonic muscovite: Genetic implications, Can. Mineral., 19, 25-34, 1981.

Miyashiro, A., Volcanic rock series in island arcs and active continental margins, Am. J. Sci., 274, 321-355, 1974.

Nakamura, N., Determination of REE, Ba, Fe, Mg, Na, and K in carbonaceous and ordinary chondrites, Geochim. Cosmochim. Acta, 38, 757-775, 1974.

Nance, W. B., and S. R. Taylor, Rare earth element patterns and crustal evolution: Australian post-Archaean sedimentary rocks, Geochim. Cosmochim. Acta, 40, 1539-1551, 1976.

Pearce, J. A., N. B. W. Harris, and A. G. Tindle, Trace element discrimination diagrams for the tectonic interpretation of granitic rocks, J. Petrol., 25, 956-983, 1984.

Pitcher, W. S., Granite type and tectonic environment, in Mountain Building Processes, edited by K. J. Hsü, pp. 19-40, Academic, Orlando, Fla., 1982.

Schmidt, D. L., and A. B. Ford, Geology of the Pensacola and Thiel Mountains, scale 1:1,000,000, in Geologic Maps of Antarctica, Antarct. Map Folio Ser., Folio 12, edited by V. C. Bushnell and C. Craddock, plate 5, American Geographical Society, New York, 1969.

Storey, B. C., and I. W. D. Dalziel, Outline of the structural and tectonic history of the Ellsworth Mountains-Thiel Mountains ridge, West Antarctica, this volume.

Streckeisen, A. L., To each plutonic rock its proper name, Earth Sci. Rev., 12, 1-33, 1976.

Stump, E., Type locality of the Ackerman Formation, La Gorce Mountains, Antarctica, in Antarctic Earth Science, edited by R. L. Oliver, P. R. James, and J. B. Jago, pp. 170-174, Australian Academy of Science, Canberra, 1983.

Taylor, S. R., Abundance of chemical elements in the continental crust: A new table, Geochim. Cosmochim. Acta, 28, 1273-1285, 1964.

Turekian, K. K., and K. H. Wedepohl, Distribution of the elements in some major units of the earth's crust, Geol. Soc. Am. Bull., 72, 1723-1728, 1961.

Vennum, W. R., and P. D. Rowley, Reconnaissance geochemistry of the Lassiter Coast Intrusive Suite, southern Antarctic Peninsula, Geol. Soc. Am. Bull., in press, 1986.

Vennum, W. R., and B. C. Storey, Correlation of gabbroic and diabasic rocks from the Ellsworth Mountains, Hart Hills, and Thiel Mountains, West Antarctica, this volume.

Vetter, U., N. W. Roland, H. Kreuzer, A. Höhndorf, H. Lenz, and C. Besang, Geochemistry, petrography, and geochronology of the Cambro-Ordovician and Devonian-Carboniferous granitoids of northern Victoria Land, Antarctica, in Antarctic Earth Science, edited by R. L. Oliver, P. R. James, and J. B. Jago, pp. 140-143, Australian Academy of Science, Canberra, 1983.

Webers, G. F., C. Craddock, M. A. Rogers, and J. J. Anderson, Geology of the Whitmore Mountains, in Antarctic Geoscience, edited by C. Craddock, pp. 841-847, University of Wisconsin Press, Madison, 1982.

Webers, G. F., C. Craddock, M. A. Rogers, and J. J. Anderson, Geology of Pagano Nunatak and the Hart Hills, in Antarctic Earth Science, edited by R. L. Oliver, P. R. James, and J. B. Jago, pp. 251-255, Australian Academy of Science, Canberra, 1983.

White, A. J. R., and B. W. Chapell, Granitoid types and their distribution in the Lachlan Fold Belt, southeastern Australia, in Circum-Pacific Plutonic Terranes, Geol. Soc. Am. Mem. 159, edited by J. A. Roddick, pp. 21-34, Geological Society of America, Boulder, Colo., 1983.

Wickham, S. M., and E. R. Oxburgh, Continental rifts as a setting for regional metamorphism, Nature, 318, 330-333, 1985.

Wildeman, T. R., and L. A. Haskin, Rare earths in Precambrian sediments, Geochim. Cosmochim. Acta, 37, 419-438, 1973.

Copyright 1987 by the American Geophysical Union.

Rb-Sr GEOCHRONOLOGY OF THE REGION BETWEEN THE ANTARCTIC PENINSULA AND THE TRANSANTARCTIC MOUNTAINS: HAAG NUNATAKS AND MESOZOIC GRANITOIDS

I. L. Millar and R. J. Pankhurst

British Antarctic Survey, Natural Environment Research Council, High Cross
Cambridge, United Kingdom

Abstract. Seventy-two new Rb-Sr whole-rock analyses are reported for Haag Nunataks, Mount Woollard, the Whitmore Mountains, the Pirrit and Nash hills, and Pagano Nunatak. For Haag Nunataks, three isochrons for gneisses and later aplogranite and microgranite sheets establish the age of crustal formation as 1000-1100 Ma. No other basement rocks of this age are known from the Antarctic Peninsula or Ellsworth Land. Results from the migmatite-pegmatite complex at Mount Woollard are inconclusive but do not suggest that this represents Precambrian crystalline basement. Provisional results for the Whitmore Mountains granites are compatible with crystallization of all components within error of a 182 ± 5 Ma isochron for fine-grained microgranite, but variation in initial $^{87}Sr/^{86}Sr$ from 0.707 for porphyritic granites to 0.722 for the microgranite rule out simple crystal fractionation models which require a common parental magma. The granites of the Ellsworth-Thiel mountains ridge are well dated as Middle Jurassic by the new data: Pirrit Hills 173 ± 3 Ma, Nash Hills 175 ± 8 Ma, and Pagano Nunatak 175 ± 8 Ma. Initial $^{87}Sr/^{86}Sr$ ratios of 0.707, 0.712, and 0.716, respectively, confirm that these are intracratonic S-type granites with a large crustal component involved in magma generation. The dolerite of Lewis Nunatak is shown by its Rb, Sr, and $^{87}Sr/^{86}Sr$ composition to be a member of the Jurassic Ferrar Supergroup.

Introduction

This paper describes the first detailed Rb-Sr whole-rock geochronological results from the "Ellsworth Mountains-Thiel Mountains Ridge" [Craddock et al., 1982] and Haag Nunataks. Samples were collected, mostly by R.J.P., during the 1983-1984 British Antarctic Survey/United States Antarctic Research Program (BAS/USARP) field expedition [Dalziel and Pankhurst, 1985].

The area of investigation is shown in Figure 1. Previous geochronological data consist mainly of K-Ar mineral ages resulting from the 1964-1965 United States expeditions by University of Minnesota geologists. Individual details are given below, but the overall impression has been formed that the granitic rocks of the area are mostly of Early to Middle Jurassic age, possibly rather older in the Whitmore Mountains. This is essentially confirmed by the present work.

New Geochronological Data

Geochronological suites of unweathered samples were collected, usually from single outcrops identified by a common BAS station number (R.22xx). Individual samples representing as much variation in igneous composition as possible were collected using 5-kg sledge hammers or, occasionally, a rock drill and wedges to split the rock. Ten kilograms was the average sample size, rather less for some aplites and dike material.

Crushed samples were analyzed at the British Geological Survey's isotope geology laboratories in London; the same techniques were used as in previous BAS work [Pankhurst, 1982, 1983]. These involve determining Rb, Sr, and Rb/Sr by X ray fluorescence spectrometry on pressed powder pellets. One-sigma errors are generally ±0.5% on Rb/Sr (and hence on $^{87}Rb/^{86}Sr$) ratios provided that Rb and Sr are both present in concentrations above 50 ppm, with fairly rapid increase in errors below this level, e.g., around 4% for samples with 4 ppm Sr. Isotopic analyses of Sr, with $^{87}Sr/^{86}Sr$ ratios normalized to $^{86}Sr/^{88}Sr = 0.1194$, were carried out on two fully automated mass spectrometers: a single-collector VG Micromass (MM) 30 and a five-collector VG 354. In the latter case the three central collectors, adjusted to a spacing of 1 amu for Sr, were used in a switched triple collection mode [Turner, 1982], achieving internal precision of around 10 ppm in 1-1.5 hours as opposed to around 50 ppm on the VG MM 30. However, because of the difficulty of reproducing results for radiogenic Sr ($^{87}Sr/^{86}Sr \geq 0.710$) on small replicate aliquots of rock powders due to inhomogeneity, an overall error of ±0.01% (1 sigma) has been assumed for all analyses. This measurement is independent of the Rb/Sr ratio determination. Results over the period concerned for NBS 987 have given 0.71034 ± 0.00003 and 0.710230 ± 0.000005 for the two machines, respectively, and comparison of samples analyzed on both justifies the adjustment of MM 30 results by -0.00010. Adjusted results are given in Tables 1

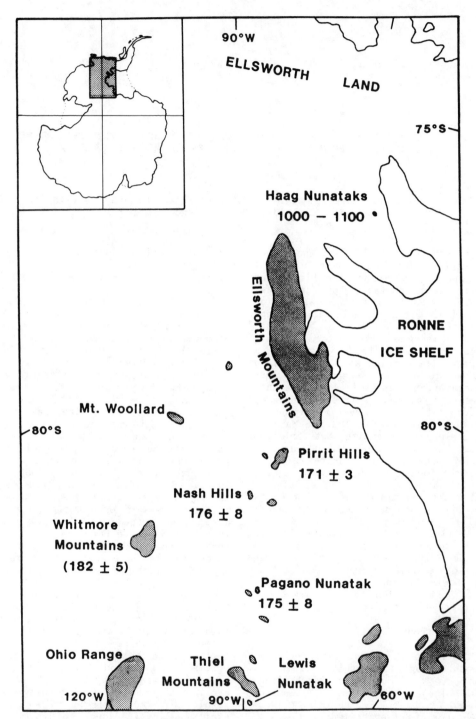

Fig. 1. Sketch map of area studied with sample localities and ages determined by this work.

and 2 and are effectively equivalent to Eimer and Amend $SrCO_3$ = 0.70800.

Isochrons have been fitted by using the York-Williamson regression [Williamson, 1968] with $\lambda Rb = 1.42 \times 10^{-11}$ a^{-1}. Where MSWD >3, indicating a high probability of scatter about the best fit line in excess of analytical errors, the error estimates have been enhanced by a factor of $\sqrt{MSWD}$ (mean square weighted deviates) in an empirical attempt to allow for nonmodel geological behavior. It should be emphasized, however, that this is not a rigorously justified procedure, and "error-

TABLE 1. Rb-Sr Data for Metamorphic Rocks

Sample Details	Rb, ppm	Sr, ppm	$^{87}Rb/^{86}Sr$	$^{87}Sr/^{86}Sr$
Haag Nunataks				
Gneisses				
R.2255.5	72	270	0.7719	0.71579
6	54	352	0.4412 (0.6)	0.71031
9	72	277	0.7524	0.71566
10	55	304	0.5200 (0.6)	0.71212
14	59	286	0.5984 (0.6)	0.71340
16[a]	72	494	0.4207	0.70884
17	54	622	0.2509 (0.6)	0.70719
18	53	306	0.5023	0.71170
Aplogranites and Pegmatites				
R.2255.1	80	318	0.7306	0.71384
4	79	392	0.5863	0.71206
8	101	166	1.7757	0.72934
11	4.5	512	0.0258 (3.8)	0.70403
13	111	123	2.6622	0.74113
Microgranites				
R.2255.2	78	413	0.5491	0.71129
7	98	346	0.8238	0.71540
12	96	508	0.5471	0.71116
15	98	404	0.7043	0.71358
Mount Woollard				
Migmatitic schists				
R.2245.2	243	134	5.235	0.73623
4	139	319	1.260	0.72353
5	116	137	2.455	0.73186
6	146	142	2.985	0.73400
11	115	139	2.408	0.73173
12	271	151	5.197	0.73590
9[b]	57	400	0.415 (0.6)	0.70512
Pegmatites				
R.2245.3A	134	144	2.696	0.72992
3B	132	131	2.919	0.72989
7	281	115	7.092	0.73939
10	194	94	6.014	0.73880
R.2251.1	91	239	1.098	0.72242
2	91	234	1.122	0.72215
Microgranite				
R.2245.1	261	128	5.892	0.73041

Errors are ±2% on Rb and Sr contents, 0.01% on $^{87}Sr/^{86}Sr$, and 0.5% on $^{87}Rb/^{86}Sr$ except where indicated in parentheses. Sample localities are as follows: R.2255 Haag Nunataks, 77°02'S, 78°20'W; R.2245 Mount Woollard, 80°33'S, 96°38'W, central portion of ridge; R.2251 Mount Woollard, 80°33'S, 96°38'W, south end of ridge.
[a] Amphibolite inclusion.
[b] Amphibolite gneiss.

chrons" with such enhanced errors must be treated with caution.

Haag Nunataks. Haag Nunataks (Figure 1) on the otherwise totally ice-covered Fowler Peninsula were discovered and sampled by a BAS airborne glaciological party in 1975. The second reported visit was by the BAS/USARP geological party, flown by Twin Otter after refueling at Siple Station on January 18, 1984. Three small (50-100 m) outcrops all exhibit the same rock types. The main component is a medium- to coarse-grained granodiorite orthogneiss, with a well-developed foliation defined by orientation of feldspar phenocrysts, amphibole, and biotite. Thin segregation pegmatites pervade most of the gneiss. This is cut by two sets of granitic sheets. The earlier set is of coarse quartz-feldspar pegmatite grading into massive aplogranite, up to 1 m thick, with wavy margins transgressive to the foliation of the gneiss. The second set (<0.5 m) is of more sharply transgressive homogeneous gray microgranite, cutting both the gneiss and pegmatite. The oldest rocks at Haag Nunataks are (?)ultramafic schistose amphibolite pods in the gneiss, and the youngest are a thin steeply dipping dike of intermediate composition. No direct age information was obtained on either of these.

The only previous geochronology for Haag Nunataks consists of K-Ar ages on separated minerals from the granodiorite gneiss collected in 1975

TABLE 2. Rb-Sr Data for Granitoids

Sample Details	Rb, ppm	Sr, ppm	$^{87}Rb/^{86}Sr$	$^{87}Sr/^{86}Sr$
Whitmore Mountains				
Mount Chapman				
R.2226.1	233	127	5.325	0.72754
2	223	104	6.191	0.72872
3	255	126	5.879	0.72659
4[a]	269	85	9.224	0.74592
5[a]	289	69	12.203	0.75405
6[a]	288	68	12.396	0.75429
R.2227.1[a]	262	97	7.876	0.74285
Mount Seelig				
R.2228.1	221	205	3.118	0.71621
R.2229.1	396	153	7.523	0.72829
2	253	167	4.348	0.71886
3	276	140	5.696	0.72328
Pirrit Hills				
R.2243.1	502	5.0	314.06 (3.1)	1.44236
2[b]	529	5.4	304.77 (2.8)	1.46637
3[b]	413	9.1	135.79 (1.7)	1.03971
4	456	42.7	31.13 (0.6)	0.78351
5	459	27.1	49.68 (0.7)	0.82837
6	601	6.2	299.09 (2.6)	1.42458
7	707	8.4	259.57 (2.0)	1.33833
9[b]	624	4.3	461.53 (3.6)	1.82979
10[b]	491	7.8	190.25 (2.1)	1.17156
11	406	13.1	92.03 (1.3)	0.93043
12	414	26.7	45.38 (0.6)	0.82047
13	368	33.5	32.05 (0.6)	0.78477
Nash Hills				
R.2242.1	321	147	6.346	0.72818
3[b]	449	28	47.484 (0.8)	0.83253
4	315	162	5.646	0.72581
5	346	125	8.060	0.73201
6	314	166	5.502	0.72639
7	310	152	5.919	0.72733
8	325	147	6.417	0.72805
9[b]	292	26	32.872 (0.8)	0.79324
Pagano Nunatak				
R.2215.1	338	78	12.600	0.74762
2	394	75	14.964	0.75223
3	325	85	11.068	0.74333
4	335	77	12.614	0.74736
5	325	78	12.137	0.74619
6	331	79	12.077	0.74618
7	345	76	13.117	0.74850
8[c]	408	6.3	197.37 (2.6)	1.19842
10[c]	341	6.8	149.31 (2.3)	1.08410
11[c]	333	13.8	70.94 (1.3)	0.89085
12[c]	425	7.6	168.64 (2.2)	1.12678
R.2256.1	332	89	10.806	0.74201
2	341	78	12.676	0.74733
4	332	88	10.997	0.74252

Errors are ±2% on Rb and Sr contents, 0.01% on $^{87}Sr/^{86}Sr$, and 0.5% on $^{87}Rb/^{86}Sr$ except where indicated in parentheses. Sample localities are as follows: R.2226 north face of Mount Chapman, 82°34′S, 105°55′W; R.2227 northwest ridge of Mount Chapman, 82°34′S, 106°00′W; R.2228 southwest of Mount Seelig (moraine), 82°28′S, 104°01′W; R.2229 west of Mount Seelig (moraine), 82°26′S, 104°06′W; R.2243 Pirrit Hills, 81°10′S, 85°00′W, north ridge of Mount Turcotte; R.2242 Nash Hills, 81°57′S, 89°44′W, prominent east-west cliff; R.2215/2256 Pagano Nunatak, 83°41′S, 87°37′W.

[a] Microgranite.
[b] Aplogranite.
[c] Aplites.

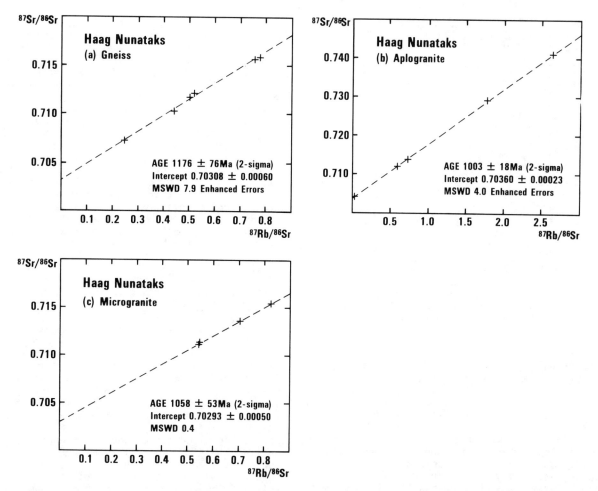

Fig. 2. Rb-Sr whole-rock isochron plots for Haag Nunataks: (a) gneisses, (b) aplogranites, and (c) microgranite. MSWD is a measure of the scatter of the data. Where MSWD <3, it is assumed that the data fit the isochron model; where MSWD >3, the calculated errors are enhanced in an attempt to allow for excess scatter (see text).

[Clarkson and Brook, 1977]. The biotites and hornblendes gave unexpectedly old ages of 991-1031 Ma, interpreted as a realistic minimum for the main amphibolite-grade metamorphism (coexisting feldspars showing significant subsequent loss of Ar). The validity of these ages and their interpretation is fully borne out by the Rb-Sr isochrons now presented (Figure 2). The late microgranites give a true isochron (i.e., MSWD <1) with a age of 1058 ± 53 Ma, the error being essentially controlled by the small range of Rb/Sr ratios of the four samples. The earlier aplogranites and pegmatites have a much more useful spread of data points which almost satisfy the criteria for an isochron, and if this small degree of residual scatter is incorporated in the error, a well-defined age of 1003 ± 18 Ma is obtained. The data for the granodiorite gneiss show rather more disturbance (MSWD = 7.9), possibly owing to its more complex history, but nevertheless give a very reasonable age of 1176 ± 76 Ma. These data do not allow any distinction between the three determined ages, although it is possible that there was a significant time lag between crystallization of the parent igneous body, or its first metamorphism, and emplacement of the aplogranite sheets. However, all three phases have indistinguishably low initial $^{87}Sr/^{86}Sr$ ratios, averaging 0.7032. This is close to estimates of ~0.703 for undepleted mantle at 1000 Ma, whereas older (e.g., Archaean) crustal rocks would have had much more radiogenic Sr at this time. This effectively precludes the possibility of any significant previous crustal history for these rocks or their source regions. The simplest interpretation of the data would involve derivation of the granodiorite magma from the mantle or a juvenile crustal source followed by essentially synchronous amphibolite-grade metamorphism and anatexis to produce the more acid material, all within a short interval of less than 100 million years.

<u>Mount Woollard</u>. This isolated mountain at 80°

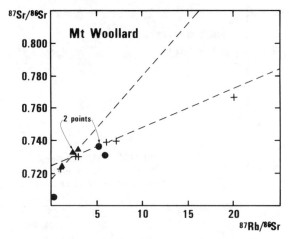

Fig. 3. Plots of $^{87}Sr/^{86}Sr$ versus $^{87}Rb/^{86}Sr$ for Mount Woollard: Triangles are migmatitic schists and pluses are pegmatites. The reference isochrons are for visual comparison and are not exact fits to the data. The steeper line corresponds to 450 Ma, and the less steep line corresponds to 200 Ma.

33°S, 96°43'W provided one of the few geological surprises of the 1983-1984 joint field season. It is depicted by Craddock [1972] as consisting of (?)Paleozoic sediments presumed equivalent to those of nearby exposures at Mount Moore, Mount Johns, and the Ellsworth Mountains. In fact it was found to consist of a schist/migmatite/pegmatite complex. The biotite and hornblende schists are migmatized by thin predeformational segregation pegmatites and are cut by thicker sheets and bodies of coarse feldspar pegmatite which are affected only by a second, open fold phase. A third set of pegmatites, like the second set up to 3 m in width and with small pink garnets, is essentially undeformed. The migmatites are also transgressed by late microgranite dikes.

The Rb-Sr whole-rock results (Figure 3) are unfortunately inconclusive. Four samples of migmatized micaschist scatter about an errorchron with an "age" of about 450 Ma, which would be consistent with a lower Paleozoic phase of metamorphism, but two further samples with a much higher proportion of igneous material plot well to the right of this line. The pegmatite samples are widely scattered and do not yield any definite age, although the general trend of the data for sample R.2245 would be consistent with a Middle Jurassic age (i.e., about 170 Ma). The scatter presumably reflects a heterogeneous local anatectic component, and the latest pegmatites at least could be related to a buried Jurassic granite. The two samples of massive pre-D2 pegmatite from sample R.2251 plot in a very ambiguous position, since there is no control over their true initial $^{87}Sr/^{86}Sr$ ratios. It is hoped that mineral dating may produce some elucidation of the ages of pegmatite emplacement, which could cover an extended time interval.

Whitmore Mountains. This 50-km-square group of spectacular mountains lies well to the west of the "Ellsworth-Thiel" topographic ridge (Figure 1). The geology has been previously described by Webers et al. [1982]. All of the main peaks are predominantly composed of a coarse porphyritic biotite-(hornblende)-granite with a variably developed micaceous foliation (named the Mount Seelig granite by Webers et al. [1982]). This is cut by a variety of later granite types including a fine-grained muscovite-bearing aplogranite developed as dikes at Linck Nunatak, aplites, pegmatites, and sheets of gray microgranite. Webers et al. [1982] considered the Linck Nunataks granite to be a late differentiate of the Mount Seelig granite and reported K-Ar biotite ages of 176 ± 5 Ma and 190 ± 8 Ma for the two granites, respectively, which they claimed to be consistent with such a relationship. However, at the same time they made the subsequently repeated assertion that the Mount Seelig granite was Late Triassic in age and the Linck Nunataks granite Early Jurassic (this ignores the errors on the determinations but even 190 Ma would be Jurassic on most currently employed time scales).

Preliminary Rb-Sr whole-rock data for the Whitmore Mountains are shown in Figure 4. So far only four samples of the very coarse porphyritic granite from the Mount Seelig area have been analyzed. In terms of Rb and Sr contents they are the least evolved of the Mesozoic granitoid rocks examined in this study (Table 2). They already show some scatter about an errorchron (MSWD = 21.3) so that the estimated age of 195 ± 21 Ma is less precise than is required for a conclusive assignment. The initial $^{87}Sr/^{86}Sr$ ratio is 0.7073 ± 0.0015. Three further samples from Mount Chapman plot well above this errorchron but do not define a line themselves. The reason for this is not clear, but the latter, although very fresh, do show signs of shearing: foliation and occasional streaked-out bands of schistose material. However, four samples of the gray microgranite from Mount Chapman, including one from a larger body comprising the northwest ridge (previously mapped by Webers et al. [1982] as metasediment), define quite a good isochron (MSWD = 3.0) corresponding to an age of 182 ± 5 Ma. This is essentially concordant with the reported K-Ar ages, and until further samples have been analyzed from the Whitmore Mountains (in progress) it would be safest to assume that all the granites here are of Middle Jurassic age. This probably includes the Mount Seelig granite, which is petrographically like the coarse granites of the Nash Hills (see below). However, high initial $^{87}Sr/^{86}Sr$ ratios of the microgranites (0.7224 ± 0.0008) preclude their origin solely by crystal fractionation from the porphyritic granite and strongly suggest upper crustal anatexis.

Pirrit Hills. There is no published map of the Pirrit Hills [Craddock, 1983], which form a spectacular group of 1000-m granite peaks 80 km southwest of the Ellsworth Mountains (Figure 1). The predominant rock type is a massive porphyritic biotite-hornblende granite which displays considerable variation in grain size and texture. There

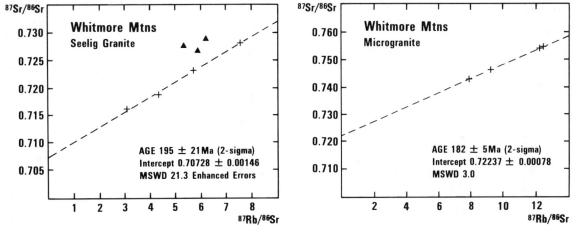

Fig. 4. Rb-Sr whole-rock isochron plots for the Whitmore Mountains. The triangles in the plot for Mount Seelig granite are samples from Mount Chapman not included in the best fit. These results are preliminary and incomplete.

are some porphyritic types with a fine-grained matrix, and both massive aplogranite and 0.5-m aplite veins are common. There is extensive late hydrothermal mineralization with amethyst and beryl-bearing pegmatites and amorphous joint fillings. A small part of the metasedimentary aureole is preserved in the southwest part of the outcrop at Bradley and Morland nunataks.

The analyzed granite samples from the northern cliffs of Mount Turcotte show a wide range of high Rb (400-700 ppm) and low Sr (4-40 ppm) contents. The Rb-Sr isotopic data (Figure 5) fit a well-defined isochron with an age of 173 ± 3, and an initial $^{87}Sr/^{86}Sr$ ratio of 0.7070 ± 0.0016. Results for the coarser granites, when regressed separately, are indistinguishable from those of the aplites and aplogranites. The age (Bathonian-Bajocian) is consistent with ages reported for the Nash Hills and Pagano Nunatak granites (see below).

Nash Hills. These hills, 50 km southwest of the Pirrit Hills, form a 30-km-long east facing scarp which again has received only the briefest mention in geological literature [Craddock, 1983]. The central and southern areas display a well-exposed, fairly homogeneous porphyritic granite intruded into a roof zone of metasediments which extend to the north. The granite is a medium- to coarse-grained biotite-hornblende granite mostly similar to the coarser types observed in the Pirrit Hills. It is often slightly altered, but fresh material was collected from the rock falls at the base of 300-m cliffs at sample R.2242. An equigranular muscovite-bearing aplogranite variety was sampled from the debris, although in situ outcrops of the 0.5-m layers in the cliffs could not be inspected closely. Craddock [1983] mentions three ages of 172-177 Ma, and Halpern [1966] reported an Rb-Sr biotite-WR age of 173 Ma (recalculated with current decay constant).

The Rb-Sr data for all samples define a reasonable isochron (Figure 5), with an age of 175 ± 8 Ma. This confirms the essential synchroneity with the Pirrit Hills granite, although at this locality the Rb and Sr contents are considerably less evolved, even the aplites containing 25-60 ppm Sr, and the initial $^{87}Sr/^{86}Sr$ ratio is significantly higher at 0.7122 ± 0.0008.

Pagano Nunatak. This small but prominent nunatak close to the metasedimentary outcrop of the Hart Hills is the most southerly exposure of the porphyritic granite type seen in the Pirrit and Nash hills. A description, five modal analyses, and one chemical analysis are given by Webers et al. [1983]. It is a medium- to coarse-grained rock, gray when freshest, with orthoclase and oligoclase phenocrysts and two micas. Muscovite is the only mica present in the 30-cm aplite and pegmatite veins. Xenoliths (of mafic amphibolite) are small and scarce, tourmaline is an accessory phase, and (?)vermiculite and epidote are developed as secondary cavity and vein minerals.

This granite, like that of the Pirrit Hills, is quite highly evolved (Rb 300-450 ppm and Sr 5-85 ppm). All data regressed together scatter somewhat (MSWD = 5.4), whereas the aplites fit a perfect isochron with an age of 175 ± 4 Ma. The initial $^{87}Sr/^{86}Sr$ ratios of the two regressions are 0.7157 ± 0.0014 for the first case and, more precisely defined, 0.7135 ± 0.0008 for the second. The agreement with the reported K-Ar biotite age of 175 ± 4 Ma [Webers et al., 1982] is remarkably good, and assuming rapid cooling of the body suggests that this firmly dates the time of intrusion.

Lewis Nunatak. Extensive collections were made in the Thiel Mountains, but these have not yet been analyzed. Previous attempts to date the porphyritic granodiorite or quartz monzonite which comprise the majority of the outcrops have been summarized by Schmidt and Ford [1969] and have clearly established a minimum age of late Precambrian (500-700 Ma). The most southerly outcrop visited, at Lewis Nunatak (85°40'S, 88°04'W), is

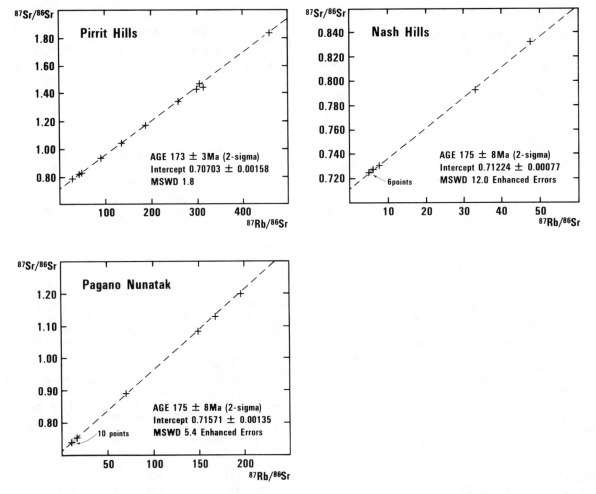

Fig. 5. Rb-Sr whole-rock isochron plots for the Pirrit Hills, Nash Hills, and Pagano Nunatak. See Figure 2 for other comments.

detached from the Thiel Mountains massif on the edge of the polar plateau scarp. It consists of an undeformed and unmetamorphosed dark brown spheroidally weathering dolerite body intruded into baked shales and sandstones probably belonging to the Beacon Supergroup. The Rb (~50 ppm) and Sr (~140 ppm) contents of the dolerite are characteristic of the Early to Middle Jurassic Ferrar Supergroup [Kyle et al., 1981] igneous rocks, and this assignment is also suggested by the anomalously high calculated initial $^{87}Sr/^{86}Sr$ ratios of 0.7105-0.7117 at 175 Ma (Table 3).

Discussion and Conclusions

This program of isotope geology is an integral part of the BAS/USARP project on the tectonic relationships of West Antarctica, but as such is only at a very early stage. We have reported our first findings here, restricted to Rb-Sr whole-rock dating of metamorphic and igneous rocks collected in the 1983-1984 field season. Even this work is incomplete in part, and in some instances has revealed the need for more detailed analytical methods in order to elucidate key points in the

TABLE 3. Rb-Sr Data for Lewis Nunatak Dolerite

Sample Details	Rb, ppm	Sr, ppm	$^{87}Rb/^{86}Sr$	$^{87}Sr/^{86}Sr$	$^{87}Sr/^{86}Sr$, 175[a]
R.2206.1	57	123	1.329	0.71384	0.71053
2	47	129	1.046	0.71349	0.71126
3	50	132	1.095	0.71407	0.71168

Errors are ±2% on Rb and Sr contents, 0.01% on $^{87}Sr/^{86}Sr$, and 0.5% on $^{87}Rb/^{86}Sr$.
[a] Calculated value of $^{87}Sr/^{86}Sr$ at 175 Ma.

geological evolution of the area. Some conclusions may already be drawn, however, both confirming previous indications of the timing of metamorphic and igneous episodes, but also revealing new information, particularly from a petrogenetic consideration of Sr-isotope compositions.

In the case of Haag Nunataks, the late Proterozoic age of this crustal fragment is confirmed in a very emphatic way. The K-Ar cooling ages previously reported are concordant with the new 1000-1100 Ma Rb-Sr whole-rock ages. Taken together with the low initial $^{87}Sr/^{86}Sr$ ratios of 0.703, these represent not merely a metamorphic event, but the primary crustal formation of these rocks at 1000-1100 Ma in a short-lived "crustal accretion and differentiation" event [Moorbath, 1977]. Furthermore, rapid cooling indicated by the K-Ar mica ages is clearly the latest significant factor in the thermal history of these rocks; they reflect nothing of the extensive Phanerozoic history represented by the Ellsworth Mountains or the rest of the study area. Although similar cratonic basement rocks could underlie much of the area, this is unproven, and in view of the geophysical discontinuities to the east of the Ellsworth Mountains, the identity of a discrete exotic "Haag" microplate cannot be ruled out. In either case, it is necessary to look beyond the Antarctic Peninsula for exposed rocks of this age. Gneissic rocks with possible ages of 1000 Ma are known at Cape Meredith in the Falkland Islands [Cingolani and Varela, 1976; Rex and Tanner, 1982], and in Heimefrontfjella, Dronning Maud Land [Juckes, 1972; P.D. Marsh and R.J. Pankhurst, BAS, unpublished data, 1984]. Unmetamorphosed igneous rocks of this age crop out closer to Haag Nunataks at the present day and at Bertrab and Littlewood nunataks, Coats Land, 1000 km east of Haag Nunataks. Eastin and Faure [1971] report an Rb-Sr isochron of 980 ± 16 Ma (recalculated by using the same decay constant as here) with an initial $^{87}Sr/^{86}Sr$ ratio of 0.704 for rhyolites and granites here which could well be equivalent to the aplogranite phase at Haag Nunataks.

The present study has failed to date the major events in the production of amphibolite-grade metamorphic and migmatitic rocks at Mount Woollard as yet, owing to the heterogeneous and polygenetic nature of the magmatic component. A possible scenario in which Phanerozoic sedimentary rocks with affinities to the lower parts of the Ellsworth Mountains succession were metamorphosed and intruded by contact migmatization in the roof of an unexposed Mesozoic pluton must be taken with a great deal of caution, since the inconclusive data are open to other interpretations. However, little in the data suggests that any of the events observed could be as old as Proterozoic; even model ages for the schists assuming an initial $^{87}Sr/^{86}Sr$ ratio of 0.703 are mostly considerably less than 1000 Ma, and these would normally be taken as likely maximum ages for the sediment provenance area. Clearly, no comparisons are to be drawn between these rocks and the gneisses and schists of Haag Nunataks. They should not be regarded, at this stage, as representing a metamorphic basement for the Ellsworth Mountains succession.

Results for the granitic rocks of the Whitmore Mountains are as yet incomplete. In particular, the predominant coarse porphyritic (Mount Seelig) variety has not yet yielded a precise age; the preliminary 195 ± 21 Ma could be as old as Triassic within error. Some uncertainty also surrounds the apparently inhomogeneous isotopic nature of this granite, especially at Mount Chapman. However, the late microgranite veins here are clearly Early to Middle Jurassic in age, as are the published K-Ar ages for both of the main granite types. It seems most likely that this pluton is synchronous with those of the Ellsworth-Thiel ridge granites, although a slightly earlier emplacement age is possible. The microgranites are S-type (2-mica) anatectic melts rather than direct differentiates of the porphyritic granite (see also Vennum and Storey [this volume]).

The granitoids of the Pirrit Hills, Nash Hills, and Pagano Nunatak form a coherent group with a well-defined Middle Jurassic age. Rb-Sr whole-rock isochrons for the granites and for late magmatic aplites give ages of 173 ± 3, 175 ± 8, and 175 ± 8 Ma, respectively, concordant with previously published K-Ar mica ages. There are, however, very distinct petrographic and geochemical differences between the three plutons, most obviously revealed in the present work by the distinct initial $^{87}Sr/^{86}Sr$ ratios of 0.7070 ± 0.0016, 0.7122 ± 0.0008, and 0.7175 ± 0.0014. The apparent progressive southerly increase should be seen as a tentative observation. Because of the high Rb/Sr ratios of the rocks, initial Sr-isotope compositions are difficult to determine precisely. However, it should be noted that the initial $^{87}Sr/^{86}Sr$ ratios do not correlate with the overall Rb/Sr ratio of each suite as should be expected if any closed system resetting had occurred. For example, the high-Rb, low-Sr rocks of the Pirrit Hills give a lower initial $^{87}Sr/^{86}Sr$ ratio than the moderate Rb and Sr granites of the Nash Hills. The rare earth element (REE) patterns for the latter are significantly light-REE enriched, however, compared to the flat patterns of the Pirrit Hills [Vennum and Storey, this volume]. This suggests that the REE patterns and initial $^{87}Sr/^{86}Sr$ ratios may be controlled by local source effects, whereas Rb/Sr may reflect a differentiation process. At present each isochron is based on a single sampling locality within each pluton, and subsequent analyses may reveal real local variation. Nevertheless, they are all relatively high values, typical of igneous rocks which are derived from, or at least extensively contaminated by, evolved continental crust. It is quite possible that the individual plutons are related at depth to a more continuous batholithic emplacement along the axis of the Ellsworth-Thiel ridge, of which they are but the high-level expression, although geophysical evidence for this is lacking [Garrett et al., this volume].

In many respects they display characteristics ascribed by Pitcher [1979] to "Hercynotype" granitoids: compositional restriction, lack of associated basic or volcanic rocks, S-type geochemistry, and emplacement into low-grade continentally derived metasediments. Such granites are usually genetically related to crustal shortening, either in an intracratonic or a continent-continent collision situation. In view of the uniformity of sediment type and structures throughout the area, the latter environment can be ruled out. There is no obvious correlation with a continent margin subduction system, but it should be noted that some early granites on the east coast of the Antarctic Peninsula (also mostly about 175 Ma) are also fairly uniformly granitic and have moderately high initial $^{87}Sr/^{86}Sr$ ratios [Pankhurst, 1982; M.J. Hole, BAS, unpublished data, 1985]. These are generally thought to be early subduction-related granites. Thus, although the results from the first stages of the BAS/USARP collaboration are already throwing new light on the geological evolution of this critical and little known area of West Antarctica, much still remains to be done. It is hoped that during the next year Rb-Sr and other isotope systematics will enable further advances, especially in the elucidation of the sparsely exposed sedimentary rocks of the area and their metamorphism.

Acknowledgments. We are indebted to USARP and the U.S. Coast Guard for their help in shipping these rock samples from the field via McMurdo Station to the Antarctic Peninsula, and to Karen Brotby, BAS, who crushed them and prepared rock powders for analysis.

References

Cingolani, C. A., and R. Varela, Investigaciones geologicas y geocronologicas en el extremo sur de la Isla Gran Malvina, Sector de Cabo Belgrano (Cape Meredith), Islas Malvinas, Actas Cong. Geol. Argent. VI, 1, 457-473, 1976.

Clarkson, P. D., and M. Brook, Age and position of the Ellsworth Mountains crustal fragment, Antarctica, Nature, 265, 615-616, 1977.

Craddock, C., Geologic map of Antarctica, scale 1:5,000,000, Am. Geogr. Soc., New York, 1972.

Craddock, C., The East-West Antarctic boundary between the ice shelves: A review, in Antarctic Earth Science, edited by R. L. Oliver, P. R. James, and J. B. Jago, pp. 94-97, Australian Academy of Science, Canberra, 1983.

Craddock, C., G. F. Webers, and J. J. Anderson, Geology of the Ellsworth Mountains-Thiel Mountains ridge (abstract), in Antarctic Geoscience, edited by C. Craddock, p. 849, University of Wisconsin Press, Madison, 1982.

Dalziel, I. W. D., and R. J. Pankhurst, Tectonics of West Antarctica and its relation to East Antarctica: Joint USARP-British Antarctic Survey geology/geophysics project, Antarct. J. U. S., in press, 1985.

Eastin, R., and G. Faure, The age of the Littlewood volcanics of Coats Land, Antarctica, J. Geol., 79, 241-245, 1971.

Garrett, S. W., L. D. B. Herrod, and D. R. Mantripp, Crustal structure of the area around Haag Nunataks, West Antarctica: New aeromagnetic and bedrock elevation data, this volume.

Halpern, M., Rubidium-strontium data from Mt. Byerly, West Antarctica, Earth Planet. Sci. Lett., 1, 455-457, 1966.

Juckes, L. M., The geology of northeastern Heimefrontfjella, Dronning Maud Land, Br. Antarct. Surv. Sci. Rep., 65, 44 pp., 1972.

Kyle, P. R., D. H. Elliot, and J. F. Sutter, Jurassic Ferrar Supergroup tholeiites from the Transantarctic Mountains, Antarctica, and their relationship to the fragmentation of Gondwana, in Gondwana Five, edited by M. M. Cresswell and P. Vella, pp. 283-287, Balkema, Rotterdam, 1981.

Moorbath, S., Ages, isotopes and evolution of Precambrian continental crust, Chem. Geol., 20, 151-187, 1977.

Pankhurst, R. J., Rb-Sr geochronology of Graham Land, Antarctica, J. Geol. Soc. London, 139, 701-712, 1982.

Pankhurst, R. J., Rb-Sr constraints on the ages of basement rocks of the Antarctic Peninsula, in Antarctic Earth Science, edited by R. L. Oliver, P. R. James, and J. B. Jago, pp. 367-371, Australian Academy of Science, Canberra, 1983.

Pitcher, W. S., Comments on the geological environments of granites, in Origin of Granite Batholiths: Geochemical Evidence, edited by M. P. Atherton and J. Tarney, pp. 1-8, Shiva Publishing, Orpington, Kent, United Kingdom, 1979.

Rex, D. C., and P. W. G. Tanner, Precambrian age for gneisses at Cape Meredith in the Falkland Islands, in Antarctic Geoscience, edited by C. Craddock, pp. 107-108, University of Wisconsin Press, Madison, 1982.

Schmidt, D. L., and A. B. Ford, Geology of the Pensacola and Thiel mountains, in Geologic Maps of Antarctica, Antarctic Map Folio Ser., folio 12, plate V, edited by V. C. Bushnell and C. Craddock, Am. Geogr. Soc., New York, 1969.

Turner, P. J., Multicollection in thermal ionization mass spectrometry, VG Isotopes Ltd., Publ. 02.485, 4 pp., VG Isotopes Ltd., Winsford, Cheshire, 1982.

Vennum, W. R., and B. C. Storey, Correlation of gabbroic and diabasic rocks from the Ellsworth Mountains, Hart Hills, and Thiel Mountains, West Antarctica, this volume, 1986.

Webers, G. F., C. Craddock, M. A. Rogers, and J. J. Anderson, Geology of the Whitmore Mountains, in Antarctic Geoscience, edited by C. Craddock, pp. 841-847, University of Wisconsin Press, Madison, 1982.

Webers, G. F., C. Craddock, M. A. Rogers, and J. J. Anderson, Geology of Pagano Nunatak and the Hart Hills, in Antarctic Earth Science, edited by R. L. Oliver, P. R. James, and J. B. Jago, pp. 251-255, Australian Academy of Science, Canberra, 1983.

Williamson, J. H., Least-squares fitting of a straight line, Can. J. Phys., 46, 1845-1847, 1968.

Copyright 1987 by the American Geophysical Union.

ELLSWORTH-WHITMORE MOUNTAINS CRUSTAL BLOCK, WESTERN ANTARCTICA: NEW PALEOMAGNETIC RESULTS AND THEIR TECTONIC SIGNIFICANCE

A. M. Grunow, I. W. D. Dalziel,[1] and D. V. Kent

Lamont-Doherty Geological Observatory of Columbia University, Palisades, New York 10964

Abstract. Preliminary paleomagnetic study of granitic and sedimentary rocks from the Ellsworth-Whitmore Mountains crustal block (EWM), West Antarctica, leads to the following conclusions: (1) The EWM has a paleopole for the Middle Jurassic located at 235°E, 41°S, (α_{95} = 5.3, N = 8 sites) assuming that no widespread regional tilting has occurred since the magnetization measured was acquired. A Middle Jurassic paleolatitude of 47°S is indicated for the sites and precludes an original location for the EWM block south of the Antarctic Peninsula crustal block (AP). (2) This pole is not significantly different from the previously published Middle Jurassic paleopole obtained from rocks of the northern Antarctic Peninsula. The combined AP-EWM paleopole, compared to the Middle Jurassic mean paleopole obtained from igneous rocks of the Ferrar Supergroup in East Antarctica, suggests about 15° tectonic clockwise rotation of the AP and EWM. Since the AP and EWM poles coincide, these two crustal blocks may have moved as one unit since the Middle Jurassic. (3) The new data are compatible with two different Gondwanaland reconstructions. The first considers the AP and EWM as separate entities. The second is based on the movement of the AP and EWM as one block. For the Middle Jurassic, both reconstructions would locate the EWM west of Coats Land and south of the Falkland Plateau, with the adjacent AP located south of southernmost South America. (4) Enigmas concerning the structural trend and isolation of the thick Ellsworth Mountains Paleozoic succession persist.

West Antarctica and the Pacific Margin of Gondwanaland

West Antarctica and New Zealand are the most difficult parts of Gondwanaland to reconstruct. This is partly due to the extensive tectonism that occurred along the margin of the Pacific Ocean during and since fragmentation of the supercontinent. It is also partly a result of the extensive ice cover in Antarctica. Yet, as has been pointed out elsewhere, the tectonic evolution of this region has important implications for understanding

---

[1]Now at Institute for Geophysics, University of Texas at Austin, Austin, Texas, 78751.

of global plate interaction, paleoclimate, and paleobiogeography. It was with this in mind that the joint U.K.-U.S. West Antarctic Tectonics Project was initiated [Dalziel and Pankhurst, this volume]. Paleomagnetic studies are clearly an essential part of such a project, especially in the light of evidence that some geologic terranes bordering the Pacific Ocean have been displaced large distances [Coney et al., 1980; Van der Voo et al., 1980; Stone et al., 1982]. Existing paleomagnetic data suggest that the four major crustal blocks of West Antarctica (Figure 1) have been in close proximity to the East Antarctic craton at least since the Late Jurassic to Early Cretaceous (for review see Dalziel and Grunow [1985]). The data base is not extensive, however, and there are indications from "overlap" in Gondwanaland reconstructions, from geologic correlation, and from some of the paleomagnetic results, that limited relative motion of these blocks and of East Antarctica has occurred (for review see Dalziel and Elliot [1982]). Radiometric ages of approximately 175 Ma reported from granites in nunataks south of the Ellsworth Mountains [Craddock, 1983] gave cause for optimism that paleomagnetic poles might be obtained for a critical time period prior to Gondwanaland breakup. At this time (Middle Jurassic [Kent and Gradstein, 1985]), Antarctica was in a middle latitude position, and hence subsequent rotations should be resolved more readily than with poles for the Cretaceous and Tertiary when Antarctica was at a very high latitude [Norton and Sclater, 1979; Delisle, 1983].

During the 1983-1984 and 1984-1985 field seasons, therefore, two of us (A.M.G. and I.W.D.D.) made extensive collections for paleomagnetic studies in the Ellsworth-Whitmore Mountains crustal block (EWM), the adjoining Thiel Mountains (part of the Transantarctic Mountains) (Figure 2), and the Thurston Island-Eights Coast crustal block (Figure 1). The collection sites were chosen on the basis of the field observations described by Storey and Dalziel [this volume]. The samples were studied in the paleomagnetic laboratory at Lamont-Doherty Geological Observatory by A.M.G. and D.V.K. Radiometric age control for the study is provided by the work of Millar and Pankhurst [this volume]. Although few of the exposed rocks are ideal for paleomagnetic study, primarily due

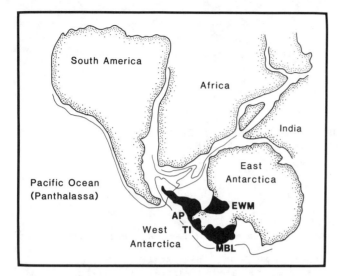

Fig. 1. Gondwanaland reconstruction of Norton and Sclater [1979]. West Antarctic microcontinents [from Dalziel and Elliot, 1982]: AP, Antarctic Peninsula crustal block; EWM, Ellsworth-Whitmore Mountains crustal block; MBL, Marie Byrd Land crustal block; and TI, Thurston Island-Eights Coast crustal block.

to the absence of paleohorizontal markers, some of the results already obtained from our first season's collecting in the EWM do provide new insights into the tectonic evolution of West Antarctica and hence of the Pacific margin of Gondwanaland. It is therefore appropriate to present here a summary of these results and our joint interpretation of them.

Ellsworth-Whitmore Mountains Crustal Block

The geology of the Ellsworth-Whitmore Mountains crustal Block is summarized in the paper by Storey and Dalziel [this volume]. The thick Paleozoic sedimentary sequence of the Ellsworth Mountains is comparable to the Gondwana craton cover exposed along the Transantarctic Mountains and the southern coast of Africa. Especially notable is the occurrence of upper Paleozoic glacial deposits, the Whiteout Conglomerate, and Glossopteris-bearing Permian strata, the Polarstar Formation [Craddock, 1969]. Together with the isolation of the mountains and their anomalous north-south structural grain, this stratigraphic comparison led Schopf [1969] to propose that the crustal block containing the Ellsworth Mountains has been displaced from an original location along the eastern margin of the Weddell Sea between the Transantarctic Mountains and the Cape Mountains of southern Africa.

The geology of the Ellsworth Mountains themselves has been recently (1979-1980) studied by a large group of scientists and is to be described in a forthcoming memoir [Craddock et al., 1986]. A paleomagnetic project was part of this effort. Preliminary results have been described by Watts and Bramall [1981]. They interpreted the data as being compatible with a 90° counterclockwise rotation of the Ellsworth Mountains relative to the East Antarctic craton since deposition of the Cambrian strata they studied. The detailed results of their extensive collecting have yet to be published. We therefore confined our work in the Ellsworth Mountains to collecting from the Permian Polarstar Formation that was not visited by Watts and Bramall.

With the exception of the Stewart Hills we collected material from all other isolated nunataks or groups of nunataks in the EWM (Figure 2). The highly deformed metasedimentary strata of the Stewart Hills were judged to be unsuitable for paleomagnetic study. This part of the collection comprises Ellsworth Mountains, Permian sedimentary rocks; Haag Nunataks, Precambrian gneiss and minor intrusions; Hart Hills, undated gabbro (hand specimens only); Martin Hills, undated metasedimentary rocks; Moreland Nunatak, undated sedimentary strata; Mount Johns, undated sedimentary strata; Mount Moore, deformed metasedimentary rocks (hand specimens only); Mount Woollard, undated gneiss, amphibolite, and pegmatites; Nash Hills, Middle Ju-

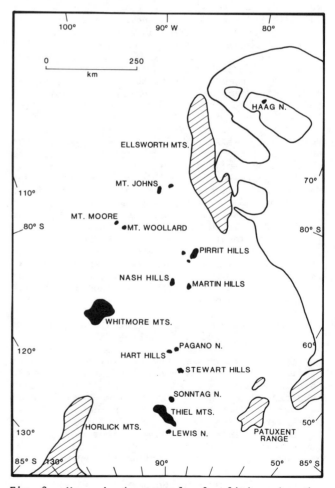

Fig. 2. Map showing sample localities in the Ellsworth-Whitmore Mountains crustal block.

rassic granite, aplite, and undated metasedimentary rocks; Pagano Nunatak, Middle Jurassic granite and aplites; Pirrit Hills, Middle Jurassic granitic rocks, aplites, and undated metasedimentary rocks; Whitmore Mountains, Early to Middle Jurassic granitic plutons, aplites, and undated metasedimentary rocks.

It should be noted that we include here the Precambrian rocks of Haag Nunataks within the EWM, although the nature of the basement in the latter region is indeterminate [Garrett et al., this volume]; see also Dalziel et al. [this volume].

Sampling and analytical procedures. A total of 480 oriented drill core samples and 24 oriented hand samples were collected from 101 sites at 12 locations in the EWM. Usually, six cores were taken at each site. At least one sun compass reading was made at each locality; the sun compass readings agreed to within 3° of magnetic readings.

Measurements of the natural remanant magnetization (NRM) of the samples were made using a cryogenic magnetometer. Pilot samples from most sites were progressively demagnetized using alternating field (AF) and/or thermal demagnetization. AF demagnetization was normally done in steps of 10 mT up to a peak of 90 mT, while in thermal demagnetization, steps of 100°C up to a temperature of 500°C were used. Beyond 500°C, smaller steps were taken up to a maximum temperature of 670°C.

Vector end-point diagrams were used to plot the information obtained from demagnetizing the pilot samples. After analysis of these diagrams, effort was concentrated on the remaining cores from the most promising and/or critical localities. All samples from the selected localities were measured with a minimum of 10 thermal or seven AF demagnetization steps and plotted on vector end-point diagrams. The component directions were calculated using linear regression analysis [Kirschvink, 1980].

Paleomagnetic results. The turbidites of the Permian Polarstar Formation of the northern Sentinel Range of the Ellsworth Mountains, although potentially critical tectonically, proved to be unsuitable for paleomagnetic study. Neither thermal nor AF demagnetization could define a consistent direction between seven sites around a fold. The samples were fairly weak, with NRM intensities between 10^{-3} and 10^{-4} A m^{-1}, and only AF demagnetization defined any type of linear trajectory. The median demagnetization field was 40 mT. Adding the bedding correction to the results did not improve the grouping of directions. The Pirrit Hills and Whitmore Mountains granites and aplites, and the Mount Woollard metamorphic rocks, proved to be magnetically unstable in that convincingly linear demagnetization trajectories were not found. The aplites from all localities were found without exception to be very weak or magnetically unstable. Studies of the Haag Nunataks, Hart Hills, Martin Hills, Moreland Nunatak, Mount Johns, and Mount Moore samples have not been completed.

Granite, aplite, and calcareous siltstone were sampled from 12 sites in the Nash Hills. The granite pluton has yielded an Rb-Sr whole-rock isochron indicating an age of 175 ± 8 Ma [Millar and Pankhurst, this volume]. The sedimentary rocks are hornfelsed and appear to be roof pendants in the granite body. Granite samples from two sites in the Nash Hills were demagnetized using AF. A single component of magnetization with a downward dipping direction to the northwest was found (Figures 3a and 4). In addition, hornfelsed calcareous red siltstone from two sites was seen to be partially magnetically overprinted by the intrusion of the granite. This overprint magnetization has a directly antiparallel direction (upward dipping to the southeast) to that of the Nash Hills granite (Figure 3b). AF demagnetization defined this secondary direction more clearly than thermal demagnetization (Figures 3c and 4). The fine-grained granite at the contact with the sedimentary country rocks at Nash Hills tended to give upward dipping directions to the southeast, but with poorly defined linear demagnetization trajectories. It seems that the chilling at the margin of the granite and the baking of the adjacent sedimentary rocks occurred during a normal polarity interval, while the main coarse-grained body of the granite intrusive recorded a later reversed polarity interval.

The magnetically overprinted metasedimentary rocks in the Nash Hills also contained a more thermally stable component of magnetization (560° to 670°) (Figures 3c and 4). These red siltstones (strike 340°, dip 45°NE) yield mean directions of D = 14.8°, I = 29.2°, α_{95} = 7.9°, and n = 12 samples resulting in a paleomagnetic pole at 285°E, 8°S, dp, dm = 4.8°, 8.7°. After tilt corrections the directions are D = 21.2°, I = 0.8°, α_{95} = 8.4°, n = 12 samples and the paleomagnetic pole is located at 292°E, 7°N, dp, dm = 4.2°, 8.4°.

At Pagano Nunatak, four sites in coarse granite and four sites in aplitic dikes were collected. The granite yielded an Rb-Sr whole-rock isochron with an age of 175 ± 8 Ma (Middle Jurassic), virtually identical to the Nash Hills granite [Millar and Pankhurst, this volume]. The four granite sites were thermally demagnetized to reveal a single very high blocking temperature component (at 580°C) with an upward dipping direction to the southeast (Figures 3d and 4). Pagano Nunatak directions are very similar to those of the hornfelsed sedimentary rocks from the Nash Hills.

The sample mean characteristic directions for the Nash Hills and Pagano Nunatak are shown in Figure 5. Samples were combined to give a site mean, and site means combined to give a unit mean. The site mean characteristic directions of the Nash Hills samples (N = 4 sites) give a unit mean of D = 137.6°, I = -64°, K = 182.5, and α_{95} = 6.8° after inverting the directions of the two granite sites (Figure 4). This corresponds to a paleomagnetic (south) pole position at 233.1°E, 39.5°S (dp, dm = 8.7°, 10.9°), and a paleolatitude for the locality of 45.7°S. The site mean directions for the Pagano Nunatak samples (N = 4 sites) have a unit mean of D = 141.6°, I = -65.5°, K = 254.6, and α_{95} = 5.8° (Figure 4). The paleomagnetic (south) pole position for Pagano Nunatak is at 237.7°E, 42.6°S (dp, dm = 7.6°, 9.4°), and the paleolatitude of Pagano Nunatak is 47.6°S.

The results for the combined Nash Hills and Pagano Nunatak site directions converted to vir-

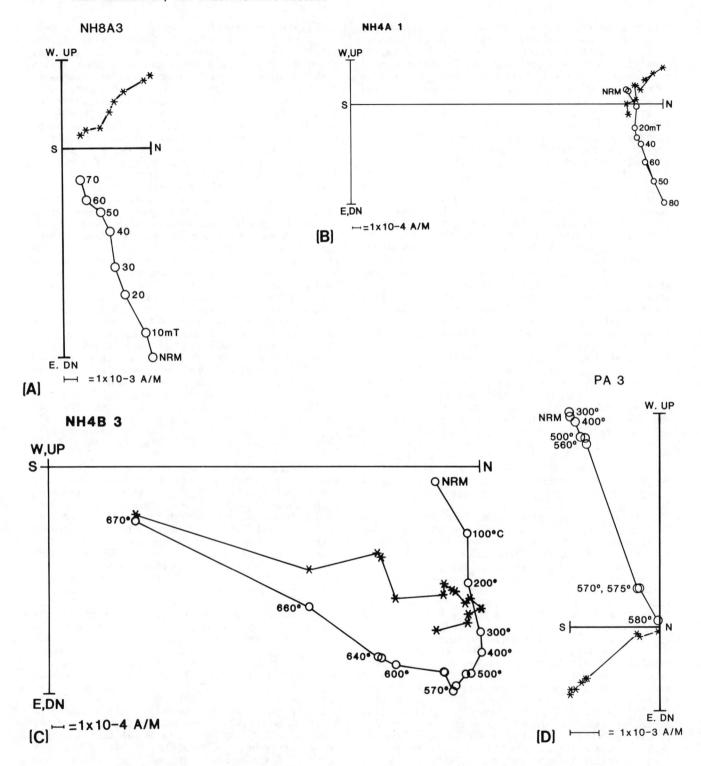

Fig. 3. Orthogonal projection of vector end points [Zijderveld, 1967] showing demagnetization behavior of samples from the Nash Hills and Pagano Nunatak. Open circles (stars) are projections on vertical (horizontal) planes at indicated levels of AF or thermal cleaning. Demagnetization fields in millitesla; temperatures in degrees Celsius. Magnetization units on axes are labeled. (a) Nash Hills granite. (b) Nash Hills baked metasediment using AF demagnetization. (c) Nash Hills baked metasediment thermally demagnetized. (d) Pagano Nunatak granite.

1. Nash Hills (81° 53'S, 89° 23'W)

Site	N/n	Lithology	Polarity	Treatment	Decl.	Incl.	K	α95	Lat.	Long.
NH1A	3/12	Granite and Metasedimentary Rock	N	AF	-	-	Unstable	-	-	-
NH1B	2/8	Granite and Metasedimentary Rock	Mixed	AF	-	-	Unstable	-	-	-
NH1C	6/6	Aplite	N	AF, TH	-	-	Unstable	-	-	-
NH1D	1/7	Granite	N	AF	-	-	Unstable	-	-	-
NH3A	6/7	Granite	R	AF	307.6	67	194.3	4.8	-44.4	224.8
NH3B	2/6	Aplite	N	AF	-	-	Unstable	-	-	-
NH3C	3/8	Aplite	N	AF	-	-	Unstable	-	-	-
NH3D	1/6	Dike	R	AF	-	-	Unstable	-	-	-
NH4A	3/6 6/6	Metasedimentary Rock Overprint (Thermal component)	N	AF, TH TH	143.6 3.9; 11.2(T)	-62.9 28.8; 9.8(T)	651.4 28.6; 94(T)	4.8 12.7; 6.9(T)	-37.6 -7.3; 3(T)	238.2 274.5; 281.8(T)
NH4B	3/6 6/6	Metasedimentary Rock Overprint (Thermal component)	N	AF, TH TH	143.6 25.3; 31(T)	-68.4 28.7;-8.2(T)	599.7 219.7;219.5(T)	5 4.5; 4.5(T)	-44.9 -7.9; 11.1(T)	239.3 295.2; 302.2(T)
NH8A	5/6	Granite	R	AF	315.8	57.1	171.2	5.9	-31.7	230.2
NH8B	1/6	Aplite	R	AF	-	-	Unstable	-	-	-

Unit mean 4/12 sites (17/25 samples, omitting unstable sites) for the Middle Jurassic:
$D = 137.6°$ $I = -64°$ $K = 182.5$ $α_{95} = 6.8°$
Pole position: 39.5° S lat., 233.1° E long., dp, dm = 8.7°, 10.9°

2. Pagano Nunatak (83° 42' S, 87° 40' W)

Site	N/n	Lithology	Polarity	Treatment	Decl.	Incl.	K	α95	Lat.	Long.
PA	4/5	Granite	N	TH	132.9	-67.9	76.6	10.6	-46.5	230.3
PB	5/5	Granite	N	TH	137.2	-62.8	65.6	9.5	-39.4	233.2
PC	2/6	Aplite	-	AF, TH	-	-	Unstable	-	-	-
PD	2/6	Aplite	-	AF, TH	-	-	Unstable	-	-	-
PE	2/6	Aplite	-	AF, TH	-	-	Unstable	-	-	-
PF	6/6	Granite	N	TH	155.4	-62.7	60.7	8.7	-38.4	249.9
PG	2/6	Aplite	-	AF, TH	-	-	Unstable	-	-	-
PH	5/6	Granite	N	TH	139.1	-67.5	85.3	8.3	-45.4	235.8

Unit mean 4/8 sites (20/22 samples, omitting unstable sites):
$D = 141.6°$ $I = -65.5°$ $K = 254.6$ $α_{95} = 5.8°$

Pole position: 42.6° S lat., 237.7° E long., dp, dm = 7.6°, 9.4°

COMBINED NASH HILLS-PAGANO NUNATAK POLE POSITION: 41.2° S lat., 235.2°E long., $A_{95} = 5.3°$, N = 8 sites

Notes: N/n = number of samples used in mean calculation/total number of samples; Treatment = demagnetization technique used; AF, alternating field and TH, thermal; K = estimate of precision parameter; $α_{95}$ = radius of circle of 95% confidence; dp and dm are the semi-axes of the oval of 95% confidence; A_{95} = radius of circle of 95% confidence for mean poleposition; T = tilt corrected.

Fig. 4. Site mean characteristic directions of the Nash Hills and Pagano Nunatak.

tual geomagnetic poles define a mean pole at 235.2°E, 41.2°S ($α_{95} = 5.3°$) (Figure 4). No tilt correction has been applied given the lack of any paleohorizontal marker. This will be discussed below.

Comparison with other results. Numerous paleomagnetic studies have been undertaken on igneous rocks of the Ferrar Supergroup in the Transantarctic Mountains. The Ferrar Supergroup comprises the Ferrar diabase sills and dikes, the Kirkpatrick basaltic lava flows, and the Dufek layered mafic igneous complex. Many of the results are of questionable validity, however, due to insufficient sampling, inadequate cleaning procedures,

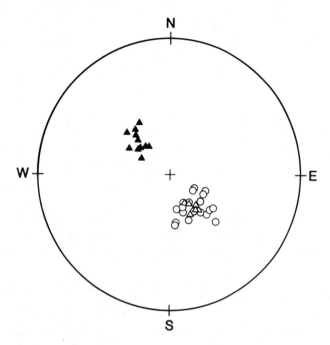

Fig. 5. Distribution of characteristic cleaned-sample directions from the Nash Hills and Pagano Nunatak. Solid (open) symbols are on lower (upper) hemisphere of equal-area projection. Circles (triangles) are samples from Pagano Nunatak granite (Nash Hills granite and baked metasediment). Ten samples rejected from 47 samples.

and incompletely published data. Nevertheless, the published poles from 15 localities along the entire length of the Transantarctic Mountains define a mean pole at 220.3°E, 54.8°S, α_{95} = 3.9° (Table 1). McIntosh et al. [1982] earlier calculated essentially the same mean pole (220°E, 55°S) from 11 localities. Individually, however, the locality poles range between 207.6°E and 231°E longitude and between 44°S and 68.6°S latitude. It is not clear whether this reflects secular variation, contaminated magnetizations, or unrecognized tectonic disturbances. Ages for the Ferrar igneous rocks based on K/Ar and Ar/Ar analyses range primarily between 160 Ma and 180 Ma [Elliot et al., 1985]. Kyle et al. [1981] believe that the main Ferrar activity was around 179 ± 7 Ma but that there may have been a younger magmatic event at 165 ± 2 Ma based on Ar/Ar analysis. Different ages of intrusion may also contribute to the scatter in the published Ferrar paleomagnetic results.

The only Early to Middle Jurassic pole from West Antarctica published prior to this study is one by Longshaw and Griffiths [1983]. They sampled acid to basic dikes and a granodiorite pluton dated by the Rb-Sr method at approximately 175 Ma and lava flows dated at approximately 175 Ma (R. J. Pankhurst, personal communication, 1985) in Graham Land, the northern part of the Antarctic Peninsula. They determined a mean paleopole at 238°E, 48°S, α_{95} = 9.5°, and N = 4 localities.

TABLE 1. Middle Jurassic South Pole Positions for East Antarctica

	n/N	Cleaning Technique	Polarity	Pole Longitude °E	Pole Latitude °S	Reference
Allan Hills	19/2	AF	N	226.8	47	Funaki [1983]
Beardmore Glacier	13/9	AF	N	221	59	Briden and Oliver [1963]
Brimstone Peak	29/12	AF	N	218	56	Cherry and Noltimier [1982]
Dufek Massif	98/-	AF	N&R	223	60	Beck et al. [1979]
Ferrar Glacier	57/5	NRM	N	218	58	Trunbull [1959]
Gorgon Peak	26/13	AF	N	230	56	Cherry and Noltimier [1982]
Mesa Ragne	60/15	AF	N	210	64	McIntosh et al. [1982]
Mount Falla	84/14	AF	N	222.6	53.8	Ostrander [1971]
Mount Fleming dike	15/1	AF	N	220.5	68.6	Funaki [1983]
Queen Alexandra Range (sills)	42/7	AF	N	220.2	54.2	Ostrander [1971]
Storm Peak	72/12	AF	N	231.5	44.1	Ostrander [1971]
Theron Mountains	8/8	NRM	N&R	224	54	Blundell and Stevenson [1959]
Wright Valley	26/1	AF	N	208	45.3	Funaki [1983]
Wright and Victoria valleys	83/46	AF	N	220	45	Bull et al. [1962]
Vestfjella dikes	109/24	AF/TH	N&R	207.6	54	Lovlie [1979]
Mean (A_{95} = 3.9; 15 studies)				220.3	54.8	

Notes: n, number of samples; N, number of sites; AF, alternating field demagnetization; TH, thermal demagnetization; A_{95}, radius of circle of 95% confidence for mean paleopole.

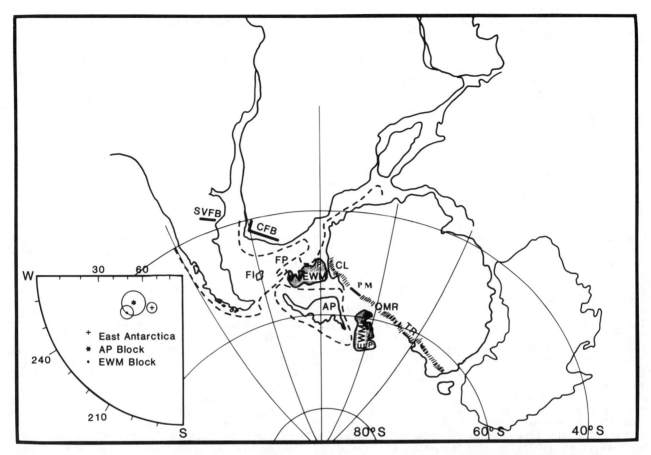

Fig. 6. A separate Ellsworth-Whitmore Mountains crustal block (EWM) located north of the Antarctic Peninsula crustal block (AP) using Norton and Sclater's [1979] reconstruction for the other Gondwanaland continents in the Early to Middle Jurassic. Thick solid lines indicate structural trends; hatched areas indicate outcrops of flat-lying Gondwana sequence cover rocks in Transantarctic Mountains. Note that the position of the EWM south of the AP [Longshaw and Griffiths, 1983] places the sample localities (P) 18° farther south than the data indicate. The equal-angle stereographic projection shows the EWM pole (small solid circle), the AP pole (star) and the East Antarctic (Ferrar) pole (plus) for the Middle Jurassic with their associated circles of 95% confidence. A 17° ± 9° counterclockwise rotation would restore the EWM pole to the Ferrar pole; a 15° counterclockwise rotation would restore the AP pole to the Ferrar pole [Longshaw and Griffiths, 1983]. PM is Pensacola Mountains; CFB, Cape Fold Belt; CL, Coats Land; FI, Falkland Islands; FP, Falkland Plateau; P, Pagano Nunatak; QMR, Queen Maud Range; SVFB, Sierra de la Ventana Fold Belt; TR, Transantarctic Range; large solid circles, Haag Nunataks; dashed lines, edges of continental shelf or inferred margin of continental block.

Longshaw and Griffiths did not apply a tilt correction to their results, as they did not have any paleohorizontal markers. Their interpretation of the findings is that an approximate 15° counterclockwise rotation about a pole at 0°E, 65°S would restore the Antarctic Peninsula to its Middle Jurassic position with respect to East Antarctica. This rotation would align their northern Antarctic Peninsula pole and the East Antarctic (Ferrar Supergroup) poles for that period of time. Longshaw and Griffiths separate the EWM from the Antarctic Peninsula crustal block (AP) and suggest a Jurassic position for the EWM that is consistent with the interpretation of Watts and Bramall [1981]. However, they move the southern part of the EWM to the west so that the Haag Nunataks are near the East Antarctic craton and the Ellsworth Mountains parallel to the Queen Maud Mountains of the Transantarctic Range (see Figure 6).

Tectonic Interpretation

Before discussing in detail the tectonic interpretation of our results, the validity of the findings must be considered because of the lack of a paleohorizontal marker. Pagano Nunatak and the

TABLE 2. Mean Middle Jurassic Paleomagnetic South Pole Positions for East and West Antarctica

	Age	Pole Longitude °E	Pole Latitude °E	A95	Reference
Antarctic Peninsula (intrusives)	175	238	48	9.5	Longshaw and Griffiths [1983]
EWM Block (intrusives and overprinted metasedimentary rocks)	175	232.2	41.2	5.3	this study
Mean AP-EWM		237	45.8	6.4	this study
East Antarctica (intrusives and extrusives)	160-180	220.3	54.8	3.9	this study

Nash Hills are located approximately 200 km apart in a north-south direction and yield the same paleomagnetic pole. Radio ice-echo sounding and aeromagnetic data indicate that the main topographic and magnetic fabric of the bedrock in this region is approximately east-west [Dalziel et al., this volume]. This fabric appears to be controlled by horsts and grabens. A depression known as the Thiel Trough extends across the EWM separating Pagano Nunatak from the Nash Hills. It is therefore unlikely that Pagano Nunatak and the Nash Hills are located on the same fault block. Yet they do yield the same paleomagnetic pole. A tilt correction of 11°S about an axis of 84°E would restore the EWM pole to the mean Ferrar pole, and an 8°S tilt correction about an axis of 91°E would restore the Antarctic Peninsula pole of Longshaw and Griffiths [1983] to the Ferrar pole. Fault blocks would therefore need to have tilted approximately the same amount and direction over a large region (the Antarctic Peninsula localities are 1500 km from the Nash Hills) to explain the coincidence of Antarctic Peninsula, Pagano Nunatak, and Nash Hills poles. This possibility certainly exists, but seems unlikely. Moreover, no regional tilting is apparent in the structural fabric of the Ellsworth domain [Storey and Dalziel, this volume]. The north to northwest trending hinge lines of upright folds in the Paleozoic succession are for the most part subhorizontal.

Proceeding, therefore, under the assumption that no tilt correction is warranted and that sufficient time is represented to average secular variation, since both normal and reversed intervals are present in the rocks (i.e., the Nash Hills and Pagano Nunatak directions are representative of the Middle Jurassic field), it now remains to determine a reasonable tectonic reconstruction consistent with the paleomagnetic, geologic, and space constraints.

We first note that the Middle Jurassic poles from the AP and the EWM are not statistically different, i.e., angular separation is under 7°, less than the α_{95} of 9.5° for the AP pole. We can therefore combine the individual locality mean poles from the AP and the EWM and calculate an overall West Antarctic Middle Jurassic pole at 237°E, 45.8°S, α_{95} = 6.4°, and N = 6 localities. Since both the AP and EWM poles differ from the Ferrar pole, it is not surprising that the combined West Antarctic pole also differs significantly from that of the Ferrar (Table 2). There is no discrepancy between the predicted and observed paleolatitude for the AP (54°S), but there is a 12° discrepancy for the EWM predicted (59°S) versus observed (47°S) paleolatitude. A 15° ± 10° counterclockwise rotation would eliminate the separation of the AP-EWM pole from the Ferrar pole. Within the paleomagnetic constraints, the AP and EWM may therefore have moved as a single or closely related unit since the Middle Jurassic with respect to East Antarctica.

The stratigraphic succession of the Ellsworth Mountains is broadly similar to that of the Gondwana craton cover in the Cape Mountains of southern Africa, the Falkland Islands and the Transantarctic Mountains [Schopf, 1969], and so it seems reasonable to conclude that the EWM should remain near the craton of Gondwanaland. Also, the timing of the Ellsworth Mountains deformation must be syn-Permian or post-Permian, which is the time of the Cape (Gondwanide) Orogeny of du Toit [1937].

Space constraints related to paleolatitudes depend on the choice of Gondwanaland reconstruction. We select the Norton and Sclater [1979] reconstruction because of their extensive use of seafloor data and the good agreement of East Antarctic and Australian Early Jurassic paleopoles with this fit [Irving and Irving, 1982]. The Middle Jurassic East Antarctic south pole used in calculating the mean Gondwanaland pole in the Norton and Sclater reconstruction differs insignificantly (5° of longitude at 55° latitude) from that determined for the Ferrar Supergroup by us from the published data. The paleomagnetically permissible locations for the EWM are (1) west of South America, (2) east of Australia, and (3) between South America, Africa, and East Antarctica.

We dismiss the first two possibilities as unrealistic. An EWM position west of South America would place the Paleozoic Gondwana craton cover succession of the Ellsworth Mountains outboard of a pre-Late Jurassic Pacific margin subduction complex [Dalziel and Forsythe, 1985]. Placing the EWM east of Australia requires a very large displacement and seems incompatible with the seafloor spreading history of the southeastern Pacif-

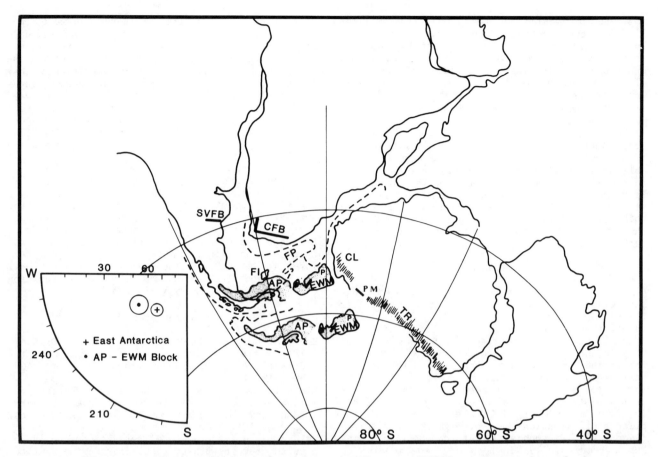

Fig. 7. A combined AP-EWM using Norton and Sclater's [1979] reconstruction of the other continents. The new paleomagnetic data predict the northern position shown for the AP-EWM. This creates an unacceptable overlap in this Gondwanaland reconstruction. Space constraints thus force an AP-EWM to the more southerly position shown on the diagram. This is 9° farther south than expected on the basis of the AP and EWM paleomagnetic results. Modifications in the Gondwanaland reconstruction and removal of the effects of extension in the EWM could eliminate the overlap. The AP-EWM (small solid circle) and East Antarctic (Ferrar) (plus) poles are plotted, with their respective circles of 95% confidence, on the equal angle stereographic projection. A 15° ± 10° counterclockwise rotation would restore the AP-EWM pole to the East Antarctic pole. Thick solid lines indicate structural trends, hatching indicates areas of flat-lying Gondwana-sequence cover rocks. See figure 6 caption for explanation of abbreviations.

ic Ocean and Tasman Sea. The third possibility results in two different tectonic reconstructions for the EWM and AP, which will be discussed below.

Within the constraints of the available paleomagnetic data, it is possible for the AP and EWM block to have moved as separate units. In support of this hypothesis is the occurrence of a major structural break along the Evans Ice Stream between the EWM block and the base of the AP [Doake et al., 1983; Garrett et al., this volume]. The reconstructed position of the AP could be as Longshaw and Griffiths [1983] proposed near the tip of South America (Figure 6). Their placement of the EWM near the Queen Maud Mountains is mainly based on space considerations and predicts an EMW paleolatitude near 65°S. However, our data show it to be near 47°S. An 18° difference in paleolatitude is beyond the errors of the analysis, and we can therefore exclude the Longshaw and Griffiths position for the EMW. Instead, we suggest that if the AP and EWM moved separately, then the AP can be positioned as suggested by Longshaw and Griffiths, but the EWM can be fitted north of the AP, into the space south of Africa, west of Coats Land, which would place the EWM at the required paleolatitude of about 47°S (Figure 6).

The alternative reconstruction, as previously mentioned, would be to keep the AP and EWM as one unit, since their poles are not significantly different. There may have been little movement of the AP relative to the EWM across the Evans Ice Stream since the Middle Jurassic. By strictly observing the rotation and paleolatitude constraints, we find that the AP overlaps southernmost South America on the Norton and Sclater [1979] reconstruction (Figure 7). The EWM is located east of the Antarctic Peninsula, south of Africa, and west of Coats Land. We can avoid gross overlap by using the outside limits of error of our paleomagnetic data and the mean Gondwanaland reference pole of Norton and Sclater [1979]; the EWM would then be near 56°S instead of our mean value of 47°S (Figure 7).

The large amount of overlap between the AP and South America shown in Figure 6 could be caused by several factors: the cumulative errors of our pole determination and the reference poles, minor motion along the East Gondwanaland-West Gondwanaland boundary, extension in South America [Gust et al., 1985] and in the AP-EWM [Garrett et al., this volume], and finally, very minor motion between the AP and EWM, i.e., not paleomagnetically discernible.

Discussion

We conclude that there are two possible reconstructions. If one takes the Norton and Sclater [1979] reconstruction and AP-EWM poles at face value, then there is no room for a combined AP-EWM because the AP would overlap with South America at 55°S in the Middle Jurassic. A separate EWM could fit in the space south of the Falkland Plateau and west of Coats Land (Figure 6). Alternatively, if the Norton and Sclater [1979] reconstruction and/or paleolatitudinal placement is modified slightly, then a combined AP-EWM, especially with the effects of extension removed, might be permissible (Figure 7). We emphasize that the position of the EWM is very similar in both of these models; we just cannot distinguish between these two reconstructions until the uncertainties in reference poles, seafloor geophysical data, and amounts of extension and tilting are better resolved.

Neither of these models disagrees with the results of Watts and Bramall [1981] if normal polarity is assumed for their Late Cambrian data. They noted that about a 30° counterclockwise rotation would then align their Ellsworth Mountains Late Cambrian pole with the early Paleozoic Gondwanaland polar wander path. If that is the case, then little rotation of the EWM from the Late Cambrian to the Middle Jurassic would be indicated. If their assumption of reversed polarity is correct, then over 100° of counterclockwise rotation would be needed between the Late Cambrian and the Middle Jurassic. It is interesting to note that our mean Nash Hills pole from the tilt-corrected thermal component of the red siltstones (292°E, 7.2°N) is virtually indistinguishable from the pole determined by Watts and Bramall [1981] in the Ellsworth Mountains (296°E, 4°N), 300 km north of the Nash Hills. If these rocks are of similar age to the Watts and Bramall samples (Late Cambrian), it would suggest that there has not been significant rotation between the Ellsworth Mountains and the Nash Hills.

The enigmas remain, then, of why the Ellsworth Mountains structural trend is at a high angle to the Cape Fold Belt-Transantarctic Mountains trends (Figures 6 and 7) and of why the thick Paleozoic succession of the Ellsworth Mountains dies out Pacificward (northward in present coordinates). The structural trends may not need to be aligned. Basement control seems likely, for example, in the case of the rocks on the Falkland Islands that change abruptly from east-west to northeast-southwest along the line of Falkland Sound [Greenway, 1972]. At present, however, there does not seem to be an obvious explanation for the "disappearance" of the thick (>10 km [Craddock, 1969]) Ellsworth Mountains succession along strike toward the Pacific.

Acknowledgments. We are grateful to our colleagues in the U.K.-U.S. West Antarctic Tectonics Project for their advice and criticism. Logistic support was provided jointly by the British Antarctic Survey (Natural Environment Research Council) and the Division of Polar Programs, National Science Foundation, Washington, D. C. (grant 82-13798 to I.W.D.D.). The efforts of the Air Unit of BAS and the Antarctic Development Squadron (VXE6) of the U.S. Navy were invaluable.

References

Beck, M. E., Jr., R. F. Burmester, and S. D. Sheriff, Field reversal and paleomagnetic pole for Jurassic Antarctica (abstract), Eos Trans. AGU, 60, 818, 1979.

Blundell, D. J., and P. J. Stevenson, Paleomagnetism of some dolerite intrusions from the Theron Mountains and Whichaway Nunataks, Antarctica, Nature, 184, 1860, 1959.

Briden, J. C., and R. L. Oliver, Paleomagnetic results from the Beardmore Glacier region, Antarctica, N. Z. J. Geol. Geophys., 6, 388-394, 1963.

Bull, C., E. Irving, and I. Willis, Further paleomagnetic results from south Victoria Land, Antarctica, Geophys. J. R. Astron. Soc., 6, 320-336, 1962.

Cherry, E. M., and H. C. Noltimier, Paleomagnetic results: Kirkpatrick Basalts at Brimstone Peak and Gordon Peak, Antarctica (abstract), Eos Trans. AGU, 63, 616, 1982.

Coney, P. J., D. L. Jones, and J. W. H. Monger, Cordilleran Suspect Terranes, Nature, 288, 329-333, 1980.

Craddock, C., Geology of the Ellsworth Mountains (Sheet 4, Ellsworth Mountains), in Geologic Maps of Antarctica, edited by V. C. Bushnell and C. Craddock, Antarctic Map Folio Series, Folio 12, Plate IV, 1969.

Craddock, C., The East Antarctic-West Antarctic boundary between the ice shelves: A review, in Antarctic Earth Science, edited by R. L. Oliver, P. R. James, and J. B. Jago, pp. 94-97, Australian Academy of Science, Canberra, 1983.

Craddock, C., J. Splettstoesser, and G. Webers, Geology and Paleontology of the Ellsworth Mountains, Antarctica, Mem. Geol. Soc. Am., in press, 1986.

Dalziel, I. W. D., and D. H. Elliot, West Antarctica: Problem child of Gondwanaland, Tectonics, 1, 3-19, 1982.

Dalziel, I. W. D., and R. F. Forsythe, Andean evolution and the terrane concept, in Tectonostratigraphic Terranes of the Circum-Pacific Regions, edited by D. G. Howell, pp. 565-581, Circum-Pacific Council for Energy and Mineral Resources, Houston, Texas, 1985.

Dalziel, I. W. D., and A. M. Grunow, The Pacific margin of Antarctica: Terranes within terranes within terranes, in Tectonostratigraphic Terranes of the Circum-Pacific Region, edited by D. G. Howell, pp. 555-581, Circum-Pacific Council for Energy and Mineral Resources, Houston, Texas, 1985.

Dalziel, I. W. D., and R. J. Pankhurst, The Joint U.K.-U.S. West Antarctic Tectonics Project: An introduction, this volume.

Dalziel, I. W. D., S. W. Garrett, A. M. Grunow, R. J. Pankhurst, B. C. Storey, and W. R. Vennum, The Ellsworth-Whitmore Mountains crustal block: Its role in the tectonic evolution of West Antarctica, this volume.

Delisle, G., Results of paleomagnetic investigations in northern Victoria Land, Antarctica, in Antarctic Earth Science, edited by R. L. Oliver, P. R. James, and J. B. Jago, pp. 146-149, Australian Academy of Science, Canberra, 1983.

Doake, C. S. M., R. D. Crabtree, and I. W. D. Dalziel, Subglacial morphology between Ellsworth Land and Antarctic Peninsula: New data and tectonic significance, in Antarctic Earth Science, edited by R. L. Oliver, P. R. James, and J. B. Jago, pp. 270-273, Australian Academy of Science, Canberra, 1983.

du Toit, A. L., Our Wandering Continents, Oliver and Boyd, 366 pp., Edinburgh, 1937.

Elliot, D. H., R. J. Fleck, and J. F. Sutter, Potassium-argon age determinations of Ferrar Group rocks, central Transantaractic Mountains, in Geology of the Central Transantarctic Mountains, Antarct. Res. Ser., vol. 36, edited by M. D. Turner and J. F. Splettstoesser, pp. 197-224, AGU, Washington, D.C., 1985.

Funaki, M., Paleomagnetic investigation of Ferrar dolerite in McMurdo Sound, Antarctica, Antarct. Record, 77, 20-32, 1982.

Garrett, S. W., L. D. B. Herrod, and D. R. Mantripp, Crustal structure of the area around Haag Nunataks, West Antarctica: New aeromagnetic and bedrock elevation data, this volume.

Greenway, M. E., The geology of the Falkland Islands, Br. Antarct. Surv. Sci. Rep., 76, 1972.

Gust, D. A., K. T. Biddle, D. W. Phelps, and M. A. Uliana, Associated Middle to Late Jurassic volcanism and extension in southern South America, Tectonophysics, 16, 223-253, 1985.

Irving, E., and G. A. Irving, Apparent polar wander paths Carboniferous through Cenozoic and the assembly of Gondwana, Geophys. Surv., 5, 141-188, 1982.

Kent, E., and F. M. Gradstein, A Cretaceous and Jurassic geochronology, Geol. Soc. Am. Bull., 96, 1419-1427, 1985.

Kirschvink, J. L., The least-square line and plane analysis of paleomagnetic data, Geophys. J. R. Astron. Soc., 62, 699-718, 1980.

Kyle, P. R., D. H. Elliot, and J. F. Sutter, Jurassic Ferrar Supergroup tholeiites from the Transantarctic Mountains, Antarctica, and their relationship to the initial fragmentation of Gondwana, in Gondwana Five, edited by L. M. Cresswell and P. Vella, pp. 283-287, A. A. Balkema, Rotterdam, 1981.

Longshaw, S. K., and D. H. Griffiths, A paleomagnetic study of Jurassic rocks from the Antarctic Peninsula and its implications, J. Geol. Soc. London, 140, 945-954, 1983.

Lovlie, R., Mesozoic paleomagnetism in Vestfjella, Dronning Maud Land, East Antarctica, Geophys. J. R. Astron. Soc., 59, 529-537, 1979.

McIntosh, W. C., P. R. Kyle, E. M. Cherry, and H. C. Noltimier, Paleomagnetic results from the Kirkpartrick Basalt Group, Victoria Land, Antarct. J. U. S., 17, 20-22, 1982.

Millar, I., and R. J. Pankhurst, Rb/Sr geochronology of the region between the Antarctic Peninsula and the Transantarctic Mountains: Haag Nunataks and Mesozoic granitoids, this volume.

Norton, I. O., and J. G. Sclater, A model for the evaluation of the Indian Ocean and the breakup of Gondwanaland, J. Geophys. Res., 84, 6803-6830, 1979.

Ostrander, J. H., Paleomagnetic investigations of the Queen Victoria Range, Antarctica, Antarct. J. U. S., 6(5), 183-185, 1971.

Schopf, J. M., Ellsworth Mountains: Position in West Antarctica due to sea-floor spreading, Science, 164, 63-66, 1969.

Stone, D. B., B. C. Panuska, and D. R. Packer, Paleolatitudes versus time for southern Alaska, J. Geophys. Res., 87, 3697-3707, 1982.

Storey, B. C., and I. W. D. Dalziel, Outline of the structural and tectonic history of the Ellsworth-Thiel Mountains ridge, West Antarctica, this volume.

Turnbull, G., Some paleomagnetic measurements in Antarctica, Arctic, 12, 151-157, 1959.

Van der Voo, R., D. L. Jones, C. S. Gromme, G. D. Eberlein, and M. Churkin, Paleozoic paleomagnetism and northward drift of the Alexander Terrane, southwestern Alaska, J. Geophys. Res., 85, 5281-5296, 1980.

Watts, D. R., and A. M. Bramall, Paleomagnetic evidence for a displaced terrain in western Antarctica, Nature, 293, 638-641, 1981.

Zijderveld, J. D., A.C. demagnetization of rocks: Analysis of results, in Methods in Paleomagnetism, edited by D. W. Collinson, K. M. Creer, and S. K. Runcorn, pp. 254-286, Elsevier, New York, 1967.

Copyright 1987 by the American Geophysical Union.

THE ELLSWORTH-WHITMORE MOUNTAINS CRUSTAL BLOCK: ITS ROLE IN THE TECTONIC EVOLUTION OF WEST ANTARCTICA

I. W. D. Dalziel,[1] S. W. Garrett,[2] A. M. Grunow,[3] R. J. Pankhurst,[2] B. C. Storey,[2] and W. R. Vennum[4]

Abstract. The 1983-1984 season of the joint British Antarctic Survey-U.S. Antarctic Research Program geology and geophysics project on the Ellsworth-Whitmore Mountains crustal block (EWM) has yielded new observations and laboratory data relevant to the geological evolution of West Antarctica and its tectonic relationship to the rest of Gondwanaland. This is a synthesis of results presented in companion papers in this volume. New paleomagnetic data favor a Jurassic reconstruction in which there has been little or no relative displacement between the EWM and the Antarctic Peninsula. They may be restored together, with a 15°-20° counterclockwise rotation and a northward translation of approximately 10° of latitude, to a position on the Pacific side of the Falkland Plateau-Cape Fold Belt-Coats Land junction. Orthogneiss exposed at Haag Nunataks represents a Proterozoic cratonization event, and aeromagnetic data demonstrate that related rocks occur beneath the ice from the northeastern edge of the Ellsworth Mountains as far as the base of the Antarctic Peninsula. Although this basement does not demonstrably extend beneath the EWM, retention of the present-day relative positions of the Antarctic Peninsula, Haag Nunataks, and the EWM is geologically compatible with the above reconstruction. To the south of the block, the igneous and sedimentary rocks of the Thiel Mountains are recognized as part of the Precambrian basement and Phanerozoic successions of the Transantarctic Mountains, geologically and geophysically distinct from the folded sedimentary succession of the EWM. The latter are mostly lithologically correlative with the lower part of the thick Cambrian-Permian Ellsworth Mountains succession, and throughout much of the area, share the same simple structural style and trend related to post-Permian folding. Discordant and more complex structures are observed at separate localities on the margins of the EWM. The youngest exposed rocks in the EWM are a suite of Middle Jurassic peraluminous "S-type" granites. These are crustal anatectic (or at least highly contaminated) melts which point to the presence of a deep continental basement beneath the EWM succession. They signify an intracontinental thermal event which, like that associated with contemporaneous magmatism in various tectonic environments throughout Gondwanaland, seems to herald the breakup of the old supercontinent. The subsequent Mesozoic-Cenozoic history of the EWM is recorded in geophysical evidence for crustal rifting and extension, some of which may be related to the migration of West Antarctic crustal blocks to their present positions, others of which are probably very young features related to Cenozoic alkali magmatism outside the EWM and the recent marked uplift of the Ellsworth Mountains.

Introduction

East Antarctica, with its peripheral but widespread exposures of Precambrian rocks, presumably continuous beneath the ice cover, is understood essentially as a large residual component of the Gondwanaland craton. It is morphologically and geologically bordered along the Ross Sea-Weddell Sea embayment by the Transantarctic Mountains. Elsewhere, it fits convincingly against the other craton fragments of South America-Africa, India, and Australia, when the oceanic lithosphere generated by Mesozoic-Cenozoic spreading is removed [Norton and Sclater, 1979]. In contrast, West Antarctica consists of at least four distinct crustal units [Jankowski and Drewry, 1981] with separately characterized and mostly younger geological histories [Dalziel and Elliot, 1982]. These units are the Antarctic Peninsula, the Thurston Island area, Marie Byrd Land, and a southern region adjoining the Transantarctic Mountains which we refer to here as the Ellsworth-Whitmore Mountains crustal block (EWM) (Figure 1).

The most obvious and best exposed portion of this latter block (Figure 2) is the Ellsworth Mountains, over 4 km above sea level and formed of a thick (>10 km [Craddock et al., 1964]) folded sequence of Paleozoic strata comparable to the Gondwanaland cratonic cover sequence of the Transantarctic Mountains (Beacon Supergroup) and the mountains of the Cape Fold Belt of southern Africa

---

[1]Institute of Geophysics, University of Texas at Austin, Austin, Texas 78715.
[2]British Antarctic Survey, High Cross, Cambridge CB3 OET, United Kingdom.
[3]Lamont-Doherty Geological Observatory of Columbia University, Palisades, New York 10964.
[4]Department of Geology, Sonoma State University, Rohnert Park, California 94928.

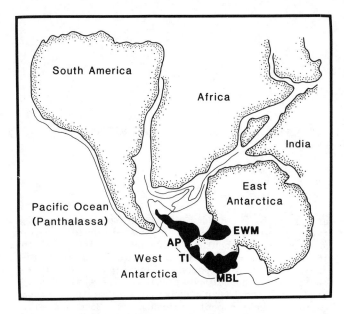

Fig. 1. Gondwanaland reconstruction of Norton and Sclater [1979] showing major continental crustal blocks of West Antarctica: AP is Antarctic Peninsula crustal block; TI, Thurston Island-Eights Coast crustal block; MBL, Marie Byrd Land crustal block and EWM, Ellsworth-Whitmore Mountains crustal block.

(Cape and Karoo supergroups). Consequently, Schopf [1969] proposed that the Ellsworth Mountains are allochthonous with respect to East Antarctica. He suggested that they were originally located along the Transantarctic margin of the craton between the Pensacola Mountains and the Cape Fold Belt mountains. Any record of this former position, and of the necessary displacement, would be obscured by ice cover at the head of the Weddell Sea.

Previous investigations in the EWM have included a detailed study of the geology of the Ellsworth Mountains [Splettstoesser and Webers, 1980], but have only included reconnaissance studies of its other widely scattered smaller nunataks (Figure 2) including the Whitmore Mountains [Craddock, 1983; Webers et al., 1982]. The first season of the British Antarctic Survey-U.S. Antarctic Research Program West Antarctic tectonics project was devoted to a study of these nunataks and adjoining parts of West and East Antarctica. We present here a summary of our preliminary results in the form of an account of the EWM and its role in the tectonic evolution of West Antarctica. Details of the data are to be found in articles by Garrett et al. [this volume], Grunow et al. [this volume], Millar and Pankhurst [this volume], Storey and Dalziel [this volume], and Vennum and Storey [this volume (a) and (b)]. This paper also contains new aeromagnetic data from the EWM; see Garrett et al. [this volume] for a description of data acquisition and processing.

Extent of the Ellsworth-Whitmore Mountains Crustal Block

The bedrock of the EWM is exposed discontinuously along a linear region of bedrock highs known as the Ellsworth Mountains-Thiel Mountains ridge [Craddock, 1983]. In this region (Figure 2) the margin of the East Antarctic craton is topographically less pronounced than elsewhere along the Transantarctic Mountains, such as the western side of the Ross Sea. The Thiel Mountains at the southern end of the ridge are geologically part of the Transantarctic Mountains. The late Precambrian porphyries and associated volcaniclastic sedimentary rocks and the early Paleozoic granites of the Thiel Mountains [Schmidt and Ford, 1969] can be correlated with the Wyatt and Ackerman formations, respectively, of the La Gorce Mountains [Stump, 1983; Storey and Dalziel, this volume] and with the Granite Harbour Intrusive Complex of Victoria Land [Gunn and Warren, 1962]. Diabases at the isolated Lewis Nunatak are petrographically and geochemically recognized as belonging to the Ferrar Supergroup [Vennum and Storey, this volume (a); Millar and Pankhurst, this volume], and we found flat-lying sedimentary strata similar to the rocks of the Beacon Supergroup beneath the contact. Along the Ellsworth Mountains-Thiel Mountains ridge to the north, sedimentary rocks of the Stewart Hills are lithologically distinct from, and significantly more deformed than, those of the Thiel Mountains [Storey and Dalziel, this volume]. Lithological and structural correlation with the upper Precambrian Patuxent Formation of the Pensacola Mountains led Craddock [1983] to conclude that the boundary of the East Antarctic craton lay to the north of the Stewart Hills. Jankowski et al. [1983] reached a similar conclusion. The rocks of the Stewart Hills are, however, structurally similar to those of Mount Moore approximately 500 km to the north [Storey and Dalziel, this volume]. We therefore suggest that the nunataks may form part of the EWM, although this does not preclude the possibility that stratigraphical correlation of the rocks of the Stewart Hills with the Patuxent Formation may also be correct. Aeromagnetic data support this interpretation. An area of 100-nT-amplitude, high-frequency magnetic anomalies reflecting shallow igneous rocks extends 50 km northward from the main escarpment of the Thiel Mountains, whereas the Stewart Hills lie within an area of more subdued magnetic activity typical of the sedimentary strata of the EWM (Figure 3). This abrupt change in profile occurs between the Stewart Hills and Sonntag Nunatak, an isolated northerly outcrop of the Thiel Mountains porphyry.

The homogenous lithology and structural geometry of sedimentary rocks from the Ellsworth domain [Storey and Dalziel, this volume], the similarity in age and petrochemistry of the granitic rocks [Millar and Pankhurst, this volume; Vennum and Storey, this volume (b)], and the identical paleomagnetic poles obtained from the granites of Nash Hills and Pagano Nunatak [Grunow et al., this volume] all make it clear that despite the likely

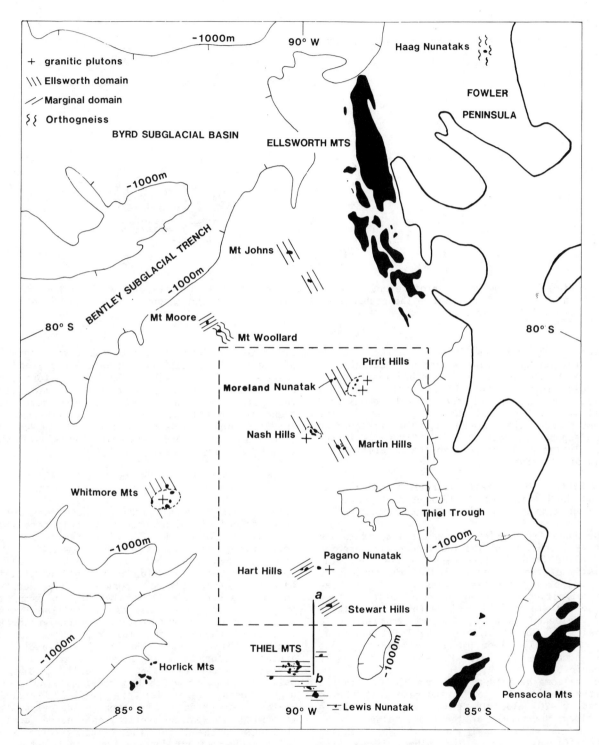

Fig. 2. Geological map of the Ellsworth-Whitmore Mountains and surrounding area showing the main tectonic provinces. Area of Figure 6 is shown by dashed outline. An aeromagnetic and topographic profile along AB is shown in Figure 3.

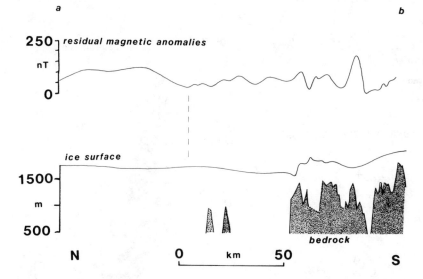

Fig. 3. Aeromagnetic profile and bedrock topography in the vicinity of the Thiel Mountains. Dashed line marks the probable northerly extent of the igneous rocks exposed in the Thiel Mountains.

presence of Mesozoic and Cenozoic faults, the EWM can be regarded as a single distinct tectonic entity. Although the nature of the basement beneath the Paleozoic succession is not yet known (see below), for purposes of continental reconstructions, we have retained the bedrock east of the Ellsworth Mountains (including the Precambrian gneiss exposed at Haag Nunataks, Figure 2) in the same relative position between the Antarctic Peninsula and the EWM.

Early History of the East Antarctic Craton Margin and the Location of the Ellsworth-Whitmore Mountains Crustal Block in Gondwanaland

The Upper Precambrian volcaniclastic strata and porphyry intrusions of the Thiel Mountains probably resulted from the same magmatic episode. The chemistry of the porphyries and of the subsequent early Paleozoic granites suggests that they reflect significant melting of continental crust [Vennum and Storey, this volume (b)]. Thus there is no evidence that the central Transantarctic Mountains in the vicinity of the Thiel Mountains were necessarily close to an active convergent plate boundary in late Precambrian to early Paleozoic times.

New paleomagnetic data from the Middle Jurassic granites of the EWM clearly indicate that it was rotated 15°-20° clockwise from its previous location along the margin of the Gondwanaland craton east of Coats Land and south of the Falkland Plateau [Grunow et al., this volume, Figure 4)]. This does not contradict the data obtained by Watts and Bramall [1981] for Cambrian rocks of the Ellsworth Mountains, which may be interpreted, assuming normal polarity, as consistent with such a position of the EWM since early Paleozoic times. This solution places the Paleozoic succession close to the Gondwanaland craton cover sequences of the Falkland Islands, southern Africa, and the Transantarctic Mountains. However, restoring the EWM 15°-20° counterclockwise in this way places the post-Permian pre-Middle Jurassic structural trend of the Ellsworth domain at a high angle to the structural trends of the Sierra de la Ventana, Cape Fold Belts, and Pensacola Mountains. This will be discussed below.

Haag Nunataks and the Basement of the Ellsworth-Whitmore Mountains Crustal Block

Arguments concerning the predrift assembly of Gondwanaland with its ancient cratonic core would be greatly clarified by a clear recognition and delimitation of the continental basement. This is hampered throughout West Antarctica by the almost complete absence of evidence for old crystalline basement. The isolated exposure at Haag Nunataks on the eastern side of the EWM is, in fact, the only demonstrable outcrop of crystalline Precambrian rock in West Antarctica. The new geochronological data confirm that this represents cratonic basement [Millar and Pankhurst, this volume]. The gneisses record a short-lived Proterozoic crust formation event resulting in rapidly cooled and uplifted basement which has remained stable at a high crustal level for the past 1000 million years. This predates all other geological history recognized in West Antarctica, and indeed, most of that demonstrated for the Transantarctic Mountains. The nearest correlatives are found on the edge of the East Antarctic shield in Coats Land and at a small exposure on the Falkland Islands [Greenway, 1972]. This is compatible with the reconstruction suggested here, which places all these occurrences in the same general area prior to Jurassic times (Figure 4).

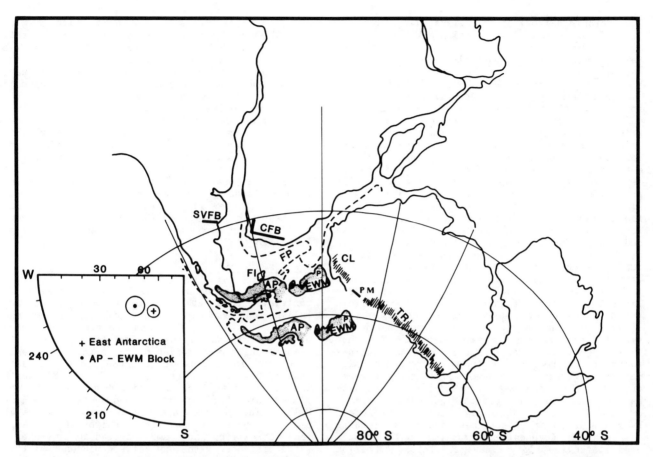

Fig. 4. Locations of the Ellsworth-Whitmore Mountains crustal block in a reconstructed Gondwanaland. It is assumed that there has been no relative motion between the AP and EWM. The more northerly position corresponds to the mean of the paleomagnetic data; the southerly one avoids overlap of AP and South America. SVFB is Sierra de la Ventana Fold Belt; FP, Falkland Plateau; FI, Falkland Islands; AP, Antarctic Peninsula crustal block; EWM, Ellsworth-Whitmore Mountains crustal block; CL, Coats Land; PM, Pensacola Mountains; TR, Transantarctic Range; and P, Pagano Nunatak.

The crucial question which must be posed is: To what extent does the tiny outcrop at Haag Nunataks represent a regional basement in West Antarctica, and particularly, Can it be inferred that it underlies the EWM? The new aeromagnetic data have convincingly shown that the gneisses outcropping at Haag Nunataks are probably present near the surface throughout the central part of the Fowler Peninsula, and that similar basement is present throughout much of the area between the base of the Antarctic Peninsula and the Ellsworth Mountains [Garrett et al., this volume]. The complete absence of this pronounced magnetic signature over the EMW places some constraints on the possible continuation of such basement. If it does indeed exist beneath the EWM, it must be very deeply buried (>20 km), and moreover, not subject to similar horst and graben tectonics of the Haag Nunataks area [Garrett et al., this volume]. Another possibility is that the basement beneath the Paleozoic succession must be of a different character.

Apart from the metasedimentary cover, the only exposed rocks in the area are Mesozoic granitoids whose chemistry and isotope geology point unambiguously to the presence of sialic continental crust at the level of anatexis, which may still be within the upper crust [Millar and Pankhurst, this volume; Vennum and Storey, this volume (b)]. They do not at present allow us to differentiate between a crystalline basement, an older metasedimentary sequence, or the base of the Paleozoic succession. However, the geophysics and geochemistry reinforce conclusions based on West Antarctic seismic refraction profiles [Bentley and Clough, 1972]. These show a layer with a P-wave velocity of 6.1 km s^{-1} close to the western edge of the block, which Bentley and Clough [1972] interpreted as part of a crystalline basement. Moreover, the Gondwana sedimentary sequences of the Falkland Islands and the Cape Fold Belt are clearly seen to unconformably overlie Precambrian or early Paleozoic crystalline basement.

Tectonic History of the Ellsworth-Whitmore Mountains Crustal Block

Paleozoic Sedimentary and Early Mesozoic Deformation

The sedimentary and structural history of the rocks investigated throughout the Ellsworth domain, the largest of two structural subdivisions (Figure 2) of the EWM [Storey and Dalziel, this volume], is comparable with part of the Cambrian-Permian sedimentary succession of the Ellsworth Mountains themselves. The rocks were mainly deposited in a shallow marine environment on a pre-Paleozoic continental basement and are most similar to the lower Paleozoic Heritage Group of the Ellsworth Mountains succession, although the facies are not identical. Deposition of these sedimentary strata was contemporaneous with some of the Gondwana cover sequences of southern Africa (Cape Supergroup), the Falkland Islands (Devono-Carboniferous Group), and the Transantarctic Mountains (Neptune Group and Beacon Supergroup). The lateral continuity of individual formations within the Gondwana cover sequence is uncertain, and it is possible that discrete depositional basins of variable orientation may have existed. The geochemical signature of the (?) Cambrian gabbroic rocks at the base of the Ellsworth Mountains succession suggests that deposition of the sedimentary strata of the EWM occurred in a subsiding basin on continental crust undergoing some crustal extension. These sedimentary strata were deformed by a single phase of northwest-southeast trending simple folds (present orientation), with associated spaced and slaty cleavage. This deformation occurred between Permian and Middle Jurassic times, probably reflecting compression of the basin. Gross structural trends within Gondwana cover sequences in the region of the predrift position suggested here for the EWM (Figure 4) show considerable variation. These trends may be controlled by preexisting basement structures and/or by the structural control of basin development. It is important to observe that abrupt changes in structural trend and style are relatively common in the Gondwanide Fold Belt; for example, in the Cape Fold Belt there is a north-south trend in the western province and an east-west trend in the southern province [Söhnge and Hälbich, 1983]. The contrast in structural trend between the rocks at the edge of the EWM (the Marginal domain of Storey and Dalziel [this volume]) and those in the Ellsworth domain may have the same cause, or possibly result from rotation of the main Ellsworth domain trends. Such rotation could have occurred either during formation of the marginal extensional rift systems discussed below or during movement of the EWM from its original position. However, it is also possible that the marginal domain reflects relics of an older orogenic belt (e.g., Ross Orogen); Craddock [1983] has compared the deformation of the Stewart Hills with that of the Patuxent Formation of the Pensacola Mountains, part of the Ross orogenic belt.

The rocks of Mount Woollard are something of a geological enigma. The structural grain of the migmatized paragneiss is concordant with that of the Ellsworth domain [Storey and Dalziel, this volume], but the higher metamorphic grade and the complex history of pegmatite emplacement are unique in this area. In view of the preliminary isotopic data [Millar and Pankhurst, this volume], we regard the gneisses as derived from Ellsworth succession sedimentary strata rather than an older crystalline basement. This implies that the complex is a deep migmatitic zone associated with Mesozoic granitic plutonism, uplifted along the margin of a later rift represented by the Bentley Subglacial Trench. However, pegmatite emplacement at Mount Woollard began at least as early as the first episode of folding undergone by the paragneisses [Storey and Dalziel, this volume], whereas the other granites of the EWM are clearly post-tectonic with respect to the folding and cleavage formation.

Igneous Activity

The igneous rocks of the EWM are predominantly restricted to a series of Middle Jurassic plutons [Millar and Pankhurst, this volume; Vennum and Storey, this volume (b)] and isolated occurrences of mafic igneous rocks. The latter include a bimodal suite of Cambrian mafic to felsic volcanic rocks and intrusions of gabbroic stocks and diabase sills into Cambrian strata in the Ellsworth Mountains as well as an undated gabbro sill emplaced into the deformed metasediments of the Hart Hills. The granites are a consanguineous suite of plutonic rocks that show great uniformity in chemical composition and a small spread in ages over a large distance, with some between-pluton variation in isotope composition. They date to 175 ± 5 Ma [Millar and Pankhurst, this volume], i.e., Middle Jurassic.

Petrologically, chemically, and isotopically, the granites strongly resemble suites of peraluminous leucogranites (S-type of White and Chappell [1983]) developed in intracontinental settings, such as the Hercynian belt of Europe, and in continental collision-type settings, such as the Bhutan and Nepalese Himalayas. A collisional environment is not obviously compatible with our knowledge of the tectonic setting of the EWM during the Middle Jurassic. These granites would appear to represent emplacement in a neutral or extensional environment following the main compressive phase that resulted in folding and cleavage development in the Paleozoic rocks of the Ellsworth domain. We ascribe their origin to melting of continental crust during a Middle Jurassic thermal event, but cannot precisely define the tectonic cause of this event. Contemporaneous magmatic activity in the Transantarctic Mountains (Ferrar Supergroup), South Africa (Karoo dolerites), and southern South America (silicic volcanic rocks of the Tobifera Group and the S-type Darwin Granite Suite) suggest a broad pattern of thermal activity (Figure 5) synchronous with, or immediately prior to, continental breakup as part of a widespread-extensional tectonic regime [Dalziel et al., 1986].

Aeromagnetic data (Figure 6) reveal a series

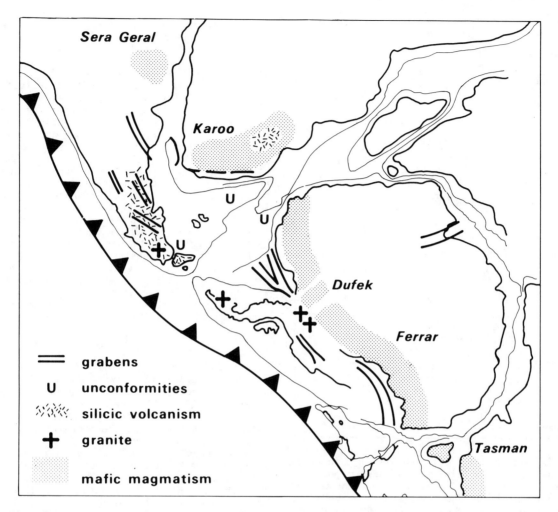

Fig. 5. Igneous rocks associated with Gondwanaland breakup [after Dalziel et al., 1986].

of low amplitude (100-300 nT) moderate wavelength (40-100 km) anomalies aligned east-west over the EWM. The anomalies are most clearly developed over exposures of granitic rocks and may represent deeper, more magnetite-rich intrusive rocks associated with the granites. Two possible tectonic controls may have resulted in the present distribution and orientation of these igneous bodies: (1) the granitic magmas were intruded as much larger masses than those presently exposed on the block and were later disrupted by faulting, or (2) the granitic magmas were emplaced in east-west orientated fractures during extensional tectonism. We know of no comparable leucocratic suites arranged in synchronous linear belts of this type. The S-type granites of the Darwin suite of the southernmost Andes were clearly emplaced during regional extension, despite the fact that they were later involved in arc-continent collision [Nelson et al., 1980]. They cut only pre-Late Jurassic basement rocks, while the contemporaneous Tobifera Group volcanic rocks were extruded onto this basement in a volcanotectonic rift zone to the east [Bruhn et al., 1978; Gust et al., 1985]. The relationship, if any, between the Ferrar, Karoo, and EWM igneous rocks regional extensional tectonism and contemporaneous Pacific margin subduction (Figure 5) remains unclear [Dalziel et al., 1986].

Mesozoic-Cenozoic Displacement and Extension

The new paleomagnetic data suggest that the Middle Jurassic paleopoles for the Antarctic Peninsula crustal block (AP) and EWM are indistinguishable [Grunow et al., this volume]. One model for post-Middle Jurassic motion is thus to keep both blocks together during a clockwise rotation of $15°-20°$ about a pole located in the vicinity of the western Drake Passage. This motion may have been accomplished by the Mesozoic seafloor spreading that formed the Weddell Sea [Barker and Jahn, 1980; LaBrecque and Barker, 1981]. The position of the AP inferred in this way creates unacceptable overlap with South America (Figure 4), although the extreme limits of paleomagnetic

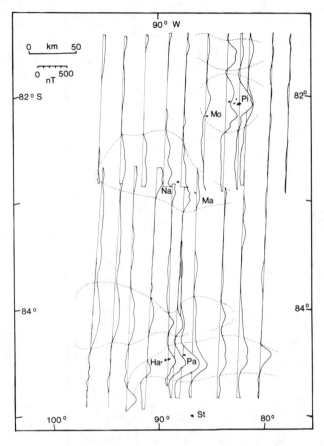

Fig. 6. Aeromagnetic profiles showing inferred distribution of granitic bodies beneath positive anomalies. Anomaly correlations are marked by dotted lines. Ha is Hart Hills; Na, Nash Hills; Ma, Martin Hills; Mo, Moreland Nunatak; Pa, Pagano Nunatak; and Pi, Pirrit Hills.

error would just allow the AP to be placed south of South America. Alternatively, the AP and EWM could have moved as separate blocks, in which case there are fewer constraints on their precise original positions [see Grunow et al., this volume].

The occurrence of rift zones throughout West Antarctica [Masolov et al., 1981] suggests that extension may have been important there both during and subsequent to displacement of the EWM, which must have been initiated after emplacement of the Middle Jurassic granites (≃175 Ma). Our new aeromagnetic data [Garrett et al., this volume] indicate that horst and graben structures between the Antarctic Peninsula and the Ellsworth Mountains are bounded by normal faults with throws of several kilometers. The Ellsworth Mountains probably developed their pronounced relief on the flanks of this rift system. The spectacular relief between the (northern) Sentinel Range, which is more than 4 km above sea level, and the bedrock beneath the Rutford Ice Stream, more than 2 km below sea level, suggests that uplift is in part of Cenozoic age. The trough between Mount Woollard and the Whitmore Mountains is 1 km deep [Jankowski and Drewry, 1981], and our new data show that it is associated with a magnetic anomaly of 750 nT in amplitude and 12 km in wavelength. This suggests that the trough is tectonic and possibly reflects extension between the two elevated areas. Another major topographic feature is the Thiel Trough (Figure 2), extending from the head of the Ronne Ice Shelf and trending east-west between Nash Hills and Pagano Nunatak. The magnetic signature over the trough and between the previously discussed anomalies associated with the granites is flat. The absence of differential tilting between the Nash Hills and Pagano Nunatak reflected in the paleomagnetic data is therefore surprising. Regardless of the nature of the Thiel Trough, it seems likely that the EWM was subject to extension during both Mesozoic and Cenozoic times. Widespread alkali basalt lavas along the Pacific margin of the continent are associated with Cenozoic extension [Garrett and Storey, 1986]. Closure of extension features in both continents may help to reduce the unacceptable overlap between the Antarctic Peninsula and South America in both this and other reconstructions of Gondwanaland in Middle Jurassic times.

Conclusion

Although laboratory analysis of the samples collected during the first season of the joint British Antarctic Survey-U.S. Antarctic Research Program (BAS-USARP) West Antarctic tectonics project is still incomplete, we believe that a new basis for understanding the tectonic evolution for West Antarctica is now emerging.

Some ideas previously based on rather sketchy data have been confirmed and amplified; for example, that the area between the Antarctic Peninsula and the Ellsworth Mountains is true Precambrian craton represented by the outcrop at Haag Nunataks. Similarly, the metasedimentary rocks of the Ellsworth Mountains-Thiel Mountains ridge and the Whitmore Mountains are all broadly correlative with the Ellsworth Mountains sequence, most probably with its lower part, the Heritage Group. The Middle Jurassic age of the granites is, in at least three cases, rigorously confirmed at about 175 Ma.

On the other hand, our observations and data have led us to challenge other well-established ideas or to develop new ones. Thus, on the basis of our paleomagnetic data, we have suggested a new type of reconstruction of Gondwanaland prior to breakup in which the AP and EWM (and the intervening cratonic basement) retain their present-day relative positions and are simply restored counterclockwise back to the margins of the Falkland Plateau and Coats Land. New geochemical and isotopic data for the Middle Jurassic granites establish their crustal anatectic origin and extend the location and style of the thermal event associated with continental breakup, variously represented by the igneous rocks of the Ferrar Supergroup, the Karoo System, the Tobifera Group, and the Darwin Granite Suite.

Further work on the samples obtained during the 1983-1984 season, and subsequently in the Jones Mountains and Thurston Island areas, may be expected to resolve some of the questions which either remain outstanding or have arisen as a result of our new ideas. For example, detailed study of the sedimentary facies, geochemistry, and isotope geology of the Paleozoic metasediments should yield information on their provenance, correlation, and depositional environment, as well as elucidate the tectonic significance of the two structural domains to which they belong. Further analysis of the Mesozoic granites (and the Mount Woollard pegmatites) will, we hope, constrain more closely their origins and, thus, the possible nature of the crust underlying the EWM. Finally, we hope to test our new hypotheses by establishing the tectonic relationship of the EWM to the Thurston Island block, one of the least understood portions of the West Antarctica mosaic.

Acknowledgments. This work is part of the joint BAS-USARP program investigating the tectonic history of West Antarctica. We are very grateful to the BAS air unit, VXE6 of the U.S. Navy, and our field assistants for their support during the field program. The U.S. side of the project was supported by the Division of Polar Programs, National Science Foundation, grant DPP 82-13798 to I. W. D. Dalziel.

References

Barker, P. F., and R. A. Jahn, A marine geophysical reconnaissance of the Weddell Sea, Geophys. J. R. Astron. Soc., 63, 271-283, 1980.

Bentley, C. R., and J. W. Clough, Antarctic subglacial structure from seismic refraction measurements, in Antarctic Geology and Geophysics, edited by R. J. Adie, pp. 683-691, Universitetsforlaget, Oslo, 1972.

Bruhn, R. L., C. R. Stern, and M. J. DeWit, Field and geochemical data bearing on the development of a Mesozoic volcano-tectonic rift zone and back-arc basin in southernmost South America, Earth Planet. Sci. Lett., 41, 32-46, 1978.

Craddock, C., The East Antarctica-West Antarctica boundary between the ice shelves: A review, in Antarctic Earth Science, edited by R. L. Oliver, P. R. James, and J. B. Jago, pp. 94-97, Australian Academy of Science, Canberra, 1983.

Craddock, C., J. J. Anderson, and G. F. Webers, Geologic outline of the Ellsworth Mountains, in Antarctic Geology, edited by R. J. Adie, pp. 155-170, North-Holland, Amsterdam, 1964.

Dalziel, I. W. D., and D. H. Elliot, West Antarctica, problem child of Gondwanaland, Tectonics, 1, 3-19, 1982.

Dalziel, I. W. D., B. C. Storey, S. W. Garrett, A. M. Grunow, L. D. B. Herrod, and R. J. Pankhurst, Extensional tectonics and the fragmentation of Gondwanaland, in Continental Extension Tectonics, edited by M. P. Coward, J. F. Dewey, and P. L. Hancock, Spec. Publ. Geol. Soc. London, in press, 1986.

Garrett, S. W., L. D. B. Herrod, and D. R. Manntripp, Crustal structure of the area around Haag Nunataks, West Antarctica: New aeromagnetic and bedrock elevation data, this volume.

Garrett, S. W., and B. C. Storey, Lithospheric extension on the Antarctic Peninsula during Cenozoic subduction, in Continental Extension Tectonics, edited by M. P. Coward, J. F. Dewey, and P. L. Hancock, Spec. Publ. Geol. Soc. London, in press, 1986.

Greenway, M. E., The geology of the Falkland Islands, Br. Antarct. Surv. Sci. Rep., 76, 42 pp., 1972.

Grunow, A. M., I. W. D. Dalziel, and D. V. Kent, Ellsworth-Whitmore Mountains crustal block, western Antarctica: New paleomagnetic results and their tectonic significance, this volume.

Gunn, G. M., and G. Warren, Geology of Victoria Land between the Mawson and Mullock glaciers, Antarctica, N. Z. Geol. Surv. Bull., 71, 157 pp., 1962.

Gust, D. A., K. T. Biddle, D. W. Phelps, and M. A. Uliana, Associated Middle to Late Jurassic volcanism and extension in southern South America, Tectonophysics, 116, 223-253, 1985.

Jankowski, E. J., and D. J. Drewry, The structure of West Antarctica from geophysical studies, Nature, 291, 17-21, 1981.

Jankowski, E. J., D. J. Drewry, and J. C. Behrendt, Magnetic studies of upper crustal structure in West Antarctica and the boundary with East Antarctica, in Antarctic Earth Science, edited by R. L. Oliver, P. R. James, and J. B. Jago, pp. 197-203, Australian Academy of Science, Canberra, 1983.

LaBrecque, J. L., and P. F. Barker, The age of the Weddell Basin, Nature, 290, 489-492, 1981.

Masolov, V. N., R. G. Kurinin, and G. E. Grikurov, Crustal structure and tectonic significance of Antarctic rift zones (from geophysical evidence), in Gondwana Five, edited by M. M. Cresswell and P. Vella, pp 303-309, A. A. Balkema, Rotterdam, 1981.

Millar, I. L., and R. J. Pankhurst, Rb-Sr geochronology of the region between the Antarctic Peninsula and the Transantarctic Mountains: Haag Nunataks and Mesozoic granitoids, this volume.

Nelson, E. P., I. W. D. Dalziel, and A. G. Milnes, Structural geology of the Cordillera Darwin--collisional style orogenesis in the southernmost Andes, Eclogae Geol. Helv., 73, 727-751, 1980.

Norton, I. O., and J. G. Sclater, A model for the evolution of the Indian Ocean and the breakup of Gondwanaland, J. Geophys. Res., 84, 6803-6830, 1979.

Schmidt, D. L., and A. B. Ford, Geology of the Pensacola and Thiel Mountains, in Geologic Maps of Antarctica, Antarctic Map Folio Ser., edited by V. C. Bushnell and C. Craddock, folio 12, plate V, American Geographical Society, New York, 1969.

Schopf, J. M., Ellsworth Mountains: Position in West Antarctica due to sea-floor spreading, Science, 164, 63-66, 1969.

Söhnge, A. P. G., and I. W. Hälbich, Geodynamics of the Cape Fold Belt, Spec. Publ. Geol. Soc. S. Afr., 12, 184 pp., 1983.

Splettstoesser, J., and G. F. Webers, Geological investigations and logistics in the Ellsworth Mountains, 1979-1980, Antarct. J. U. S., 15, 36-39, 1980.

Storey, B. C., and I. W. D. Dalziel, Outline of the structural and tectonic history of the Ellsworth Mountains-Thiel Mountains ridge, West Antarctica, this volume.

Stump, E., Type locality of the Ackerman Formation, La Gorce Mountains, Antarctica, in *Antarctic Earth Science*, edited by R. L. Oliver, P. R. James, and J. B. Jago, pp. 170-174, Australian Academy of Science, Canberra, 1983.

Vennum, W. R., and B. C. Storey, Correlation of gabbroic and diabasic rocks from the Ellsworth Mountains, Hart Hills, and Thiel Mountains, West Antarctica, this volume (a).

Vennum, W. R., and B. C. Storey, Petrology, geochemistry, and tectonic setting of granitic rocks from the Ellsworth-Whitmore Mountains crustal block and Thiel Mountains, West Antarctica, this volume (b).

Watts, D. R., and A. M. Bramall, Palaeomagnetic evidence for a displaced terrain in western Antarctica, *Nature*, 293, 638-642, 1981.

Webers, G. F., C. Craddock, A. M. Rogers, and J. J. Anderson, Geology of the Whitmore Mountains, in *Antarctic Geoscience*, edited by C. Craddock, University of Wisconsin Press, Madison, 1982.

White, A. J. R., and B. W. Chappell, Granitoid types and their distribution in the Lachlan Fold Belt, southeastern Australia, *Mem. Geol. Soc. Am.*, 159, 21-34, 1983.

Copyright 1987 by the American Geophysical Union.

SEDIMENTARY ROCKS OF THE ENGLISH COAST, EASTERN ELLSWORTH LAND, ANTARCTICA

T. S. Laudon

Department of Geology, University of Wisconsin-Oshkosh, Oshkosh, Wisconsin 54901

D. J. Lidke

U.S. Geological Survey, Denver, Colorado 80225

T. Delevoryas and C. T. Gee

Department of Botany, University of Texas at Austin, Austin, Texas 78712

Abstract. Nunataks scattered over 16,000 km^2 of the western English Coast along the Bellingshausen Sea are the westernmost rock exposures of the Antarctic Peninsula tectonic province. The nunataks are composed of sedimentary rocks of probable Paleozoic, Mesozoic, and probable Mesozoic age, volcanic and plutonic rocks of probable Mesozoic age, and basaltic volcanic and volcaniclastic rocks of probable late Cenozoic age. Most rocks exposed on the English Coast are correlatable with Mesozoic and Cenozoic units which occur in the Orville Coast and the Lassiter Coast regions to the east. One outcrop in the English Coast consists of sedimentary rocks containing fossil plants including abundant Glossopteris leaves and fragments of Phyllotheca and Equisetum. The age of the rocks is probably Permian, significantly older than the oldest previously dated rocks (Middle Jurassic) from the southern Antarctic Peninsula. This is the first reported occurrence of Glossopteris from the Antarctic Peninsula province and the second reported occurrence from West Antarctica. Permian(?) sedimentary rocks suggest that the English Coast was probably near the Pacific Coast of Gondwana. Sandstone, shale, and conglomerate at several other localities in the English Coast are correlated with the Jurassic Latady Formation of the Orville Coast. Sedimentary rocks of unknown age at two other nunataks include metamorphosed quartz-sandstone and cross-bedded sandstone.

Introduction

Sixteen small nunataks and nunatak groups are scattered over more than 16,000 km^2 in the western English Coast of the Bellingshausen Sea (Figure 1). Fourteen of these were visited for the first time during the 1984-1985 field season. These exposures, in northeastern Ellsworth Land, are the westernmost exposures of the southern Antarctic Peninsula tectonic province, one of four-or-more discrete lithospheric fragments (sometimes called microplates) which apparently moved in relation to one another and to the East Antarctic craton during the breakup of Gondwana.

The Antarctic Peninsula tectonic province consists mostly of a Middle Jurassic to lower Tertiary magmatic arc, which developed above a south- to east-dipping subduction zone on the Pacific Ocean (Bellingshausen Sea) side of the region [Rowley et al., 1983]. Igneous rocks of the magmatic arc extend eastward from the English Coast through other parts of eastern Ellsworth Land and the Orville Coast, then northward through the Lassiter and Black coasts, and the northern part of the Antarctic Peninsula. An associated subduction complex (the LeMay Group [Burn, 1984]) and fore-arc-basin deposits (the Fossil Bluff Formation [Thomson, 1982]) occur on Alexander Island, and backarc-basin deposits (the Latady Formation [Laudon et al., 1983]) occur in the Orville and Lassiter Coast regions (Figure 1). Rocks assigned to the pre-Jurassic "basement" in the northern part of the province have been interpreted as part of an earlier, upper Paleozoic to lower Mesozoic subducted Pacific margin of Gondwana, which was uplifted and eroded following deformation and metamorphism in an early Mesozoic Gondwanide Orogeny [e.g., Burn, 1984; Dalziel, 1982, 1983, 1984; Dalziel and Elliot, 1982; Smellie, 1981; Thomson et al., 1983].

Within the English Coast, Rowley et al. [1985] noted calc-alkalic volcanic rocks of probable Mesozoic age at FitzGerald Bluffs, Marshall Nunatak, Schwartz Peak, Mount Southern, Mount Harry, Mount Peterson, and Mount Rex and plutonic rocks of probable Mesozoic age at FitzGerald Bluffs and Mount Harry (Figure 1). Calc-alkalic volcanic rocks in the English Coast are similar in lithology and field occurrence to the Mount Poster Formation [Rowley et al., 1982] of the Antarctic Peninsula Volcanic Group [Thomson and Pankhurst, 1983] in the Orville and Lassiter Coast regions to the east. Jurassic age for the Mount Poster Formation is based on its intertonguing relationships

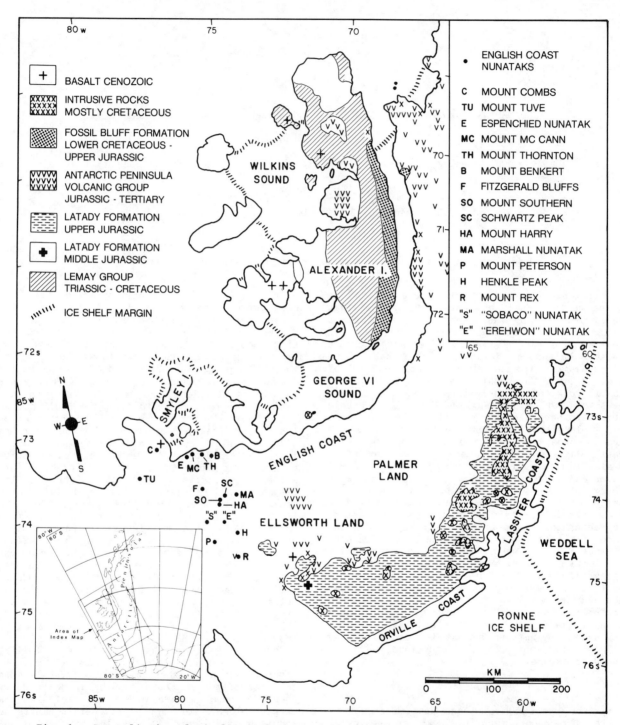

Fig. 1. Generalized geological map of eastern Ellsworth Land, southern Palmer Land, and Alexander Island showing locations of outcrops in the English Coast. In the Orville and Lassiter coasts, rocks of the Antarctic Peninsula Volcanic Group belong to the Jurassic Mount Poster Formation, and intrusive rocks belong to the Lassiter Coast Intrusive Suite.

with ammonite-bearing marine strata of the Jurassic Latady Formation, and the fact that both the Latady Formation and the Mount Poster Formation were deformed prior to emplacement of Cretaceous plutons [Laudon et al., 1983; Rowley et al., 1983]. However, Thomson and Pankhurst [1983] have pointed out that in the absence of paleontologic or radiometric evidence, no outcrop of the Antarctic Peninsula Volcanic Group can be dated more closely than Jurassic-Tertiary. Plutonic rocks in the English Coast are probably equivilant to the Cretaceous Lassiter Coast Intrusive Suite [Rowley et al., 1983].

Basaltic volcanic and volcaniclastic rocks of probable late Cenozoic age occur in the English Coast at Mount Benkert, Mount Thornton, Mount McCann and Espenchied Nunatak (Figure 1) [O'Neill and Thomson, 1985].

Laudon et al. [1985] briefly described sedimentary rocks which are exposed at nine localities in the English Coast. Sedimentary rocks at "Erehwon" Nunatak contain plant fossils including Glossopteris leaves of probable Permian, or possibly Triassic age, considerably older than the oldest previously known rocks (Middle Jurassic) from the southern Antarctic Peninsula. This discovery, the most important of the field season, provides the oldest known age of basement rocks which underlie and perhaps locally protruded through the Mesozoic magmatic arc of the Antarctic Peninsula province. Sedimentary rocks at Henkle Peak, Mount Peterson, Mount Rex, and Marshall Nunatak are probably equivilant to the Jurassic Latady Formation. Sedimentary rocks at FitzGerald Bluffs and "Sobaco" Nunatak are of unknown age.

Structural attitudes of sedimentary rocks (including Paleozoic(?) rocks) in the English Coast are generally accordant, with northwesterly strikes and southwesterly dips. They appear to be continuations of Lower Cretaceous structural trends of the Orville Coast.

We describe and interpret the sedimentary rocks of the English Coast below.

Sedimentary Rocks of "Erehwon" Nunatak

The field name Erehwon was used for an unnamed snow-capped nunatak located at approximately 74°31'S, 76°25'W (Figure 1). It is approximately 6 m high, but is not shown on any maps, and is not visible on available tri-camera photography. Two small outcrops on its eastern side are composed of dark gray, thinly laminated, interbedded mudstone, siltstone, and fine-grained sandstone, in an approximately 2-m-thick section that strikes N75°W and dips S25°. Some beds consist of upward fining sequences from 2 to 10 cm thick. Lenticular sandstone beds as thick as 6 cm and as wide as 50 cm are present.

One bed, approximately 10 cm thick, contains abundant leaves of Glossopteris (Figures 2a and b) associated with fragments of two sphenophyte genera, Phyllotheca (Figure 2c) and Equisetum. Glossopteris fossils include vegetative leaves and scale leaves. Variation in the form of vegetative leaves is within the limits of specific variation. They closely resemble G. zeilleri, known from the Kharharbari Stage (Early Permian) of the Giridih Coalfield in the Indian Gondwanas [Chandrah and Surange, 1979]. Phyllotheca is most abundant in the Permian, although its fossil record extends from the Pennsylvanian to the Cretaceous [Delevoryas, 1962]. Equisetum is an extant genus which first appeared in the Carboniferous. The presence of Glossopteris and Phyllotheca strongly suggests that the rocks at Erehwon Nunatak are of Permian age, although they might be as young as Triassic [Schopf and Askin, 1980]. This is the first reported occurrence of Glossopteris from the Antarctic Peninsula tectonic province and only the second related occurrence from West Antarctica.

Abundant Glossopteris leaves, together with the fact that many are intact, suggest that transport distance of the leaves was limited. Some low-energy terrestrial environment of deposition, perhaps a swamp, seems most likely for these rocks, but field evidence is meager, and quiet paralic environments cannot be ruled out.

Sedimentary Rocks Correlated With the Latady Formation

Sedimentary rocks exposed in the English Coast at Henkle Peak, Mount Peterson, Mount Rex, and Marshall Nunatak are tentatively correlated with the Middle and Upper Jurassic Latady Formation of the Orville and Lassiter Coast regions of eastern Ellsworth Land and southern Palmer Land (Figure 1). The Latady Formation consists of thick structurally complex sequences of sandstone, siltstone, and shale with locally interbedded volcanic rocks, coal, and limestone of Bajocian to Kimmeridgian age. Most of the Latady was deposited in shallow marine environments in a backarc basin on the Atlantic side of the Mesozoic magmatic arc. In the interior of the Orville and Lassiter coasts, sedimentary rocks of the Latady Formation were deposited in both marine and terrestrial environments, and they intertongue to the north with volcanic rocks of the Mount Poster Formation of the Antarctic Peninsula Volcanic Group [Laudon et al., 1983].

Henkle Peak. Henkle Peak (Figure 1) is a mostly snow covered northwest trending ridge about 1 km long and 100 m high. On its southwest side, black carbonaceous conglomerate, sandstone, and siltstone, with abundant plant debris, are exposed in a 30-m-thick section that strikes N30°W and dips SW15°. A thicker but inaccessible section that includes the same beds is exposed on the southeast end of the ridge. Small outcrops of similar rocks with strike N74°W and dip SW24° are intruded by a mafic sill on the northeast side.

A measured stratigraphic section on the southwest side of the ridge begins with 10.7 m of conglomerate, overlain by 8.8 m of sandstone, 1 m of conglomerate, and 9.4 m of siltstone. All of the lithologies weather massively, but contacts between them are sharp. Conglomerate is thick bedded to massive. Sandstone and siltstone units contain internal laminations and cross-laminations. Many bedding planes in conglomerate are conspicuously limonite stained.

All of the rocks are black and carbonaceous,

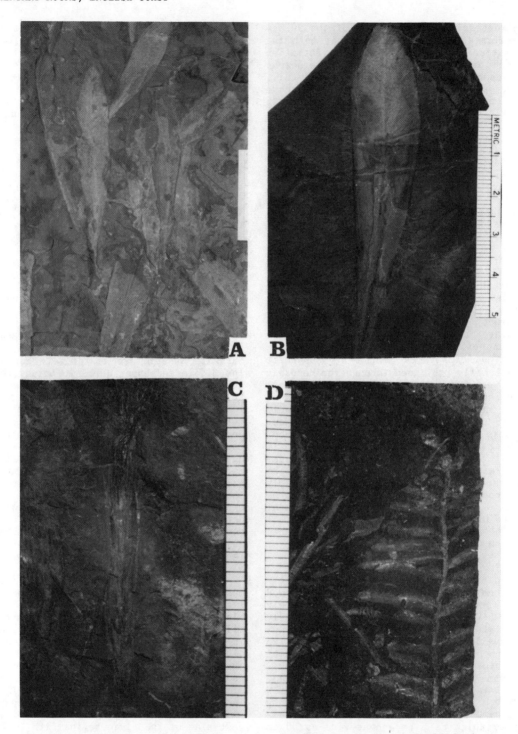

Fig. 2. Fossil plants from the English Coast. Specimen numbers are field designators. Specimens are presently housed at the Department of Botany, University of Texas at Austin. (a) Glossopteris leaf litter from Erehwon Nunatak, specimen DL15b. (b) Compression of Glossopteris leaf from Erehwon Nunatak, specimen DL15b. Midrib is composed of numerous veins, some of which separate and extend into the lamina. Venation is reticulate; veins both dichotomize and anastomose. (c) Fragment of Phyllotheca stem from Erehwon Nunatak, specimen DL15b. (d) Leafy twig of Elatocladus planus from Henkle Peak, specimen L107E.

and they contain abundant plant fragments as large as 10 cm wide and 40 cm long. Clast-supported conglomerate, with well-rounded pebbles as long as 7 cm, appears to be intraformational. Both pebbles and matrix are composed of fine-grained carbonaceous clastic material. Fluvial environments of deposition, perhaps associated with anastomosing streams, seem likely for the rocks at Henkle Peak.

Leafy twigs of Elatocladus planus (Figure 2d) were recovered from siltstone near the top of the measured stratigraphic section. Elatocladus planus is an arborescent conifer commonly found in Jurassic rocks, although its known range extends from Upper Triassic to Lower Cretaceous. It has been found in Queensland, Tasmania, New South Wales, India, and Antarctica [Hill et al., 1966], including the Jurassic Latady Formation in the Orville and Lassiter coasts [Gee, 1984].

Mount Peterson. Mount Peterson (Figure 1) is a northeast trending ridge approximately 1 km long and 100 m high. Outcrops on its southeast side are composed mostly of red and green clastic sedimentary rocks interbedded with and overlain by volcanic rocks. The ridge is broken by at least three faults with moderate to steep southerly dips. Sedimentary strata strike between N60°E and N48°W with southerly dips from 10° to 24°. Exposed stratigraphic sections, in different fault blocks, have (from north to south) the following thicknesses: 14 m, 32 m, and 54 m.

Sedimentary rocks in the two northern sections are predominantly green, whereas those in the southern section are mostly red. Mudstone, siltstone, sandstone, and conglomerate are interbedded in lenticular units. Mudstone units are as thin as 25 cm; conglomerate units are as thick as 12 m. Sedimentary structures include laminations, cross-laminations, convolute laminations, intraclasts, ripples, and raindrop prints. Many sedimentary units are upward fining sequences. No fossils were found at Mount Peterson.

Conglomerate consists of granules, pebbles, cobbles, and boulders in arkose matrix. Lenticular beds of laminated siltstone and sandstone are common in thicker conglomerate units. Large clasts are predominantly volcanic rocks of many different textures and compositions. Felsic plutonic rocks and fine-grained clastic sedimentary rocks are minor constituents.

The central part of the ridge is capped by at least 23 m of fine-grained vitric tuff or flow rock, apparently in conformable stratigraphic contact with underlying sedimentary rocks. Similar volcanic rocks occur rarely as thin beds in predominantly sedimentary sequences lower on the ridge. The southern fault block is composed of volcanic rocks with lenticular beds of red siltstone.

Sedimentary rocks at Mount Peterson are interpreted as alluvial fan deposits that formed near an elevated terrane consisting predominantly of volcanic rocks, with smaller plutonic and sedimentary components. Interbedded flows indicate that there was volcanic activity in the region during formation of the fans.

That the volcanic rocks at Mount Peterson are similar to the Jurassic Mount Poster Formation (Antarctic Peninsula Volcanic Group) [Rowley et al., 1982] in the nearby Orville Coast region suggests that associated sedimentary rocks may be terrestrial facies of the Jurassic Latady Formation (Figure 1). However, no other alluvial fan deposits are known from the Latady Formation. Elsewhere in the Orville and Lassiter Coast regions, and at nearby Henkle Peak, terrestrial facies of the Latady Formation were deposited in low-energy swamp, lacustrine, or fluvial environments [Laudon et al., 1983].

Mount Rex. Mount Rex (Figure 1) consists of four separate peaks composed mostly of porphyritic dacite [Laudon et al., 1964], which has been correlated with the Jurassic Mount Poster Formation of the Antarctic Peninsula Volcanic Group [Rowley et al., 1985]. The east face of the easternmost peak consists of at least three massive, cliff-forming sequences of flow rocks, separated by at least two recessively weathering sequences of sandstone and siltstone, which strike N65°W and dip SW27°. The lower clastic sequence is about 28 m thick and the upper about 22 m thick. A snow-covered ramp near the top of the face is probably underlain by a younger clastic sequence.

Sedimentary rocks are dark colored and contain abundant feldspar and lithic grains. Location (see Figure 1), composition, interbedding with volcanic rocks which have been correlated with the Jurassic Mount Poster Formation, and structural attitude suggest that they are part of the Jurassic Latady Formation.

Marshall Nunatak. Marshall Nunatak (Figure 1) is about 1 km long, with a precipitous western face about 100 m high, and a gentle, mainly snow covered eastern slope. It is composed predominantly of felsic volcanic rocks similar to the Jurassic Mount Poster Formation. Thin sequences of fine-grained clastic sedimentary rocks occur within the volcanic sequence. Several small outcrops immediately east of the main outcrop area, but separated from it by snow cover, are composed of boulder conglomerate, which also occurs as lenticular bodies within the volcanic sequence. The largest exposed conglomerate lens is about 15 m long and 4 m thick.

Conglomerate consists of granules, pebbles, cobbles, and boulders in a matrix of volcanic rock. Most conglomerate is matrix supported, but some is clast supported. Most matrix is black and fine grained, but some consists of porphyritic dacite. The largest clast observed is 53 cm long; clasts from 30 to 45 cm long are common. Clast composition is approximately 55% volcanic and 45% sedimentary and metasedimentary. Boulders and cobbles of cross-laminated metaquartzite, similar to quartzite exposed at FitzGerald Bluffs, make up approximately 25% of the clasts.

Conglomerate at Marshall Nunatak is interpreted as representing either volcanic flows or mudflows. Interbedding with volcanic rocks similar to the Mount Poster Formation suggests that its age is Jurassic.

Sedimentary Rocks Whose Ages Are Unknown

"Sobaco" Nunatak. The field name Sobaco was used for a small (approximately 20 m high) unnamed nunatak, located near 74°29.5'S, 77°12'W (Figure 1). On its north side, a 28-m-thick section of gray sandstone is exposed, which strikes N66°W and dips SW32°. The bottom 1.5 m of the exposed section consists of interbedded siltstone and fine-grained sandstone, with lenses of black shale, laminations, cross-laminations, and ripples. This is overlain by 18.6 m of medium-bedded, dark gray, fine-grained sandstone containing lenses of coarse-grained light gray laminated sandstone. The upper 7.9 m of the section consists of medium-bedded, medium-grained, gray sandstone that is strongly laminated and cross-laminated. Near the base of the upper unit there is a 15-cm-thick zone containing intraclasts, apparently derived from the underlying fine-grained unit. Cross-laminations are accentuated by fine-grained carbonaceous material.

Although field evidence is scant, occurrence of planar and trough cross-laminations, intraclasts, and fine-grained carbonaceous material suggests fluvial or paralic environments of deposition for these rocks. No fossils were found, and the age of the rocks is not known.

FitzGerald Bluffs. FitzGerald Bluffs (Figure 1) are part of a north-facing east-west trending escarpment that is about 12 km long. Most outcrops are of felsic plutonic rocks that are similar in composition and field occurrence to the calc-alkalic Lassiter Coast Intrusive Suite of Cretaceous age in the Lassiter and Orville coasts (Figure 1). Volcanic rocks similar to the Jurassic Mount Poster Formation are also present [Rowley et al., 1985]. The highest part of the ridge, near its western end, is underlain by bedded quartzite containing minor silty and argillaceous interbeds, in a nearly vertical escarpment about 300 m high capped by overhanging ice cliffs. It was possible to examine only a small portion of the outcrop at two places. About 100 boulders that had fallen from the cliff were also examined. At the base of the cliff the quartzite has been intruded and hornfelsed by porphyritic granite and aplite. The intrusives and host rock have been sheared.

The quartzite appears to be metamorphosed quartz sandstone. It is mostly medium bedded, and is laminated and cross-laminated. Laminae are accentuated by fine-grained dark minerals that have been chloritized. The age of the quartzite is not known, but it is older than plutonic rocks intruding it which are of probable Cretaceous age [Rowley et al., 1985].

Mount Harry and Mount Southern. Thin sequences of fine-grained sedimentary rocks occur as interbeds in predominantly volcanic rock sequences at Mount Southern and Mount Harry (Figure 1).

Discussion

Glossopteris-bearing clastic sedimentary rocks from Erehwon Nunatak are of probable Permian age. They are the oldest dated rocks from the Antarctic Peninsula tectonic province and are the only rocks older than Middle Jurassic that are known from eastern Ellsworth Land and southern Palmer Land. The occurrence of Glossopteris at Erehwon Nunatak is very significant and indicates that the upper Paleozoic Gondwana floral province includes the Pacific Coast of the southern Antarctic Peninsula tectonic province. Although Glossopteris is widespread in Permian rocks of Gondwana, including the Transantarctic Mountains, its only previously reported occurrence in West Antarctica is in the Polarstar Formation of the Ellsworth Mountains (south of the locations shown in Figure 1 but not included), approximately 350 km south of the English Coast [Craddock et al., 1965].

Taylor and Smoot [1986] suggest middle to Late Permian age for the Polarstar flora, which apparently does not include the species found in the English Coast. Collinson et al. [1985] suggest that the Polarstar Formation was deposited in a basin which was bounded on its Pacific margin by volcanoes, and which had only limited passages to the Pacific Ocean. If this model is accurate, Glossopteris-bearing rocks in the English Coast may have been deposited in the Polarstar basin near its Pacific margin. Unfortunately, the scant field evidence available at Erehwon Nunatak provides little basis for stratigraphic or paleogeographic correlation.

Carbonaceous clastic sedimentary rocks containing Elatocladus planus from Henkle Peak are of probable Jurassic age, and they are probably terrestrial facies of the Latady Formation. Clastic sedimentary rocks at Mount Rex, Mount Peterson, and Marshall Nunatak are interpreted to be part of the Latady Formation, primarily on the basis of interbedding with volcanic rocks which have been correlated with the Mount Poster Formation of the Antarctic Peninsula Volcanic Group. However, because neither paleontologic nor radiometric evidence regarding their ages is available, they may be as young as Tertiary.

Sedimentary rocks of unknown age in the English Coast promote further speculation regarding the pre-Mesozoic nature of the English Coast. Conspicuously cross-bedded quartzite at FitzGerald Bluffs appears to be metamorphosed quartz sandstone, suggesting cratonic provenance for the sand. Large clasts of similar-appearing quartzite are abundant (approximately 25%) in volcanic conglomerate at Marshall Nunatak. The nearest described unit with similar lithology is the Devonian Crashsite Quartzite of the Ellsworth Mountains [Craddock et al., 1964]. Sobaco Nunatak is composed of sandstone, which is similar in appearance, but may be compositionally more complex.

Large clasts in conglomerate at both Mount Peterson and Marshall Nunatak include a variety of volcanic, plutonic, sedimentary, and metamorphic rocks, many of which appear to be lithologically distinct from known Mesozoic bedrock units in the region. Petrologic and geochemical analyses of these may yield additional information regarding the pre-Jurassic geology of the English Coast.

Acknowledgments. Much of this report is based on observations and notes of the other members of the English Coast field party, P. D. Rowley, party leader, K. S. Kellogg, J. M. O'Neill, J. W. Thomson, and W. R. Vennum. We thank them for this and for their help in many ways. We thank the

National Science Foundation Division of Polar Programs, U.S. Navy, and ITT Antarctic Services for essential assistance. This work was supported by National Science Foundation grant DPP-8318183 to the U.S. Geological Survey, and grant DPP-8319569 to the University of Wisconsin-Oshkosh, and by the University of Wisconsin-Oshkosh Faculty Development Program.

References

Burn, R. W., The geology of the LeMay Group, Alexander Island, Br. Antarct. Surv. Sci. Rep. 109, 65 pp., 1984.

Chandrah, S., and K. R. Surange, Revision of the Indian Species of Glossopteris, Monogr. Ser., vol. 2, 291 pp., Birbahl Sahni Institute of Paleobotany, Lucknow, India, 1979.

Collinson, J. W., C. L. Vavra, and J. L. Zawiskie, Sedimentology of Polarstar Formation, Permian, Ellsworth Mountains, Antarctica, in Abstracts, Sixth Gondwana Symposium, Inst. of Polar Stud. Misc. Publ. 231, p. 20, The Ohio State University, Columbus, 1985.

Craddock, C., J. J. Anderson, and G. F. Webers, Geologic outline of the Ellsworth Mountains, in Antarctic Geology, edited by R. J. Adie, pp. 155-170, North-Holland, Amsterdam, 1964.

Craddock, C., T. W. Bastien, R. H. Rutford, and J. J. Anderson, Glossopteris discovered in West Antarctica, Science, 148, 634-637, 1965.

Dalziel, I. W. D., The early (pre-Middle Jurassic) history of the Scotia Arc region: A review and progress report, in Antarctic Geoscience, edited by C. Craddock, pp. 111-126, University of Wisconsin Press, Madison, 1982.

Dalziel, I. W. D., The evolution of the Scotia Arc: A review, in Antarctic Earth Science, edited by R. L. Oliver, P. R. James, and J. B. Jago, pp. 283-288, Australian Academy of Science, Canberra, 1983.

Dalziel, I. W. D., Tectonic evolution of a forearc terrane, southern Scotia Ridge, Antarctica, Spec. Pap. Geol. Soc. Am., 200, 32 pp., 1984.

Dalziel, I. W. D., and D. H. Elliot, West Antarctica: Problem child of Gondwanaland, Tectonics, I, 3-19, 1982.

Delevoryas, T., Morphology and Evolution of Fossil Plants, 189 pp., Holt, Rinehart, and Winston, New York, 1962.

Gee, C. T., Preliminary studies of a fossil flora from the Orville Coast, eastern Ellsworth Land, Antarctic Peninsula, Antarct. J. U. S., 19, 36-37, 1984.

Hill, D., G. Playford, and J. T. Woods, Jurassic Fossils of Queensland, pp. j1-j32, Queensland Paleontographic Society, Brisbane, 1966.

Laudon, T. S., J. C. Behrendt, and N. J. Christensen, Petrology of rocks collected on the Antarctic Peninsula Traverse, J. Sediment. Petrol., 34, 360-364, 1964.

Laudon, T. S., M. R. A. Thomson, P. L. Williams, K. L. Milliken, P. D. Rowley, and J. M. Boyles, The Jurassic Latady Formation, southern Antarctic Peninsula, in Antarctic Earth Science, edited by R. L. Oliver, P. R. James, and J. B. Jago, pp. 308-314, Australian Academy of Science, Canberra, 1983.

Laudon, T. S., D. J. Lidke, T. Delevoryas, and C. T. Gee, Sedimentary rocks of the English Coast, eastern Ellsworth Land, Antarct. J. U. S., 20, 38-40, 1985.

O'Neill, J. M., and J. W. Thomson, Tertiary mafic volcanic and volcaniclastic rocks of the English Coast, Antarctica, Antarct. J. U. S., 20, 36-38, 1985.

Rowley, P. D., D. L. Schmidt, and P. L. Williams, The Mount Poster Formation, southern Antarctic Peninsula, Antarct. J. U. S., 17, 38-39, 1982.

Rowley, P. D., W. R. Vennum, K. S. Kellogg, T. S. Laudon, P. E. Carrara, J. M. Boyles, and M. R. A. Thomson, Geology and plate tectonic setting of the Orville Coast and eastern Ellsworth Land, Antarctica, in Antarctic Earth Science, edited by R. L. Oliver, P. R. James, and J. B. Jago, pp. 245-250, Australian Academy of Science, Canberra, 1983.

Rowley, P. D., K. S. Kellogg, and W. R. Vennum, Geologic studies in the English Coast, eastern Ellsworth Land, Antarctica, Antarct. J. U. S., 20, 34-36, 1985.

Schopf, J. M., and R. A. Askin, Permian and Triassic floral biostratigraphic zones of southern land masses, in Biostratigraphy of Fossil Plants, edited by D. L. Dilcher and T. N. Taylor, pp. 119-152, Dowden, Hutchinson and Ross, Stroudsburg, Penn., 1980.

Smellie, J. L., A complete arc-trench system recognized in Gondwana sequences of the Antarctic Peninsula region, Geol. Mag., 118, 139-159, 1981.

Taylor, T. N., and E. L. Smoot, Permian plants from the Ellsworth Mountains, West Antarctica, in Geology and Paleontology of the Ellsworth Mountains, edited by C. Craddock, J. F. Splettsoesser and G. F. Webers, Mem. Geol. Soc. Am., in press, 1986.

Thomson, M. R. A., Mesozoic paleogeography of West Antarctica, in Antarctic Geoscience, edited by C. Craddock, pp. 331-338, University of Wisconsin Press, Madison, 1982.

Thomson, M. R. A., and R. J. Pankhurst, Age of post-Gondwanian calc-alkaline volcanism in the Antarctic Peninsula region, in Antarctic Earth Science, edited by R. L. Oliver, P. R. James, and J. B. Jago, pp. 328-333, Australian Academy of Science, Canberra, 1983.

Thomson, M. R. A., R. J. Pankhurst, and P. D. Clarkson, The Antarctic Peninsula--A late Mesozoic-Cenozoic arc (review), in Antarctic Earth Science, edited by R. L. Oliver, P. R. James, and J. B. Jago, pp. 289-294, Australian Academy of Science, Canberra, 1983.

THE GONDWANIAN OROGENY WITHIN THE ANTARCTIC PENINSULA: A DISCUSSION

B. C. Storey, M. R. A. Thomson, and A. W. Meneilly

British Antarctic Survey, Natural Environment Research Council
High Cross, Cambridge CB3 0ET, United Kingdom

Abstract. The relevance of the term "Gondwanian orogeny" to deformation within upper Paleozoic to Mesozoic rocks of the Antarctic Peninsula is reviewed, and the tectogenesis of deformation within this time frame is discussed. The term, originally applied by du Toit to deformation of middle Paleozoic to lower Mesozoic rocks in the Samfrau geosyncline, has been used extensively in Antarctic literature. However, the Gondwanian fold belt includes deformed regions of significantly different ages. For example, in southern Africa the Gondwanide Cape folding is a single-phase multiple-event period of deformation which spanned the beginning of the Permian to the Middle Triassic, a period of approximately 50 m.y. In the Antarctic Peninsula, although data are insufficient to reconstruct a full tectonic history, the Gondwanian orogen is now known to include rocks and structures of a younger age. The only ages which presently fall within the above time frame are some estimates of a period of metamorphism whose tectonic significance is uncertain. The authors favor the use of the term "Peninsula orogeny" for deformation of Triassic and Early Jurassic rocks below the "Peninsula unconformity" in preference to the term Gondwanian orogeny.

Introduction

The term "Gondwanide orogeny" was introduced by du Toit [1937] to describe a phase of late Paleozoic deformation, recognized in the Carboniferous and older rocks of the Samfrau geosyncline marginal to Gondwanaland. It is widely used to describe deformation within the Cape Fold Belt, and also deformation of Paleozoic metasedimentary rocks in parts of South America and the Falkland Islands. On du Toit's [1937, Figure 7] map, the Antarctic Peninsula was not specifically included, although it was said to belong to the outer side of the same geosyncline [du Toit, 1937, p. 63]. However, in the form "Gondwanian orogeny," the term has also been used to describe deformation of (?) late Paleozoic to early Mesozoic sedimentary and metasedimentary rocks observed below a (?) middle Mesozoic unconformity in the Antarctic Peninsula. The unconformity is present at the base of the Antarctic Peninsula Volcanic Group and was believed to separate Gondwanian deformation from Andean volcanic and tectonic events [e.g., Dalziel and Elliot, 1971, 1973; Elliot, 1975; Smellie, 1981; Dalziel, 1982; Thomson and Pankhurst, 1983]. Evidence for this comes primarily from the South Orkney Islands and northern Antarctic Peninsula where Late Jurassic-Early Cretaceous marine rocks rest unconformably on the Scotia metamorphic complex [Thomson, 1981] and ?Early Cretaceous nonmarine conglomerates unconformably overlie deformed graywackes of the Trinity Peninsula Group [Farquharson, 1984]. It may also be present in central Alexander Island where Edwards [1979] reported an unconformable contact between the Late Jurassic-Early Cretaceous Fossil Bluff Formation and metasedimentary rocks of the LeMay Group. The deformed rocks below the unconformity are commonly referred to as the basement to the Andean system [Dalziel and Elliot, 1971; Pankhurst, 1983], and include subduction complex and forearc sedimentary rocks [Smellie, 1981] that accumulated along an Andean-type proto-Pacific margin of Gondwanaland. They are believed to have been deformed before breakup of the supercontinent during a period referred to as the Gondwanian orogeny [Dalziel, 1982]. However, Miller [1983] has suggested that folding, metamorphism, and plutonism at Triassic-Jurassic boundary times within the Antarctic Peninsula should be referred to as the Peninsula Orogeny. Analogous field relations also occur in the southern Andes where Late Jurassic-Early Cretaceous marine and volcanic sequences rest unconformably on deformed graywackes and metasedimentary rocks of presumed late Paleozoic age [e.g., Riccardi, 1971; Caminos, 1980].

Deformation of Paleozoic sedimentary rocks in the Ellsworth and Pensacola Mountains of West Antarctica has been referred to as the Ellsworth and Weddell orogenies, respectively [Craddock, 1972, 1975; Ford, 1972]. Both have been equated with the Gondwanian orogeny [Dalziel and Elliot, 1982] and their deformed rocks are considered with the Cape Fold Belt of southern Africa, the Sierra de La Ventana Fold Belt of Argentina, and the Falkland Islands to constitute part of the backarc basin or foredeep of the Gondwanian arc and forearc system [see De Wit, 1977; Dalziel and Elliot, 1982], a concept in accordance with du Toit's [1937] original idea.

This paper reviews the known deformation, sedi-

Copyright 1987 by the American Geophysical Union.

mentation, and magmatic history within the Antarctic Peninsula (Figure 1) and, by briefly reviewing Gondwanian events and their correlatives within the rest of Gondwanaland, assesses the relevance of the term Gondwanian orogeny to known events within the Antarctic Peninsula. Evidence for interpreting Gondwanian events in relation to a forearc, arc, and backarc system is also considered. In keeping with present geological usage, a common term should only be used for spatially linked and broadly temporal deformational, metamorphic, and magmatic events, and it should be independent of the tectonic processes involved. The spatially linked criteria are not discussed in detail in this paper as the true position of the Antarctic Peninsula and other parts of the Gondwanian system within Gondwanaland prior to breakup is still uncertain [Dalziel and Elliot, 1982]. New paleomagnetic data, however, suggest that they may be linked [Grunow et al., this volume].

Antarctic Peninsula

Within the Antarctic Peninsula the concept of a twofold division of the tectonic history into Andean and Gondwanian events has been challenged [Storey and Garrett, 1985]. Recent advances in geochronology have blurred the distinction between Gondwanian and Andean events, and it has been suggested that subduction-related processes, including accretion, magmatism, and extension, were virtually continuous since at least early Mesozoic times. These processes led to the development of a complex Mesozoic magmatic arc, which formed on Paleozoic basement marginal to Gondwanaland. During development of the arc a narrow segment of continental crust, containing the active arc ("Antarctic Peninsula"), split off from the rest of Gondwanaland and became part of West Antarctica.

As the times of sedimentation, deformation, and magmatism are critical to understanding the tectonic history, we would at this stage emphasize the uncertainties surrounding some of these. From our present knowledge, however, the main stratotectonic events are summarized in Figure 1 and briefly discussed below.

Basement to the Mesozoic Arc

A fundamental problem in interpreting the early tectonic history is concerned with determining the age of widespread orthogneiss, amphibolite, paragneiss, and migmatite mapped as metamorphic basement or complex within the arc terrane. As a result of isotopic dating, many areas of "ancient" metamorphic complex are now known to be sheared Mesozoic plutons [Gledhill et al., 1982; Pankhurst, 1983]. Some foliated granite gneisses, which were syntectonically emplaced in a regional stress field, give ages of crystallization of 180-200 Ma [Pankhurst, 1983], and at Engel Peaks, northeastern Palmer Land, a mid-Cretaceous pluton is deformed in a major shear zone to form mylonites and cataclastites that had previously been interpreted as Paleozoic sialic basement [Meneilly et al., this volume]. However, at Target Hill, eastern Graham Land, granite sheets in a banded migmatite gave an apparent Rb/Sr whole-rock age of 336 ± 34 Ma [Pankhurst, 1983]. This overlaps with an age of 386 ± 40 Ma, obtained from granitic cobbles from part of the Trinity Peninsula Group, and suggests the presence of an older Paleozoic basement to the Mesozoic arc [Storey and Garrett, 1985]. Although the details of the basement history are not known, an early Triassic metamorphism ($\approx$ 245 Ma) overprints the older deformation history [Pankhurst, 1983]. The tectonic significance of the metamorphism is not clearly understood.

Evolution of the Mesozoic Magmatic Arc

Magmatism and sedimentation related to the subduction of proto-Pacific Ocean lithosphere were virtually continuous throughout different parts of the Antarctic Peninsula since at least Late Triassic times (Figure 1) [Thomson and Pankhurst, 1983, p. 330]. The time of initiation of the Mesozoic subduction system is uncertain, but there is no clear evidence of subduction prior to the Triassic. Subduction complex rocks within part of the Scotia metamorphic complex contain Triassic microfossils [Dalziel et al., 1981] and were deformed and accreted prior to or during the Early Jurassic, with a thermal event at 190 Ma representing rapid cooling after the main prograde metamorphism [Tanner et al., 1982]. The metamorphic and structural history of part of this complex is consistent with deformation within an accretionary prism [Storey and Meneilly, 1985]. Sedimentary rocks within the forearc region also contain Triassic fossils (Legoupil Formation, Trinity Peninsula Group [Thomson, 1975]) and Upper Jurassic fossils (LeMay Group, Alexander Island [Thomson and Tranter, 1986]) which place a maximum age on subduction-related deformation for these rocks. However, some sedimentary rocks within the Trinity Peninsula Group may be significantly older than this. Mudstones within the Hope Bay Formation, Trinity Peninsula Group [Hyden and Tanner, 1981] give a Rb/Sr whole-rock age of 281 ± 16 Ma (Early Permian), which has been interpreted as the age of diagenesis [Pankhurst, 1983]. The stratigraphy and tectonic setting of the Trinity Peninsula Group are critical to understanding the early history of arc development. Smellie [this volume] recognized two distinct petrofacies within the group. One may be derived from an active magmatic arc, whereas the origin of the other, part of which is Mesozoic in age, is more ambiguous, and it may be derived from an older deeply dissected arc and deposited on a passive margin; the View Point Formation contains granitic cobbles at least 100 m.y. older than the depositional age. Some of the deformation of these rocks is syn-Triassic or post-Triassic in age.

During development of the magmatic arc, discrete periods of accretion, magmatism, uplift, compression, and extension of the arc can be recognized [Storey and Garrett, 1985; Meneilly et al., this volume]. These processes were not mutually exclusive during the history of arc evolution. Although magmatism was virtually continuous from the Late Triassic, pulses of increased activ-

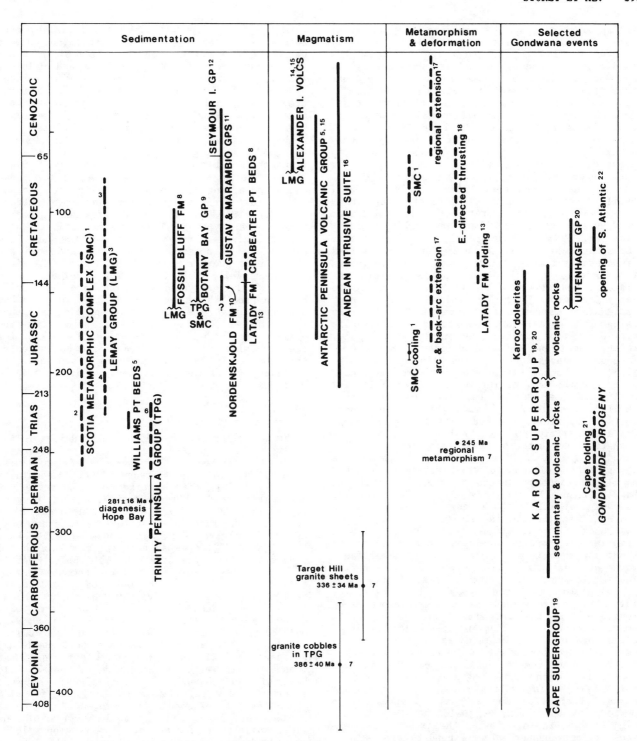

Fig. 1. Timing of major geological events in the Antarctic Peninsula compared with selected Gondwana events. Data sources are as follows: (1) Tanner et al. [1982]; (2) Dalziel et al. [1981]; (3) Burn [1984]; (4) Thomson, new data; (5) Smellie et al. [1984]; (6) Thomson [1975]; (7) Pankhurst [1983]; (8) Taylor et al. [1979]; (9) Farquharson [1984]; (10) Farquharson [1983]; (11) Ineson et al. [1986]; (12) Zinsmeister [1982]; (13) Rowley et al. [1983]; (14) Burn [1981]; (15) Thomson and Pankhurst [1983]; (16) Pankhurst [1982]; 17) Storey and Garrett, 1985; 18) Meneilly et al. [this volume]; (19) SACS [1980]; (20) Dingle et al. [1983]; (21) Söhnge and Hälbich [1983]; and (22) Sclater et al. [1977]. The time scale is taken from Harland et al. [1982].

ity apparently occurred during the Early Jurassic, Early and Late Cretaceous, and early Tertiary [Pankhurst, 1982]. Although the magmatism has traditionally been considered as subduction related, some of the Early Jurassic east coast plutons have high initial $^{87}Sr/^{86}Sr$ ratios (0.721 [Meneilly et al., this volume]). These may have formed entirely by partial melting of the continental crust and may be related to either subduction or high geothermal gradients prior to breakup of Gondwanaland. Deformation and moderate- to high-pressure metamorphism within accretionary prism rocks occurred during the Early Jurassic ($\simeq$ 190 Ma) and Late Cretaceous [Tanner et al., 1982]. Uplifted subduction complex rocks are overlain by Late Jurassic-Early Cretaceous volcaniclastic sedimentary rocks and Tertiary volcanic rocks [Thomson, 1981; Burn, 1981]. Synkinematic gneissose fabrics within Lower Jurassic granites are consistent with contemporaneous deformation in the accretionary prism and emplacement in a regional stress field. During Middle Jurassic to Early Cretaceous, marginal basins filled by thick sedimentary sequences formed behind the arc. In Palmer Land a basin may have formed as an ensialic rift during the Middle Jurassic [Meneilly et al., this volume] and deformed during an Early Cretaceous compressional period [Rowley et al., 1983]. In Graham Land the basin formed as a half graben on the edge of the Weddell Sea. Following uplift of the arc in the Late Jurassic, down-faulted extensional basins within the arc were infilled by fluviatile sediments during the Early Cretaceous [Farquharson, 1983]. Within the arc, forearc, and backarc, east directed thrusting occurred during the Cretaceous. A large ductile shear zone related to this movement deformed middle Cretaceous plutons on the east coast of Palmer Land [Meneilly et al., this volume]. However, not all of the movement in ductile shear zones within the arc is related to compressional tectonics. In the Marguerite Bay area, foliated granite and granite gneisses were deformed in shear zones with up to 1 km dip-slip normal fault movement [Matthews, 1983], and ductile normal faults also occur in northeast Palmer Land.

The history of arc development illustrates the interaction of compressional and extensional tectonics within the arc and emphasizes how late Mesozoic ductile shearing produced highly deformed rocks, previously interpreted as metamorphic basement deformed during the Gondwanian orogeny. The interaction of compressional and extensional tectonics is common within magmatic arcs and may be related to variation in subduction or spreading rates. Periods of slow subduction correspond to periods of accretion, extension of the arc and migration of the arc toward the trench, and periods of fast subduction correspond to low angle of the subduction zone, magmatism, and compressional tectonism.

The Gondwanian Orogeny Within Other Parts of Gondwanaland

South Africa

In southern South Africa, Paleozoic and early Mesozoic platform and molasse sediments of the Cape and Karoo supergroups were deformed by four paroxysms, producing coaxial structures over a period of 48 m.y.: the Cape Fold Belt of the Gondwanide orogeny, 230-278 Ma [Söhnge and Hälbich, 1983]. It is unconformably overlain by Lebombo Group volcanic rocks (Early Jurassic) and Early Cretaceous conglomerates (Enon Formation, Uitenhage Group) [Dingle et al., 1983]. The deformation of the Cape Fold Belt has been related to a number of tectonic causes: gravity sliding induced by block faulting [Newton, 1973], collision of lithospheric plates [De Swart et al., 1974], docking of subduction complex against Gondwanaland, and advance of the Andean island arc against the proto-African plate during flat-plate subduction [Lock, 1980]. It has been further suggested that Andean-type subduction accounted for the entire Gondwanian orogeny, and that the Cape Fold Belt could be considered as a foreland fold and thrust belt [De Wit, 1977; Smellie, 1981; Dalziel, 1982; Dalziel and Elliot, 1982]. However, in a study of the geodynamics of the Cape Fold Belt, Hälbich [see Söhnge and Hälbich, 1983, p. 182] has concluded that there is no compelling evidence that the fold belt has had a subduction-type origin. Estimated movements are much too slow for the flat-plate subduction model, and he suggested that the fold belt bears more resemblance to ensialic Pan-African belts and their associated platforms.

West Antarctica

In the Ellsworth Mountains, thick Paleozoic and early Mesozoic platform and basinal sedimentary rocks, correlated with an undeformed Gondwanian succession in the Transantarctic Mountains [Elliot, 1975], were deformed into a series of asymmetric folds during the Ellsworth Orogeny [Craddock, 1972]. The age of this deformation is not well constrained. It affects strata as young as Permian, gives K-Ar age ranges of 235-254 Ma [Yoshida, 1983], and has been correlated with the Gondwanian orogeny of the Cape Fold Belt. Within the Pensacola Mountains a complex orogenic history exists, and F3 post-Permian and pre-Jurassic deformation of an Ordovician-Devonian succession, termed the Weddell Orogeny, is correlated with Gondwanian and Ellsworth orogenies [Ford, 1972; Dalziel and Elliot, 1973]. Dalziel and Elliot [1982] postulated that the Cape Fold Belt and Pensacola and Ellsworth Mountains formed a linear mountain chain with the Sierra de La Ventana Fold Belt of east Argentina [Coates, 1969] that was deformed in a foreland fold and thrust belt during the Andean-type Gondwanian orogeny, prior to breakup of Gondwanaland and dispersion of the microcontinental fragments. Harrington [1970] considered the Sierra de La Ventana Fold Belt as an aulacogenic chain.

Falkland Islands

Within the Falkland Islands, Paleozoic marine and terrestrial sedimentary rocks of the Devono-Carboniferous Group and Lafonian Supergroup (Upper Carboniferous to Permo-Triassic) were deformed about discordant fold trends [Greenway, 1972]. Greenway attributed the deformation to block faulting in the basement during the Late Jurassic to Early Cretaceous, whereas others [e.g., De Wit, 1977; Dalziel and Elliot, 1982] have attributed the deformation to the Gondwanian orogeny.

New Zealand

New Zealand probably formed part of the proto-Pacific margin of Gondwanaland during the Mesozoic. Complex stratotectonic units of a Carboniferous-Cretaceous arc and forearc sequence were deformed during episodes of the Rangitata Orogeny (for review see Bradshaw et al. [1981]; and Adams et al. [1985]). Unlike the Antarctic Peninsula, in New Zealand there is a Permian arc sequence and Triassic-Cretaceous basinal sediments of a forearc accretionary prism. The forearc was deformed during an orogenic episode at the end of the Triassic (Rangitata I), with accretion of the older part of a wedge to the trench slope suite. Younger parts of the clastic wedge were deformed during an Early Cretaceous main phase of the orogeny (Rangitata II).

South America

The southern Andes formed part of an active plate margin from Middle Devonian to Triassic times, with accretion of a thick wedge of sedimentary rocks, pillow basalt, bedded chert, and calcareous rocks (for review see Forsythe [1982]). Some of the limestones contain Early Permian fusulines. The accretionary prism was deformed during the Gondwanian orogeny and was unconformably overlain by Middle to Upper Jurassic volcanic rocks of the Tobifera Formation [Dalziel and Elliot, 1973]. These volcanic rocks formed in an extensional volcanotectonic rift zone [Bruhn et al., 1978], prior to formation and closure of a Late Jurassic-Early Cretaceous backarc basin system [Dalziel, 1981]. The backarc basin was deformed during mid-Cretaceous closure of the basin.

Discussion

There is considerable variation in the timing and type of tectonic and magmatic events along the proto-Pacific margin and within Gondwanaland during the late Paleozoic and early Mesozoic. These include (1) the Permian-Cretaceous subduction-accretion processes of New Zealand, (2) Permo-Triassic accretion in the southern Andes followed by a middle Jurassic-Cretaceous extensional rift system, and (3) a Permo-Triassic folding of a Paleozic succession in southern Africa, Ellsworth Mountains, and Pensacola Mountains.

At present the geological history and tectonic setting of the Antarctic Peninsula during the late Paleozoic and early Mesozoic is unclear. Sedimentation in parts of the accretionary prism was under way by Triassic or possibly late Paleozoic times, and the earliest known subduction-related plutonism is also Triassic. The deformation of part of the accretionary prism below the unconformity involves rocks as young as Early Jurassic and Triassic. Much of the deformation was consequently younger than the main deformation period within the Cape Fold Belt. The only events correlated with the Gondwanide deformation proper are the regional metamorphism at ~245 Ma and sedimentation within parts of the Trinity Peninsula Group. The Hope Bay Formation, which gave the Rb/Sr whole-rock age of 281 ± 16 Ma may be derived from contemporaneous arc activity. Other parts are younger, contain Triassic fossils, and may be derived from an older dissected arc terrane and deposited in a passive margin setting.

Although a widespread unconformity occurs below late Mesozoic strata in the Antarctic Peninsula region and has been used to separate Gondwanian and Andean events, it does not appear either to represent a significant break in plate margin processes or to be a consistent time plane along the Pacific margin in general. The magmatic history within the Antarctic Peninsula spans the possible age range of the unconformity which may be related to uplift associated with the breakup of Gondwanaland [Dalziel, 1982]. Although this unconformity may be a consistent time plane within the south Atlantic region, this cannot hold along the whole length of the proto-Pacific margin because New Zealand started to separate from Antarctica during the Late Cretaceous [Hayes and Ringis, 1973]. Within the Antarctic Peninsula region it is by no means certain that the unconformity is the same age throughout. In Alexander Island both Late Jurassic volcaniclastic sedimentary rocks and Tertiary calc-alkaline volcanic rocks directly overlie subduction complex rocks with strong unconformity.

A fuller understanding of the Gondwanide fold belt as a whole and its interaction with plate margin processes is partly dependent on the correct reassembly of the component parts prior to breakup of the supercontinent. It will only then be possible to assess if the component parts were spatially linked parts of the same orogenic belt and if a Gondwanian arc, forearc, and backarc system existed. There does, however, appear to be some variation in timing between the deformation within the Cape Fold Belt and Pacific margin processes in the Antarctic Peninsula sector. Furthermore, there is some uncertainty concerning the tectonic setting of the folding in the Cape Orogeny (for a review see Hälbich [Söhnge and Hälbich, 1983, pp. 177-184). Opposing viewpoints are (1) that the Cape Fold Belt was part of a backarc basin or foredeep involving a flat-plate subduction model [Lock, 1980; Dalziel and Elliot, 1982] or (2) that it bears more resemblance to Pan-African belts and was independent of plate margin processes [Söhnge and Hälbich, 1983]. There appears to be little compelling evidence to support either theory. The similarity of the Cape Fold Belt to the Laramide orogen in the western United States [Lock, 1980] and the occurrence of volcanic strata within the Permian [Elliot and Watts, 1974] are commonly sited as a link between the Cape Fold Belt and plate margin processes. However, there is as yet no detailed geochemistry of the volcanic rocks to support this, and it is also possible that the volcanism which occurs close to the axis of the Karoo Basin could be related to ensialic rifting. A curious aspect of the Gondwanian cover sequence in the South Atlantic-Weddell Sea region is the lack of deformation in the Beacon Supergroup of the Transantarctic Mountains, which is stratigraphically equivalent to part of the folded strata of the Cape Fold Belt and the Ellsworth Mountains. If a flat-plate subduction model applied, one might expect deformation throughout this region. Sedimentological studies [Loock and Visser, 1985] also suggested that one continuous fold belt may not have existed, and that a series of unconnected, enclosed, or partly enclosed basins existed in the southern continents during Paleozic to early Mesozoic times. Their suggestion is supported by differences in fossil assemblages, radiometric ages, and paleomagnetic data. Deformation of these basins may have resulted in the geometry and fold pattern seen in the Cape Fold Belt and Pensacola and Ellsworth Mountains and may be controlled by the orientation of the preexisting basin and/or basement structures. An analogous situation may exist in West Antarctica today, where a complex series of sedimentary basins are separated by topographic highs.

Conclusions

1. The timing of supposed Gondwanian events within the Antarctic Peninsula does not correspond with the known tectonic and sedimentation events within the classic Gondwanide succession of the Cape Fold Belt, and much of the previously assumed Gondwanide deformation in the Antarctic Peninsula is much younger.
2. In view of the timing problems, the authors favor the use of the term "Peninsula orogeny" for deformation of Triassic and Early Jurassic rocks below the unconformity and that the unconformity be referred to as the "Peninsula unconformity."
3. The significance and spatial link of the ~245-Ma metamorphism and the early history of the Trinity Peninsula Group to the classical Gondwanide orogen remains unknown.
4. The future possibility of identifying events that can be more directly correlated with the Gondwanian orogeny is not ruled out.
5. The tectonic relationship between folding in the Cape Fold Belt and Pacific margin processes during the Permo-Triassic remains unproven.

Acknowledgments. We wish to thank all our colleagues at the British Antarctic Survey for many useful discussions and two anonymous referees for greatly improving an earlier draft of the manuscript.

References

Adams, C. J., D. G. Bishop, and J. E. Gabites, Potassium-argon age studies of a low-grade, progressively metamorphosed greywacke sequence, Dansey Pass, South Island, New Zealand, J. Geol. Soc. London, 142, 339-349, 1985.

Bradshaw, J. D., P. D. Andrews, and C. J. Adams, Carboniferous to Cretaceous of the Pacific margin of Gondwana: The Rangitata phase of New Zealand, in Gondwana Five, edited by M. M. Cresswell and P. Vella, pp. 217-221, A. A. Balkema, Rotterdam, 1981.

Bruhn, R. L., C. R. Stern, and M. J. De Wit, Field and geochemical data bearing on the development of a Mesozoic volcanotectonic rift zone and backarc basin in southernmost South America, Earth Planet. Sci. Lett., 41, 32-46, 1978.

Burn, R. W., Early Tertiary calc-alkaline volcanism on Alexander Island, Br. Antarct. Surv. Bull., 53, 175-193, 1981.

Burn, R. W., The geology of the LeMay Group, Alexander Island, Br. Antarct. Surv. Sci Rep., 109, 65 pp., 1984.

Caminos, R., Cordillera fuegina, in Geologia Regional Argentina, vol. 2, coordinated by J. C. M. Turner, pp. 1463-1501, Academia Nacional de Ciencias, Cordoba, 1980.

Coates, D. A., Stratigraphy and sedimentation of the Sauce Grande Formation, Sierra de La Ventana, southern Buenos Aires Province, Argentina, in Gondwana Stratigraphy, International Union of Geological Sciences Symposium, Buenos Aires, 1967, vol. 2, edited by A. J. Amos, pp. 799-819, UNESCO, Paris, 1969.

Craddock, C., Antarctic tectonics, in Antarctic Geology and Geophysics, edited by R. J. Adie, pp. 449-455, Universitetsforlaget, Oslo, 1972.

Craddock, C., Tectonic evolution of the Pacific margin of Gondwanaland, in Gondwana Geology, edited by K. S. W. Campbell, pp. 609-618, Australian National University Press, Canberra, 1975.

Dalziel, I. W. D., Backarc extension in the southern Andes: A review and critical reappraisal, Philos. Trans. R. Soc. London, Ser. A, 300, 319-335, 1981.

Dalziel, I. W. D., The early (pre-Middle Jurassic) history of the Scotia arc region: A review and progress report, in Antarctic Geoscience, edited by C. Craddock, pp. 111-126, University of Wisconsin Press, Madison, 1982.

Dalziel, I. W. D., and D. H. Elliot, Evolution of the Scotia arc, Nature London, 233, 246-252, 1971.

Dalziel, I. W. D., and D. H. Elliot, The Scotia arc and Antarctic margin, in The South Atlantic, the Ocean Basins and Margins, edited by A. E. M. Nairn and F. G. Stehli, pp. 171-245, Plenum, New York, 1973.

Dalziel, I. W. D., and D. H. Elliot, West Antarctica: Problem child of Gondwanaland, Tectonics, 1, 3-19, 1982.

Dalziel, I. W. D., D. H. Elliot, D. L. Jones, J. W. Thomson, M. R. A. Thomson, N. A. Wells, and W. J. Zinsmeister, The geological significance of some Triassic microfossils from the South Orkney Islands, Scotia Ridge, Geol. Mag., 118, 15-25, 1981.

De Swart, A. M. J., O. Fletcher, and P. Toschek, Note on orogenic style in the Cape Fold Belt by I. W. Hälbich, Trans. Geol. Soc. S. Afr., 77, 53-58, 1974.

De Wit, M. J., The evolution of the Scotia arc as a key to the reconstruction of southwestern Gondwanaland, Tectonophysics, 37, 53-81, 1977.

Dingle, R. V., W. G. Siesser, and A. R. Newton,

Mesozoic and Tertiary Geology of Southern Africa, 375 pp., A. A. Balkema, Rotterdam, 1983.
du Toit, A. L., Our Wandering Continents, 366 pp., Oliver & Boyd, Edinburgh, 1937.
Edwards, C. W., New evidence of major faulting on Alexander Island, Br. Antarct. Surv. Bull., 49, 15-20, 1979.
Elliot, D. H., Tectonics of Antarctica: A review, Am. J. Sci., 275, 45-106, 1975.
Elliot, D. H., and D. R. Watts, Nature and origin of volcaniclastic material in some Karoo and Beacon rocks, Trans. Geol. Soc. S. Afr., 77, 105-108, 1974.
Farquharson, G. W., The Nordenskjold Formation of the northern Antarctic Peninsula: An Upper Jurassic radiolarian mudstone and tuff sequence, Br. Antarct. Surv. Bull., 60, 1-22, 1983.
Farquharson, G. W., Late Mesozoic, nonmarine conglomeratic sequences of northern Antarctic Peninsula (Botany Bay Group), Br. Antarct. Surv. Bull., 65, 1-32, 1984.
Ford, A. B., Weddell Orogeny-latest Permian to early Mesozoic deformation at the Weddell Sea margin of the Transantarctic Mountains, in Antarctic Geology and Geophysics, edited by R. J. Adie, pp. 419-425, Universitetsforlaget, Oslo, 1972.
Forsythe, R., The late Palaeozoic to early Mesozoic evolution of southern South America: A plate tectonic interpretation, J. Geol. Soc. London, 139, 671-682, 1982.
Gledhill, A., D. C. Rex, and P. W. G. Tanner, Rb-Sr and K-Ar geochronology of rocks from the Antarctic Peninsula between Anvers Island and Marguerite Bay, in Antarctic Geoscience, edited by C. Craddock, pp. 315-323, University of Wisconsin Press, Madison, 1982.
Greenway, M. E., The geology of the Falkland Islands, Br. Antarct. Surv. Sci. Rep., 76, 42 pp., 1972.
Grunow, A. M., I. W. D. Dalziel, and D. V. Kent, Ellsworth-Whitmore Mountains crustal block, Western Antarctica: New paleomagnetic results and their tectonic significance, this volume.
Harland, W. B., A. V. Cox, P. G. Llewellyn, C. A. G. Pickton, A. G. Smith, and R. Walters, A Geological Time Scale, 131 pp., Cambridge University Press, New York, 1982.
Harrington, H. J., Las Sierras Australes de Buenos Aires, Republica Argentina: Cadena aulacogenica, Rev. Asoc. Geol. Argent., 25, 151-181, 1970.
Hayes, D. W., and J. Ringis, Seafloor spreading in the Tasman Sea, Nature London, 243, 454-458, 1973.
Hyden, G., and P. W. G. Tanner, Late Palaeozoic-early Mesozoic forearc basin sedimentary rocks at the Pacific margin in western Antarctica, Geol. Rundsch., 70, 529-541, 1981.
Ineson, J. R., J. A. Crame, and M. R. A. Thomson, Lithostratigraphy of the Cretaceous strata of West James Ross Island, Antarctica, Cretaceous Res., 7, 141-159, 1986.
Lock, B. E., Flat plate subduction and the Cape Fold Belt of South Africa, Geology, 8, 35-39, 1980.
Loock, J. C., and J. N. J. Visser, A Gondwana sequence deposited on a supercontinent: Fact or fantasy? (abstract), in Sixth Gondwana Symposium, p. 61, Institute of Polar Studies, Misc. Publ. 231, Ohio State University, Columbus, 1985.
Matthews, D. W., The geology of Horseshoe and Lagotellerie islands, Marguerite Bay, Graham Land, Br. Antarct. Surv. Bull., 52, 125-154, 1983.
Meneilly, A. W., S. M. Harrison, B. A. Piercy, and B. C. Storey, Structural evolution of the magmatic arc in northern Palmer Land, Antarctic Peninsula, this volume.
Miller, H., The position of Antarctica within Gondwana in the light of Palaeozoic orogenic development, in Antarctic Earth Science, edited by R. L. Oliver, P. R. James, and J. B. Jago, pp. 579-581, Australian Academy of Science, Canberra, 1983.
Newton, A. R., A gravity-folding model for the Cape Fold Belt, Trans. Geol. Soc. S. Afr., 76, 145-152, 1973.
Pankhurst, R. J., Rb-Sr geochronology of Graham Land, Antarctica, J. Geol. Soc. London, 139, 701-711, 1982.
Pankhurst, R. J., Rb-Sr constraints on the ages of basement rocks of the Antarctic Peninsula, in Antarctic Earth Science, edited by R. L. Oliver, P. R. James, and J. B. Jago, pp. 367-371, Australian Academy of Science, Canberra, 1983.
Riccardi, A. C., Estratigrafia en el oriente de la Bahia de la Lancha, Lago San Martin, Santa Cruz, Argentina, Rev. Mus. La Plata Secc. Geol., 7, Geologia 61, 245-318, 1971.
Rowley, P. D., W. R. Vennum, K. S. Kellogg, T. S. Laudon, P. E. Carrara, J. M. Boyles, and M. R. A. Thomson, Geology and plate tectonic setting of the Orville Coast and eastern Ellsworth Land, Antarctica, in Antarctic Earth Science, edited by R. L. Oliver, P. R. James, and J. B. Jago, pp. 245-250, Australian Academy of Science, Canberra, 1983.
SACS, (South African Committee for Stratigraphy) Stratigraphy of South Africa, Part 1, compiled by L. E. Kent, Hand. Geol. Surv. S. Afr., 8, 690 pp., 1980.
Sclater, J. G., S. Hellinger, and C. Tapscott, The paleobathymetry of the Atlantic Ocean from the Jurassic to the present, J. Geol., 85, 509-552, 1977.
Smellie, J. L. S., A complete arc-trench system recognized in Gondwana sequences of the Antarctic Peninsula region, Geol. Mag., 118, 139-159, 1981.
Smellie, J. L., Sandstone detrital modes and basinal setting of the Trinity Peninsula Group, northern Graham Land, Antarctic Peninsula: A preliminary survey, this volume.
Smellie, J. L., R. J. Pankhurst, M. R. A. Thomson, and R. E. S. Davies, The geology of the South Shetland Islands, VI, Stratigraphy, geochemistry and evolution, Br. Antarct. Surv. Sci. Rep., 87, 85 pp., 1984.
Söhnge, A. P. G., and I. W. Hälbich (Eds.), Geodynamics of the Cape Fold Belt, Spec. Publ. Geol. Soc. S. Afr., 12, 184 pp., 1983.
Storey, B. C., and S. W. Garrett, Crustal growth of the Antarctic Peninsula by accretion, magma-

tism, and extension, Geol. Mag., 122, 5-14, 1985.

Storey, B. C., and A. W. Meneilly, Petrogenesis of metamorphic rocks within a subduction-accretion terrane, Signy Island, South Orkney Islands, J. Metamorph. Geol., 3, 21-42, 1985.

Tanner, P. W. G., R. J. Pankhurst, and G. Hyden, Radiometric evidence for the age of the subduction complex in the South Orkney and South Shetland islands, West Antarctica, J. Geol. Soc. London, 139, 683-690, 1982.

Taylor, B. J., M. R. A. Thomson, and L. E. Willey, The geology of the Ablation Point-Keystone Cliffs area, Alexander Island, Br. Antarct. Surv. Sci. Rep., 82, 65 pp., 1979.

Thomson, M. R. A., New palaeontological and lithological observations on the Legoupil Formation, north-west Antarctic Peninsula, Br. Antarct. Surv. Bull., 41/42, 169-185, 1975.

Thomson, M. R. A., Late Mesozoic stratigraphy and invertebrate palaeontology of the South Orkney Islands, Br. Antarct. Surv. Bull., 54, 65-83, 1981.

Thomson, M. R. A., and R. J. Pankhurst, Age of post-Gondwanian calc-alkaline volcanism in the Antarctic Peninsula region, in Antarctic Earth Science, edited by R. L. Oliver, P. R. James, and J. G. Jago, pp. 328-333, Australian Academy of Science, Canberra, 1983.

Thomson, M. R. A., and T. H. Tranter, Early Jurassic fossils from central Alexander Island and their geological setting, Br. Antarct. Surv. Bull., 70, 23-39, 1986.

Yoshida, M., Structural and metamorphic history of the Ellsworth Mountains, West Antarctica, in Antarctic Earth Science, edited by R. L. Oliver, P. R. James and J. B. Jago, pp. 266-269, Australian Academy of Science, Canberra, 1983.

Zinsmeister, W. J., Review of Upper Cretaceous-Lower Tertiary sequence of Seymour Island, Antarctica, J. Geol. Soc. London, 139, 779-785, 1982.

Copyright 1987 by the American Geophysical Union.

SANDSTONE DETRITAL MODES AND BASINAL SETTING OF THE TRINITY PENINSULA GROUP, NORTHERN GRAHAM LAND, ANTARCTIC PENINSULA: A PRELIMINARY SURVEY

J. L. Smellie

British Antarctic Survey, Natural Environment Research Council, High Cross, Cambridge
United Kingdom CB3 0ET

Abstract. Sandstone detrital modes for a representative sample of the Trinity Peninsula Group in northern Graham Land are described and assessed. Whereas the volumetrically dominant quartz and feldspar were derived principally from erosion of a plutonic and high-rank metamorphic terrane, the lithic population was derived mainly from a volcanic cover. The data clearly indicate the presence of two major sandstone suites (petrofacies I and II) with distinctive and probably separate provenances. Further scope for subdivision is limited by the small sample set, but four petrofacies (Ia, Ib, IIa, and IIb) may be present, three of which correspond with previously described lithostratigraphical units (Legoupil, Hope Bay, and View Point formations). The sample distribution and detrital modes enable approximate geographical limits to be assigned to each petrofacies for the first time, although the nature of the boundaries (stratigraphical or structural) is unknown. Petrofacies II could have been derived from an active magmatic arc and deposited in a forearc basin (sensu lato) or series of basins at a major consuming margin. Petrofacies I is a much more quartzose suite, although otherwise petrographically very similar to petrofacies II. Its depositional setting is ambiguous on the basis of the data presently available, and deposition can only be said to have occurred at either an active or a passive continental margin. Finally, there is the possibility that strike-slip faulting has structurally shuffled the Trinity Peninsula Group, causing the pronounced age and compositional contrasts observed.

Introduction

The Trinity Peninsula Group (TPG) encompasses monotonous, polydeformed, siliciclastic turbidite strata of ill-defined age (Permo-Carboniferous-Triassic?) [Thomson, 1975; Pankhurst, 1983] which crop out over wide areas in the Antarctic Peninsula region (Figure 1). Together they comprise the most extensive exposures of pre-Late Jurassic sedimentary rocks in the Scotia Arc-Antarctic Peninsula region.

The stratotectonic position of the TPG is poorly understood because the paleogeographical setting of the Pacific margin of Gondwana, its constituent terranes, and their translations and rotations are obscure. Moreover, the duration of turbidite deposition is poorly defined. As a result, we do not know the probable sites of sediment efflux from the Gondwana craton, or the positions through time of the TPG relative to the same, or the nature of the Gondwana Pacific margin (active or passive) during sedimentation.

Previous authors have envisaged deposition of the TPG as a clastic wedge at a passive or active continental margin or at a margin initially passive; then active [Dalziel and Elliot, 1973; Smellie, 1981; Burn, 1984; Storey and Garrett, 1985]. Within the active margin setting, deposition within upper-slope [Smellie, 1981] and trench-slope [Storey and Garrett, 1985] basins have been postulated.

A serious shortcoming affecting all of these studies is the lack of knowledge of the stratigraphy of the TPG and, in particular, stratigraphically controlled provenance variations which would enable a more rigorous stratotectonic appraisal. Therefore, the purpose of this study is to examine, by means of sandstone petrography, the detrital constitution of TPG sandstones in northern Graham Land (where they are least altered and deformed), to test for internally consistent mappable provenance-related variations that can be used as a stratigraphical guide [cf. Ingersoll, 1978, 1983; Floyd, 1982], to document more precisely the nature of the provenance terrane, and to interpret all these aspects in terms of a basin setting within a Gondwana framework.

Stratigraphy. Aitkenhead [1975] subdivided the Trinity Peninsula Group in northern Graham Land on an areal basis. Six divisions were established, each of which was defined as a natural group with common affinities, such as "common topographical expression or similar strike direction, lithology or metamorphic grade." The scheme was devised for descriptive purposes only, and no stratigraphical implications were suggested. By contrast, Hyden and Tanner [1981] distinguished three lithological associations, each of which was regarded as a distinct mappable unit of formation status extending over a significant area and recognized mainly by the ratios of coarse- to fine-grained rocks and

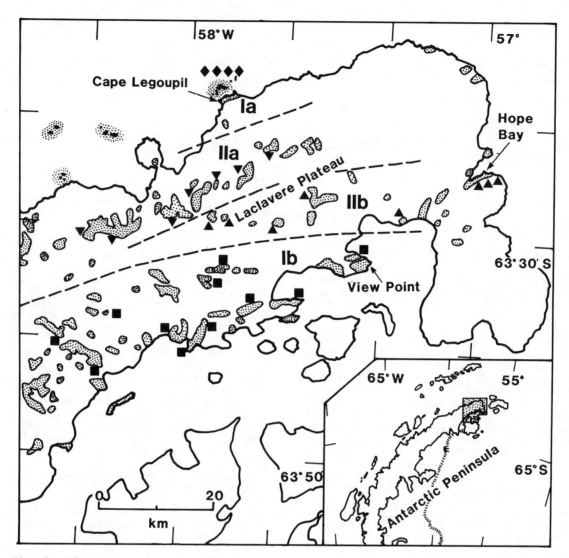

Fig. 1. Sketch map of Trinity Peninsula, northern Graham Land, showing outcrops of the Trinity Peninsula Group, sample localities, and approximate distribution of the four petrofacies described in this paper.

the presence of unusual minor lithologies. No attempt was made to map the formation boundaries, but the following criteria were listed. The Hope Bay Formation is correct way up and gently to moderately dipping, comprising subequal proportions of massive and thinly interbedded sandstones and mudstones. The View Point Formation is characterized by massive and thinly interbedded siliceous sandstones and mudstones; less than 30% of exposed strata are sandstones, and the beds are generally overturned. The Legoupil Formation [Halpern, 1965] is also mudstone dominated but has thinner massive sandstones than the other formations. Massive graded sandstones up to 2 m thick are common in the Hope Bay and View Point formations. Cleavage is well developed in mudstones of the View Point and Legoupil formations but is poorly developed in the Hope Bay Formation.

Petrography. Previous workers who have studied the petrology of sandstones in northern Graham Land include Aitkenhead [1975] and Elliot [1965]. However, only geographically restricted parts of the entire sequence exposed in Trinity Peninsula were examined by these workers who used a traditional quartz-feldspar rock fragments (QFR) method of point counting in which detrital modes are strongly influenced by grain-size differences between samples [e.g., Basu, 1976; Suttner et al., 1981; Ingersoll et al., 1984]. Nevertheless, Aitkenhead [1975] was able to demonstrate a broad bimodal distribution pattern for his data; quartzose/feldspar-poor rocks occurring on southeast Trinity Peninsula and a feldspar-rich suite restricted mainly to central Trinity Peninsula. Specimens examined by Elliot [1965] fall mainly within a feldspar-rich group. The quartz-rich samples were also characterized by proportionately greater counts for allanite, tourmaline, and garnet, and lesser epidote and sphene; igneous rock fragments were said to be more abundant in the central group of feldspar-rich sandstones.

For this study, thin sections were made from 30 sandstones selected from collections housed at the British Antarctic Survey. Initially, only medium sandstones were chosen in order to ease point-counting procedures and to achieve the greatest uniformity of results. However, it was found necessary to include fine (12 samples) and coarse (1 sample) sandstones to maintain an even and widespread areal coverage (Figure 1). Each section was stained for both plagioclase and potassium feldspar by using the methods of Houghton [1980]. The Gazzi-Dickinson point-counting method was chosen to minimize grain-size variation effects on detrital modes [cf. Ingersoll et al., 1984] and to ensure a standardized modern technique comparable with techniques of workers elsewhere in the world. In particular, the technique has been used extensively to categorize modern basinal sands in terms of plate tectonic settings [e.g., Crook, 1974; Dickinson and Suczek, 1979; Valloni and Maynard, 1981].

There are no along-strike differences in detrital modes in the present data set, but cross-strike variation is evident and was tested by projecting sample locations along strike (northeast-southwest) to intersect an arbitrarily positioned strike-perpendicular line. Modal parameters (Table 1) were then plotted along this line to yield a pseudotransect some 40 km long. This, in combination with triangular and binary variation diagrams (Figures 2 and 3), indicates that at least two, and possibly up to four, natural groups can be distinguished. They are referred to here as petrofacies because they are distinguished primarily on the basis of their petrography, and to a lesser extent, field appearance, their form, boundaries, or mutual relations not being known.

Texture. With the exception of sandstones from the Duroch Islands (near Cape Legoupil, Figure 1), which are poorly to moderately sorted and bimodal, the sandstones examined from each petrofacies are typically poorly sorted and matrix poor. Grains of all types are typically angular to subangular, but a significant proportion are subrounded to well rounded in petrofacies I, a feature that is most noticeable in the quartz grains. Some of the softer clasts have been shape modified owing to compaction against more competent grains, an effect particularly noticeable in the micas, which are generally bent and crushed. "Matrix" principally consists of minor phyllosilicate cement which occurs as discontinuous to complete radial and concentric rims of fine crystals, accompanied, in the matrix-rich rocks, by finely comminuted detritus. Minor to advanced recrystallization has occurred in a few sandstones (not considered here) from widely scattered localities involving partial polygonization of quartz grains and development of a coarse commonly aligned sericitic and/or quartzofeldspathic groundmass. These are the most intense dynamothermal effects of a regional metamorphism which generally only reached prehnite-pumpellyite facies [Hyden and Tanner, 1981].

Composition. The sandstones are predominantly subquartzosefeldspathic and lithofeldspathic arenites, with a few feldspatholithic and quartzose arenites. No grain types are exclusive to any petrofacies, although their relative abundances often vary strikingly between petrofacies (Table 1). Quartz (Q) and feldspar (F) are predominant, with substantial regional variation in Q/F ratios (Figures 2 and 3). Quartz occurs in several forms [cf. Elliot, 1965], but no attempt was made here to classify the different types except in terms of undulosity. Slight to marked undulose extinction is very common, but nonundulose quartz (sensu Basu et al. [1975]) is always more abundant than undulose quartz. Quartz, with numerous tiny acicular crystallite inclusions of rutile (?) [cf. Aitkenhead, 1975], is virtually restricted to petrofacies I. Plagioclase (albite-oligoclase [Elliot, 1965; Aitkenhead, 1975]) is predominant among the feldspars, occurring as twinned and untwinned grains that are rarely zoned; most show slight to extensive sericitization, but except for rare checkerboard albite and a few grains containing epidote or prehnite, there is little evidence to suggest that much of the albite formed from more calcic plagioclase by diagenetic alteration. A small proportion (generally <10%) of potassium feldspar (orthoclase [Elliot, 1965; Aitkenhead, 1975]) occurs in most of the sandstones, although it is typically absent from sandstones of petrofacies I. It is characteristically untwinned and unaltered or, at most, carries a slight brownish dusting. String, rod, and patch microperthites are common, whereas microcline is uncommon.

Rock fragments are usually less common than monominerallic clasts (Table 1). Four major types are recognized. Volcanic clasts, always the most abundant (Figures 3c, d), are mainly fine-grained felsitic and quartzofeldspathic mosaic types, with a range of phenocrysts or microphenocrysts (plagioclase, quartz, potassium feldspar, biotite, amphibole, apatite, pyroxene, and opaque ore) and occasional spherulitic potassium feldspar textures indicative of an acidic-intermediate (rhyolite-andesite) volcanic provenance. Other volcanic fragments, probably intermediate or basic in composition, are less common and include intersertal to trachytic-textured plagioclase-rich types and completely altered texturally destroyed clasts dominated by chlorite (and/or epidote and sphene) that are particularly characteristic of petrofacies II; hypabyssal (volcanic) types are coarser-grained hypidiomorphic igneous rocks composed of plagioclase, potassium feldspar, and quartz, some with myrmekitic and graphic textures. Sedimentary clasts are mainly dark penecontemporaneous fine-grained siltstones and shales which lack tectonic fabrics. Metasedimentary fragments are fine-grained aggregates of quartz and mica, containing occasional traces of plagioclase or chlorite, that are almost invariably foliated. The extremely sparse meta-igneous clasts are dominated by feldspar, quartz, and/or chlorite, and may or may not be foliated.

Accessory framework minerals include biotite, muscovite, hornblende, epidote, sphene (or rutile), opaque ore, garnet, zircon, allanite, tourmaline, and apatite.

Petrofacies characteristics. The grouping of

TABLE 1. Modal Data Used in the Present Study: Means and (in Brackets) Standard Deviations

Petrofacies	QFL[a]				FRMWK[a] M,%	P/F[a]	Q/F[a]	Q_p/Q[a]	L_{vm}/L[b]
	Q%	Q_m%	F%	L%	L_t%				
Ia	52.5 (7.1)	50.2 (7.7)	35.0 (6.1)	12.6 (3.0)	14.9 (3.5)	0.86 (0.78)	0.82 (0.15)	1.56 (0.50)	0.05 (0.02)	0.81 (0.10)
Ib	53.5 (21.2)	48.2 (19.1)	23.3 (7.4)	23.2 (17.6)	28.6 (16.4)	0.44 (0.46)	0.98 (0.04)	2.88 (2.43)	0.09 (0.06)	0.58 (0.07)
IIa	29.4 (4.5)	27.5 (5.2)	56.3 (9.8)	14.4 (7.1)	16.2 (6.7)	2.42 (1.62)	0.89 (0.04)	0.55 (0.20)	0.07 (0.04)	0.68 (0.08)
IIb	24.6 (3.3)	22.9 (3.4)	52.9 (7.0)	22.5 (5.3)	24.2 (5.2)	1.72 (0.69)	0.79 (0.03)	0.48 (0.12)	0.07 (0.05)	0.80 (0.07)

Abbreviation	Definition
Q	Total quartzose grains = $Q_m + Q_p$, where Q_m is monocrystalline quartz grains and Q_p is polycrystalline quartz grains.
F	Total feldspar grains = P + K, where P is plagioclase feldspar grains and K is potassium feldspar grains.
L_t	Total aphanitic lithic grains = L + Qp, where L is total unstable lithic grains.
L	$L_m + L_v + L_s$, where L_m is metamorphic aphanitic lithic grains, L_v is volcanic and hypabyssal aphanitic lithic grains, and L_s is sedimentary aphanitic lithic grains.
	$L_{vm} + L_{sm}$, where L_{vm} is volcanic-hypabyssal and metavolcanic aphanitic lithic grains and L_{sm} is sedimentary and metasedimentary aphanitic lithic grains.
FRMWK	Q + F + L + M + Mc, where M is monocrystalline phyllosilicate grains and M_c is miscellaneous and unidentified framework grains.

Abbreviations are from Ingersoll and Suczek [1979].
[a] Data obtained from counts of 500 points per thin section, using a point count interval of 0.3 mm and a maximum grid spacing that resulted in coverage of the entire section.
[b] Data obtained from counts of 200 points per thin section, for lithic clasts only, in 17 of the freshest samples (5 samples from petrofacies IIa; 4 samples from each of the other petrofacies).

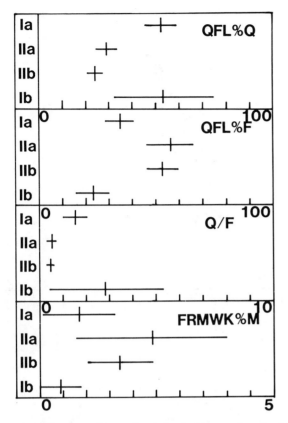

Fig. 2. Means and standard deviations of selected key parameters used to distinguish petrofacies in the Trinity Peninsula Group. The length of each horizontal line is equivalent to 1 standard deviation on either side of the mean.

petrofacies was established by a combination of characteristics: gross lithological similarity, geographical location, and in particular, differences in detrital framework modes (Figures 2 and 3). Once the groupings had been established, means and standard deviations were used to indicate approximate field boundaries (Table 1; Figure 3) based on the 2σ method of Ingersoll [1978]. It is evident from Figures 2 and 3 that no single parameter is sufficient to uniquely identify any petrofacies. However, certain key parameters (QFL%Q, QFL%F, and Q/F, Figure 2; see Table 1 for parameter abbreviations) enable the two main petrofacies to be distinguished with a minimum of effort.

The distinctive characteristics of petrofacies I are generally high QFL%Q and Q/F, low QFL%F, FRMWK%M, and K, and a large range of QFL%L (Table 1; Figures 2 and 3). The sandstones are often siliceous in appearance, and this petrofacies contains the sole recorded occurrences of polymict conglomerate, red and green chert, pillow lava, and hyaloclastite. Although quantitative data are lacking at present, the petrofacies also contains consistently greater proportions of allanite, garnet, zircon, and tourmaline than petrofacies II, which is characterized by higher hornblende, sphene, and epidote. The conglomerates are rich in cobbles and pebbles of igneous and high-rank metamorphic rocks (mainly acid volcanic rocks, granite, orthogneiss, and paragneiss) and sedimentary material (mainly quartzite, arkose, siltstone, and mudstone); schist, phyllite, and other low-rank metamorphic detritus are minor [Aitkenhead, 1975; Thomson 1975]. Within this petrofacies, scope for subdivision is restricted, although the significance is uncertain as very few samples are currently available from the Cape Legoupil area (petrofacies Ia, Figure 1). Despite major overlap in the compositions of the two sets of samples, sandstones from the Cape Legoupil area are somewhat more feldspathic and less lithic, with low P/F, Q/F, and Q_p/Q ratios and higher L_{vm}/L; and potassium feldspar, typically absent in petrofacies Ib sandstones, is present in three of the four petrofacies Ia samples analyzed (Table 1). Petrofacies Ia and Ib may correspond to the Legoupil and View Point formations of Hyden and Tanner [1981].

Petrofacies II, with its notably lower QFL%Q, higher QFL%F, low Q/F, and moderate to high L_{vm}/L ratios, and presence of basic volcanic clasts, can also be subdivided. Samples from the northwestern flanks of Laclavere and Louis-Philippe plateaux (petrofacies IIa) are consistently lithic-poor, with noticeably lower L_{vm}/L and higher P/F ratios (reflecting lower K contents) (Table 1). Despite the large overlap in the compositional data, the two populations are statistically separable (Mann-Whitney U test (one tailed), p = 0.001-0.242 for $Q_mFLt\%Lt$, L_{vm}/L, and P/F). Moreover, the northwestern samples often contain garnet (virtually absent in samples of petrofacies IIb to the southeast) and a greater proportion of opaque ore. Petrofacies IIa has no published counterpart of formational status, whereas petrofacies IIb may correspond to the Hope Bay Formation of Hyden and Tanner [1981].

Discussion

Potassium feldspar variation. If the virtual absence of potassium feldspar in petrofacies I is an initial depositional property, these sandstones clearly did not have the same provenance as sandstones of petrofacies II. Scarcity or absence of detrital potassium feldspar has also been observed in numerous other forearc suites [cf. Middleton, 1972; Moore, 1979; Dickinson et al., 1982] but is an unreliable provenance discriminant. Its distribution is usually ascribed to control by source rocks or to removal of potassium feldspar by post-depositional alteration or solution. On the Q_mFL_t diagram (Figure 3), sandstones of petrofacies I have a nearly constant Q/F ratio falling between 1 and 2 (much higher than the 0.3-0.6 range of values characteristic of petrofacies II), with a residual trend to higher values in the most quartz-rich samples. It is significant that the few samples that do contain detrital potassium feldspar also have Q/F values within the "normal" range for the petrofacies, suggesting that the potassium feldspar was lost diagenetically. Al-

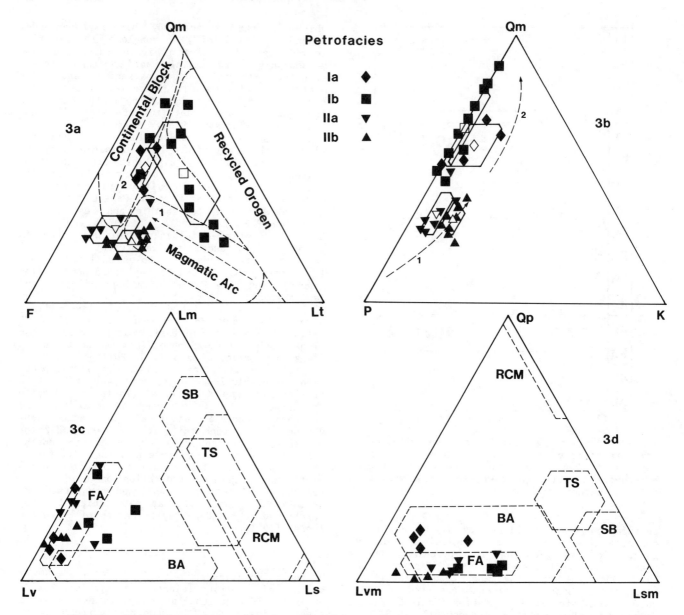

Fig. 3. Triangular plots of selected modal parameters for Trinity Peninsula Group petrofacies; dashed-line fields and trends for modern basinal sands taken from Dickinson and Suczek [1979] and Ingersoll and Suczek [1979]. Solid-line polygons are 1 standard deviation on either side of the mean [cf. Ingersoll, 1978]. Open symbols represent mean values for each petrofacies. In Figures 3a and b, trend 1 represents an increasing ratio of plutonic:volcanic components in magmatic arc provenances; trend 2 depicts increasing maturity or stability from continental block provenances. RCM is rifted continental margins; SB, suture belts; TS, mixed magmatic arc and subduction complexes (trench-slope basins with detritus reworked from subduction complex); FA, magmatic arcs, forearc areas; BA, mixed magmatic arcs and rifted margins (backarc basins).

ternatively, some of this loss may have occurred prior to deposition, as suggested by granite-gneiss clasts in conglomerate from View Point, which show potassium feldspar in various stages of replacement by checkerboard albite and the presence of the latter as a detrital mineral in the sandstones [Aitkenhead, 1975]. The residual trend to higher Q/F values could have been caused by removal of original potassium feldspar by dissolution or clay alteration.

Internal cross-strike variation. Within petrofacies II, the petrographical differences are very small and of uncertain significance, consisting mainly of an increased proportion of lithic fragments (mainly volcanic) in petrofacies IIb relative to IIa. This variation may be due partly to grain size differences between the two sample sets (petrofacies IIa samples are generally finer than petrofacies IIb samples), but other explanations are possible (e.g., time-dependent changes in the detritus as the provenance evolved, derivation from different geographical segments of that terrain, or complex recycling processes between source and basin). Insufficient data are available for petrofacies Ia, but striking modal variation exists within petrofacies Ib, which ranges progressively from lithic-rich to quartz-rich compositions (Figure 3a); the more mature quartzose samples are also characterized by the presence of stable and ultrastable accessories such as zircon and tourmaline. Again, the distribution of fine-grained and medium-grained specimens in the sample set suggests some grain-size control on the variation, but the large compositional range strongly suggests that other factors were more important. For example, such a trend could result from normal sedimentary differentiation as characterized by modern big river sands [cf. Potter, 1978; Franzinelli and Potter, 1983], which become progressively depleted in lithic and other labile clasts with increasing distance from source. Other factors, such as climate, sedimentation rate, and residence time in diverse intermediary depositional environments, may also significantly modify the provenance-inherited imprint [cf. Suttner et al., 1981].

Provenance. The two major petrofacies share many modal characteristics that suggest a provenance composed of coarse-grained plutonic and high-rank metamorphic rocks: low proportion of lithic clasts, dominantly quartzofeldspathic composition, scarcity of "volcanic" quartz, unzoned or slightly zoned, sometimes myrmekitic or graphic-textured plagioclase of albitic or oligoclase composition, perthitic potassium feldspar, presence of blue tourmaline, garnet, and zircon, biotite consistently more abundant than muscovite, and the relative proportions of nonundulose to undulose monocrystalline quartz and polycrystalline quartz. In addition, the lithic populations are overwhelmingly dominated by andesite-rhyolite volcanic fragments, although these are usually volumetrically subordinate to the monomineralic clasts. Considered together, the types of clast present and their relative proportions in petrofacies II (Figure 3) strongly suggest a provenance for this petrofacies composed of unroofed highly dissected batholiths and their metamorphic envelope, capped locally by isolated volcanoes. For petrofacies I, the high quartz content, presence of rounded and well-rounded (second cycle) quartz grains, and in the Legoupil Formation, better sorting, indicate that mature sedimentary rocks also formed a significant part of the provenance and may now be represented by abundant clasts of quartzose sandstone in conglomerates at View Point [cf. Aitkenhead, 1975]. Moreover, unlike petrofacies II, the relative proportions of the different clast types strongly suggest recycled orogen or possibly continental block provenances (Figures 3a, b). Given the abundance of plutonic and volcanic detritus, the lithological constitution of the source area for petrofacies I was evidently similar to a magmatic arc, although the dissimilarities are significant. Petrofacies I also shows a trend toward lithic-rich compositions comparable with volcaniclastic sandstones (Figure 3a). However, although data are still sparse, no compelling evidence suggests that this trend is due to an increased influx of volcanic detritus (e.g., there is no obvious increase in the proportion of volcanic lithic fragments, calcic plagioclase, "volcanic" quartz, or unstable heavy minerals).

Basinal setting. Impressed by the presence of quartzose turbidites in the TPG, Dalziel and Elliot [1973] suggested deposition as a geosynclinal wedge in one or more discrete troughs along an essentially passive continental margin, subsequent compression and accompanying minor subduction resulting in deformation and low-grade metamorphism. By contrast, Smellie [1981] proposed deposition in a forearc (upper-slope) basin, a model based largely on the possible presence of older sialic crust beneath the TPG (in contrast to its apparent absence beneath coeval subduction complex deposits in Alexander Island and South Shetland and South Orkney islands). A reappraisal by Burn [1984] indicated that in structural style, metamorphism, and apparent scarcity of volcaniclastic detritus, a subduction zone setting was more appropriate, and this suggestion was endorsed by Storey and Garrett [1985], who assigned the TPG mainly to a subduction complex (trench slope basin), but did not exclude a passive margin setting for early parts of the succession.

This study has demonstrated that the two major TPG petrofacies identified in northern Graham Land were derived wholly (petrofacies II), or possibly in part (petrofacies I), from magmatic arcs. Age control is poor, but given a Permo-Carboniferous age for at least part of petrofacies II [Pankhurst, 1983] and a Triassic or broadly "Mesozoic" age for petrofacies I [Thomson, 1975; Hyden and Tanner, 1981], the apparent contrasts in age and detrital modes between the two petrofacies render it unlikely that they were derived from the same arc segment or possibly even the same magmatic arc.

Although it is difficult to prove, arc magmatism may have been active during deposition of petrofacies II, which has the least conflicting modal characteristics. The arc was highly dissected, and a forearc (upper/lower slope or trench basin) depositional setting is suggested (Figure 3). By contrast, interpretation of the detrital modes of petrofacies I is more ambiguous, and at least three principal settings are compatible with the data. (1) One compatible setting is a passive margin, shedding quartzose sediments but including an ancient arc orogen. In this context, the arc-related basin setting suggested by the lithic population (Figures 3c and d) is anomalous, although it accurately reflects the petrological character of the source rocks. Later subduction at this continental margin would result in the incorporation of these sediments in a subduction complex, consistent with the structural interpretations [Burn, 1984; Storey and Garrett, 1985; cf. Dalziel and Elliot, 1973]. (2) A consuming margin, in which erosion of an uplifted nonvolcanic frontal arc, such as characterizes some "Andean" arc systems [Moore, 1979], could also create the ambiguous modal characteristics. In this case, the lithic population accurately reflects both the petrological character and tectonic setting of the source rocks. (3) Another compatible setting is a consuming margin in which quartzose sediments were derived outside of the arc-trench system, introduced sedimentologically (e.g., as abyssal plain turbidites [Dickinson, 1982]), and subsequently accreted [cf. Ingersoll and Suczek, 1979; Velbel, 1980].

Finally, given the pronounced age and compositional contrasts between petrofacies I and II, and our lack of knowledge of the early structural history of the region, structural shuffling by strike-slip faulting must also remain a possible additional complication in the history of these rocks.

Conclusions

1. The Trinity Peninsula Group in northern Graham Land contains at least two major sandstone suites identified by their sandstone detrital modes (petrofacies I and II).

2. Each suite can be further subdivided to yield a total of four petrofacies (Ia, Ib, IIa, and IIb), three of which correspond with previously identified stratigraphical units (Legoupil, Hope Bay, and View Point formations).

3. Approximate geographical limits can be placed on each petrofacies for the first time, although the nature of the boundaries (stratigraphical or structural) is unknown.

4. Petrofacies I and II had disparate provenances, but they share many features which indicate that the provenances contained coarse-grained plutonic and high-rank metamorphic rocks, responsible for the volumetrically predominant quartzofeldspathic detritus, and a few mainly andesite-rhyolite volcanoes from which most of the lithic clasts were derived. Quartzose sedimentary rocks also formed a significant part of the provenance of petrofacies I.

5. For petrofacies II, a deeply dissected magmatic arc provenance is indicated, formed of unroofed batholiths, their metamorphic envelope, and isolated volcanoes. The arc may have been active, and deposition took place in a forearc (trench or upper-slope) basin or series of basins.

6. Interpretation of the depositional setting of petrofacies I is ambiguous and is not fully constrained by any information currently available. At least three principal models can be proposed and indicate that deposition could have occurred at either an active or a passive margin.

7. The striking age and compositional contrasts between petrofacies I and II suggest that some structural transposition by strike-slip faulting cannot be discounted.

References

Aitkenhead, N., The geology of the Duse Bay-Larsen Inlet area, northeast Graham Land (with particular reference to the Trinity Peninsula Series), Br. Antarct. Surv. Sci. Rep., 51, 62 pp., 1975.

Basu A., Petrology of Holocene fluvial sand derived from plutonic source rocks: Implications to paleoclimatic interpretation, J. Sediment. Petrol., 46, 694-709, 1976.

Basu, A., S. W. Young, L. J. Suttner, W. C. James, and G. H. Mack, Re-evaluation of the use of undulatory extinction and polycrystallinity in detrital quartz for provenance interpretation, J. Sediment. Petrol., 45, 873-882, 1975.

Burn, R. W., The geology of the LeMay Group, Alexander Island, Br. Antarct. Surv. Sci. Rep., 109, 64 pp., 1984.

Crook, K. A. W., Lithogenesis and geotectonics; the significance of compositional variation in flysch arenites (graywackes), in Modern and Ancient Geosynclinal Sedimentation, Spec. Publ. 19, edited by R. H. Shaver, pp. 304-310, Society of Economic Paleontologists and Mineralogists, Tulsa, Okla., 1974.

Dalziel, I. W. D., and D. H. Elliot, The Scotia Arc and Antarctic margin, in The Ocean Basins and Their Margins, vol. 1, The South Atlantic, edited by F. G. Stehli and A. E. M. Nairn, pp. 171-246, Plenum, New York, 1973.

Dickinson, W.R., Compositions of sandstones in circum-Pacific subduction complexes and forearc basins, Am. Assoc. Pet. Geol. Bull., 66, 121-137, 1982.

Dickinson, W. R., and C. A. Suczek, Plate tectonics and sandstone compositions, Am. Assoc. Pet. Geol. Bull., 63, 2164-2182, 1979.

Dickinson, W. R., R. V. Ingersoll, D. S. Cowan, K. P. Helmold, and C. A. Suczek, Provenance of Franciscan graywackes in coastal California, Geol. Soc. Am. Bull., 93, 95-107, 1982.

Elliot, D. H., Geology of northwest Trinity Peninsula, Graham Land, Br. Antarct. Surv. Bull., 7, 1-24, 1965.

Floyd, J. D., Stratigraphy of a flysch succession: The Ordovician of W Nithsdale, SW Scotland, Trans. R. Soc. Edinburgh, 73, 1-9, 1982.

Franzinelli, E., and P. E. Potter, Petrology, chemistry, and texture of modern river sands, Amazon River system, J. Geol., 91, 23-39, 1983.

Halpern, M., The geology of the General Bernardo O'Higgins area, northwest Antarctic Peninsula, in Geology and Paleontology of the Antarctic, edited by J. B. Hadley, pp. 177-209, AGU, Washington, D.C., 1965.

Houghton, H. F., Refined techniques for staining plagioclase and alkali feldspars in thin section, J. Sediment. Petrol., 50, 629-631, 1980.

Hyden, G., and P. W. G. Tanner, Late Palaeozoic-early Mesozoic forearc basin sedimentary rocks at the Pacific margin in western Antarctica, Geol. Rundsch., 70, 529-41, 1981.

Ingersoll, R. V., Petrofacies and petrologic evolution of the Late Cretaceous forearc basin, northern and central California, J. Geol., 86, 335-352, 1978.

Ingersoll, R. V., Petrofacies and provenance of late Mesozoic forearc basin, northern and cenral California, Am. Assoc. Pet. Geol. Bull., 67, 1125-1142, 1983.

Ingersoll, R. V., and C. A. Suczek, Petrology and provenance of Neogene sand from Nicobar and Bengal fans, Deep Sea Drilling Project sites 211 and 218, J. Sediment. Petrol., 49, 1217-1228, 1979.

Ingersoll, R. V., T. F. Bullard, R. L. Ford, J. P. Grimm, J. D. Pickle, and S. W. Sares, The effect of grain size on detrital modes: A test of the Gazzi-Dickinson point-counting method, J. Sediment. Petrol., 54, 103-116, 1984.

Middleton, G. V., Albite of secondary origin in Charny sandstones, Quebec, J. Sediment. Petrol., 42, 341-349, 1972.

Moore, G. F., Petrography of subduction zone sandstones from Nias Island, Indonesia, J. Sediment. Petrol., 49, 71-84, 1979.

Pankhurst, R. J., Rb-Sr constraints on the ages of basement rocks of the Antarctic Peninsula, in Antarctic Earth Science, edited by R. L. Oliver, P. R. James, and J. B. Jago, pp. 367-371, Australian Academy of Science, Canberra, 1983.

Potter, P. E., Petrology and chemistry of modern big river sands, J. Geol., 86, 423-449, 1978.

Smellie, J. L., A complete arc-trench system recognized in Gondwana sequences of the Antarctic Peninsula region, Geol. Mag., 118, 139-159, 1981.

Storey, B. C., and S. W. Garrett, Crustal growth of the Antarctic Peninsula by accretion, magmatism, and extension, Geol. Mag., 122, 5-14, 1985.

Suttner, L. J., A. Basu, and G. H. Mack, Climate and the origin of quartz arenites, J. Sediment. Petrol., 51, 1235-1246, 1981.

Thomson, M. R. A., New palaeontological and lithological observations on the Legoupil Formation, northwest Antarctic Peninsula, Br. Antarct. Surv. Bull., 41/42, 169-185, 1975.

Valloni, R., and J. B. Maynard, Detrital modes of recent deep-sea sands and their relation to tectonic setting: A first approximation, Sedimentology, 28, 75-83, 1981.

Velbel, M. A., Petrography of subduction zone sandstones: A discussion, J. Sediment. Petrol., 50, 303-304, 1980.

STRUCTURAL EVOLUTION OF THE MAGMATIC ARC IN NORTHERN PALMER LAND, ANTARCTIC PENINSULA

A. W. Meneilly, S. M. Harrison, B. A. Piercy, and B. C. Storey

British Antarctic Survey, Natural Environment Research Council, High Cross, Cambridge
CB3 OET, United Kingdom

Abstract. In northern Palmer Land, Antarctic Peninsula, the magmatic arc has developed by the interaction of compressional and extensional tectonics. Periods of magmatism and arc compression in the Early Jurassic and Early to middle Cretaceous separate periods of arc and backarc extension in the Middle to Late Jurassic and Tertiary. Sheeted migmatite complexes and orthogneiss are mostly Jurassic arc plutons emplaced into older paragneiss. Heterogeneous regional deformation produced major shear zones, and the orientation of shear zones, foliation, and banding describe a fan diverging upward from the center of the arc. Mafic dikes and steep extensional shear zones throughout northern Palmer Land and thick amygdaloidal basalt sheets in northeastern Palmer Land are related to Jurassic extension of the arc and formation of a backarc basin filled by thick clastic sediments (Mount Hill and Latady formations). The arc and backarc were intensely deformed during the Cretaceous. Penetrative deformation of the backarc basin sedimentary rocks produced east verging folds; whereas the arc was deformed by discrete east directed thrust zones as much as 1 km thick. Previously, the metamorphic rocks in northern Palmer Land were considered as part of a pre-late Paleozoic basement or as the roots of the Gondwanian orogen. Most of them are now shown to be deformed plutons of the Mesozoic magmatic arc. The deformation on the west side of the peninsula may be due to trench-directed thrusting; whereas the deformation on the east side may be the result of backarc basin closure.

Introduction

Subduction of Pacific and proto-Pacific ocean crust beneath the Antarctic Peninsula margin of Gondwanaland has taken place since at least early Mesozoic time [Smellie, 1981; Dalziel, 1982; Storey and Garrett, 1985]. Subduction has resulted in accretion to the Antarctic Peninsula continental fragment, both before and after it separated from the East Antarctic craton during Jurassic or Early Cretaceous time to form the Weddell Sea basin [Barker and Griffiths, 1977; La Brecque and Barker, 1981]. Recent studies in geochronology [Pankhurst, 1982] suggest that subduction has been virtually continuous during this time, and the distinction between Gondwanian and Andean orogenic events has become blurred [Storey and Garrett, 1985]. Models for the evolution of the magmatic arc on the Antarctic Peninsula have been based mainly on the geology of its northern part (Graham Land). Until recently, little was known about the deformed gneiss and metasedimentary rocks of northern Palmer Land (Figure 1) and whether they were part of the ancient cratonic basement or were formed during subduction.

This paper presents new data on the field relations, structure, and age of the rocks in northern Palmer Land and discusses their relations to the tectonic events associated with evolution of the magmatic arc and breakup of Gondwanaland. Particular attention is given to major shear zones and faults in eastern Palmer Land revealed by recent mapping [Meneilly, 1983]. The magnitude, sense of displacement, and strain are known from some of these fault zones and give a better understanding of the kinematics of arc evolution. The paper is based on a full analysis of eastern Palmer Land field data by one author (A. W. M., 1980-1981, 1982-1983) and only a preliminary assessment of more recent western Palmer Land field data by the authors (B. A. P. and S. M. H., 1983-1984, 1984-1985).

Rock Units

The rocks in northern Palmer Land are divided into four main units: (1) paragneiss, some of which may be pre-Mesozoic continental crust, (2) plutonic rocks of the magmatic arc, many of them deformed, ranging in age from Early Jurassic to at least Late Cretaceous, (3) backarc basin metasedimentary rocks, and (4) volcanic rocks of the Antarctic Peninsula Volcanic Group (APVG) [Thomson, 1982]. The age ranges of these tectonostratigraphic units are summarized in Figure 2 and the field relations are illustrated schematically in Figure 3.

Paragneiss. Scattered exposures of paragneiss are most common in northwest Palmer Land and occur locally in northeast Palmer Land. The rocks are garnetiferous, quartzofeldspathic, or mafic biotite gneiss containing sillimanite and andalusite, and migmatitic paragneiss with granitic segregations and veins. At Auriga Nunataks (Figure 1) a 250-m-thick marble layer is present, affirming a sedimentary origin at this locality.

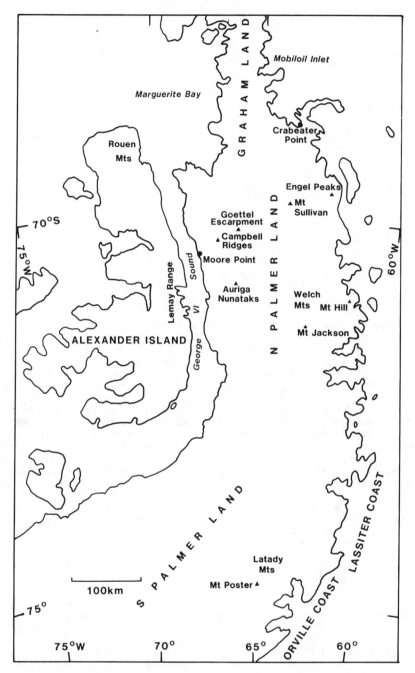

Fig. 1. Location map of Palmer Land.

The age of the paragneiss is uncertain. At Auriga Nunataks (Figure 1) it is cut by orthogneiss and Jurassic amphibolite dikes (see below), and near Mount Jackson (Figure 1) the gneiss appears to result from contact metamorphism by (?)Early Jurassic orthogneiss. Thus, some of the paragneiss in northern Palmer Land is at least as old as Early Jurassic.

The paragneiss could be part of an Early Jurassic or older subduction complex formed by accretion at deep levels and subsequently intruded by arc plutons. This possibility is supported by the presence of the thick marble layer, a lithology common in parts of the Scotia metamorphic complex, a Mesozoic accretionary complex in the southern Scotia arc [Tanner et al., 1982; Storey and Meneilly, 1985]. The only other significant exposure of marble noted from the Antarctic Peninsula is a small outcrop in eastern Graham Land speculatively assigned to the (?)Carboniferous to Triassic Trinity Peninsula Group [Fleet, 1965]. Alternatively, the paragneiss and marble may not be re-

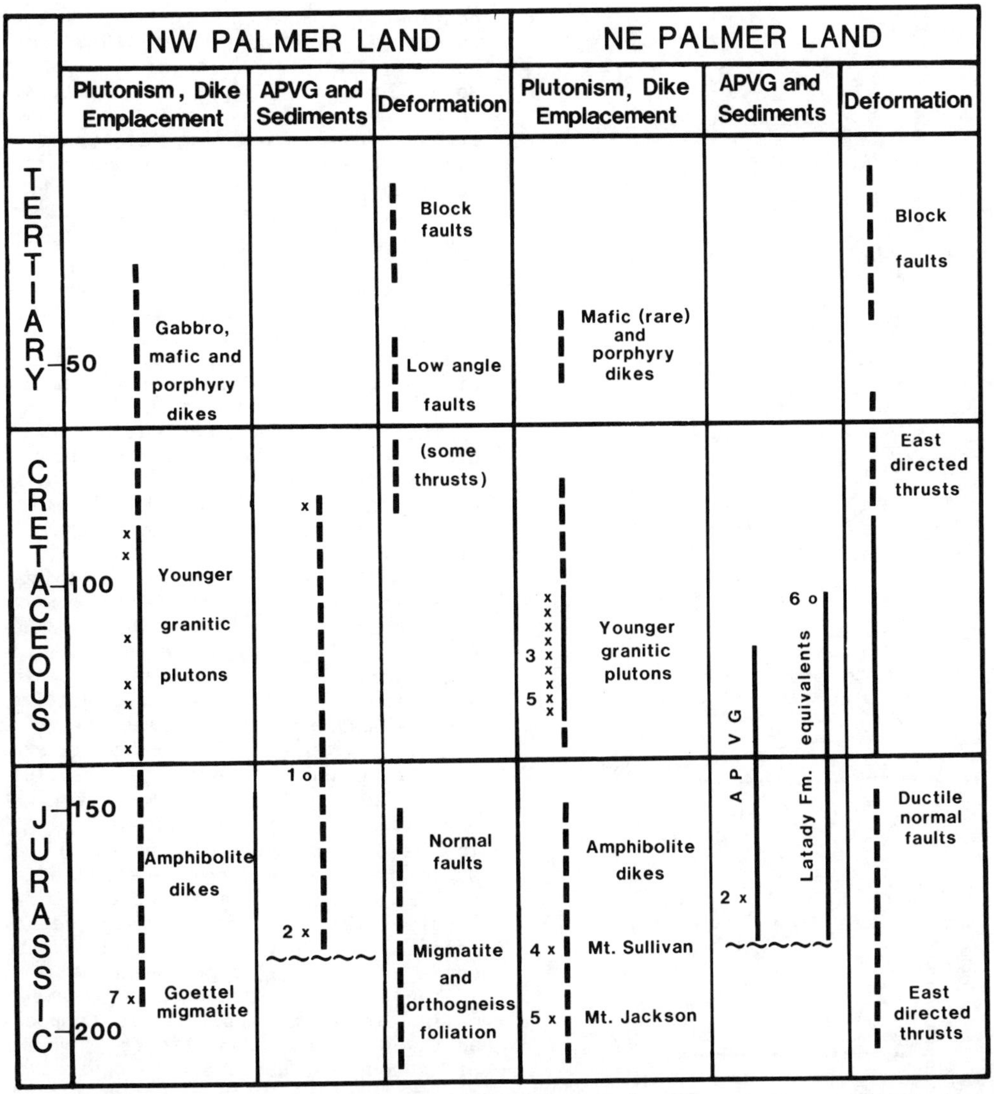

	NW PALMER LAND			NE PALMER LAND		
	Plutonism, Dike Emplacement	APVG and Sediments	Deformation	Plutonism, Dike Emplacement	APVG and Sediments	Deformation
TERTIARY (−50)	Gabbro, mafic and porphyry dikes		Block faults; Low angle faults	Mafic (rare) and porphyry dikes		Block faults
CRETACEOUS (−100)	Younger granitic plutons (x x x x x x)	x	(some thrusts)	Younger granitic plutons (3 x x x x x 5 x x)	6 o (APVG equivalents)	East directed thrusts
JURASSIC (−150 to −200)	Amphibolite dikes; 7 x Goettel migmatite	1 o; 2 x	Normal faults; Migmatite and orthogneiss foliation	Amphibolite dikes; 4 x Mt. Sullivan; 5 x Mt. Jackson	2 x Latady Fm.	Ductile normal faults; East directed thrusts

Fig. 2. Timing of geological events in Palmer Land. Data are from British Antarctic Survey [1982] and (1) Thomson, 1975; (2) Thomson and Pankhurst, 1983; (3) Rowley et al., 1983; (4) Pankhurst, 1983; (5) Meneilly, Millar, and Pankhurst, unpublished data, 1985; (6) Crame, 1985. (7) Piercy, Miller, and Pankhurst, unpublished data, 1985. x denotes radiometric age determination, and o denotes fossil age. Solid lines indicate probable extent of activity, and broken lines indicate inferred activity. Wavy lines are possible stratigraphic positions of unconformities.

lated to subduction but may be part of much older continental crust on which the magmatic arc was founded. Within the Transantarctic Mountains, the Nelson Limestone of the Pensacola Mountains contains Middle Cambrian fossils [Schmidt et al., 1965], and a similar age for the Palmer Land marble would be consistent with its interpretation as a continental basement to the arc. Another possibility is that some of the paragneiss is Early Jurassic or older volcanic rocks forming the oldest part of the APVG.

Plutonic rocks. About 60% of the exposed rock in northern Palmer Land consists of orthogneiss, migmatitic orthogneiss, and plutons of the magmatic arc. There is a wide range in composition and in degree of deformation of plutonic rocks ranging in age from at least Early Jurassic to Late Cretaceous or even Cenozoic. To date, there are no consistent field, petrographic, or geochemical criteria to distinguish different plutonic events and to categorize every exposure of plutonic rock in the area. The history of the magmatic arc presented here (Figure 2) is based on the field relations at several key exposures and on, as yet, sparse radiometric ages.

The highly deformed nature of many of the plutonic rocks has prompted some workers to suggest that the orthogneiss represents the crystalline

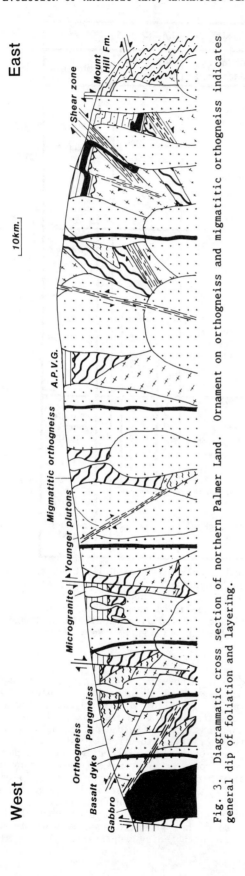

Fig. 3. Diagrammatic cross section of northern Palmer Land. Ornament on orthogneiss and migmatitic orthogneiss indicates general dip of foliation and layering.

continental basement to the Carboniferous and younger magmatic arc [Adie, 1954; Smellie, 1981]. However, Early Jurassic ages of orthogneiss from Mount Sullivan (Figure 1) [Pankhurst, 1983] and Mount Jackson (see Figure 1 for reference) and from sheeted migmatite at Goettel Escarpment (see Figure 1 for reference) suggest that many are deformed early Mesozoic arc plutonic rocks. This well-developed episode of emplacement of plutons and sheeted migmatite in the Early to Middle Jurassic is correlated with the plutonic event of the same age recognized in eastern Graham Land by Pankhurst [1982]. However, Paleozoic orthogneiss has been recognized in eastern Graham Land [Pankhurst, 1983], and the possibility that some of the gneiss in northern Palmer Land is of a similar age is not ruled out.

The orthogneiss and migmatite are cut by mafic dikes which are themselves deformed before intrusion of Cretaceous plutons. At two localities in northwest Palmer Land, orthogneiss is unconformably overlain by undeformed volcanic rocks of the APVG.

A second major phase of plutonism, during the Early to middle Cretaceous, is also widespread in both northern and southern Palmer Land [Farrar and Rowley, 1980]. In northern Palmer Land, mainly granodioritic and granophyric plutons cut the paragneiss, orthogneiss and migmatite, the APVG, and the backarc basin sedimentary rocks. Gray and pink microgranite and porphyry dikes may represent the last phase of the younger granitoid plutonism.

The latest phase of igneous activity in northern Palmer Land is the intrusion of gabbro plutons in the extreme west (Moore Point) and of related mafic dikes in western Palmer Land. These late basalt dikes are rare on the east side of the peninsula. The dikes occur in two sets: a major set trending north-south, parallel to the peninsula, and a minor one at right angles to this.

The sequence of pluton emplacement and deformation is well displayed at Campbell Ridges in northwest Palmer Land (Figure 1) where five phases of igneous intrusion are recognized. Coarse biotite-orthogneiss with a well-developed but irregularly oriented foliation is cut by more massive granodiorite and diorite. This is followed by a dense suite of mafic and intermediate dikes, striking 60°-120°, as much as 10 m in thickness and associated with normal faulting. The granodiorite, diorite, and dikes were partially recrystallized (amphibolite-grade metamorphism) but with no regional deformation. Undeformed granodiorite plutons, petrographically very similar to the widespread Cretaceous plutons found throughout Palmer Land, are the next phase of intrusion, and are, in turn, cut by late basalt dikes. These latter dikes are less dense than the earlier suite and rarely exceed 3 m in thickness.

Migmatitic orthogneiss is exposed mainly along the western flank of the central spine of northern Palmer Land. At Goettel Escarpment (Figure 1), gabbro and diorite are intimately veined by granodiorite, which has given a rubidium-strontium (Rb-Sr) whole-rock date of approximately 200 Ma (B. A. Piercy, I. L. Millar, and R. J. Pankhurst, unpublished data, 1985). In places, deformation has produced a gently east dipping gneissosity and rotated and folded granitic veins to form a banded, foliated migmatitic gneiss with relict pods of gabbro. At other exposures of migmatite in northwest Palmer, undeformed granitoid sheets, as much as 200 m thick, intrude mafic plutonic rock and dip consistently eastward. This regional easterly dip of sheeting and foliation suggests a regional tectonic control on emplacement of the migmatite which was maintained during subsequent deformation. The origin of the mafic plutonic rock into which the granitoids were emplaced is uncertain, but it is probably the less differentiated component of the early Mesozoic arc plutonism.

The ages of pluton intrusion and deformation are constrained at only one locality in eastern Palmer Land. One kilometer northeast of Mount Jackson (Figure 1), a plagioclase and K-feldspar porphyroblastic granite, cut by mafic dikes, yields a Rb-Sr whole-rock date of magmatic crystallization of 199 Ma (A. W. Meneilly, I. L. Millar, and R. J. Pankhurst, unpublished data, 1985). Granite and mafic dikes are deformed and then cut by Early Cretaceous granodiorite [Singleton, 1980].

Backarc basin metasedimentary rocks. Singleton [1980] described a sequence of unfossiliferous metapelitic rocks from northeast Palmer Land (Mount Hill Formation) and correlated them on lithological grounds with the Middle to Upper Jurassic backarc sedimentary rocks of the Latady Formation in southeast Palmer Land [Laudon et al., 1983]. Belemnite and plant fragments collected by one of us (A. W. M.) from Mount Hill support this correlation. At Mount Hill, meta-siltstone and mudstone have a steep west-northwest slaty cleavage, axial-planar to steeply west plunging folds. Mineral stretching lineations pitch at around 90° on cleavage. The Mount Hill Formation is intruded and hornfelsed by Cretaceous plutons [Singleton, 1980]. Andalusite porphyroblasts overprint the slaty cleavage but predate a crenulation cleavage. In northeast Palmer Land, deformed Lower Cretaceous sedimentary rocks at Crabeater Point [Thomson, 1967] may be Albian in age [Crame, 1985]. Various sedimentary rocks in the Mobiloil Inlet area were considered to be correlatives of the (?)upper Paleozoic Trinity Peninsula Group by Fraser and Grimley [1972]. However, it is more likely that they, and the Crabeater Point sedimentary rocks, are lateral equivalents of the Mesozoic arc and backarc basin sedimentary rocks as they are interbedded with volcanic rocks and amygdaloidal basalt, similar to rocks found within the APVG on the eastern flank of the magmatic arc in northern Palmer Land (see below).

Antarctic Peninsula Volcanic Group. Calc-alkaline volcanic rocks of Early Jurassic (possibly even Triassic) to early Tertiary age are widespread on the Antarctic Peninsula [Thomson and Pankhurst, 1983]. Middle to Late Jurassic representatives of the APVG in southern Palmer Land have been named the Mount Poster Formation by Rowley et al. [1982]. A K-Ar whole-rock age of 88 Ma [Rex, 1976] and a general trend of about 175 Ma for most of the northern Palmer Land volcanic rocks [Thomson and Pankhurst, 1983] suggest that the APVG in the area is mainly Early-Middle Jurassic but ranges to at least Early Cretaceous. The volcanic rocks unconformably overlie the orthogneiss and are intruded by Cretaceous plutons. The volcanic rocks are undeformed in western Palmer Land but must predate east dipping shear zones which cut Cretaceous plutons. In contrast, on the eastern side of Palmer Land the APVG is commonly weakly schistose and deformed by the west dipping shear zones which cut the Cretaceous plutons. Another striking difference between the two sides of Palmer Land is the presence within the volcanic rocks in northeast Palmer Land of isolated, but very thick (as much as 1 km) sheets of amygdaloidal basalt; it is uncertain whether these are composite lava flows, dikes, or sills. To the south of Mobiloil Inlet there are thick sequences of amygdaloidal lava flows [Fraser and Grimley, 1972].

Metavolcanic rocks around Mobiloil Inlet and in northeast Palmer Land have previously been considered to represent the frontal edge of a (?) Carboniferous-Triassic volcanic arc [Smellie, 1981]. It is more likely that they are part of the Jurassic APVG and that their metamorphism and deformation is of Cretaceous age [Thomson and Pankhurst, 1983]. Recent fieldwork (D. I. M. Macdonald, personal communication, 1985) has indicated that a thick sequence of volcanic rocks is interbedded with sedimentary rocks of Albian strata at Crabeater Point in northeast Palmer Land.

Deformation in the Magmatic Arc: Major Shear Zones and Faults

Deformation affected a large part of the plutonic rocks and is characterized by heterogeneous shear zones and retrogression to the greenschist facies. Although the timing of deformation events is not well constrained, three distinct phases can be recognized.

The first event recorded in the orthogneiss and migmatite produced a regional foliation which diverges across the arc (Figure 3), dipping to the east on the west side and to the west on the east side of northern Palmer Land. The foliation at some localities has a consistent dip and is penetrative over large areas, but generally this regional deformation is very heterogeneous and is concentrated into major shear zones. Shear sense can be recognized in northeast Palmer Land and indicates east directed thrusting. The next event involved steep shear zones with normal fault displacement and downthrow toward the center of the arc. Finally, at several localities the fabric in these ductile normal faults is cut by dextral shear zones. Mafic dikes cut the early regional foliation but are deformed in the ductile normal faults and dextral shear zones.

Although the Cretaceous plutons lack a regional foliation, many have deformed, banded, and migmatitic margins and are cut by brittle and ductile thrust faults. In northeast Palmer Land the thrusts are east directed, and in northwest Palmer Land they are west directed. Deformed Cretaceous plutons can be impossible to distinguish from

Jurassic plutons, and large areas of gneissic granite in northern Palmer Land are of uncertain age.

The relations between several phases of magmatic intrusion and deformation are illustrated by the following examples. (1) The Early Jurassic granite at Mount Jackson (Figure 1) has a variably developed gneissosity dipping moderately to the west, and shear sense criteria in mylonite zones indicate that this fabric is part of an east directed ductile thrust. Within the mylonite zones, the granite has suffered retrogression and deformation to a porphyroclastic biotite-muscovite gneiss with recrystallized quartz ribbons. In places the gneissosity is cut by narrow (<4 m) steep shear zones with a normal fault displacement and downthrow to the west. (2) In the eastern Welch Mountains (Figure 1) a foliated granodiorite of unknown age has good textural evidence indicating that the west dipping foliation developed during east directed ductile thrusting: C and S planes were recognized where the C planes are parallel to the shear direction and the S planes are parallel to the XY plane of the strain ellipsoid [Berthé et al., 1979]. Farther west, granodiorite is cut by a steep north striking ductile shear zone with normal fault displacement and downthrow to the west. K-feldspar porphyroblast growth occurred within the shear zone predating the main deformation which involved greenschist facies retrogression to micaceous augengneiss and banded mica-schist. The steep west dipping foliation, with stretching lineation pitching at right angles, is rotated clockwise, folded about steeply plunging axes, cut by minor dextral shear zones, and the stretching lineation rotated to plunge moderately northwest. These structures can be explained by superimposing major dextral shear on the west dipping foliation. (3) The Engel Peaks granophyre in northeast Palmer Land (Figure 1) was emplaced into volcanic rocks of the APVG at about 113 Ma (A. W. Meneilly, I. L. Millar, and R. J. Pankhurst, unpublished data, 1985). This pluton (formerly thought to be early Paleozoic because of a supposed unconformity with Carboniferous sedimentary rocks [Davies, 1984]) is cut by a steep north striking shear zone at least 1 km thick [Meneilly, 1983]. Kinematic analysis indicates that the displacement on the shear zone is that of an east directed thrust. Both brittle and ductile deformation mechanisms and retrogression reduced the granophyre to muscovite-epidote mylonite. Amygdaloidal basalt within the volcanic rocks is involved in the shearing, and strain analysis on the deformed amygdales gives a minimum estimate of 3.5 for the shear strain. Parallelism of long axes of amygdales and approximately constant axial ratios on each of the three principal planes of the strain ellipsoid indicates all amygdales had a very similar initial shape fabric. Assuming initial sphericity, arithmetic means of the small range of axial ratios were used as an estimate for the strain ratio. Shear strain was calculated by assuming simple shear and using the graphical method of Ramsay [1967, figure 3.21]. This gives a displacement of at least 3.5 km, and in the steep shear zone, a relative vertical movement of about 3.2 km. Since the deformation in the basalt appears much less intense than in the granite, the actual displacement may be much greater.

Arc Evolution in Northern Palmer Land

The maximum known age ranges of magmatism, sedimentation, and deformation are summarized in Figure 2. It is clear that these three processes were not mutually exclusive during the history of arc evolution. In Early Cretaceous time for instance, sedimentation or volcanism took place throughout northern Palmer Land; while in southeast Palmer Land, the Latady and Mount Poster formations were being deformed and in western Palmer Land granitic plutonism was active. Extensive radiometric dating in Graham Land has revealed a pattern of continuous plutonism and volcanism from Late Triassic to Tertiary with pulses of activity during the Early Jurassic, Early Cretaceous, Late Cretaceous, and Tertiary [Pankhurst, 1982]. Although similar age control is not available for Palmer Land, the preliminary data suggest that here too, magmatism was continuous from at least the Early Jurassic until middle Cretaceous and spanned periods of backarc sedimentation and of deformation across the arc. In southern Chile, Hervé et al. [1984] recognized plutons intruded at the site of the marginal basin and possibly while it was being filled by turbidites.

Many attempts have been made recently to relate the tectonic and lithological elements of the Antarctic Peninsula to processes of Pacific plate subduction and Gondwanaland breakup drawing mainly on field relations and data from Graham Land [Suárez, 1976; Smellie, 1981; Barker, 1982; Dalziel, 1982, 1983; Elliot, 1983; Thomson et al., 1983; Storey and Garrett, 1985]. The following tectonic events are thought to have played a major role in controlling the evolution of Palmer Land: (1) continuous subduction from the Triassic until the early Tertiary; (2) breakup of Gondwanaland which commenced in the Early Jurassic; (3) extension and backarc basin formation in the Middle to Late Jurassic; (4) arc-continent compression and closure of the backarc basin in the Early to middle Cretaceous; and (5) Cenozoic ridge crest-trench collision and cessation of subduction.

The evolution of northern Palmer Land in periods marked by one or more of these major events is discussed and illustrated by schematic cross sections of the arc in Figure 4. In the absence of good paleomagnetic control, the collision and accretion of exotic continental fragments are not discussed. However, they are not ruled out as possible mechanisms for deformation within the magmatic arc and forearc and backarc sequences within this part of the proto-Pacific margin. The interpretation of the tectonic history of western Palmer Land is based mainly on field data and may be modified by subsequent laboratory studies.

Basement to the magmatic arc. The paragneiss is stratigraphically the oldest rock unit in Palmer Land. Its absolute age is unknown but it could be part of the continental basement into which the arc plutonic rocks were emplaced or from which components of the plutonic rock were derived by partial melting. Similarly, some of the orthogneiss in northern Palmer Land may be as old as the Paleozoic orthogneiss of eastern Graham Land [Pankhurst, 1983] and may also be part of the pre-Mesozoic continental crust.

(?)Triassic-Middle Jurassic: Accretion and early arc plutonism. Subduction of Pacific and proto-Pacific Ocean crust beneath the Gondwanaland craton may have begun in the Triassic. There is no evidence at present within the Antarctic Peninsula of subduction prior to this [Storey and Garrett, 1985]. The earliest record of subduction in northern Palmer Land may be marked by the Early-Middle Jurassic arc plutonism (Figure 4a). Granitic intrusion in eastern Palmer Land produced discrete plutons, while in western Palmer Land gabbro and diorite was invaded by sheets and veins of granitic material, possibly at a higher structural level. A heterogeneous regional foliation cuts the early arc plutons and dips toward the center of the arc (Figure 4a).

The Early to Middle Jurassic plutonism in eastern Palmer Land, and also in eastern Graham Land [Pankhurst, 1983], coincides with the uplift and erosion of the magmatic arc over the whole of West Antarctica and the southern Andes that may be related to the initial stages of breakup of Gondwanaland [Dalziel, 1982]. Some of the granitic plutons with high initial $^{87}Sr/^{86}Sr$ ratios (A. W. Meneilly, I. L. Millar, R. J. Pankhurst, and B. A. Piercy, unpublished data, 1985) may be formed entirely by partial melting of the crust; they may not be subduction related but may be due to high geothermal gradients during the breakup of Gondwanaland.

Middle Jurassic-earliest Cretaceous: Arc and backarc extension. Sediments of the Mount Hill Formation and sediments around Mobiloil Inlet were deposited in backarc basins (Figure 4b) which extended northward into eastern Graham Land [Farquharson et al., 1984] and southward into the Lassiter and Orville coasts (Latady Formation), [Rowley et al., 1983]. These form part of a series of backarc basins present along the Pacific margin of South America and Antarctica [Súarez, 1976] during the Jurassic and Cretaceous. There is considerable variation in the timing of development and filling of these basins. Opening of the basins in the southern Antarctic Peninsula began during the early stages of the breakup of Gondwanaland (Early to Middle Jurassic). They may represent a response to this regional lithospheric extension or may be directly related to plate margin processes. The basins in eastern Graham Land were filled during the Cretaceous and Tertiary [Farquharson et al., 1984].

The thick amygdaloidal basalt sheets in eastern Graham Land may be products of continental rifting during backarc extension or during breakup of Gondwanaland. A similar origin has been suggested for the Jurassic dolerite sheets of the Ferrar Supergroup of the Transantarctic Mountains by Elliot [1975]. That basalt is absent from the APVG in southern Palmer Land [Rowley et al., 1982] suggests that crustal extension was less marked here than in northern Palmer Land. Thick quartz-dolerite sills were also emplaced into the backarc sedimentary rocks in the southern Andes before middle Cretaceous deformation [Katz and Watters, 1966]. Basalt dikes and sills within Jurassic arc plutons in northern Palmer Land strike parallel to the peninsula and indicate widespread extension of the arc as well as the backarc at this time. It is possible that some of the Jurassic granitoid plutons and some of the granitic sheets within the migmatites were emplaced during this arc extension. The ductile normal faults with downthrows toward the center of the arc (Figure 4b) affecting Jurassic plutons and mafic dikes are probably due to the same arc extension. Dikes in eastern Graham Land are also Middle Jurassic in age [Pankhurst, 1982]. In the southern Andes, Hervé et al. [1984] decribed a mafic dike swarm parallel to the regional tectonic strike and cutting arc plutonic rocks prior to backarc deformation.

Early-middle Cretaceous: Arc-continent compression and backarc closure. The magmatic arc and backarc basin sedimentary rocks in Palmer Land were intensely deformed during the Early and middle Cretaceous, contemporaneous with magmatism and sedimentation (Figure 4c). In southeast Palmer Land the backarc basin sedimentary rocks of the Latady and Mount Hill formations were deformed in the Early Cretaceous [Rowley et al., 1983]. In northeast Palmer Land, deformed sedimentary rocks are probably as young as Albian, and deformation of one part of the backarc basin had started while sedimentation was still taking place. Deformation lasted until at least Albian time. Plutonism continued during the Early Cretaceous compression, and ages from east Palmer Land overlap the period of probable deformation of backarc sedimentary rocks in northeast Palmer Land.

There is marked contrast in deformation styles between the arc and backarc regions. Penetrative deformation in the backarc region produced folds and thrusts in the Latady and Mount Poster formations verging eastward away from the arc [Rowley et al., 1983], and west plunging stretching lineations in the Mount Hill Formation are consistent with east directed transport. In contrast, the eastern margin of the arc was subjected to heterogeneous deformation, producing the major east directed ductile thrust zones described above. A similar contrast in deformation styles between homogeneous folding of the backarc sediments and inhomogeneous shearing of the arc batholith and basin floor has been noted in the southern Andes by Dalziel et al. [1974] and Hervé et al. [1984]. Backarc basin sedimentary rocks along the peninsula only show deformation in the southern part (Palmer Land). This may reflect the contrasting types of sedimentary basins within this backarc region. The basins in the northern part (Graham Land) may have developed as half grabens on one limb of the opening Weddell Sea, whereas the basins in Palmer Land may be ensialic. The latter were probably deformed by closure against continental crust in the south Weddell Sea whereas those in Graham Land were backed by oceanic crust and did not suffer compressional deformation [Elliot, 1983].

Middle Cretaceous: Plutonic maximum, thrusting, and uplift. A pulse of plutonism in Palmer Land,

A TRIASSIC - MID JURASSIC
Accretion and early arc plutonism

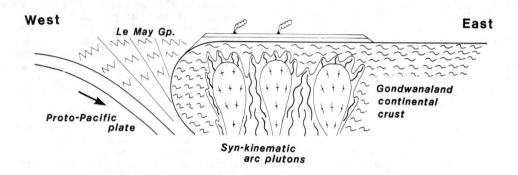

B MID JURASSIC - EARLIEST CRETACEOUS
Arc and back-arc extension

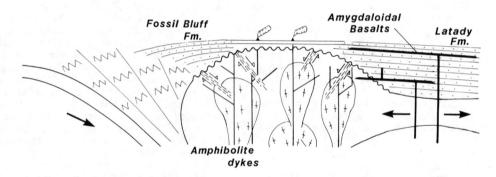

C EARLY - MID CRETACEOUS
Arc-continent compression and back-arc closure

Fig. 4. Schematic cross sections illustrating the evolution of the magmatic arc in northern Palmer Land. Ornament as in Figure 3. Bold arrows indicate direction of compression, extension, or subduction.

beginning in the Early Cretaceous and peaking in middle Cretaceous, could be attributed to a middle Cretaceous global increase in sea-floor spreading rates [Larson and Pitman, 1972]. Although these younger plutons are generally undeformed or have only minor, possibly intrusion-related marginal deformation, they are cut by shear zones throughout northern Palmer Land. These are particularly well developed in eastern Palmer Land where many are east directed thrusts similar to those in older plutons (Figure 4d). This east directed shearing, initiated during arc-continent compression and backarc deformation in the Early Cretaceous, may be continuous through middle Cretaceous or later and may span the plutonic peak. Thus, during the Early and middle Cretaceous, the con-

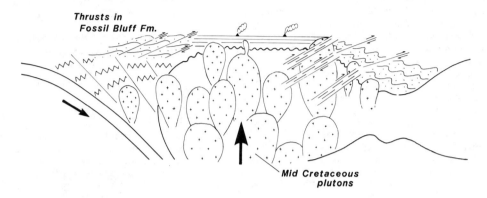

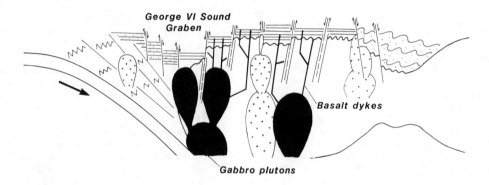

Fig. 4. (continued)

currence of increased spreading rate and arc-continent compression resulted in increased magmatism, compression, and uplift of the arc. The middle Cretaceous uplift is recorded in northern Graham Land by a planar unconformity in the back-arc basin stratigraphical record [Farquharson et al., 1984] and in southern South America by uplift and deformation of the Rocas Verdes marginal basin [Dalziel, 1986].

Post-Albian east directed thrusts are also present within the Late Jurassic-Early Cretaceous forearc basin sedimentary rocks of the Fossil Bluff Formation [Horne, 1967]. The significance of these east directed thrusts is uncertain at present; arc-directed thrusts are common in forearc regions and may be due to thin-skinned back thrusting within the forearc region [Bally, 1975] or due to cessation of subduction and uplift of the accretionary prism. Alternatively, they may be part of the same tectonic event responsible for east directed thrusting within the arc and back-arc regions during the Cretaceous.

Late Cretaceous-Tertiary: Intra-arc extension. The last major tectonic event in Palmer Land was a (?)Tertiary intra-arc extensional regime (Figure 4e). The main features associated with this extensional period are formation of the George VI Sound graben, normal and block faulting, and emplacement of late gabbro plutons and mafic dikes. Block faulting is conspicuous when juxtaposing orthogneiss or migmatite with undeformed volcanic rocks, although fault traces can rarely be seen in the field. Storey and Garrett [1985] suggested that this extension was caused by slowing and cessation of subduction due to arrival of the ridge crest at the trench. Mafic dikes are very common in western Palmer Land and are mainly parallel or normal to George VI Sound. Similar dikes are rare in eastern Palmer Land. Subduction continued until at least 40 Ma with intrusion of the Rouen Mountains Batholith into the LeMay Group on Alexander Island [Care, 1983] and probably with the late microgranite and porphyry intrusions in Palmer Land. Absolute dating of shear zone rocks is lacking, and it is possible that thrusts in east and west Palmer Land are as young as Tertiary.

Conclusions

The geology of northern Palmer Land can be explained by continuous subduction and magmatism during the Mesozoic with peaks of magmatism and arc compression in the Early Jurassic and Early to middle Cretaceous, separating a period of arc and backarc extension in the Middle to Late Jurassic. The large areas of migmatite and orthogneiss exposed in northern Palmer Land are probably mostly Jurassic (with possibly some Cretaceous) syntectonic arc plutons. Confirmation of the presence of pre-Mesozoic continental basement awaits temperature-pressure studies on metamorphic assemblages in the paragneiss of western Palmer Land and a comprehensive radiometric dating survey. The widespread exposure of gneiss in northern Palmer Land contrasts with the absence of such rocks in southern Palmer Land [Rowley and Williams, 1982]. This is probably due to the higher structural level of exposure in the south where undeformed Cretaceous plutons, but no Jurassic plutons, are exposed.

The main phase of arc and backarc extension took place in the Middle to Late Jurassic. Arc-continent compression and closure of the backarc basin commenced in the earliest Cretaceous and initiated deformation and thrusting in eastern Palmer Land that lasted through much of the Cretaceous. Sedimentation, due to renewed arc uplift, continued during deformation, at least in the Mobiloil Inlet area, until Albian time. The middle Cretaceous plutonic maximum overlapped with backarc sedimentation and deformation.

During arc compression, the orientation of foliation, migmatite and orthogneiss banding, and shear zones fan across the arc, accommodated in the plutonic rocks by major divergent brittle and ductile thrusts. This fanning could be due to, or enhanced by, Pacific floor subduction on the west and possibly also by overriding of the thinned continental crust or mafic basin floor by the arc on the east of the peninsula.

The interaction of compressional and extensional tectonics within magmatic arcs which may be due to variation in subduction rates has also been documented within other parts of the proto-Pacific margin (e.g., Peruvian Andes), [Bussell, 1983; Pitcher et al., 1984]. However, within the Antarctic Peninsula it seems unlikely that the large amount of compressive shortening seen in the arc and backarc regions is due to subduction processes; most modern magmatic arcs display extensional features due to rise and spread of large volumes of magma. The amount of lithospheric shortening is more likely to be due to some form of collision tectonics perhaps during closure of the backarc basin and collision of the arc against the continental basement or due to docking of subduction complex rocks or continental fragments within the forearc region, or to ridge crest-trench collisions.

Acknowledgments. We thank our many colleagues in the British Antarctic Survey who assisted with the fieldwork and discussed the geology with us.

References

Adie, R. J., The petrology of Graham Land, 1, The basement complex; Early Paleozoic plutonic and volcanic rocks, Br. Antarct. Surv. Sci. Rep., 11, 22 pp., 1954.

Bally, A. W., A geodynamic scenario for hydrocarbon occurrences, Proc. World Pet. Congr., 9, 33-44, 1975.

Barker, P. F., The Cenozoic subduction history of the Pacific margin of the Antarctic Peninsula: Ridge crest-trench interaction, J. Geol. Soc. London, 139, 787-801, 1982.

Barker, P. F., and D. H. Griffiths, Towards a more certain reconstruction of Gondwanaland, Philos. Trans. R. Soc. London, Ser. B., 279, 143-159, 1977.

Berthé, D., P. Choukroune, and P. Jegouzo, Orthogneiss, mylonite, and noncoaxial deformation of granites; the example of the South American Shear Zone, J. Struct. Geol., 1, 31-42, 1979.

British Antarctic Survey, British Antarctic Territory geological map, sheet 5, 1:50,000, BAS 500G Ser., compiled by J. W. Thomson and J. S. Harris, British Antarctic Survey, Cambridge, 1982.

Bussell, M. A., Timing of tectonic and magmatic events in the central Andes of Peru, J. Geol. Soc. London, 140, 279-286, 1983.

Care, B. W., The petrology of the Rouen Mountains, northern Alexander Island, Br. Antarct. Surv. Bull., 52, 63-86, 1983.

Crame, J. A., Lower Cretaceous inoceramid bivalves from the Antarctic Peninsula region, Palaeontology, 28, 475-525, 1985.

Dalziel, I. W. D., The early (pre-Middle Jurassic) history of the Scotia Arc region: a review and progress report, in Antarctic Geoscience, edited by C. Craddock, pp. 111-126, University of Wisconsin Press, Madison, 1982.

Dalziel, I. W. D., The evolution of the Scotia Arc: a review, in Antarctic Earth Science, edited by R. L. Oliver, P. R. James, and J. B. Jago, pp. 283-288, Australian Academy of Science, Canberra, 1983.

Dalziel, I. W. D., Collision and cordilleran orogenesis: An Andean perspective, in Collision Tectonics, edited by J. G. Ramsay, M. P. Coward, and A. C. Ries, pp. 389-404, Blackwell Scientific, Oxford, London, 1986.

Dalziel, I. W. D., M. J. de Wit, and K. F. Palmer, Fossil marginal basin in the southern Andes, Nature, 250, 291-294, 1974.

Davies, T. G., The geology of part of northern Palmer Land, Br. Antarct. Surv. Sci Rep., 103, 46 pp., 1984.

Elliot, D. H., Tectonics of Antarctica: a review, Am. J. Sci., 275-A, 45-106, 1975.

Elliot, D. H., The mid-Mesozoic to mid-Cenozoic active plate margin of the Antarctic Peninsula, in Antarctic Earth Science, edited by R. L. Oliver, P. R. James, and J. B. Jago, pp. 347-351, Australian Academy of Science, Canberra, 1983.

Farquharson, G. W., R. D. Hamer, and J. R. Ineson, Proximal volcaniclastic sedimentation in a Cretaceous back-arc basin, northern Antarctic Peninsula, in Marginal Basin Geology, Spec. Publ. 16, edited by B. P. Kokelaar, and M. P. Howells, pp. 219-229, Blackwell Scientific, Oxford, London, 1984.

Farrar, E., and P. D. Rowley, Potassium-argon ages of Upper Cretaceous plutonic rocks of Orville Coast and eastern Ellsworth Land, Antarct. J. U. S., 15, 26-28, 1980.

Fleet, M., Metamorphosed limestone in the Trinity Peninsula Series of Graham Land, Br. Antarct. Surv. Bull., 7, 73-76, 1965.

Fraser, A. G., and P. H. Grimley, The geology of parts of the Bowman and Wilkins coasts, Antarctic Peninsula, Br. Antarct. Surv. Sci. Rep., 67, 59 pp., 1972.

Hervé, M., M. Súarez, and A. Puig, The Patagonian Batholith S. of Tierra del Fuego, Chile: timing and tectonic implications, J. Geol. Soc. London, 141, 877-884, 1984.

Horne, R. R., Structural geology of part of southeastern Alexander Island, Br. Antarct. Surv. Bull., 11, 1-22, 1967.

Katz, H. R., and W. A. Watters, Geological investigations of the Yaghan Formation and associated igneous rocks of Navarino Island, southern Chile, N. Z. J. Geol. Geophys., 9, 323-359, 1966.

La Brecque, J. L., and P. F. Barker, The age of the Weddell Basin, Nature, 290, 489-492, 1981.

Larson, R. L., and W. C. Pitmann III, World-wide correlation of Mesozoic magnetic anomalies, and its implications, Geol. Soc. Am. Bull., 83, 3645-3662, 1972.

Laudon, T. S., M. R. A. Thomson, P. L. Williams, K. L. Milliken, P. D. Rowley, and J. M. Boyles, The Jurassic Latady Formation, southern Antarctic Peninsula, in Antarctic Earth Science, edited by R. L. Oliver, P. R. James, and J. B. Jago, pp. 308-314, Australian Academy of Science, Canberra, 1983.

Meneilly, A. W., Deformation of granitic plutons in eastern Palmer Land, Br. Antarct. Surv. Bull., 61, 75-79, 1983.

Pankhurst, R.J., Rb-Sr geochronology of Graham Land, Antarctica, J. Geol. Soc. London, 139, 701-711, 1982.

Pankhurst, R.J., Rb-Sr constraints on the ages of basement rocks on the Antarctic Peninsula, in Antarctic Earth Science, edited by R. L. Oliver, P. R. James, and J. B. Jago, pp. 367-371, Australian Academy of Science, Canberra, 1983.

Pitcher, W. S., M. P. Atherton, E. J. Cobbing, and R. D. Beckinsale, (editors), Magmatism at a Plate Edge: The Peruvian Andes, 328 pp., Blackie, Glasgow, 1984.

Ramsay, J. G., Folding and Fracturing of Rocks, 568 pp., McGraw-Hill, New York, 1967.

Rex, D. C., Geochronology in relation to the stratigraphy of the Antarctic Peninsula, Br. Antarct. Surv. Bull., 43, 49-58, 1976.

Rowley, P. D., D. L. Schmidt, and P. L. Williams, Mount Poster Formation, southern Antarctic Peninsula and eastern Ellsworth Land, Antarct. J. U. S., 17, 38-39, 1982.

Rowley, P. D., W. R. Vennum, K. S. Kellogg, T. S. Laudon, P. E. Carrara, J. M. Boyles, and M. R. A. Thomson, Geology and plate tectonic setting of the Orville Coast and Eastern Ellsworth Land, Antarctica, in Antarctic Earth Science, edited by R. L. Oliver, P. R. James, and J. B. Jago, pp. 245-250, Australian Academy of Science, Canberra, 1983.

Rowley, R. D., and P. L. Williams, Geology of the northern Lassiter Coast and southern Black Coast, Antarctic Peninsula, in Antarctic Geoscience, edited by C. Craddock, pp. 339-348, University of Wisconsin Press, Madison, 1982.

Schmidt, D. L., P. L. Williams, W. H. Nelson, and J. R. Ege, Upper Precambrian and Paleozoic stratigraphy and structure of the Neptune Range, Antarctica, Geol. Surv. Prof. Pap. U. S., 525-D, 112-119, 1965.

Singleton, D. G., The geology of the central Black Coast, Palmer Land, Brit. Antarct. Surv. Sci Rep., 102, 50 pp., 1980.

Smellie, J. L., A complete arc-trench system recognised in Gondwana sequences of the Antarctic Peninsula region, Geol. Mag., 118, 139-159, 1981.

Storey, B. C., and S. W. Garrett, Crustal growth of the Antarctic Peninsula by accretion, magmatism, and extension, Geol. Mag., 122, 5-14, 1985.

Storey, B. C., and A. W. Meneilly, Petrogenesis of metamorphic rocks within a subduction-accretion terrane, Signy Island, South Orkney Islands, J. Metamorphic Geol., 3, 21-42, 1985.

Súarez, M., Plate tectonic model for southern Antarctic Peninsula and its relation to southern Andes, Geology, 4, 211-214, 1976.

Tanner, P. W. G., R. J. Pankhurst, and G. Hyden, Radiometric evidence for the age of the subduction complex in the South Orkney and South Shetland islands, West Antarctica, J. Geol. Soc. London, 683-690, 1982.

Thomson, M. R. A., A probable Cretaceous invertebrate fauna from Crabeater Point, Bowman Coast, Graham Land, Br. Antarct. Surv. Bull., 14, 1-14, 1967.

Thomson, M. R. A., Upper Jurassic mollusca from Carse Point, Palmer Land, Br. Antarct. Surv. Bull., 41/42, 23-30, 1975.

Thomson, M. R. A., Mesozoic paleogeography of Western Antarctica, in Antarctic Geoscience, edited by C. Craddock, pp. 331-338, University of Wisconsin Press, Madison, 1982.

Thomson, M. R. A., and R. J. Pankhurst, Age of post-Gondwanian calc-alkaline volcanism in the Antarctic Peninsula region, in Antarctic Earth Science, edited by R. L. Oliver, P. R. James, and J. B. Jago, pp. 328-333, Australian Academy of Science, Canberra, 1983.

Thomson, M. R. A., R. J. Pankhurst, and P. D. Clarkson, The Antarctic Peninsula--A late Mesorzoic-Cenozoic arc (review), in Antarctic Earth Science, edited by R. L. Oliver, P. R. James, and J. B. Jago, pp. 289-294, Australian Academy of Science, Canberra, 1983.

LATE PALEOZOIC ACCRETIONARY COMPLEXES ON THE GONDWANA MARGIN OF SOUTHERN CHILE:
EVIDENCE FROM THE CHONOS ARCHIPELAGO

John Davidson,[1] Constantino Mpodozis,[1] Estanislao Godoy,[2] Francisco Hervé,[2]
Robert Pankhurst,[3] and Maureen Brook[4]

Abstract. The late Paleozoic "basement" rocks that crop out along the Pacific side of the Chonos Archipelago (44°-46°S) can be divided into two north-south trending belts: (1) an eastern belt formed of submarine fan-turbidites and subordinate pelagic cherts, each containing well-preserved primary sedimentary structures, and (2) a western belt, mainly formed by strongly foliated mica schists and greenschists. Trace element contents in the cherts and greenschists indicate rocks of oceanic affinity. The structures present within the eastern rock suite are principally subisoclinal folds (with tectonic imbrication) and locally developed zones of broken formation. The transition from these rocks into the foliated schists appears to be related to a progressive increase in metamorphism and strain associated with the development of westward verging recumbent folds and a flat-lying crenulation cleavage. It is inferred that these structures developed during the construction of a Late Carboniferous-Early Permian accretionary prism (about 260 Ma Rb-Sr ages), although sedimentation may have taken place throughout the upper Paleozoic. Rb-Sr whole-rock isochrons giving Late Jurassic to Early Cretaceous ages for some localities may indicate much later development of S_2 structures. Alternatively, they may represent isotopic resetting by hydrothermal effects during the emplacement of transgressive Early Cretaceous granites, one of which gives a new Rb-Sr isochron age of 125 ± 2 Ma. This overall scenario seems to be consistent with that reported in the slightly older coastal metamorphic basement north of 34°S and equivalent or younger complexes farther south in the Madre de Dios Archipelago.

Introduction

Low-grade metamorphic rocks within the Chonos Archipelago in southern Chile (44°-46°S), belong to a long belt of a late Paleozoic forearc assemblage extending continuously south of 34°S in the Coastal Range and sporadically from this latitude to the north. They have been interpreted as a subduction complex accreted to the ancestral Pacific margin of Gondwana [Hervé et al., 1981; Forsythe, 1982]. The presence of metamorphic rocks in the Chonos Archipelago was first noted by Darwin [1846]. Stiefel [1970] attempted lithological subdivisions of the basement rocks, and Miller [1973, 1979] made numerous structural observations and found Devonian fossils at one locality. Subsequent work by Hervé and collaborators [e.g., Godoy et al., 1984] has demonstrated that the rocks in the area have suffered a progressive westward increase in strain and metamorphism.

The purpose of this paper is to describe the lithology, age, metamorphism, and structure of the metamorphic rocks of the archipelago. These features are then related to the construction of a late Paleozoic accretionary prism along the southern South American sector of the Pacific margin of Gondwana.

Lithology

The rocks in the Chonos Archipelago form two adjacent north-south trending belts (Figure 1). The eastern belt consists of submarine fan-turbidites and subordinate pelagic cherts, both of which contain well-preserved primary sedimentary structures. Bands of sheared sandstones, mudstones, and conglomerate (broken formation) are locally developed. The western belt consists largely of strongly foliated mica schists and greenschists. They have a pronounced metamorphic fabric, primary structures are seldom preserved, and some schists contain high-pressure minerals such as glaucophanic amphibole.

Eastern Belt (EB)

Rocks within this belt form two major lithological units (Teresa and Potranca units), both in gradational contact toward the west with the western metamorphic belt. To the east they are intruded by Mesozoic granitoids or faulted against Mesozoic volcanic rocks.

Teresa unit. This unit crops out north of Canal Ninualac (Figure 1) and consists predominantly of medium- to coarse-grained, thick-bedded

---

[1] Servicio Nacional de Geología y Minería, Santiago de Chile.
[2] Departamento de Geología, Universidad de Chile, Santiago de Chile.
[3] British Antarctic Survey, Natural Environment Research Council, Cambridge, United Kindgom.
[4] British Geological Survey, Natural Environment Research Council, London, United Kingdom.

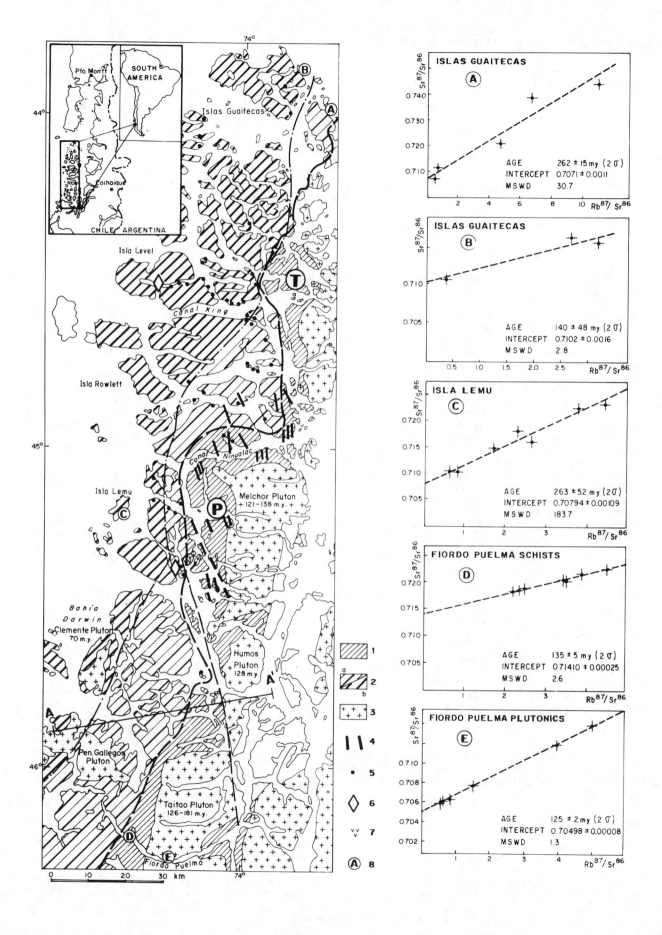

(up to 200 cm) quartzofeldspathic sandstones. It also contains thick intervals (100-200 cm) largely of mudstones and thin-bedded quartzofeldspathic sandstones: "classical" turbidites. The rest of the clastic succession is formed by disrupted mudstone and thin-bedded sandstone (broken formation), which also includes incoherent meter-size sandstone blocks.

Ten-meter-thick sequences of radiolarian-banded cherts are common, sometimes as part of the broken formation (Isla Teresa). Major and transition element geochemistry of the cherts has demonstrated [Godoy et al., 1984] that they represent deposits similar to pelagic oceanic sediments.

Within the context of a submarine fan model for the Teresa unit, the lithofacies association indicates a middle-to-suprafan setting, probably built over oceanic crust. Interbedded chert occurrences could represent periods of diminished to nonexistent clastic deposition. It is also possible that the chert beds correspond to more distal areas of sedimentation, far from turbidity current flows, and that they were tectonically interleaved during accretion.

Potranca unit. The unit crops out south of Canal Ninualac (Figure 1) and is made up predominantly of thick-bedded to massive coarse-grained sandstones, lithic arkoses, and pebbly conglomerates. On Isla Potranca, bed thickness reaches more than 10 m. Amalgamation of beds is very common, as well as rip-up intraclasts. The rest of the unit consists of thin-bedded, fine- to medium-grained sandstones with pelitic intervals. The sand/pelite ratio is close to unity, and bed thickness ranges between 1 and 10 cm. Exposures of broken formation are commonly associated with the above-mentioned rock sequences. The Potranca unit lithofacies specifies an environment of principally middle-to-supra submarine fan setting with episodes of high-energy turbidite deposition feeder channels. The inferred channelized sequences are coarser grained than other pebbly sandstones in the eastern belt terranes, indicating that the Potranca unit is probably the more proximal.

Western Belt (WB)

This is composed mainly of muscovite schist, minor greenschist, and sporadic metacherts and meta-andesites. The mica schists bear albite porphyroblasts in the westernmost islands. Muscovite schists are composed of interlayered quartz-albite bands and muscovite-chlorite bands with intrafolial folds. At Islas Guaitecas, muscovite is phengitic (Fe total = 2-6%, MgO = 2%), and chlorite is ferrimagnesic. Albite poikiloblasts include sphene, clinozoisite, garnet, and pistacite, which are altered and fractured when in the groundmass. Common accessory minerals are calcite, apatite, zircon, and tourmaline.

The metacherts usually contain abundant garnet, epidote, magnetite, Mn-rich stilpnomelane (parsentensite), and occasionally glaucophanic amphibole.

Greenschists are widely distributed. The common mineral assemblages are quartz ± albite ± chlorite ± calcite ± muscovite ± garnet ± stilpnomelane ± actinolite ± biotite ± glaucophanic amphibole. The last mineral has only been found at the eastern shore of Isla Venezia, where it has actinolitic nuclei. Amphibole is commonly actinolitic, sometimes with hornblendic cores. Epidote is zoned, with Fe-rich cores. Major [Godoy, 1980] and trace element chemistry [Godoy et al., 1984] indicates that they have mid-ocean ridge basaltic composition in the usual discriminant diagrams. The previously described mineral assemblages are indicative of low-grade metamorphism, transitional from greenschist to blueschist facies.

At Isla Dring and Isla Italia there are massive, pale gray rocks with cumulate diabasic relic textures which constitute small stocks or diabase sheets.

Structure

Structural features of the southern half of the archipelago are presented in Figure 2, which includes a transverse profile and stereoplots of foliation, fold axis, and lineations.

Eastern Belt

Strong lithological control of deformation is observed in this part of the complex. The massive sandstone sequences are homoclinal and uncleaved. Rhythmic turbidites show gently plunging, subisoclinal-to-chevron folds, trending north-northwest and overturned to the southwest. Their axial plane cleavage (S_1) is coaxially refolded by decimeter-size recumbent chevron-like folds, which are associated with a subhorizontal crenulation cleavage (S_2). Both cleavages are inhomogeneously developed; the second was recognized only in the southern part of the area (Islas Caiquenes). Finally, pelite-rich horizons, which probably include olistostromes, have been sheared off into mélange-like selvages, where quartz-rich rootless folds may locally be observed.

Western Belt

Westward coarsening of the crenulation cleavage fabric occurs in this belt. It relates to progressively increasing strain, associated with pervasive microscopic refolding and pressure solution of an also coarsened S_1. Early (D_1) isoclinal

Fig. 1. (Opposite) Generalized geological map of the Chonos Archipelago, southern Chile. (1) Eastern belt: T, Teresa unit; P, Potranca unit. (2) Western belt: 2a, schists; 2b, albite-bearing schists. (3) Jurassic-Cretaceous granitoids. (4) Broken formations. 5) Greenschists. 6) Glaucophanic amphibole locality. 7) Meta-andesites. (8) Rb-Sr whole-rock sampling sites keyed to isochron diagrams. Later geological units are left blank.

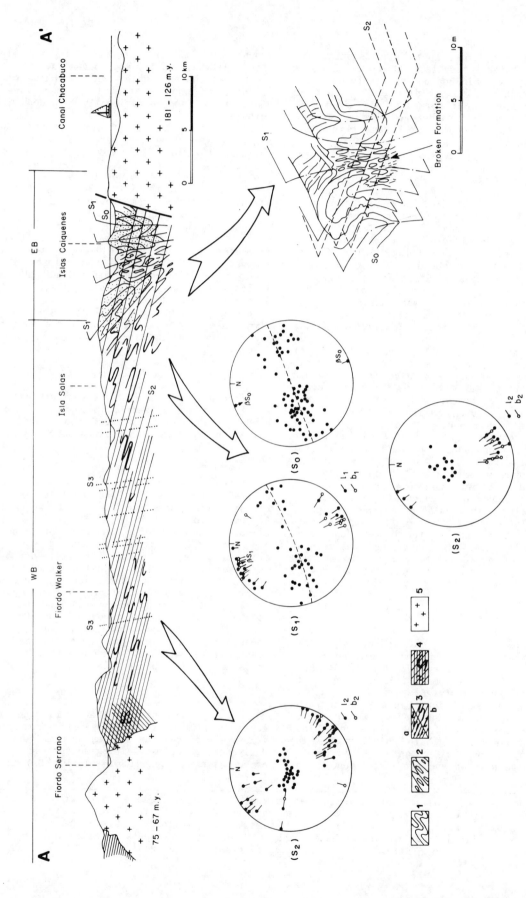

Fig. 2. Cross section AA' between Fiordo Serrano and Canal Chacabuco (see Figure 1 for location) showing stereoplots of stratification and cleavages, fold axis, and lineations from the eastern belt (right) and the western belt (left). (1) Eastern belt sandstones. (2) Broken formations. (3) Western belt schists: 3a, albite-bearing schists; 3b, schists and phyllites. (4) Contact metamorphic rocks. (5) Jurassic and Cretaceous granitoids.

folding is preserved only inside albite porphyroblasts. S_0 is preserved only in thick metachert-metatuff sequences, such as the one cropping out on Isla Rowlett [Miller, 1979, p. 436]. The main foliation (S_2) is flat lying to gently northeast dipping, but late warping and kinking, also recorded in the EB, locally modify this regional dip. Quartz rods defining L_2 are very abundant in some areas.

Geochronology

Only one fossil locality, on Isla Potranca in the eastern belt, is known for the "basement" rocks in this area. Miller and Sprechmann [1978] described a brachiopod fauna indicating a Devonian age of sedimentation. The locality was revisited in 1985, and further specimens were collected from the base of a thick turbidite flow, but as yet no specific identifications have been made.

Metasedimentary rocks of this type are difficult to date meaningfully by radiometric methods, but Rb-Sr whole-rock ages have been attempted from several sites, mostly within the WB, with varying degrees of S_2 development. Unmetamorphosed fine-grained sediments can yield diagenetic ages, close to that of deposition, using this technique, but even quite poorly developed low-grade recrystallization can result in resetting of the Rb-Sr systems. Thus the significance of results can be ambiguous. In general, a well-defined isochron mean square of weighted deviates (MSWD $\leq$ 3) in which the data fit a straight line within the limits of analytical error can be taken as recording a discrete geological event. Excess scatter, indicated by MSWD > 3, results in an errorchron where the model is not met. In such cases, errors are usually enhanced (by a factor of $\sqrt{MSWD}$), but it must be remembered that this may not adequately allow for isotopic disturbance resulting from partial rehomogenization.

Two ages have previously been reported from the area of Islas Guaitecas at the northern end of the archipelago [Godoy et al., 1984]. The first, from Isla Leucayec, is based on a 5-point errorchron giving 262 ± 15 Ma with initial ration (IR = 0.7071 (locality A in Figure 1). The samples, collected over a distance of 10 km, are fine-grained banded quartz-mica schists in a chlorite-calcite zone of metamorphism. Sedimentary structures are partly preserved, and the main foliation is S_1: a subvertical S_2 fabric is poorly developed. The second age, 148 ± 48 Ma with IR = 0.7101, is based on three coarse-grained semischists from Isla Ascension in the WB, about 10 km northwest of locality A. These rocks are in a higher (actinolite) zone of metamorphism and exhibit a strong subhorizontal S_2 foliation. Each of these ages is supported by concordant K-Ar dating of samples from the two localities [Godoy et al., 1984], so that there is a clear relationship of the younger age to the intensity of S_2.

The Permian age indicated for locality A has now been confirmed by a second errorchron, giving 263 ± 52 Ma, for seven samples of coarse-grained micaschists from Isla Lemu, Isla Kent, and Isla Dring (locality C; data in Table 1). These mostly possess a penetrative S_2 crenulation cleavage, sometimes folded by a further event, and four contain Ab porphyroblasts. The IR of 0.7079 ± 0.0011 is also similar to that for locality A, although because of the very high MSWD the error estimates on this and the age are rather unreliable. In such a situation the age would normally be considered a minimum for the event which produced S_2, but since the samples were taken over a large area (20 x 10 km), it is possible that they represent, imperfectly, a Rb-Sr system which remained essentially closed to isotopic migration during this event. Finally, an almost perfect isochron of 135 ± 35 Ma has been obtained as part of the present work for semischists in Fiordo Puelma (locality D). These rocks occur within the area mapped as EB but close to the WB boundary, and they do have a well-developed flat-lying S_2 cleavage. In this case the sampling area was much smaller (about 30 m beach section), and it is much more likely that metamorphic resetting has occurred. This age is very close to the Jurassic/Cretaceous boundary. The initial $^{87}Sr/^{86}Sr$ ratio is relatively high at 0.7141 ± 0.0003. This is consistent with 100-150 Ma for closed-system growth of radiogenic ^{87}Sr in rocks such as those which have given the Permian ages (i.e. with an initial value of about 0.707 and a mean $^{87}Rb/^{86}Sr$ ratio of about 3). This suggests that the younger Late Jurassic-Early Cretaceous isochrons represent isotopic homogenization during a discrete event affecting Rb-Sr systems such as those observed at localities A and C.

The geological interpretation of these two possible events is ambiguous. The Permian ages could in principle represent sedimentation or diagenesis, since the IR values of 0.707 and 0.708 are quite reasonable for continentally derived graywackes, comprising a mixture of old sialic detritus and juvenile volcanic products. However, this hypothesis is not consistent with the paleontological estimate of a Devonian age, which is probably also within the range of possibilities allowed by the isotope systems if it is assumed that on sedimentation they had a lower IR (e.g., 0.704) or that the Rb/Sr ratios were originally significantly lower and that they were increased by metasomatism at or prior to Permian times. Further work may clarify this possibility, but it is suggested that the 260 Ma ages are minima for the constructive phase of accretionary growth, since the albite porphyroblasts in the Isla Lemu area contain relict S_1 structures and probably represent a thermal event sufficiently intense to disturb Rb-Sr systematics significantly.

The younger ages are more problematic. The only deformational event with which they could be correlated would be the formation of S_2 during the "destructive" tectonic phase in the accretionary complex. However, although this could well be an inhomogenously and diachronously developed set of structures of similar appearance, we would prefer to consider that they are associated with underthrusting caused by continued subduction of oceanic crust consequent upon (and closely following in time) the formation of S_1 structures. This would require that the event at about 135 Ma was a later "cryptic" resetting with no obvious structural

TABLE 1. New Rb-Sr Whole-Rock Data

	Rb, ppm	Sr, ppm	$^{87}Rb/^{86}Sr$	$^{87}Sr/^{86}Sr$
Isla Lemu-Kent-Dring Schists				
SLL 215	73	255	0.834	0.71040
SLL 220	53	242	0.629	0.71042
SLL 226	88	96	2.647	0.71633
SLL 229	65	109	1.734	0.71501
SLL 231	218	142	4.441	0.72364
SLL 232	195	143	3.791	0.72255
SLL 256	74	92	2.312	0.71825
Fiordo Puelma Schists				
SLL 300A	59	72	2.367	0.71855
B	188	120	4.531	0.72273
C	65	76	2.485	0.71885
E	160	119	3.895	0.72170
F	72	59	3.506	0.72058
G	46	60	2.219	0.71842
H	102	86	3.432	0.72061
Fiordo Puelma Granite				
SLL 298A	53	287	0.535	0.70595
B	56	279	0.582	0.70609
SLL 299	109	199	1.526	0.70773
SLL 306B	79	280	0.818	0.70632
C	57	305	0.542	0.70595
E	134	98	3.986	0.71196
F	132	77	4.999	0.71393
SLL 294	71	242	0.857	0.70673
SLL 295	84	228	1.067	0.70713
SLL 296	71	258	0.801	0.70679
SLL 297	60	269	0.652	0.70614

Determined at British Geological Survey, London [see Millar and Pankhurst, this volume]: Rb/Sr ± 0.5%; $^{87}Sr/^{86}Sr$ ± 0.01% (1 sigma).

expression. One possibility would be thermal resetting during the emplacement of transgressive granitoids, as explained below.

High-level plutonic rocks also define two belts in the archipelago. The Melchor, Humos, and Taitao plutons (Figure 1) bound the metamorphic rocks to the east. Their Early Cretaceous cooling ages (see Godoy et al. [1984, Table 1] for a review of the ages) are here confirmed by a good isochron of 125 ± 2 Ma (MSWD = 1.3), obtained in Fiordo Puelma. The low IR of 0.7050 for rocks with moderately high Rb/Sr ratios suggests that this age represents igneous crystallization. Although it is clearly a minimum age for all the deformation of the metasediments, and it is just distinguishable from the age obtained for the Puelma schists, their statistical and geographical proximity would nevertheless be consistent with resetting of the Rb-Sr systems in the latter during batholithic emplacement. The Peninsula Gallegos and Isla Clemente plutons, on the other hand, give well-defined 67-75 Ma K-Ar ages for the western intrusive belt.

Discussion

The lithology and structure of the Chonos Archipelago metamorphic rocks are compatible with the following development. During Devonian to late Paleozoic times, a pelagic turbiditic suite with oceanic affinities was progressively accumulated along this part of the South American margin of Gondwana. It appears to have been accreted onto the continent, perhaps during Permian times, in a "constructive" stage of tectonism. Broken formations originated between wedges of proximal to middle fan turbidites and pelagic cherts of the EB. At the same time, oceanic basalt slices were stacked oceanward or slid into soft sediments [Ozawa, 1985] of the WB. First-generation southwest-overturned subisoclinal folds are most likely related to this Permian stage. Similar features have been described for simple accretionary prisms elsewhere [Seely et al., 1974; Karig, 1974; Moore and Karig, 1976; Cowan and Silling, 1978].

Possibly also starting in Permian times, but perhaps extending locally into the Early Cretaceous, a "destructive" stage of postaccretionary tectonism is recognized in the WB. It was responsible for the strongly penetrative west-prograding second foliation and the growth of glaucophanic amphibole found in some of the greenschists. Following Moore [1979], the destructive character of this event could be related to an increase in shear strain along the base of the accretionary prism. This phenomenon could be enhanced by a faster convergence rate, a reduced sedimentary input, or a decrease in pore factors, resulting in a strong mechanical coupling between the oceanic plate and the accretionary wedge.

Extensive outcrops of metamorphic rocks east of the Patagonian Batholith in the mainland of southern Chile and Argentina have been compared with those of the Chonos Archipelago [Forsythe, 1982]. They include turbidites with minor marble, greenschists, and cherts [Ramos, 1979]. However, the greenschist trace element contents and the uplifted craton provenance of the metasandstones distinguish these rocks from those of the Chonos Archipelago and suggest a different tectonic scenario [Godoy et al., 1984]. The metamorphic basement north of Chonos Archipelago shares similar characteristics with that described here, although metamorphism of its northernmost outcrops (34°-30°S) seems to have taken place during the Late Carboniferous [Hervé et al., 1982].

A more direct comparison may be made with the area south of Golfo de Penas, where accretion of a compound prism which includes exotic Permian limestones and radiolarian cherts took place in the late Paleozoic to early Mesozoic [Forsythe and Mpodozis, 1983]. Many of these rocks are only metamorphosed to low grade, which relates to a higher structural level than the Chonos accretionary wedge.

Acknowledgments. The Chilean contribution to this paper is a joint research effort between Universidad de Chile (DIB grants E1300, E083, E886, and E1702) and Servicio Nacional de Geología y Minería. M. Brook and R. J. Pankhurst contributed with the permission of the directors of British Geological Survey and the British Antarctic Survey, respectively, as part of a research project funded by the Overseas Development Administration of the Foreign and Commonwealth Office of Great Britain. We are grateful to Eugenia Pirzio-Biroli, Town Mayor of Puerto Cisnes, for invaluable maritime logistic support. This paper is a contribution to IGCP projects 193 and 211.

References

Cowan, D. F., and R. M. Silling, A dynamic, scaled model of accretion at trenches and its implications for the tectonic evolution of subduction complexes, J. Geophys. Res., 83, 5389-5396, 1978.

Darwin, C., Geological Observations on South America, 279 pp., Smith Elder, London, 1846.

Forsythe, R., The late Palaeozoic to early Mesozoic evolution of southern South America: A plate tectonic interpretation, J. Geol. Soc. London, 139, 671-682, 1982.

Forsythe, R., and C. Mpodozis, Geología del basamento pre-jurásico superior en el archipiélago Madre de Dios, Magallanes, Chile, Bol. Serv. Nac. Geol. Mineral., 39(63), Santiago, 1983.

Godoy, E., Zur Geochemie der Grünschiefer des Grundgebirges in Chile, Muenstersche Forsch. Geol. Palaeontol., 51, 161-182, 1980.

Godoy, E., J. Davidson, F. Hervé, C. Mpodozis, and K. Kawashita, Deformación sobreimpuesta y metamorfismo progresivo en un prisma de acreción paleozoico: Archipiélago de los Chonos, Aysén, Chile, Actas. Congr. Geol. Argent. IX, 4, 211-232, 1984.

Hervé, F., J. Davidson, E. Godoy, C. Mpodozis, and V. Covacevich, The late Palaeozoic in Chile, stratigraphy, structure and possible tectonic framework, Rev. Acad. Bras. Cienc., 53, 362-373, 1981.

Hervé, F., K. Kawashita, F. Munizaga, and M. Bassei, Edades Rb-Sr de los cinturones pareados de Chile Central, Actas Congr. Geol. Chileno 3rd, 2, D116-D135, 1982.

Karig, D. E., Evolution of arc systems in the western Pacific, Annu. Rev. Earth Planet. Sci., 2, 51-75, 1974.

Millar, I. L., and R. J. Pankhurst, Rb-Sr geochronology of the region between the Antarctic Peninsula and the Transantarctic Mountains: Haag Nunataks and Mesozoic granitoids, this volume.

Miller, H., Características estructurales del basamento geológico chileno, Actas Congr. Geol. Argent. V, 4, 101-115, 1973.

Miller, H., Das Grundgebirge der Anden im Chonos-Archipel, Region Aisén, Chile, Geol. Rundsch., 68(2), 428-456, 1979.

Miller, H., and P. Sprechmann, Eine devonische Faunula aus dem Chonos-Archipel Region Aisén, Chile, und ihre stratigraphische Bedeutung, Geol. Jahrb., Reihe B, 28, 37-45, 1978.

Moore, J. C., Variations in strain and strain rate during underthrusting of trench deposits, Geology, 7, 185-188, 1979.

Moore, G. F., and D. E. Karig, Development of sedimentary basins on the lower trench slope, Geology, 4, 693-697, 1976.

Ozawa, Y., Variety of subduction and accretion processes in Cretaceous to recent plate boundaries around southwest and central Japan, Tectonophysics, 112, 493-518, 1985.

Ramos, V., Tectónica de la región del río y lago Belgrano, Cordillera Patagónica, Argentina, Actas Cong. Geol. Chileno 2, 1, B1-B32, 1979.

Seely, D. R., P. R. Vail, and G. G. Walton, Trench slope model, in The Geology of Continental Margins, edited by C. A. Burke and C. L. Drake, pp. 249-260, Springer-Verlag, New York, 1974.

Stiefel, J., Das Andenprofil im Bereich des 45 sudlichen Breitengrades, Geol. Rundsch., 59, 961-797, 1970.

Copyright 1987 by the American Geophysical Union.

EARLY PALEOZOIC STRUCTURAL DEVELOPMENT IN THE NW ARGENTINE BASEMENT OF THE ANDES
AND ITS IMPLICATION FOR GEODYNAMIC RECONSTRUCTIONS

A. P. Willner

Institut für Mineralogie, Ruhr-Universität Bochum, Federal Republic of Germany

U. S. Lottner

Geologisch-Paläontologisches Institut, Westfälische Wilhelms-Universität Münster
Federal Republic of Germany

H. Miller

Institut für Allgemeine und Angewandte Geologie, Ludwig-Maximilans-Universität München
Federal Republic of Germany

Abstract. Tectonometamorphic zones were defined within the lower Paleozoic basement of the NW Argentine Andes in a transitional zone between two Andean segments of different geotectonic evolution. In the Cambrian, the Pacific edge of Gondwana changed from a passive to an active continental margin. This event began with folding of a Vendian/Eocambrian sediment wedge (Puncoviscana Formation and equivalents). The effects can be traced progressively over all structural levels with exposed depth increasing from north to south. Phenomena of a second deformation are of different nature and age but mostly characterized by shear belts causing large-scale crustal imbrication. In the lower tectonic levels this phase coincides with subduction-related magmatism of Ordovician age. A flat subduction slab is supposed, somewhat steeper in the northern than in the southern segment. The following anatectic-granitic magmatism and weak deformation in the Devonian may have marked a new change to passive margin conditions.

Introduction

Different patterns of geodynamic evolution of various segments of the Andes, especially during the Mesozoic and Cenozoic, have been known for a long time [e.g., Gansser, 1973]. Jordan et al. [1983] and Allmendinger et al. [1983] attribute the Neogene tectonic segmentation of the central Andes to a segmentation of the underthrusting oceanic plate, to the difference in dip of the subduction slab, and to preexisting crustal heterogeneities. Such differences in evolutionary processes can also be detected in the lower Paleozoic of Bolivia, north Chile, and NW Argentina where four principal segments of different orogenic and paleogeographical evolution may be identified for that time.

The segment north of the Argentine-Bolivian border is characterized by a huge intracontinental trough filled with more than 10,000 m of clastic sediments (Bolivian eastern Cordillera and Subandean Sierras), which was continuously subsiding from the Cambrian to the Devonian. It was underlain by continental crust between a Precambrian Arequipa Massif in the west and the Brazilian shield in the east [Dalmayrac et al., 1980]. Deformation and magmatism occurred only in the Arequipa Massif during the early Paleozoic (Ordovician Atico event [Shackleton et al., 1979]). The lower Paleozoic sediments were only weakly folded during the Variscan orogeny.

In the segment between 22°S and approximately 26°S (Figure 1; segment A) the following early Paleozoic paleogeographical elements may be identified. In the Subandean Sierras and below the Chaco-Pampas Plain, continuous Late Cambrian to Devonian shelf sediments unconformably overlie the weakly metamorphosed and moderately deformed graywackes and pelites of the Puncoviscana Formation (upper Precambrian-Lower Cambrian) which contain few Cambrian granitoids [Aceñolaza and Toselli, 1981a]. West of the basement, Ordovician volcanics and plutonic rocks of the "faja eruptiva de la Puna" [Méndez et al., 1972] form the eastern margin of a vast forearc basin of Ordovician flysch sediments [Coira et al. 1982] extending across the Chilean border. Devonian sandstones overlie a second belt of Ordovician plutons in Chile [Mpodozis et al., 1983], yet flysch predominates farther west. Breitkreuz and Zeil [1984] suppose a continuation of the Arequipa Massif as far south as the Peninsula Mejillones (23°S), but a Precambrian age has not been proven. Older continental crust might, however, have underlain the lower Paleozoic basins as far west as the Argentine Puna, where some scattered basement outcrops are known [Allmendinger et al., 1983].

A third segment (Figure 1; segment B) farther

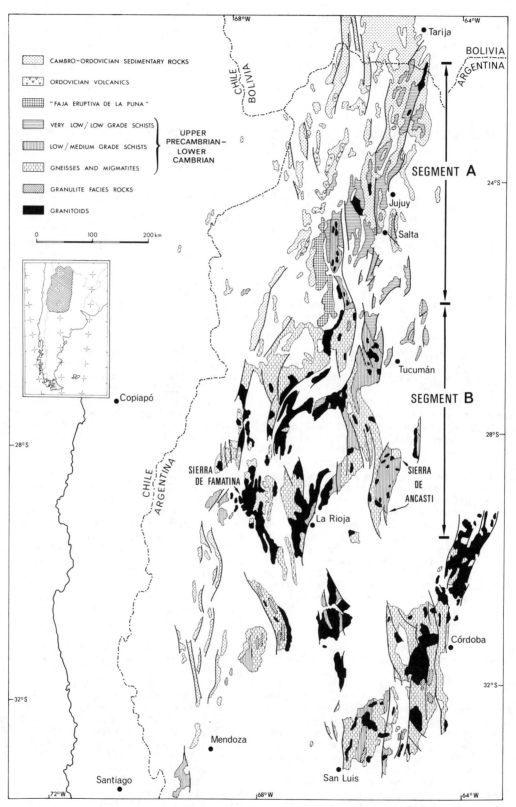

Fig. 1. The lower Paleozoic basement of the NW Argentine Andes; Andean segments A and B studied in detail.

south (26°S to 30°S) shows considerably larger outcrops of medium- to high-grade metamorphic equivalents of the Puncoviscana Formation. The metamorphism of this unit occurred mainly in the Cambrian and Ordovician [Bachmann et al. 1986]. Calcalkaline plutonism in this segment took place from the Late Cambrian to the Early Carboniferous [Knüver and Miller, 1982]. Deposition of Ordovician clastic sediments and volcanics in the Sierra de Famatina farther west was synchronous with strong magmatic and tectonic activity in the east. West of the Famatina area, basement rocks with intermediate-pressure metamorphism and basic to ultrabasic intrusives of late Precambrian to early Paleozoic age occur [Villar, 1975]. Devonian flysch sediments are known from the northern Precordillera to the present Pacific coast [Hervé et al., 1981].

In a fourth segment south of 30°S, deeper tectonic levels are exposed in the southern Pampean Ranges with an even broader outcrop width of basement rocks. Granulite belts with Precambrian radiometric ages are found in gneissic and migmatitic zones that were affected by substantial Paleozoic magmatism [Caminos et al., 1982]. They might be interpreted as relicts of an older continental crust. Farther west, the deeply eroded basement is abruptly confined by the Precordilleran trough filled with Cambrian to Devonian sediments. The nature of contact, however, is unknown so far. Farther west of the Precordillera an allochthonous terrain is supposed in Chile by indirect evidence [Ramos et al., 1984].

The basement of the NW Argentine Andes essentially consists of of the metagraywackes and metapelites of the Puncoviscana Formation, which can be traced into equivalents of higher metamorphic grade, as shown by structural continuity [Willner and Miller, 1986] and geochemical similarity [Willner et al., 1985]. The extreme homogeneity of these turbiditic sediments over huge areas is very useful in defining tectonometamorphic zones (Figure 2). As the most critical area for detailed studies, a transition zone of as yet unknown nature between the mentioned segments A and B was chosen in the provinces of Salta and Tucumán.

The scope of this work is to show the nature of crustal thickening during the early Paleozoic in the Andean basement and its role in the development of an early convergent margin of the Cordilleran type in the selected segments A and B.

Tectonometamorphic Zonation in the
Basement Between Salta and Tucumán

Figure 3 summarizes all features of the tectonometamorphic zones in the studied part of the basement (Figure 2) between the mainly very low grade metamorphic area in the north (segment A) and the medium- to high-grade metamorphic areas in the south (segment B). The characteristics of the first two superimposed deformation phases and accompanying thermal events have been studied in detail, as they produced the strongest imprints everywhere and as they are most important in defining a model for the structural evolution of the basement. Only structures of the first folding phase may be traced over all structural levels, and these structures progressively increase in intensity toward the south. They may be attributed to the Pampean orogeny (according to Aceñolaza and Toselli [1981a]). Its approximately Lower to Middle Cambrian age is based on (1) trace fossil occurrences of Early Cambrian age in the Puncoviscana Formation [Aceñolaza and Toselli, 1981a], (2) a few K-Ar ages for the first synkinematic metamorphism in zone II (512 Ma in the Sierra San Javier; 570 Ma in the Sierra de Medina [cf. Aceñolaza and Toselli, 1981b], (3) well-proven early Rb-Sr thin slab-isochrone ages of 557-569 Ma in zone IV, synchronous with the formation of an s_1 banding [Bachmann et al., 1986], and (4) so far inconsistent Rb-Sr ages of 717 ± 19 Ma [Omarini et al., 1985], 497-601 Ma, and K-Ar ages of 489-530 Ma of syntectonic to posttectonic granitoid intrusions in the uppermost structural levels (summarized by Toselli and Aceñolaza [1978]).

The progressive development of the s_1 foliation and tightening of F_1 folds with synkinematic increasing metamorphism defines the tectonic levels. Thus a tectonic banding is produced by microscopically well visible pressure solution along an axial plane cleavage in psammites. It first appears in zone Ib and becomes pronounced in zone II. It is developed as a separation into 1- to 5-cm-thick quartz-rich laminae and 0.5 thick laminae relatively enriched in micas and heavy minerals. In the deeper tectonic levels (zones III to VI), gliding on the banding planes destroys the F_1-fold hinges, and thus the banding is found subparallel to the bedding. Stronger pressure solution makes the banding more pronounced; most of the dissolved quartz and feldspar leaves the primary rock to form s-parallel segregations. These are isoclinally folded or deformed to pinch and swell structures by the transposition process. The "transposition isotect" marks a most important transition within the basement. It should be noted that, with increasing depth of the tectonic level, a progressive inclination of the first foliation surfaces, known from other orogens as, for example, the mid-European Variscides, is not observed. The main strike of the F_1 structures is approximately meridional with a deviation to the northeast from the southern Cumbres Calchaquíes to the Sierra de Candelaria.

The effects of a second deformation, mainly characterized by dm- to m-scale drag folds refolding the s_1 banding, are of various origin and age in different zones. For example, in some regions (zones Ia, IIIb, and IV), local steeply plunging fold axes are observed, which could be interpreted as being caused by horizontal slip movements along nearby deep-seated early Paleozoic wrench faults. Two major terrains with more widespread horizontal F_2-drag folds of different ages are distinguished.

Within zone II, F_2-drag folds and a flat west-dipping s_2 cleavage keep a consistent shear sense over considerable distances, even where the dip of banding locally changes (Figure 3). The drag folds are not associated with major synclines and anticlines, but are products of differential move-

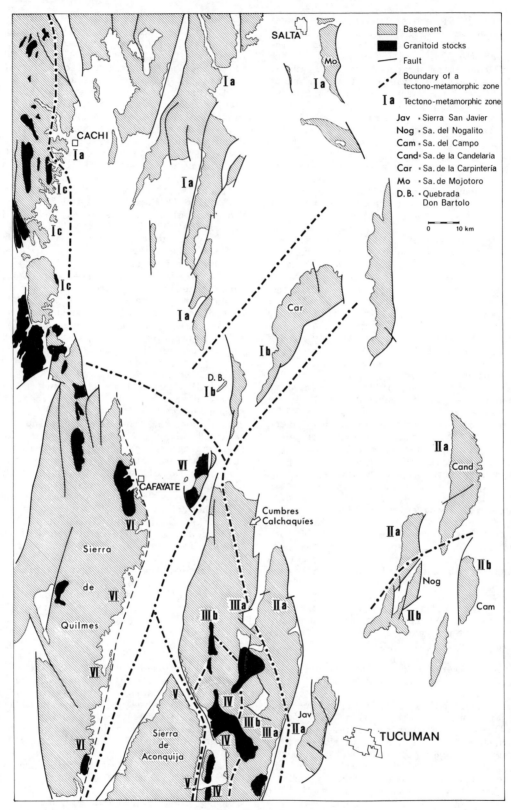

Fig. 2. Tectonometamorphic zonation in the basement between Tucumán and Salta.

ZONE	Ia	Ib	Ic	IIa	IIb	IIIa	IIIb	IV	V	VI
	← UPPER TECTONIC LEVEL →		SHEAR ZONE	← INTERMEDIATE TECTONIC LEVEL →	SHEAR ZONE	← TRANSITIONAL ZONE →		← LOWER TECTONIC LEVEL →		SHEAR ZONE
F_1-FOLDS	OPEN TO CLOSED, dm-Dm SCALE, VARIABLE STRIKE INTENSITY, MORPHOLOGY & CYLINDRICITY	CLOSED CHEVRON FOLDS, m-Hm SCALE, MERIDIONAL STRIKE GAINS CONSTANCY	RARELY DETECTABLE, SAME AS IA	CYLINDROIDAL CHEVRON FOLDS m-Hm SCALE	SUBHORIZONTAL TIGHT, RARELY PRESERVED	TRANSPOSED FOLDS: DESTRUCTION OF F_1-FOLD HINGES BY GLIDING ALONG BANDING PLANES; CONVERGENT LITHOLOGICAL BOUNDARIES; s_1-PARALLEL QUARTZ-SEGREGATIONS REFOLDED (F_{1B}-FOLDS) →				
s_1-FOLIATION	CLEAVAGE IN PELITES ONLY	WEAK INCIPIENT BANDING IN PSAMMITES	CLEAVAGE IN PELITES ONLY	BANDING IN PSAMMITES AS AXIAL PLANE FOLIATION	BANDING IN PSAMMITES	BANDING BECOMES MORE PRONOUNCED BY STRONGER FLATTENING; QUARTZ SEGREGATIONS	BANDING TRANSPOSED SUBPARALLEL TO BEDDING (SEDIMENTARY FEATURES DESTROYED) →			
M_1-METAMORPHISM SYNKINEMATIC -------- CHARACTERIST. MICROFABRIC	VERY LOW GRADE ORIGINAL CLASTIC FABRIC "SPINY" WHITE MICA CEMENT, QUARTZ OVERGROWTH	← (VERY LOW GRADE) →		LOW GRADE NEW CHLORITE & WHITE MICA STRONG MATRIX RECRYSTALLIZATION REDUCTION OF CLAST SIZE BY PRESSURE SOLUTION	SOME INCIPIENT BIOTITE	LOW GRADE (CHLORITE ZONE) PRESSURE SOLUTION MORE PRONOUNCED	← TRANSPOSITION ISOTECT (LOW GRADE) ALMANDINE BLASTESIS →			
F_2-FOLDS	LOCAL, dm-m SCALE, STEEPLY DIPPING AXES	LOCAL, OPEN, dm-m SCALE, SUBHORIZONTAL AXES	RARE	DRAG FOLDS; SUBHORIZONTAL, SLIGHTLY CURVED AXIS, CONCENTRIC-SIMILAR	INTRAFOLIAL DRAG FOLDS, cm-dm SCALE CONCENTRIC-SIMILAR, EXTREMELY HARMONIC	SAME AS IN IIA F_2-DRAG FOLDS IN IB, II, IIIA RELATED TO LARGE SCALE SHEARING IN ZONE IIB	OPEN TO CLOSED DRAG FOLDS, VERTICAL OR SUBHORIZONTAL AXES, NON-CYLINDRICAL		HIGH INTENSITY OF IRREGULAR FLOW FOLDS	dm-m SCALE, FLAT W-VERGENT DRAG FOLDS WITH LOWER LIMBS SHEARED; STRETCHING LINEATIONS PROMINENT
s_2-FOLIATION	RARE	RARE	PENETRATIVE TRANSPOSITION CLEAVAGE	CRENULATION CLEAVAGE IN PELITES ("COLD" DEFORMATION)	PENETRATIVE FLAT, E-VERGENT SHEAR PLANES; STRETCHING LINEATIONS PROMINENT	SAME AS IN IIA	AXIAL PLANE SCHISTOSITY (CRENULATION) →			SUBPARALLEL TO s_1; SHEAR PLANES AND AXIAL PLANE SCHISTOSITY
M_2-METAMORPHISM -------- CHARACTERIST. MICROFABRIC	NONE	NONE	PROGRADE, LOW TO HIGH GRADE METAMORPHISM OF ABUKUMA-TYPE (PRE-F_2)	NONE	DYNAMIC RECRYSTALLIZATION OF QUARTZ, SOME BIOTITE, EXTREME PRESSURE SOLUTION	LOW GRADE INCIPIENT BIOTITE; STRONGER RECRYSTALLIZATION	LOW GRADE - MEDIUM GRADE STRONG BIOTITE BLASTESIS; STILLPRIM. CHLORITE METIC MICAS ORIGINAL CLASTIC FABRIC DESTROYED	STATIC POSTKINEMATIC HEATING: STAUROLITE, ANDALUSITE;	INCIPIENT SILLIMANITE; GNEISSES STRONG INCREASE IN GRAIN SIZE	SYN- TO POSTKINEMATIC: GRANULITES; PROGRESSIVE BARROW TYPE (CY); MIGMATITES (CORD/ALM : 6kb)
MAGMATISM	NONE	NONE	POSTTECTONIC (F_1) TONALITES & GRANODIORITES	NONE	NONE	NONE	POST-F_2-GRANODIORITE STOCKS		SYNTECTONIC (F_2) TONALITIC TO GRANITIC INTRUSIVES	SYNTECTONIC (F_2 AND 2) MAGMATISM METASOMATISM POSTTECTONIC STOCKS
M_3-METAMORPHISM	NONE	NONE	LOW TO MEDIUM GRADE; STATIC RECRYSTALIZATION	NONE	NONE	FIRST RETROGRADE CHLORITE; NEW CHLORITE GROWTH	MUSCOVITE/CHLORITE ON EXPENSE OF STAUROLITE	← RETROGRADE EFFECTS CHLORITIZATION OF ALMANDINE →		
LATE DEFORMATION EFFECTS			KINKBANDS AND CATACLASITES		LARGE SCALE WEAK REFOLDING OF S_2		REJUVENATED MOVEMENTS ALONG s_1- AND s_2-FOLIATIONS LOCAL PSEUDOTACHYLITES			SOME CROSS FOLDING

Fig. 3. Characteristics of the tectonometamorphic zones of the basement between Tucumán and Salta (compare with Figure 2).

ment along banding planes caused by external rotation due to intensive shear movements in certain regions of the basement [Willner and Miller, 1986]. Subzone IIb represents such a shear belt in the Sierra del Nogalito. Monotonous subhorizontal planes with extremely harmonic intrafolial folds and thinned banding with strong pressure solution effects on internal clasts may be best explained by simple shear with synchronous flattening.

In the nearby Sierra de Candelaria, zone IIa type rocks are unconformably overlain by the Upper Cambrian Meson Group. This means that in this region intermediate levels rapidly emerged to the surface owing to strong crustal thickening by imbrication. Thus, deformation in zones Ib and II had ended in the Late Cambrian.

In zone III, within the Cumbres Calchaquíes (Figure 2), a major change of dip of the banding planes occurs, extending to the northwest and south (as far as the Sierra de Ancasti; Figure 1), and is accompanied by a change of vergence and morphology of F_2-drag folds, whereas the central zone III is hardly affected by F_2. Similar to the previously described eastern F_2-fold terrain, the F_2-drag folds in the western Cumbres Calchaquíes are related to movements in a major west-vergent shear belt (zone VI) in the Sierra de Quilmes. The rocks of this sierra were affected by intermediate- to high-pressure metamorphism [Toselli et al., 1978] in a notably close proximity to the low-pressure zone of the Cumbres Calchaquíes. It is also characterized by monotonous flat-lying s surfaces and parallel subhorizontal irregular drag folds with lower limbs sheared off and exhibit incipient anatexis. Considerable metasomatic phenomena are prominent. All K-Ar data compiled by Toselli et al. [1978] point to an Ordovician syntectonic metamorphism. Hence F_2 folding in this terrain is much younger than in the described region farther east (zone II).

Near the border of zone III, biotite starts to grow across an F_2-deformed M_1 fabric, which is rapidly obliterated by complete recrystallization during progressive metamorphism of the low-pressure type (staurolite in zone IV; incipient sillimanite in zone V). This post-F_2 static heating M_2 (459-470 Ma according to Bachmann et al. [1986]) is most characteristic for the deeper tectonic levels. It is related to huge thermal domes and may be traced with similar characteristics at least as far south as the Sierra de Ancasti [Willner, 1983, Figure 1]. In close relation to this static heating, syntectonic granitoids formed in zone V, where F_2-drag folds show some relation to a forceful magma injection and large posttectonic stocks (tonalites to granodiorites) developed in zones III and IV. Hence the most important transition within the NW Argentine basement between the Puncoviscana Formation in the north and its higher metamorphic equivalents to the south occurs within the Cumbres Calchaquíes. It is a two-stage deformation: the first is linked to a Middle Cambrian F_1 deformation, the second is linked to a steep metamorphic gradient during an Ordovician posttectonic static heating.

A third locally restricted shear belt within an upper tectonic level is represented by subzone Ic, where post-F_1 granitoid stocks and their low- to high-grade contact metamorphosed wall rocks are penetrated by a subvertical s_2 foliation. An Ordovician thermal overprint produced later recrystallization of quartz (cf. K-Ar data mentioned by Toselli and Aceñolaza [1978]).

The Character of Crustal Thickening

In summary, four successive processes of continuous crustal thickening may be identified in the basement. Deposition of a huge pile of clastic sediments on a passive margin was followed by strong folding and flattening. Deformation style then changed to imbrication of crustal blocks, while synchronous to subsequent mantle-derived magmatic rocks were intruded. Intensity of all four processes strongly diminishes from south to north.

Sediment accumulation. During late Precambrian to Early Cambrian time the continental margin in the considered sector of the Andes was a passive one [Ježek et al., 1985]. An enormous continuously subsiding trough was separated toward the east from the intracontinental Brazilian mobile belts far in the hinterland by a belt of older Precambrian cratons. The uplift of the Brazilides may have provided the material influx for the graywackes of the several thousand-meter-thick Puncoviscana Formation.

Folding and flattening. The change from a passive to an active continental margin took place in the Middle Cambrian, perhaps as early as the late Precambrian in the southern Pampean Ranges. During this Pampean orogeny [Aceñolaza and Toselli, 1981a], strong shortening and flattening occurred within the entire basement, especially in the deeper tectonic levels. Toward the east (zone IIb), F_1 folds continuously tighten as their axial planes are rotated from an upright into a subhorizontal position during increasing shear strain (Figure 4). Finally the s_1 banding was thinned and later crenulated as shear strain continued. The F_2-drag folds in zone II are related to this shear movement which ended in the Late Cambrian. Their orientation points to a clockwise external rotation of initially subvertical s_1 planes into a more inclined position. About 300 km farther south, a continuation of this shear zone appears in the Sierra de Ancasti (Figure 1) in deeper structural levels (Sierra Brava Complex), where shear movements were synchronous with the transposition process [Willner, 1983]. Early basic magma intrusions of mantle provenance are associated with this shear belt [Lottner, 1985]. Farther south (Sierras of Córdoba) they may continue as a belt of basic to ultrabasic rocks [Villar, 1975]. A similar belt described by Villar [1975] farther west (west of the Sierra de Famatina, Figure 1) might be a time equivalent.

Crustal imbrication. In the Ordovician a considerable change in the type of crustal thickening is noted. Within the area studied in detail, a large-scale shear belt of greater depth and different vergence developed in the west (zone VI). Simple shear caused an anticlockwise rotation of

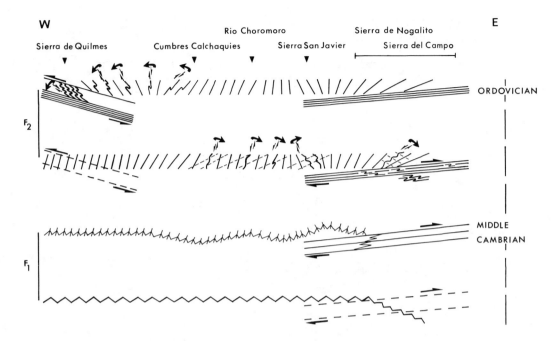

Fig. 4. Schematic structural evolution model along a section between the Sierra de Quilmes and the Sierra del Campo (Tucumán province).

the banding in the neighboring Cumbres Calchaquíes (Figure 4), as deduced from the orientation of F_2-drag folds in this area. This shear zone continues farther south along the western Sierra del Aconquija to the sierras immediately adjacent to the Sierra de Ancasti (Figure 1). A similar external rotation from an east- to a west-vergent position of the banding planes caused by a large-scale imbrication of crustal blocks can also be observed in the Sierra de Ancasti [Willner, 1983]. General westward vergence also dominates farther west, and several repetitions of this type of shear belt to the west are supposed, though this has not been proven yet owing to lack of structural data.

Magmatic accretion. During and after these processes, considerable magmatic accretion contributed to crustal thickening, e.g., from Cambrian to Early Carboniferous in the Sierra de Ancasti [Knüver and Reissinger, 1982]. The magmatic evolution in the Sierra de Ancasti may be taken as typical for the segment B. Its evolution was summarized by Lottner [Lottner, 1985] and Lottner and Miller [1986]. Initially, a tholeiitic magma was derived from partial melting of mantle peridotite due to dehydration of a subducting oceanic slab. By gravitational fractionation of cumulates and due to hornblende, plagioclase, and biotite fractionation, this magma gave rise to a basic to intermediate sequence of an increasingly calc-alkaline character (gabbro, diorite, quartz diorite, tonalite, and trondhjemite). Tonalites and trondhjemites were strongly contaminated during their synkinematic ascent along shear belts in the Early Ordovician. Nearly contamination-free acid end-members with initial Sr ratios around 0.705 (monzogranite and muscovite granite) intruded at that time as well.

This early mantle-derived sequence probably caused contemporary static heating which gave rise to a crustal anatectic-granitic magma sequence (granodiorite, biotite granite, muscovite granite, and pegmatites). These magmas intruded from the Late Ordovician to the Early Carboniferous mainly, as late tectonic to posttectonic stocks along early Paleozoic wrench faults. It is notable that both suites developed continuously, but only once. No repetitions are observed.

Proposed Comprehensive
Geodynamic Interpretation

The geodynamic evolution may be demonstrated by the models proposed in Figure 5 on two representative sections selected at 25°S and 29°S.

Sandstone provenance in the Puncoviscana trough [Ježek et al., 1985] and the lack of any proven evidence of a continental land mass to the west during the early Paleozoic, at least south of 22°S, support a passive continental margin position for the area under consideration during late Precambrian-Early Cambrian time. The existence of a directly bordering proto-Pacific ocean has also been presumed for the late Precambrian by Shackleton et al. [1979]. No true ophiolites of the late Precambrian/Cambrian age have been detected yet in the studied area. Those reported from the Argentine Puna by Coira et al. [1982] are interpreted as later orogenic intrusions according to Aceñolaza and Toselli [1984].

During the Cambrian orogeny a symmetrical configuration of the orogen is probable, at least in

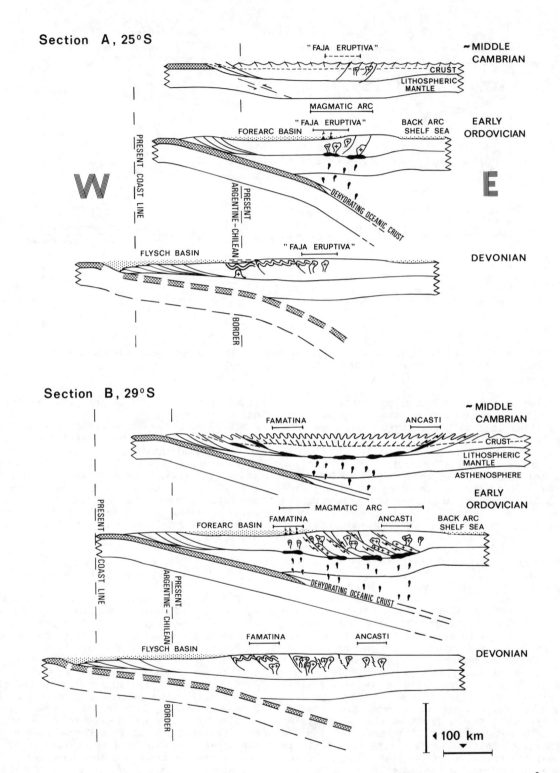

Fig. 5. Geodynamic evolution of the Andes in the lower Paleozoic on sections at 25°S and 29°S.

segment B (29°S). On the eastern flank of the orogen, the discussed east-vergent shear belt with a few basic intrusions developed. In most of the central part of the orogen the s_1 banding was subvertical. In the west, subhorizontal first folds were developed and ultrabasic to basic intrusions of late Precambrian to early Paleozoic age are reported from the basement west of the Sierra de Famatina [Kilmurray, 1969; Villar, 1975].

At the beginning of this Pampean orogeny the subcrustal mantle was continuously downwarped by the convergence of oceanic and continental plates, producing a concomitant symmetrical shortening of the overlying crust with its thick terrigenous sediment pile. This downwarping probably began during sedimentation. Subduction of oceanic crust started in the Cambrian. When it had reached the critical depth for dehydration, the first basic to ultrabasic partial melts formed, which migrated upward from the crust/mantle border along the flat-lying shear zones.

Subduction of a flat continentward dipping oceanic slab slowly continued; the Early Ordovician seems to be the first time when all paleogeographical elements typical for such a geotectonic setting are present. The Lower Ordovician calc-alkaline plutonism in the Sierra de Ancasti is typical for subduction processes [Lottner, 1985]; the Lower Ordovician synsedimentary volcanism in the Argentine Puna starts with basic rocks of chemical characteristics typical for low-K arc tholeiites and changes upward into calc-alkaline series [cf. Coira et al., 1982]. The Lower Ordovician graywacke-filled basin west of this arc may be interpreted as a forearc basin, while the shelf sediments east of the uplifting basement may be seen as a backarc shelf sea according to the nomenclature of Dickinson and Seely [1979].

We have tried to reconstruct the Early Ordovician geotectonic setting using typical thicknesses for the crust and the lithospheric mantle. According to Dickinson and Seely [1979], dehydration of the cold subducted oceanic crust occurs between 100 and 150 km in depth (below 80 km according to most authors). This means that we have to assume a fairly flat subduction slab in segment B (29°S) to cause the broad magmatic arc known to exist in this segment during the time under consideration. In segment A (25°S) we have to assume a steeper slab dip, as the magmatic arc is much more narrow. Still farther north at the Peruvian coast, the slab might have been even steeper. Shackleton et al. [1979] detected an Ordovician Atico event in the Arequipa Massif, while a long-lived subsiding backarc trough existed to the east (Bolivian eastern Cordillera).

There is a remarkable coincidence of the known structural evolution with the constructed geotectonic setting. In segment B, after extreme flattening and east-vergent shearing during the Pampean orogeny, large scale west-vergent crustal imbrication over a wide region became prevalent in the Early Ordovician. This imbrication is restricted to the magmatic arc region. Steepening of the subducting slab in segment A also coincides with decreased deformation and crustal thickening. These differences between both segments may be the cause for the crustal heterogeneities that influenced Neogene Andean tectonics [Jordan et al., 1983; Allmendinger et al., 1983]. Orogenic activity is known to move continuously westward during the Paleozoic [Miller, 1984]. Thus, during the Famatinian orogeny of Silurian to Early Devonian time, the forearc basin sediments were folded, while in the basement only minor late deformation effects occurred at that time (e.g., in the Sierra de Ancasti, cf. Willner [1983]). After a probable gap in forearc basin sedimentation in the Silurian, when major uplift prevailed and deposition mainly occurred in the backarc and trench area, a vast flysch basin developed farther west in the Devonian in both segments [Hervé et al., 1981].

Devonian magmatism in the east seems to be entirely of anatectic origin, and crustal uplift was accompanied by strong mylonitization [Caminos, 1979; Willner, 1983]. In the west, i.e., along the present-day Chilean coast, no subduction-related magmatism is documented for the Devonian. It seems that subduction slowly ceased at that time. A stable continental margin setting was apparently created. The absence of mantle-derived magmas may also be the result of a final depletion in partial mantle melts above the Benioff zone.

In the late Paleozoic, considerable magmatism is known from the Chilean coast [e.g., Drake et al., 1982; Hervé et al., 1981; Halpern, 1978], as well as a local low-temperature/high-pressure zone south of the area under consideration [Hervé et al., 1974]. Hence a new rather steeply dipping subduction zone must have been formed at that time causing considerable magmatism and metamorphism near the present coast. This style is contrary to the situation during the Ordovician and Silurian when the proposed dip angle of the subducting slab was fairly shallow.

Conclusions

Tectonic segmentation in the Andean chain has been identified in the early Paleozoic. Subdivision and detailed studies within the NW Argentine basement reveal a transition between two structural levels or Andean segments and make apparent a striking increase in crustal thickening from north to south.

After a change from a passive continental margin with thick turbiditic sedimentation to an active margin in the Middle Cambrian, crustal thickening in the early Paleozoic is mainly characterized by an initial tight folding and strong flattening, followed by large-scale imbrication of crustal blocks. Features of the first deformation event mainly define the structural levels with increasing exposed depths from north to south. A transposition isotect, where F_1-fold hinges are destroyed, is the most important transition in the basement. It roughly coincides with the onset of progressive posttectonic static heating, characteristic for the deeper tectonic levels.

During the Early Ordovician, imbrication of crustal blocks occurs within a wide magmatic arc including mantle-derived plutonic rocks. A flat subduction zone is assumed to have caused this arc. The slab steepens in the north, where the arc is considerably smaller and where deformation and crustal thickening diminish.

After a westward shift of deformation during the Ordovician and Silurian, subduction seems to have ceased in the Devonian, when anatectic magmas prevailed and deformation decreased. A new onset of subducting with a steep slab is proposed for the late Paleozoic.

Acknowledgments. This work was executed in the course of the scientific cooperation convention signed between the universities of Münster and Tucumán. It was supported by the Deutsche Forschungsgemeinschaft, the Deutscher Akademischer Austauschdienst, and the Stiftung Volkswagenwerk.

References

Aceñolaza, F. G., and A. J. Toselli, Geología del noroeste Argentino, Fac. Cienc. Nat., Esp. Publ., Universidad Nacional de Tucumán, 1287, 212 pp., 1981a.

Aceñolaza, F. G., and A. J. Toselli, The Precambrian-Lower Cambrian formations of northwestern Argentina, in Short Papers of the 2nd International Symposium Cambrian Systems, Open File Rep. 81-743, pp. 1-4, edited by M. Taylor, U.S. Geological Survey, Reston, Va., 1981b.

Aceñolaza, F. G., and A. J. Toselli, Lower Ordovician volcanism in northwest Argentina, in Aspects of the Ordovician System, edited by D. L. Brunton, 295 pp., Universitetsforlaget, Oslo, 1984.

Allmendinger, R. W., V. A. Ramos, T. E. Jordan, M. Palma, and B. Isacks, Paleogeography and Andean structural geometry, northwest Argentina, Tectonics, 2(1), 1-16, 1983.

Bachmann, G., B. Grauert, and H. Miller, Isotopic dating of polymetamorphic metasediments from NW-Argentina, Zentralbl. Geol. Palaeontol., Teil I, 1985, 1257-1268, 1986.

Breitkreuz, C., and W. Zeil, Geodynamic and magmatic stages on a traverse through the Andes between 20°S and 24°S (north Chile, south Bolivia, northwest Argentina), J. Geol. Soc. London, 141, 861-868, 1984.

Caminos, R., Sierras Pampeanas noroccidentales. Salta, Tucumán, Catamarca, La Rioja y San Juan, in Segundo simposio de geologia regional Argentina, pp. 225-291, Academia Nacional de Ciencias, Córdova, 1979.

Caminos, R., C. A. Congolani, F. Hervé, and E. Linares, Geochronology of the pre-Andean metamorphism and magmatism in the Andean Cordillera between latitudes 30°S and 36°S, Earth Sci. Rev., 18, 333-352, 1982.

Coira, B., J. Davidson, C. Mpodozis, and V. Ramos, Tectonics and magmatic evolution of the Andes of northern Argentina and Chile, Earth Sci. Rev., 18, 303-332, 1982.

Dalmayrac, B., G. Lambacher, R. Marocco, C. Martinez, and B. Tomasi, La chaîne hercynienne d'Amérique du Sud--Structure et évolution d'orogène intracratonique, Geol. Rundsch., 69(1), 1-21, 1980.

Dickinson, W. R., and D. R. Seely, Structure and stratigraphy of forearc regions, Am. Assoc. Pet. Geol. Bull., 61(1), 1-31, 1979.

Drake, R., M. Vergara, F. Munizaga, and J. C. Vicente, Geochronology of Mesozoic-Cenozoic magmatism in central Chile, lat. 31°S-36°S, Earth Sci. Rev., 18, 365-393, 1982.

Gansser, A., Facts and theories on the Andes, J. Geol. Soc. London, 129, 93-131, 1973.

Halpern, M., Geological significance of Rb/Sr isotopic data of northern Chilean crystalline rocks of the Andean orogen between latitudes 23°S and 27°S, Geol. Soc. Am. Bull., 89, 522-537, 1978.

Hervé, F., F. Munizaga, E. Godoy, and L. Aguirre, Late Paleozoic K/Ar ages of blueschists from Pichilemu, central Chile, Earth Planet. Sci. Lett., 23, 261-267, 1974.

Hervé, F., J. Davidson, E. Godoy, C. Mpodozis, and V. Covacevich, The late Paleozoic in Chile: Stratigraphy, structure and possible tectonic framework, An. Acad. Bras. Cienc., 54(2), 361-373, 1981.

Ježek, P., A. P. Willner, F. G. Aceñolaza, and H. Miller, The Puncoviscana trough--A large basin of late Precambrian to Early Cambrian age on the Pacific edge of the Brazilian shield, Geol. Rundsch., 74(3), 573-584, 1985.

Jordan, T. E., B. L. Isacks, R. W. Allmendinger, J. A. Brewer, V. Ramos, and C. J. Ando, Andean tectonics related to geometry of subducted Nazca plate, Geol. Soc. Am. Bull., 94, 341-361, 1983.

Kilmurray, J. O., Petrología de las rocas metamórficas del sector noreste de la Sierra de Maz, Villa Unión, provincia de La Rioja (estudio preliminar), Actas Congr. Geol. Argent. 4, 1, 409-428, 1969.

Knüver, M., and H. Miller, Rb/Sr-geochronology of the Sierra de Ancasti (Pampean Ranges, NW-Argentina), Actas Congr. Latinoamer. Geol. Buenos Aires V, 3, 457-471, 1982.

Knüver, M., and M. Reissinger, The plutonic and metamorphic history of the Sierra de Ancasti (Catamarca province, Argentina), Zentralbl. Geol. Palaeontol., Teil I, 1981(3/4), 285-294, 1982.

Lottner, U.S., Strukturgebundene Magmenentwicklung im altpäozoischen Grundgebirge NW-Argentiniens am Beispiel des Südteils der Sierra de Ancasti, PhD thesis, 170 pp., Univ. Münster, Germany, 1985.

Lottner, U. S., and H. Miller, The Sierra de Ancasti as an example of the structurally controlled magmatic evolution in the lower Paleozoic basement of the NW-Argentine Andes, Zentralbl. Geol. Palaeontol., Teil I, 1985, 1269-1281, 1986.

Méndez, J., A. Navarini, D. Plaza, and V. Viera, Faja eruptiva de la Puna oriental, Actas Congr. Geol. Argent. Buenos Aires V, 4, 83-100, 1972.

Miller, H., Orogenic development of the Argentine/Chilean Andes during the Paleozoic, J. Geol. Soc. London, 141, 885-892, 1984.

Mpodozis, C., F. Hervé, J. Davidson, and S. Rivano, Los granitoides de Cerros de Lila, manifestaciones de un episodio intrusivo y termal del Paleozoico Inferior en los Andes del norte de Chile, Rev. Geol. Chile, 18, 3-14, 1983.

Omarini, R. H., A. Aparicio Yague, C. Párica, S. Pichowiak, L. García, K. W. Damm, J. G. Viramonte, J. A. Salfity, and R. N. Alonso, New geochronologicial data on the Precambrian age of the Puncoviscana Formation, northwestern Argentina, Comun. Dep. Geol. Univ. Chile, 35, 181-183, 1985.

Ramos, V. A., T. E. Jordan, W. Allmendinger, S. M. Kay, J. M. Cortés, and M. A. Palma, Chilenia: Un terreno alóctono en la evolución Paleozoica de los Andes centrales, Actas Congr. Geol. Argent. IX, 2, 84-106, 1984.

Shackleton, R. M., A. C. Ries, M. P. Coward, and P. R. Cobbold, Structure, metamorphism, and geochronology of the Arequipa Massif of coastal Peru, J. Geol. Soc. London, 136, 195-214, 1979.

Toselli, A. J., and F. G. Aceñolaza, Geochronología de las Formaciones Puncoviscana y Suncho, Provincias de Salta y Catamarca, Asoc. Geol. Argent. Rev., 33(1), 76-80, 1978.

Toselli, A. J., J. N. Rossi de Toselli, and C. W. Rapela, El basamento metamórfico de la Sierra de Quilmes, República Argentina, Asoc. Geol. Argent. Rev., 33(2), 105-121, 1978.

Villar, M., Las fajas y otras manifestaciones untrabásicas en la República Argentina y su significado metalogenético, Actas Congr. Ibero Am. Geol. Econ. II, 3, 133-155, 1975.

Willner, A. P., Mehrphasige deformation und metamorphose im altpaäozoischen Grundgebirge des Nordteils der Sierra de Ancasti (Prov. Catamarca, NW-Argentinien), PhD thesis, 203 pp., Univ. Münster, Germany, 1983.

Willner, A. P., and H. Miller, Structural division and evolution of the lower Paleozoic basement in the NW-Argentine Andes, Zentralbl. Geol. Palaeontol., Teil. I, 1985, 1245-1255, 1986.

Willner, A. P., H. Miller, and P. Jezek, Geochemical features of an upper Precambrian-Lower Cambrian greywacke/pelite sequence (Puncoviscana trough) from the basement of the NW-Argentine Andes, Neues Jahrb. Geol. Palaeontol. Monatsh., 1985(8), 498-512, 1985.

Copyright 1987 by the American Geophysical Union.

PALEOMAGNETISM OF PERMIAN AND TRIASSIC ROCKS, CENTRAL CHILEAN ANDES

Randall D. Forsythe

Department of Geological Sciences, Rutgers University, New Brunswick, New Jersey 08903
Lamont-Doherty Geological Observatory, Palisades, New York 10964

Dennis V. Kent

Lamont-Doherty Geological Observatory and Department of Geological Sciences
Columbia University, Palisades, New York 10964

Constantino Mpodozis and John Davidson

Servicio Nacional de Geología y Minería, Casilla 10465, Santiago, Chile

Abstract. The first paleomagnetic data from Permian and Triassic formations west of the Andean divide are presented. Four formations of Permian or Triassic age in the central Chilean Andes have been investigated: two are located in the coastal ranges, and two are in the main cordillera. Of the formations in the main cordillera (Pastos Blancos and Matahuaico formations), only the Pastos Blancos Formation has yielded characteristic directions. While a fold test is absent, magnetizations are most likely secondary and yield pretilt corrected concordant inclinations, but yield declinations discordant 30° clockwise in comparison to the South American apparent polar wander path. Both formations from the coastal ranges (Cifuncho and Pichidangui formations) yielded stable directions. Postfolding magnetizations in the Cifuncho Formation also show declinations discordant 30° clockwise and concordant inclinations. The Pichidangui Formation has two stable components: one of postfolding age is concordant to apparent polar wander path data, and one of probable prefolding (Late Triassic) age is concordant in declination, but discordant in inclination. Further work is needed to better define the prefolding magnetizations in the Pichidangui Formation, but at present these preliminary results are the first paleomagnetic signs of displaced terranes along the Pacific margin of Chile. If correct, the results suggest that the Pichidangui Formation was some 15° of latitude farther south during the Late Triassic and had likely moved northward to its present latitudinal position with respect to cratonic South America by Middle to Late Jurassic.

Introduction

Through paleomagnetic studies of late Paleozoic and Mesozoic sequences from the more stable platform or cratonic regions of South America [e.g., Creer, 1970; Valencio and Vilas, 1972], one can argue that much of the South American landmass was an integral portion of the Gondwana supercontinent during the late Paleozoic and early Mesozoic, until the opening of the South Atlantic in the Early Cretaceous. Thus far, paleomagnetic studies of rocks from the Andean Cordillera have yielded results both concordant and discordant with the South American and Gondwana apparent polar wander (APW) paths [e.g., Palmer et al., 1980a, b; Heki et al., 1983]. Most of the studies made thus far, especially in sequences exposed west of the Pacific/Atlantic drainage divide, have been in units of post-Early Jurassic age. Here, we report on magnetizations acquired by principally Permian and Triassic formations from the western flanks of the Andes between 25°S and 31°S in Chile (Figure 1).

The paleomagnetism of rocks west of the Andean divide, particularly in the coast ranges and from units of pre-Jurassic age, is important for two reasons. First, in recent years, the geologic community has become increasingly aware of the unique character of the western forearc region of the central Peruvian and Chilean Andes, which is dominated by metamorphic and plutonic rocks of Precambrian to late Paleozoic age. The proximity of these pre-Jurassic crystalline rocks to the Peru-Chile trench has led to speculations that the forearc either was truncated via a cryptic process of "subduction erosion" [Rutland, 1971; Scholl et al., 1977; Kulm et al., 1977; and von Huene et al., 1985] or was the site of the accretion of exotic microplates [Nur and Ben-Avraham, 1978, 1982; Nur, 1983]. Acquiring paleogeographic data is an important step for testing such hypotheses. Second, attempts to make Paleozoic paleogeographic reconstructions for the regions of South America affected by later "Andean" orogenic events must rely on isolated outcrops of pre-Jurassic rocks that are often poorly fossiliferous and variably affected by Andean orogenic events.

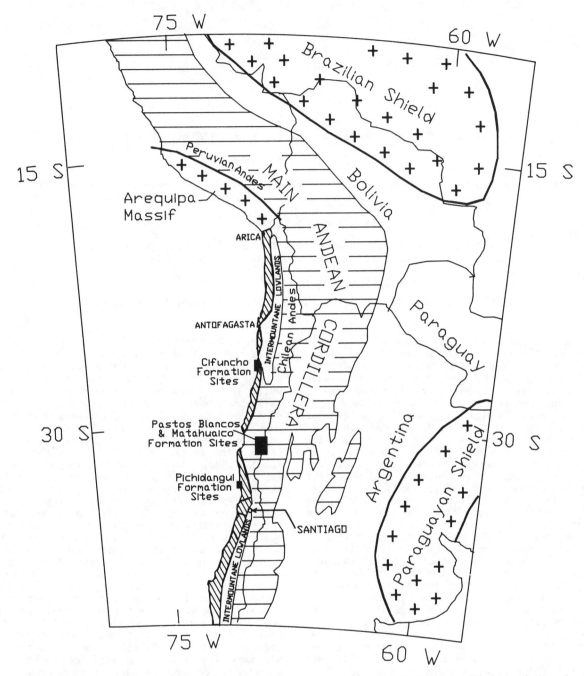

Fig. 1. Location map of the three sampling areas (enlarged in Figures 2, 3, and 4) of Permian and Triassic rocks in north central Chile.

Paleomagnetic data for these isolated complexes, as well as younger units, are necessary to build confidence that such complexes form part of a unified history of the southwest margin of Gondwanaland, rather than a collage of "terrane" elements which might have been brought into juxtaposition by the "docking" of various exotic continental or oceanic microplates during the Mesozoic and Cenozoic.

Outcrops of pre-Jurassic rocks in central and northern Chile are found either in the coastal cordillera or in the western ranges of the main cordillera that form the border between Argentina and Chile. In the coast ranges, rocks of Triassic age tend to be of distinctly different facies from those in the main cordillera, and most of the coast ranges are separated from the main cordillera by fault systems. Thus paleogeographic at-

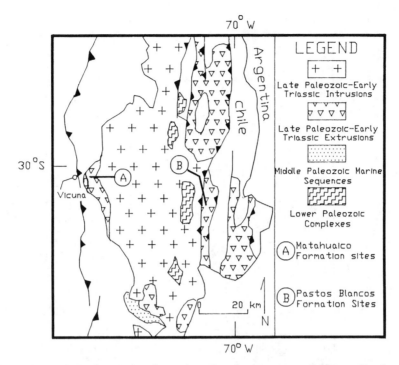

Fig. 2. Paleozoic and Early Mesozoic units in the western Chilean flanks of the main cordillera east of Vicuña. The two sampling localities in the Pastos Blancos and Matahuaico formations are shown.

tempts to reconstruct the pre-Jurassic framework of the central Andes are tenuous on either stratigraphic or tectonic grounds. To determine the tectonic relationships between pre-Jurassic units of the western flanks of the main cordillera and those of the coastal cordillera, we have sampled a number of late Paleozoic, Triassic, and Jurassic units between latitudes 25°S and 31°S. Here, we report on results from four formations: two in the main cordillera and two in the coastal cordillera. We briefly discuss first the geology of the four units sampled in the main and coastal cordilleras and then the paleomagnetic data and their ramifications.

Geology of the Units Sampled

Main Cordillera

Two formations of late Paleozoic to possibly Triassic age were sampled in the main cordillera in the upper reaches of the Elqui River valley (31°S). These are the Pastos Blancos Formation, sampled in the La Laguna River area, and the Matahuaico Formation, sampled east of Vicuña (Figure 2).

Pastos Blancos Formation. The Pastos Blancos Formation [after Thiele, 1964; Maksaev et al., 1984; and Mpodozis and Cornejo, 1986 is a sequence, thousands of meters thick, of rhyolitic ignimbrites and lavas, pyroclastic beds, a few intercalations of andesite, and continental sedimentary rocks. It is thought to be equivalent to the Choiyoi Group of the Frontal Cordillera of Mendoza and San Juan, Argentina [Rolleri and Criado Roque, 1969; Caminos, 1979] that forms an extensive magmatic belt from the North Patagonian Massif [Llambias et al., 1984] toward Iquique, Chile [Coira et al., 1982]. In the La Laguna river valley the Pastos Blancos Formation is formed of ignimbrites, breccias, and rhyolitic lavas intercalated with red continental sediments. The exact age of the formation is uncertain, and the base is not exposed in this area. In the Hurtado River valley, 30 km to the southwest, it rests with angular discordance over the Hurtado Formation, a sequence of graywacke and shale of probable Devonian age [Mpodozis and Cornejo, 1986]. It is intruded by numerous plutons (leucocratic granites) of Permian and Triassic age [Cornejo et al., 1984]. The oldest date obtained is 276 ± 4 Ma (biotite-K/Ar) for a pluton sampled south of the La Laguna reservoir. This date suggests that the Pastos Blancos Formation in the La Laguna area is no younger than Early Permian, but it could be substantially older (Carboniferous). The Choiyoi Group, exposed in Argentina, has always been considered younger, i.e., post-Early Permian, since it rests in the Frontal Cordillera over Upper Carboniferous to Lower Permian marine strata (Tupungato, El Plata, and Cerro Agua Negra formations [Polanski, 1970]). In the valley of Huasco in Chile (29°S), the Pastos Blancos Formation rests over the Las Placetas Formation, which contains Carboniferous plant remains [Reutter, 1974; Nasi et al., 1986]. It seems most likely that the Pastos Blancos Formation is Carboniferous

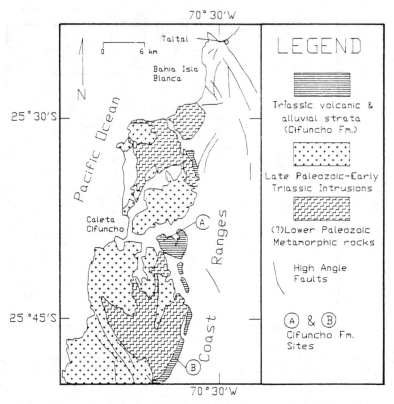

Fig. 3. Paleozoic and lower Mesozoic units in the coast range south of Taltal, near Caleta Cifuncho, with sampling localities indicated.

to Lower Permian. In the sampling area the sequence is homoclinally dipping about 60° to the north-northwest; thus no fold tests were feasible.

Matahuaico Formation. The Matahuaico Formation [after Dedios, 1967], exposed in the Elqui River valley east of Vicuña, is formed of acidic pyroclastic breccias, vitreous lavas, finely laminated tuff, and red sandstones and conglomerates. Its base is not exposed. It is covered by Upper Triassic continental sequences and intruded by leucocratic granite bodies of Permian and Triassic ages [Cornejo et al., 1984]. It can be considered the equivalent to the Pastos Blancos Formation. According to Letelier [1977], it contains plant remains (Cordaites hislopi and Noeggerathiopsis hislopii) attributed to the Permian. The formation is folded at the sampling locality into an upright syncline, with a hinge plunging approximately 40° to the west.

Coastal Cordillera

Two units of Triassic age, the Cifuncho Formation and the Pichidangui Formation, were sampled in the coast ranges (Figures 3 and 4).

Cifuncho Formation. The Cifuncho Formation (25°30'S; after García [1967]) rests with angular discordance over the Paleozoic basement of the coast ranges which here is composed of a deformed sequence of turbidites thought to be of forearc origin [Bell, 1982]. The Cifuncho Formation is concordantly overlain by fossiliferous marine strata of Hettangian age. Following Suarez et al. [1984], the formation is a thick sequence of red conglomerates, sandstones, and shales that were deposited in alluvial fans and braided streams within basins limited by faults in the beginning stages of the "Andean Cycle." It includes scarce remains of plants similar to the classic "flora de la Ternera" attributed to the Upper Triassic [Naranjo and Puig, 1984]. Its precise pre-Jurassic age is uncertain. It is the only Triassic sequence known in the coast ranges of the Antofagasta region of Chile.

Pichidangui Formation. The Pichidangui Formation (latitude 32°S, longitude 71°30'W [after Vicente, 1976]) is composed of 4000 to 5000 m of breccias, tuffs, and keratophyric lavas that were deposited in fluctuating subaquatic continental conditions. The formation extends in the coast from Caleta el Quereo in the north, south of Los Vilos, to the village of Los Molles (Figure 4). It also probably includes some associated intrusions and sills.

The base of the formation rests over Anisian sandstones of the El Quereo Formation with Daonella fauna and scarce ammonites (Gymnites, Ptychites, Sturia, Ceratites, Gryphoceras, and Trematoceras) [Cecioni and Westermann, 1968]. The upper part of the Pichidangui Formation contains intercalations of black shales carrying Esteria and rich in plant remains with Dicroidium, Yabeil-

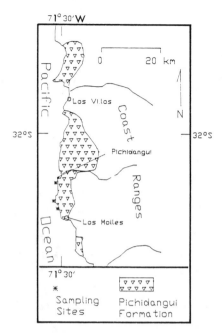

Fig. 4. Distribution of the Late Triassic Pichidangui Formation in the coast ranges near Los Molles along with the sampling locations.

la, Podazamites, Chiropteris, and Neoggerathiopsis attributed to the upper Carnian [Fuenzalida, 1937, 1938; Cecioni and Westermann, 1968].

Near the village of Los Molles, the formation is covered by black shales with plant remains and ammonites (Sandlingites) attributed to the upper Norian. Over these beds lies the Los Molles Formation with a complete sequence passing from the Triassic to the Jurassic in marine facies.

Paleomagnetic Results

Samples oriented by magnetic compass were obtained in the field with a gasoline-powered drill. Sampling was arranged according to sites, each represented by three or more independently oriented samples. At least one 2.2-cm-long specimen core was prepared from each core sample for magnetic measurement.

Natural remanent magnetization (NRM) was measured on either a computer-interfaced two-axis superconducting magnetometer [Goree and Fuller, 1976] or a flux-gate spinner magnetometer [Molyneux, 1971]. Detailed alternating field (AF) and thermal demagnetization studies were conducted initially on a pilot sample from each site. The sample demagnetization data were analyzed on orthogonal projections of vector end-point diagrams, and a progressive demagnetization schedule at a minimum of six levels was designed to isolate significant magnetization components for the remaining available samples from each site. Magnetization components were calculated by least-squares line fitting to the demagnetization trajectories [Kirschvink, 1980]; component mean directions were averaged according to standard paleomagnetic statistical analysis [Fisher, 1953; Watson and Irving, 1957].

Main Cordillera, Pastos Blancos Formation

A total of 39 samples from 10 sites were available for paleomagnetic study. The NRM directions tend to fall in the northeast quadrant with negative inclination, away from the present-day field direction. The inferred stability of the magnetizations is confirmed by demagnetization studies. AF treatment generally defines a demagnetization trajectory that converges toward the origin (Figure 5a). Up to 60% of the NRM, however, can remain after 100 mT, and subsequent thermal treatment to 670°C and higher is required to remove the remaining fraction. Evidently, high-coercivity, high-blocking-temperature hematite is an important remanence carrier in these rocks. The sample magnetizations can, in any case, be regarded as essentially univectorial, origin-seeking trajectories defining characteristic directions that are usually upward and northerly (e.g., Figure 5a), but in two sites are more southerly and downward.

Characteristic directions were isolated in 33 samples representing all 10 sites (Table 1). The northerly and upward (normal polarity) directions from eight sites are well clustered, with a mean of declination (D) of 33°, a mean of inclination (I) of -44.4°, and an alpha 95 of 6.9°, before correction for tectonic tilt (Table 1). After correction for tilt, the directions become northwesterly and shallower. These eight sites represent a variety of rock types, including rhyolites, welded tuffs, and fine-grained dikes.

The downward directions from the two rhyolite flows can be interpreted as of reversed polarity, but they neither are closely grouped (even though from adjacent lava beds) nor collectively or individually appear to be antipodal to the eight normal-polarity site directions. The age of magnetization of the poorly represented reversed polarity directions may therefore differ significantly from that of the normal polarity directions. Although no fold test was possible at the sampling locality, we note that the reversed polarity directions go from westerly to the more expected southerly declination with application of tilt correction.

Main Cordillera, Matahuaico Formation

Of the 24 samples from six sites available for paleomagnetic study from the Matahuaico Formation, 18 were subjected to demagnetization treatments. The sampling locality of this unit, thought to be correlative to the Pastos Blancos Formation, afforded the opportunity of a fold test to help constrain the age of magnetization. Upon demagnetization, most samples showed evidence of multicomponent magnetizations, often with seriously overlapping stability spectra. Despite attempts to isolate characteristic directions for each site (with a preference for the high-temperature component that converged on the origin), site directions failed to group either before or after tilt correction. Unfortunately, the magnetizations of the Matahuaico Formation at the sites sampled are

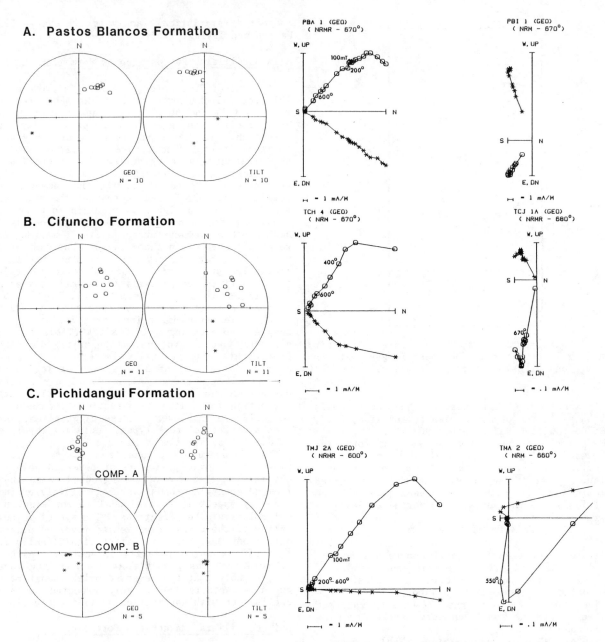

Fig. 5. Paleomagnetic data from the (a) Pastos Blancos, (b) Cifuncho, and (c) Pichidangui formations. At left are equal area plots of site mean directions in geographic (GEO) and tilt corrected coordinates (TILT) for the components isolated in each rock unit. At right are vector end-point diagrams of principally thermal (in degrees Celsius) sometimes preceded by AF demagnetization for representative specimens from each formation. Open circles (stars) are projections on vertical (horizontal) planes.

very complex with no convincing definition of characteristic directions for the formation.

Coastal Cordillera, Cifuncho Formation

The sample collection included sandstones (six sites), basic dikes (three sites), and lava flows (three sites) for a total of 53 samples. Redbed samples were thermally demagnetized up to 680°C, and dike and lava samples AF demagnetized up to 100 mT, although some of the lava samples required subsequent thermal treatment up to 680°C for complete demagnetization. Complex demagnetization trajectories were commonly encountered, although a sufficient number of sites yielded relatively straightforward univectorial decays of consistent

TABLE 1. Paleomagnetic Data From Permian and Triassic Rocks, Chile

			Before Tilt Correction				After Tilt Correction			
Sites	Samples	Polarity	Declination	Inclination	k	alpha 95	Declination	Inclination	k	alpha 95
Pastos Blancos Formation (Late Carboniferous-Early Permian, 31.1°S, 70.1°W)										
8	29	N	33.1[a]	-44.4	65	6.9	344.7	-29.8	65	6.9
2	4	R	272.4	36.3	--	---	174.9	71.2	--	---
Cifuncho Formation (Late Triassic, 25.6°S, 70.6°W)										
9	28	N	40.1	-47.6	29	9.8	50.7	50.8	16	13.3
2	5	R	196.0	59.0	--	---	163.6	52.2	--	----
11	33	N, R	36.6[b]	-50.0	24	9.6	40.3	-53.8	10	15.1
Pichidangui Formation (Late Triassic, 31.2°S, 71.5°W)										
10(A)	33	N	353.1[c]	-50.0	69	5.8	341.5	-40.9	24	10.1
5(B)	14	R	242.7	71.1	50	11.0	191.7[d]	74.6	134	6.6

N is normal; R is reversed. Dashes indicate only marginally significant data.
[a] The pole is 60.7°S, 200.3°E; dp = 5.5°, dm = 8.7°.
[b] The pole is 57.5°S, 217.1°E; dp = 8.6°, dm = 12.8°.
[c] The pole is 84.1°S, 20.7°E; dp = 5.2°, dm = 7.8°.
[d] The pole is 59.0°S, 277.5°E; dp = 10.9°, dm = 12°.

orientation from sample to sample, as illustrated in Figure 5b, to allow meaningful averaging of what we regard as the characteristic direction. One lava site, however, failed to yield at least one interpretable sample demagnetization diagram.

The site-mean characteristic directions fall predominantly into a normal polarity, northwesterly, and up grouping (nine sites), but two sites (sediment and a lava) have nearly antipodal (reversed polarity), southerly, and down directions (Table 1). After tectonic tilt corrections (assuming the dikes are feeders for the lavas), scatter in the site-mean directions increases. The precision parameter decreases with tilt correction by a factor of 2.4, which is significant at the 95% confidence level. It therefore appears that the directions are of postfolding origin, and the in situ mean for the 11 sites (D=36.6°, I=-50°, and alpha 95=9.6°) can be regarded as a good estimate of the geomagnetic field in which the secondary magnetization was acquired.

Coastal Cordillera, Pichidangui Formation

Basaltic and rhyolitic lavas and tuffs were sampled at 13 sites (47 samples). The NRM directions tend to fall near the present geomagnetic field. AF and thermal demagnetization reveal the presence of two components of magnetization. The most common and dominant component (labeled A) is northerly and up and can be isolated in 10 sites. In six of these sites, it is the only consistent component of magnetization present as shown by linear trajectories that converge toward the origin by 100 mT AF or 600°C thermal treatment (Figure 5c). In four sites, however, the A component trajectory does not go toward the origin, and further temperature treatment up to 680°C reveals a southerly and downward magnetization (component B) of high stability (Figure 5c). One additional site shows the B component even though the A component cannot be well resolved.

Component A fails the fold test at the 95% confidence level; for these normal polarity sites the in situ mean is D=353.1°, I=-50°, with alpha 95=5.8° (Table 1). The precision parameter improves by a factor of 2.7 after tilt correction for the B component, and therefore this reversed polarity direction may be of pre-folding origin. For the five sites, the mean B component direction after tilt correction is D=191.7° and I=74.6°, with alpha 95=6.6° (Table 1), which may possibly represent a Late Triassic direction.

Discussion

For comparison with the paleomagnetic directions reported here from northern Chile, we use the apparent polar wander path for South America determined by Irving and Irving [1982]. They compiled 34 paleopoles representing the time interval 380 Ma to the present. These paleopoles, thought to provide reference directions for the stable interior of South America, were averaged through a 30 Ma time window, moved in 10 Ma steps, and assigned confidence circles. These poles are used to define a 300 to 120 Ma polar wander path shown in Figure 6 along with the results obtained from the two coastal localities.

We note that for virtually all of Mesozoic and Cenozoic time, and even in the Late Permian (i.e., 260 Ma to the present), the mean paleopoles for South America as calculated by Irving and Irving [1982] are not significantly different from the present dipole field (geographic axis). Only a Late Triassic (200 Ma) mean shows a departure (16°) which exceeds the associated 95% confidence level (11°). Therefore, if these references poles are taken at face value, predicted directions from 60 Ma to the present should have declinations near

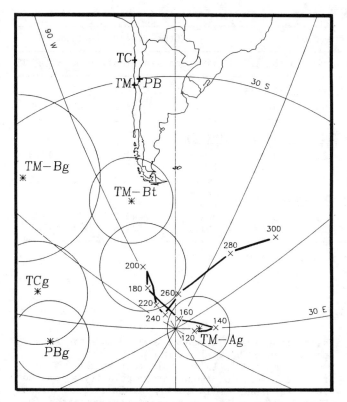

Fig. 6. Averaged APW path for South America [Irving and Irving, 1982] along with the poles and their 95% confidence limits for the localities reported here (TCg is Cifuncho Formation without tilt correction; TM-Bg and TM-Bt are Pichidangui Formation B component, without and with tilt correction, respectively, TM-Ag is Pichidangui Formation A component without tilt correction; and PBg is Pastos Blancos Formation without tilt correction). The confidence circle for the 200 Ma position of the APW is shown for comparison with the poles derived from the sampled formations. Formation confidence circles are calculated from virtual geomagnetic poles determined for each site mean.

0° (180° for reversed polarity) and inclinations (negative for normal, positive for reverse polarity) conforming to the present latitude of the sampling locality according to the dipole formula. For Late Carboniferous and Early Permian times (300 Ma to 270 Ma), the mean paleopoles for South America show significant departures from the present field axis, and would predict in north-central Chile (~31°S, 71°W) northwesterly declinations (e.g., 322° for a 300 Ma mean pole) and inclinations somewhat steeper than at present (e.g., -58.5° for a 300 Ma mean pole, compared to -50° for the present dipole field).

The reference pole data set for South America is admittedly not very satisfactory because of the general absence of fold and other field tests to demonstrate that the magnetizations are not seriously contaminated by more recent overprints. However, some support for the hypothesis that can be drawn from these data, that South America in fact occupied, since about the late Permian, an orientation not very dissimilar from today's position, comes from comparison of paleopoles from Africa and the other southern continents when they are reconstructed to their past positions relative to South America [e.g., Van der Voo and French, 1974; Irving and Irving, 1982; Norton and Sclater, 1979].

In the absence of a fold test, it is uncertain whether the paleomagnetic results from the Pastos Blancos Formation represent primary or secondary magnetizations. We believe that at least the normal polarity directions that are dominant in our collection do not represent primary magnetizations because this would conflict with the predominantly reversed polarity Kiaman Interval of Late Carboniferous and Permian age [Irving and Pullaiah, 1976], the time interval in which the Pastos Blancos Formation is thought to have formed. If the normal polarity magnetizations are of secondary origin, we find that the mean northeasterly declination calculated without tilt correction does not correspond to predicted Mesozoic or Cenozoic field directions, even though there is no notable discrepancy in the mean inclination. Therefore the data would suggest a clockwise rotation of about 30°, but no discernible latitudinal motion of the Pastos Blancos Formation since acquisition of its postfolding characteristic magnetizations. If the secondary magnetizations were acquired after the Kiaman Interval but were still prefolding, we must conclude the opposite: no significant discordance in declinations but more than 15° of latitudinal discrepancy, a result placing the Pastos Blancos Formation at much lower latitudes with respect to cratonic South America. As is discussed below, the former alternative implies a rotation that is consistent with that seen in the Cifuncho Formation, while the latter implies a latitudinal shift that is inconsistent with that seen in the Pichidangui Formation. Thus secondary acquisition of the characteristic directions in the Pastos Blancos Formation appears to be the more reasonable alternative. Unfortunately, one cannot exclude the possibility that the directions may have been subsequently affected by local tilting. The poorly defined reversed directions at two sites in the formation might record an earlier, possibly primary, magnetization.

The stable directions from 11 sites in the Cifuncho Formation are apparently of secondary origin because of a negative fold test, even though two of the sites have reversed polarity. As for the Pastos Blancos Formation secondary magnetizations, the Cifuncho Formation mean direction without correction for bedding tilts is rotated by about 30° clockwise, but there is no significant discrepancy in the mean inclination with respect to available post-Late Triassic reference directions for South America. The clockwise discordance is significant regardless of what available Late Triassic or younger reference direction is chosen for comparison. Of course, one cannot exclude the possibility that there may be errors in the APW data or, for that matter, that small amounts of block tilting about horizontal axes could be present. However, below we show that the clockwise discordance seen here is part of a broad regional pattern along the western flank of the Chilean Andes, leaving the alternative of local block tilting seemingly ad hoc.

The paleomagnetic results from the Pichidangui Formation show two components of magnetization. The normal polarity A component fails the fold test. However, unlike the secondary magnetizations in the Pastos Blancos and Cifuncho formations, the A component mean direction conforms in both declination and inclination with a younger (post-Late Triassic) paleomagnetic field direction for South America. We do not know the age of the secondary magnetizations, but we know that it must be younger than the ages of the rock units and their time(s) of deformation. If the secondary magnetizations of the Pastos Blancos, Cifuncho, and Pichidangui formations are all of the same age, the Chilean Andes have a complex regional pattern of apparent rotation, with clockwise rotation in the main cordillera and the Cifuncho region, but apparently no rotation in the more southerly Pichidangui coastal area. Alternatively, the A component of the Pichidangui Formation may be significantly younger, acquired after the clockwise rotation was recorded in the Pastos Blancos and Cifuncho formations.

The reversed polarity B component in the Pichidangui Formation may represent an original (i.e., Late Triassic) magnetization. It is the only magnetization component found in any of the rock units reported here which shows better grouping after tilt corrections. The south pole position for the tilt-corrected direction falls at latitude 59.0°S, longitude 82.5°W (dp=10.9°, dm=12°). This is closest to the 200 Ma (Late Triassic reference pole of Irving and Irving [1982], but still some 15° of arc different. The 200 Ma reference pole predicts a declination of 187° and an inclination of 64.5° at the Pichidangui sites; in contrast, the observed declination and inclination were 191.7° and 74.6°, respectively. The good agreement between predicted and observed declinations suggests that there have been few significant rotations; however, the difference between predicted and observed inclinations is statistically significant and implies that almost 15° of northward latitudinal transport occurred sometime after the Late Triassic. Because most samples were dominated by the secondary A component, additional sampling will be necessary before the B component can be better documented. If the secondary A component was acquired as a thermochemical overprint during the peak of arc activity in the Pichidangui area in the Late Jurassic or Early Cretaceous, the concordance of the A component with the South American reference APW path would suggest that the apparent northward translation of this coastal block was completed in the Jurassic. Note that by considering the B component to also be secondary, the amount of directional discordance increases in comparison with younger reference poles, so that about 30° of latitudinal motion would be required to reconcile the observed with predicted paleomagnetic directions.

Recently, paleomagnetic results from Jurassic and Early Cretaceous intrusions in this coastal region, including bodies which cut the Pichidangui Formation, have been reported [Irwin et al., 1985] and show no major discordance with respect to Jurassic or Cretaceous reference positions. These results clarify interpretations of the magnetizations recorded in the Pichidangui Formation. First, since intrusive units of Jurassic age that intrude the Pichidangui Formation show no apparent rotations with respect to reference directions, folding in the Pichidangui Formation must have been completed prior to emplacement of the intrusions. This suggests that a period of deformation occurred not previously reported elsewhere in Chile and that the B component of magnetization in the Pichidangui Formation is likely of pre-Late Jurassic age. Since the directions of the A component are consistent with the directions recorded in the Jurassic and Early Cretaceous intrusions, it is not unreasonable to suppose that the A component was acquired as a secondary thermochemical overprint during emplacement and initial cooling of the intrusions. While further work is clearly needed, it appears that structural and magnetic data support the existence of a microplate of Late Triassic age represented by the Pichidangui Formation, which had moved northward with respect to cratonic South America and been deformed (during "docking"?) between the Late Triassic and Late Jurassic, after which it was intruded by the initial plutonic elements of the Andean arc system in the Late Jurassic and Early Cretaceous.

Further comparison of results obtained in Pastos Blancos, Cifuncho, and Pichidangui formations to other published results from the Andes between 10°S and 32°S brings out several interesting relationships. In Figures 7a and 7b, observed declinations and inclinations from these studies (including the four reported here) are plotted against predicted declinations and inclinations for the various sampling sites from regions north of 19°S and regions south of 19°S, respectively (assuming sites are fixed with respect to stable South America). Looking first at results reported from the western flank of the Andes in Peru, we find that observed inclinations show no consistent patterns of discordance with respect to the expected (cratonic reference) inclinations. While one could say that the majority of the observed results are consistent with no latitudinal discordance of coastal terranes, some of the inclinations are too steep and others are too shallow. That inconsistency arises even in units of similar age sampled from one general region leads to the suspicion that errors may be present. However, the observed declinations plotted against expected declinations show a surprising consistency (given the inconsistency in the inclinations) in their general counterclockwise discordance. This pattern has been noted by Heki et al. [1983], May and Butler [1985], and Beck [1985]. Heki et al. have pointed out, following the suggestion of Carey [1955], that it may reflect oroclinal bending from Peru to Chile (Arica orocline). In Figure 7b, data for the western flank of the Andes south of 19°S is plotted. First, note that with the excep-

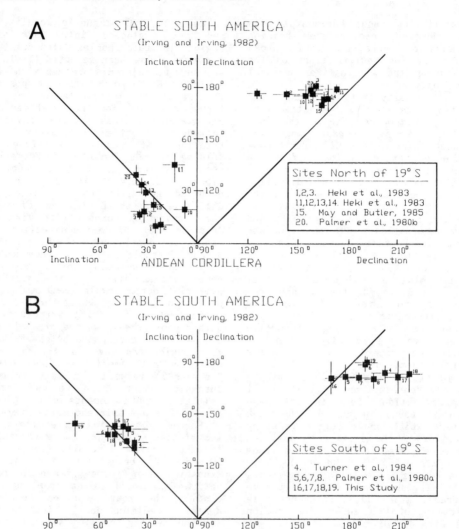

Fig. 7. Predicted versus observed declinations and inclinations for a number of units that have been sampled between 10°S and 32°S in the Andes: (a) sites north of 19°S and (b) sites south of 19°S. The predicted values for each locality assume no relative motion between the Andean ranges and "stable" or cratonic South America when APW data published by Irving and Irving [1982] are used. Vertical bars represent a range of predicted values based on the uncertainty of the age of the units sampled or the time of NRM aquisition. The horizontal bars reflect confidence limits of mean directions determined for the various units. All reported paleomagnetizations are believed to be Jurassic or younger with the exception of the B component from the Pichidangui Formation (site 19) reported here.

tion of the B component of the Pichidangui Formation (which is the only magnetization direction plotted of likely pre-Jurassic age), none of the reported inclinations are significantly discordant from their expected inclinations. Thus we first conclude that these data contain no evidence of latitudinally displaced terranes of Late Jurassic or younger ages. Second, observed declinations show the opposite pattern of discordance from that recorded from the Peruvian Andes. While the tectonic mechanism responsible for this systematic shift or discordance is debatable, we note that simple hypotheses of oroclinal bending are unlikely to explain the observed variations in amounts of discordance, for example, that seen between the Cifuncho and Pichidangui formations of the Chilean coast ranges. Other hypotheses, such as systematic block rotations of border terranes in association with oblique plate convergence [Beck, 1980], are difficult to reconcile with rotations seen in the units from the main Cordillera, e.g., Pastos Blancos Formation.

Much more work, including better refining of the APW data for stable South America, is needed to define the extent and timing of the apparent rotations before a tectonic agent can be isolated.

Finally, evidence for latitudinal displacements revealed in the paleomagnetic results from the Late Triassic Pichidangui Formation emphasizes the need to isolate primary components in pre-Jurassic units of the Andes.

Acknowledgments. Support for this work comes from National Science Foundation grants INT 83-00521, EAR 82-12549, and EAR 85-07046, and from the Servicio Nacional de Geología y Minería. We thank reviewers for their comments and criticisms. This is Lamont-Doherty Geological Observatory Contribution No. 4075.

References

Beck, M. E., Jr., Paleomagnetic record of plate-margin tectonic processes along the western edge of North America, J. Geophys. Res., 85, 7115-7131, 1980.

Beck, M. E., Jr., Tectonic significance of paleomagnetic studies in the Andes, Comunicaciones, 35, 15-18, 1985.

Bell, C. M., The lower Paleozoic metasedimentary basement of the coastal range of Chile between 25°30'S and 27°S, Rev. Geol. Chile, 17, 339-348, 1982.

Caminos, R., Cordillera Frontal, in Geología Regional Argentina, edited by A. F. Leanza, pp. 305-343, Academia Nacional de Ciencias, Cordoba, 1979.

Carey, S. W., The orocline concept in geotectonics, R. Soc. Tasmania Proc., 89, 255-288, 1955.

Cecioni, G., and T. E. Westermann, The Triassic-Jurassic marine transition of coastal-central Chile, Pac. Geol., 1, 41-75, 1968.

Coira, B., J. Davidson, C. Mpodozis, and V. Ramos, Tectonic and magmatic evolution of the Andes of northern Argentina and Chile, Earth Sci. Rev., 18, 303-332, 1982.

Cornejo, P., C. Nasi, and C. Mpodozis, La Alta Cordillera entre Copiapo y Ovalle, in Seminario Geología de Chile, Apuntes, Miscelania, pp. H1-H45, Servicio Nacional de Geología y Minería, Santiago, Chile, 1984.

Creer, K. M., A paleomagnetic survey of South American rock formations, Philos. Trans. R. Soc. London Ser. A, 267, 457-558, 1970.

Dedios, P., Caudrángulo Vicuña, provincia de Coquimbo, in Carta Geologica de Chile, 16, 65 pp., Instituto de Investigaciones Geologicas, Santiago, Chile, 1967.

Fisher, R. A., Dispersion on a sphere, Proc. R. Soc. London, Ser. A, 217, 295-305, 1953.

Fuenzalida, H., El Rético de la costa de Chile central, Bol. Minas Pet., 65, 1-11, 1937.

Fuenzalida, H., El Rético de la costa de Chile central, Bol. Mus. Hist. Nac., 10, 66-98, 1938.

García, F., Geología del Norte Grande de Chile, in Symposium Sobre el Geosinclinal Andino, Soc. Geol., Chile Publ., vol. 3, 183 pp., Sociedad Geologico de Chile, Santiago, 1967.

Goree, W. S., and M. Fuller, Magnetometers using RF-driven squids and their applications in rock magnetism and paleomagnetism, Rev. Geophys., 14, 591-608, 1976.

Heki, K., Y. Hamono, and M. Kono, Rotation of the Peruvian block from paleomagnetic studies of the central Andes, Nature, 305, 514-516, 1983.

Irving, E., and G. A. Irving, Apparent polar wander paths Carboniferous through Cenozoic and the assembly of Gondwana, Geophys. Surv., 5, 141-188, 1982.

Irving, E., and G. Pullaiah, Reversals of the geomagnetic field, magnetostratigraphy, and relative magnitude of paleosecular variation in the Phanerozoic, Earth Sci. Rev., 12, 35-64, 1976.

Irwin, J. J., R. E. Drake, and W. D. Sharp, Some paleomagnetic constraints on the tectonic evolution of the coastal Cordillera of central Chile, Eos Trans. AGU, 66, 866, 1985.

Kirschvink, J. L., The least-squares line and plane and the analysis of paleomagnetic data, Geophys. J. R. Astron. Soc., 62, 699-718, 1980.

Kulm, L. D., W. J. Schweller, and A. Masías, A preliminary analysis of the subduction process along the Andean continental margin 6° to 45°S, in Island Arcs, Deep Sea Trenches, and Back-Arc Basins, Maurice Ewing Ser., vol. 1, edited by M. Talwani and W. C. Pitman III, pp. 285-302, AGU, Washington D. C., 1977.

Letelier, M., Petrología, ambiente de depositación y estructura de las formaciones Matahuaico, Las Breas, Tres Cruces sensu lato, e intrusivos hipabisales Permo-Triásicos, IV región, Chile, para título de Geólogia (for obtaining the title of Geologist), 131 pp., Dep. de Geol. de Chile, University of Santiago, Santiago, 1977.

Llambias, E., R. Caminos, and C. Rapela, Las plutonitas y volcanitas del ciclo eruptivo Gondvánico, Congr. Geol. Argent. Bariloche Relat., 9th, 85-117, 1984.

Maksaev, V., R. Moscoso, C. Mpodozis, and C. Nasi, Las unidades volcánicas y plutónicas del Cenozoico superior en la Alta Cordillera del Norte Chico (29°-31°S), Geología, alteración hidrotermal y mineralización, Rev. Geol. Chile, 21, 1151, 1984.

May, S. R., and R. F. Butler, Paleomagnetism of the Puente Piedra Formation, central Peru, Earth Planet. Sci. Lett., 72, 205-218, 1985.

Molyneux, L., A complete result magnetometer for measuring the remanent magnetization of rocks, Geophys. J. R. Astron. Soc., 10, 429, 1971.

Mpodozis, C., and P. Cornejo, Geología de la Hoja Pisco Elqui, in Carta Geologica de Chile, scale 1:250,000, Servicio Nacional de Geología y Minería, Santiago, Chile, in press, 1986.

Naranjo, J. A., and A. Puig, Geología de las Hojas Taltal y Chañaral, in Carta Geologica de Chile, scale 1:250,000, 62-63, 140 p., Servicio Nacional de Geología y Minería, Santiago, Chile, 1984.

Nasi, C., R. Moscoso, and V. Maksaev, Geología de la Hoja Guanta, in Carta Geologia de Chile, scale 1:250,000, Servicio Nacional de Geología y Minería, Santiago, Chile, in press, 1986.

Norton, I. O., and J. G. Sclater, A model for the evolution of the Indian Ocean and the breakup of Gondwanaland, J. Geophys. Res., 84, 6803-6830, 1979.

Nur, A., Accreted terranes, Rev. Geophy., 21, 1779-1785, 1983.

Nur, A., and Z. Ben-Avraham, Speculations on mountain building and the lost Pacific continent, J. Phys. Earth, Suppl. 26, 21-37, 1978.

Nur, A., and Z. Ben-Avraham, Oceanic plateaus, the fragmentation of continents and mountain building, J. Geophys. Res., 87, 3644-3661, 1982.

Palmer, H. C., A. Hayatsu, and W. D. MacDonald, The Middle Jurassic Camaraca Formation, Arica, Chile: Paleomagnetism, K-Ar age dating and tectonic implications, Geophys. J. R. Astron. Soc., 62, 155-172, 1980a.

Palmer, H. C., A. Hayatsu, and W. D. MacDonald, Paleomagnetic and K-Ar age studies of a 6 km-thick Cretaceous section from the Chilean Andes, Geophys. J. R. Astron. Soc., 62, 133-153, 1980b.

Polanski, J., Carbónico y Pérmico de la Argentina, 216 pp., EUDEBA, State University of Buenos Aires, 1970.

Reutter, K. J., Entwicklung und Bauplan der chilenischen Hochkordillere im Bereich 29°S sudlicher Breite, Neues Jahrb. Geol. Palaeontol. Abh., 146, 153-178, 1974.

Rolleri, E., and P. Criado Roque, Geología de la provincia de Mendoza, Actas Congr. Geol. Argent. 4, 2, 1-60, 1969.

Rutland, R. W. R., Andean orogeny and ocean flow spreading, Nature, 233, 252-255, 1971.

Scholl, D. W., M. S. Marlow, and A. K. Cooper, Sediment subduction and offscraping at Pacific Margins, in Island Arcs, Deep Sea Trenches, and Back-Arc Basins, Maurice Ewing Ser., vol. 1, edited by M. Talwani and W. C. Pitman III, pp. 199-210, AGU, Washington, D. C., 1977.

Suarez, M., J. A. Naranjo, and A. Puig, Estratigrafia de la Cordillera de la Costa al sur de Taltal, etapas iniciales de la evolución andina, Rev. Geol. Chile, 24, 19-28, 1984.

Thiele, R., Reconocimiento geológico de la Alta Cordillera de Elqui, Geol. Publ. 27, 73 pp., Univ. de Chile, Santiago, 1964.

Turner, P., H. Chemmey, and S. Flint, Paleomagnetic studies of a Cretaceous molasse sequence in the central Andes (Coloso Formation, northern Chile), J. Geol. Soc. London, 141, 869-876, 1984.

Valencio, D. A., and J. F. Vilas, Paleomagnetism of late Paleozoic and early Mesozoic rocks of South America, Earth Planet Sci. Lett., 15, 75-85, 1972.

Van der Voo, R., and R. B. French, Apparent polar wandering for the Atlantic bordering continents: Late Carboniferous to Eocene, Earth Sci. Rev., 10, 99-119, 1974.

Vicente, J. C., Exemple de volcanisme initial euliminaire: Les complexes albitopyriques neotriasiques et meso Jurassiques du secteur cotier des Andes meridionales centrales 32° à 33° L. sud), Spec. Ser. Int. Assoc. Volcanol. Chem. Earth's Inter., 267-329, 1976.

von Huene, R., L. D. Kulm, and J. Miller, Structure of the frontal part of the Andean convergent margin, J. Geophys. Res., 90, 5429-5442, 1985.

Watson, G. S., E. Irving, Statistical methods in rock magnetism, Mon. Not. R. Astron. Soc., Geophys. Suppl., 7, 289-300, 1957.

Copyright 1987 by the American Geophysical Union.

LATE PALEOZOIC PSEUDOALBAILLELLID RADIOLARIANS FROM SOUTHERNMOST CHILE
AND THEIR GEOLOGICAL SIGNIFICANCE

Hsin Yi Ling

Department of Geology, Northern Illinois University, DeKalb, Illinois 60115

Randall D. Forsythe

Department of Geological Sciences, Rutgers University, New Brunswick, New Jersey 08854
Lamont-Doherty Geological Observatory, Columbia University, Palisades, New York 10964

Abstract. Details of late Paleozoic (Late Carboniferous and Early Permian) radiolarians from Isla Madre de Dios and Isla Regalada of southernmost Chile are presented for the first time, and their faunal compositions compared with those of the published records. One new species, Pseudoalbaillella chilensis, and one new subspecies, Ps. u-forma reflexa are proposed. Ages suggested by these findings, having older (Late Carboniferous) age limits for the Madre de Dios Archipelago and younger (Early Permian) for Regalada Island, support a S to SE younging pattern for the timing of accretion and the material accreted for the forearc terrane. The trend may be continued into the Antarctic Peninsula, where even younger Triassic radiolarian chert has been reported.

Introduction

The successful recovery of well-preserved and abundant radiolarians and fusulinids of latest Carboniferous and early Early Permian ages from the Madre de Dios Archipelago provided the age constraints necessary for the tectonic interpretation of the Chilean part of the Gondwanaland forearc [Ling et al., 1985]. We have subsequently identified the earliest Permian radiolarian cherts from Isla Regalada (Figure 1). The finding of slightly younger radiolarian assemblages from Isla Regalada, located south of Isla Madre de Dios, agrees well with the southward accretional interpretation for the Gondwanaland forearc regional development previously suggested [Ling et al., 1985].

Sample Location

Radiolarian-bearing samples from Isla Madre de Dios of the Madre de Dios Archipelago have already been discussed [Ling et al., 1985, Figure 1]. A preliminary examination of Isla Madre de Dios samples revealed two radiolarian faunas belonging to Carboniferous and Permian periods. Approximately equal amounts of the samples were treated for equal amounts of time so that the populations for the samples and the relative abundances of respective taxa within the assemblages could be compared. The samples from Isla Regalada (Figure 1) were collected by one of us (R.D.F.) during the 1984 austral summer season. All samples from the island yielded abundant radiolarians. Unfortunately, the area was tectonically disturbed, thus hampering efforts to decipher their stratigraphic relationships. The samples were treated with dilute hydrofluoric acid at different concentrations and for varying durations in order to collect at least 100 specimens from each sample.

Age Determination and Discussion
of Radiolarian Assemblages

Although the beginning of Paleozoic radiolarian study can be traced back to an analysis by Rüst [1892], or more recently by Deflandre [1952], investigative efforts have intensified only in the last decade. This has resulted in a preliminary zonation for Upper Devonian through Permian from the United States [Holdsworth and Jones, 1980], an Upper Carboniferous through Permian biostratigraphy from Japan [Ishiga, 1982; Ishiga and Imoto, 1980; Ishiga et al., 1982a, b, c; 1984], and a comparison of Upper Carboniferous through Permian between the Urals and Texas [Nazarov and Ormiston, 1985; Cornell and Simpson, 1985]. The fact that the radiolarian faunas from southernmost Chile consist almost exclusively of species from the genus Pseudoalbaillella permits close comparison with faunas of Japanese occurrence and, to a lesser extent, to the fauna of the Urals.

In Japan, a radiolarian faunal succession encompassing Late Carboniferous through Early to Middle Permian was initially established by Ishiga [1982] and Ishiga and his associates [Ishiga and Imoto, 1980; Ishiga et. al, 1982a, b, c] in ascending order as Ps. bulbosa assemblage for the latest Carboniferous and Ps. u-forma - Ps. elegans and Ps. lomentaria assemblages for early to middle Early Permian. Subsequently realizing that Ps.

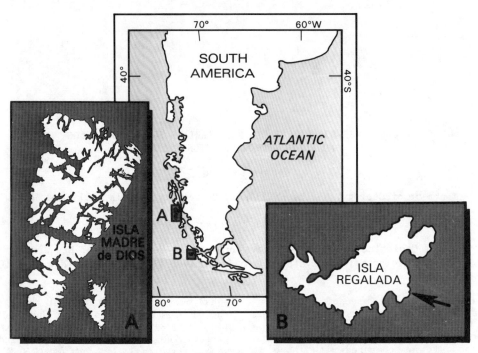

Fig. 1. Index map showing (a) Isla Madre de Dios (see Ling et al. [1985, Figure 1] for details) and (b) Isla Regalada (see arrow) where studied samples were collected.

u-forma can be differentiated into two closely related but stratigraphically separated morphotypes, Ishiga et al. [1984] added a new assemblage at the basal part of the Lower Permian section. Their revised Early Permian succession is thus, in ascending order, Ps. annulatus - Ps. u-forma morphotype I (= u-forma u-forma in this paper), Ps. u-forma morphotype II (= u-forma reflxa n. subsp. in this paper) - Ps. elegans, and Ps. lomentaria assemblages (Figure 2).

It should be mentioned here that although spumellian radiolarians were also observed in the samples along with sponge spicules [Ling et al., 1985, Figure 1 P-T], they are less diversified and far fewer in number. Given the restraints for this article, they are thus excluded from the following discussion.

Isla Madre de Dios assemblages. As presented briefly in an earlier paper [Ling et al., 1985], two distinct radiolarian assemblages are identified from the island (Figure 2). The B1 and B3 sample assemblages were characterized by the predominance of the rather limited geological range of Pseudoalbaillella sp. cf. Ps. bulbosa, similar to the latest Carboniferous "Ps. bulbosa assemblage" recorded from Japan by Ishiga [1982], or, similar to the assemblage of Gzhelian age of the southern Urals by the occurrence of Haplodiacanthus circinatus by Nazarow and Ormiston [1985]. However, the complete absence of Ps. bulbosa and the dominance of Ps. simplex led us to assign the A6 samples to the Early Permian Ps. u-forma morphotype II (= u-forma reflexa) - Ps. elegans assemblage [Ishiga, 1982; Ishiga and Imoto, 1980; Ishiga et al., 1982a, 1982b, 1984], although

specimens of Ps. u-forma reflexa were not observed in the A6 sample.

Isla Regalada assemblages. The occurrence and relative abundance of radiolarians in six (IR 1-6) samples are shown in Figure 3. All samples contain Ps. chilensis n.sp., Ps. simplex, and Ps. annulatus, and, in addition, samples 1 and 3 yielded Ps. u-forma u-forma (= u-forma morphotype I of Ishiga et al. [1984]).

The initial appearance of of Ps. u-forma (morphotype I = u-forma u-forma) was selected as the base of the Ps. u-forma - Ps. elegans assemblage of the basal Wolfcampian [Holdsworth and Jones, 1980; Ishiga and Imoto, 1980; Ishiga et al., 1982b, 1984]. In the original presentation of this assemblage from the Sasayama area, Ps. simplex is dominant over Ps. u-forma (= u-forma morphotype II = u-forma reflexa) in the lowest sample (55). In the next sample (50), less than 1 m above, Ps. elegans was present in place of Ps. simplex, but radiolarians were rare. A similar radiolarian fauna persisted in the next samples (27-38), which were 2 m stratigraphically higher. Subsequent samples, above the 5.5-m stratigraphic interval, contained the Ps. lomentaria assemblage.

In a recent study from the Ohmori area, as previously discussed, Ishiga et al. [1984] recognized two morphotypes within a species of Ps. u-forma and consequently established three assemblages within the Wolfcampian. The lowest, Ps. annulata - Ps. u-forma morphotype I (= u-forma u-forma) assemblage (block A), is characterized by the co-occurrence of the named species, but without Ps. simplex. These two taxa became extinct before reaching the next higher zone (block B) of Ps.

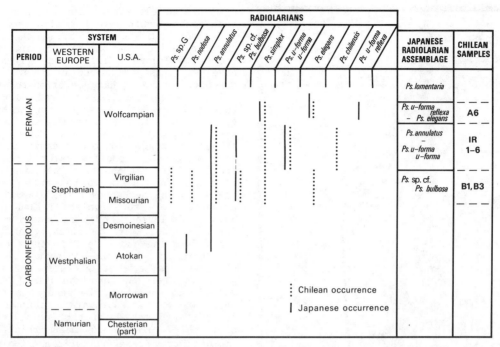

Fig. 2. Comparison of pseudoalbaillellid radiolarian occurrences between Chile and Japan. Japanese ranges are after Ishiga [1982] and Ishiga et al. [1984]. Chilean samples A and B are from Isla Madre de Dios; IR, from Isla Regalada.

u-forma morphotype II (= u-forma reflexa n. subsp. herein) - Ps. elegans where the local range of Ps. simplex was observed. Ps. u-forma morphotype II (= u-forma reflexa) made its initial appearance about 40 cm above the first occurrence of Ps. simplex. In the next stratigraphically higher sample, Ps. elegans showed its initial appearance, and finally, Ps. sp. aff. Ps. scalprata appeared in the slightly higher samples. The Ps. lomentaria assemblage is characterized by the occurrence of Ps. lomentaria and Ps. sakmarensis, but these two species were never observed in samples from southernmost Chile.

In comparison with these Japanese occurrences, the Isla Regalada samples can be grouped into two faunas: fauna I for samples 1 and 3 below and fauna II for the remaining samples above, based on the presence or absence of Ps. u-forma u-forma (u-forma morphotype I of Ishiga et al. [1984]). The relative stratigraphic position of individual samples within respective assemblages is not possible, as stated earlier.

Fauna I is characterized by the occurrence of Ps. u-forma u-forma and Ps. annulatus. Occurrence of Ps. simplex in the Chilean samples but absence from the correlative Japanese samples ("Block A" of Ishiga et al. [1984]) is puzzling. However, this can be explained by allowing its earlier geological appearance in Chile. Fauna II is recognized by the absence of Ps. u-forma u-forma. In the Ohmori section of Japan [Ishiga et al., 1984], there is a radiolarian faunal gap between the last occurrence of Ps. u-forma u-forma and the initial appearance of Ps. simplex in the uppermost part of block A and the lowermost part of block B. Unfortunately, one major and one minor fault were also recognized in this interval.

With these considerations, sample A6 of Isla Madre de Dios can then be interpreted as slightly younger than fauna II of Isla Regalada.

The sharpest contrast between the Chilean and Japanese faunas (as well as the Russian fauna) is the common cooccurrence of conodonts in Japanese

Sample No.	Taxa			
	Ps. u-forma u-forma	Ps. annulatus	Ps. chilensis	Ps. simplex
2		10	69	21
4		15	71	14
5		7	60	33
6			85	15
1	54	3	5	38
3	47	16	8	30

Fig. 3. Distribution and relative abundance (in percentage) of pseudoalbaillellid radiolarians from the Isla Regalada samples.

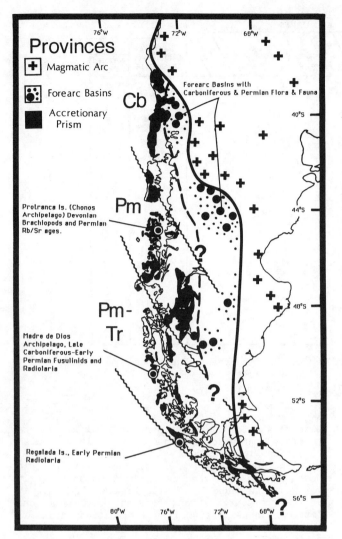

Fig. 4. Distribution of forearc complexes in southernmost South America with fossiliferous localities shown. General provinces are after Forsythe and Mpodozis [1983]. Age relationships suggest a S to SW younging pattern of accretion and accreted material from the Carboniferous (Cb) and Permian (Pm) to Permo-Triassic (Pm-Tr) times.

samples; yet only one or two broken denticles of conodonts were observed in the Chilean samples examined.

Regional Significance

The overall distribution and possible extension of the South American sector of the Gondwana forearc is shown in Figure 4. Only the three indicated localities at the western margin of the forearc have reported fossils that are, with some degree of confidence, situated within the accretionary belt of the forearc. Other localities to the east could be either forearc basin (in situ) or accreted elements. The northern locality in the Chonos Archipelago has also been the subject of recent radioisotopic dating [Davidson et al., this volume]. The brachiopod findings from the Protranca Island [Miller and Sprechmann, 1978] have been assigned Devonian ages. Rb/Sr dates suggest, however, that accretion-related metamorphism and deformation occurred during the Late Carboniferous-Early Permian. In the Madre de Dios Archipelago, latest Carboniferous-Early Permian ages [Douglass and Nestell, 1976; Ling et al., 1985] are suggested for oceanic assemblages of both pelagic and shallow water facies. Ages of accreted material appear to be younger and timing of accretion appears to be earlier here than to the north in the Chonos Archipelago. The results presented here for Regalada Island indicate oceanic (radiolarian chert) sequences no older than Early Permian, which is slightly younger than for those from the Madre de Dios Archipelago. Thus, in general, we can conclude that a simple south to southeast pattern of accretion, initially suggested by Forsythe and Mpodozis [1983], remains valid. In addition, the general suggestion that there may be a southward continuation of this forearc into the Antarctic Peninsula [Grunow et al., this volume] is also reinforced, since radiolarian chert of Triassic age [Dalziel et al., 1981] from the South Orkney Islands would lie to the south of Regalada Island in most reconstructions.

Systematic Paleontology

All examinations and photomicrography of the radiolarians were made under a Leitz stereomicroscope and JEOL JSM 50-A scanning electron microscope. The holotype and paratypes of the new species Ps. chilensis are deposited in the collection of the U.S. National Museum, Washington, D.C., and are referred to by USNM number.

Subclass R A D I O L A R I A Müller 1858
Order POLYCYSTINA Ehrenberg 1838, emend. Riedel 1967
Suborder ALBAILLELLARIA Deflandre 1953, emend. Holdsworth 1969
Family ALBAILLELLIDAE Deflandre 1953, emend. Holdsworth 1969
Genus Pseudoalbaillella Holdsworth and Jones, 1980

Psedudoalbaillella annulatus Ishiga (Plate 1, Figure 1). Pseudoalbaillella annulatus [Ishiga et al., 1984, pp. 48-49, Plate 1, Figures 6-11]. Psudoalbaillella nodosa [Ling et al. (not Ishiga), 1985, p. 358, Figure 3 K]. Remarks: An overall slender conical outline with a slightly lobated pseudoabdominal margin due to diagonal segments characterizes the present species.

In a previous article, Ling et al. [1985] incorrectly assigned these Chilean speciemens to Ps. nodosa [Ishiga, 1982, p. 28, Plate 1, Figures 3-7] due to the presence of diagonal segments. Although not abundant, more specimens were recovered from the area. They are considered as conspecific with Ps. annulatus, although clear horizontal bands in the apical cone and any extended features beyond the pseudoabdominal segments were not observed in Chilean specimens.

Pseudoalbaillella sp. cf. Ps. bulbosa Ishiga (Plate 1, Figures 2-4). Pseudoalbaillella bulbosa [Ishiga, 1982, p. 335, Plate 1, Figures 8-13, 16, 17]. Haplodiacanthus circinatus [Nazarov and Ormiston 1985, pp. 47-48, Plate 6, Figures 9, 10; Text Figure 11]. Pseudoalbaillella sp. cf. Ps. bulbosa Ishiga [Ling et al., 1985, p. 358, Figures 3 G-I]. The present species is closely related to those reported from Japan by Ishiga [1982]. Furthermore, it is considered as conspecific with the Russian species, Haplodiacanthus circinatus, proposed by Nazarov and Ormiston [1985]. However, the reasons for not adopting the later generic name Haplodiacanthus at present are as follows.

According to Nazarov and Rudenko [1981], specimens assignable to their new genus Haplodiacanthus are characterized by the presence of two columellae along the ventral and dorsal side of the skeleton and by apparent segmentation ("band" of some authors) in both pseudocephalic and pseudoabdominal section, as described and illustrated by its type species, H. anfractus [Nazarov and Rudenko, 1981, p. 33, Figures 5-7]. This concept is generally followed for another new species, H. circinatus, by Nazarov and Ormiston [1985].

Such a columellar element was not observed in the studied Chilean specimens when they were examined under a conventional light transmission microscope and with a phase contrast attachment after being mounted with Canada balsam on a microslide. Segmentations are also absent in these radiolarian. A detailed discussion of these observations including a photomicrographic presentation is currently being prepared. Therefore, although the Chilean specimens are also identical to the Russian species even with the characteristic basal prolongation but to a lesser extent, they are classified under the present taxon.

Pseudoalbaillella sp. cf. Ps. bulbosa Ishiga is the most dominant and characteristic species in the B1 and B3 samples from Isla Madre de Dios. The limited geological occurrence for this species has been clearly documented from Missourian to Wolfcampian in Japan [Ishiga, 1982] and almost exclusively in Gzhelian C from Russia [Nazarov and Ormiston, 1985], as stated earlier.

Pseudoalbaillella chilensis Ling and Forsythe, n. sp. (Plate 1, Figures 5-8). Pseudoalbaillela sp. [Ishiga, 1982, Plate 1, Figures 14, 20]. Description: skeleton of moderate size with a graceful sinus figure to a slight hook at the distal end. The pseudocephalic segment is long and conical, smoothly tapered toward the distal end. The pseudothoracic part is slightly expanded to spherical with two lateral wings. The diameter of the pseudoabdominal segment is either uniform or gradually increasing distally; the distal third of the skeleton shows the slight curved shape of a fishing hook. Basal prolongations (spines) are very short or completely absent. The distal end (pylome) is open. Derivation of name: species name referring to their abundant occurrence in the Chilean samples studied. Measurements: maximum height (measurement "H" of Nazarov and Ormiston [1985]) 250 to 400 um (based on 30 specimens). Remarks: the rather gracefully curved outline into a "fishing hook" characterizes the present species, which can be readily differentiated from specimens of related forms, such as Ps. u-forma u-forma, by the degree of curvature in the distal third of the skeleton. It is highly conceivable that the present species could be considered a phylogenic link from Ps. elegans to Ps. u-forma u-forma to Ps. u-forma reflexa.

Pseudoalbaillella elegans Ishiga and Imoto (Plate 1, Figure 9). Pseudoalbaillella elegans [Ishiga and Imoto, 1980, p. 337, Plate 1, Figures 9-12; Ishiga et al., 1982b, Plate 1, Figures 2, 3; Ishiga et al., 1984, pp. 49-50, Plate 1, Figures 12-16; Ling et al., 1985, p. 358, Figure 3 J]. Remarks: A cylindrical and occasionally slightly curved pseudoabdomen characterizes the present species.

Pseudoalbaillella simplex Ishiga and Imoto (Plate 1, Figures 10, 11). Pseudoalbaillella simplex [Ishiga and Imoto, 1980, p. 337, Plate 1, Figures 13-18; Ishiga et al, 1984, p. 49, Plate 1, Figures 17-22; Ling et al., 1985, p. 358, Figures 3 L, M.; Hattori and Yoshimura, 1982, p. 113, Plate 1, Figure 2]. Remarks: the Chilean specimens generally agree well with the Japanese specimens except for the complete absence of a pseudoabdominal flap. However, the presence of such a skeletal element is hardly discernable from the specimens illustrated, including the holotype and some of the paratypes.

Pseudoalbaillella u-forma u-forma Holdsworth and Jones (Plate 1, Figures 12-14). Pseudoalbaillella u-forma [Holdsworth and Jones, 1980, p. 285, Figures 1 C, F; Ujiie and Hashimoto, 1983, p. 711, Figure 54]. Pseudoalbaillella sp. aff. Ps. u-forma [Holdsworth and Jones, 1980; Ishiga, 1982, Plate 1, Figures 18, 19]. Pseudoalbaillella u-forma morphotype I [Ishiga et al., 1984, p. 48, Plate 1, Figures 1-4]. Remarks: among radiolarians whose pseudoabdominal segment shows a distinct curved shape, two morphologic types were recognized from Japan, morphotype I, which shows a definite bending into the letter "L", and morphotype II, which resembles the letter "U". Furthermore, type I occurs earlier than type II, appearing to limit the lower part of the Wolfcampian. Ishiga et al. [1984] postulated that the occurrence of this subspecies seems to limit the lower part of the Wolfcampian, possibly below the "Ps. u-forma - Ps. elegans" zone of Ishiga et al. [1982a].

Pseudoalbaillella u-forma reflexa Ling and Forsythe, n. ssp., not illustrated. Pseudoalbaillella u-forma [Holdsworth and Jones, 1980; Ishiga and Imoto, 1980, Plate 1, Figures 6-8; Ishiga et al., 1982b, Plate 1, Figure 1; Hattori and Yoshimura, 1982, Plate 1, Figure 1]. Pseudoalbaillella u-forma morphotype II [Ishiga et al., 1984, pp. 48-49, Plate 1, Figure 2]. Description: differentiated from the above type subspecies by the more bent pseudoabdominal distal part, thus forming the letter "U". Derivation of name: from the Latin reflexus, bent or turned back, characterizing the

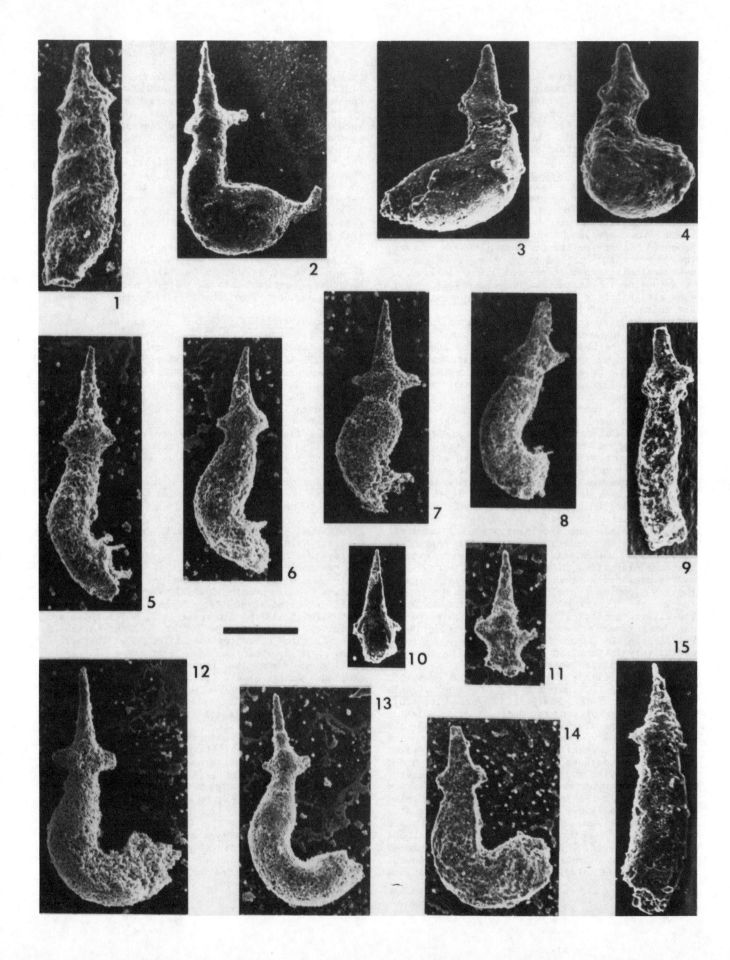

nature of the U-shaped pseudoabdominal section. Holotype [Ishiga and Imoto, 1980, Plate 1, Figure 6 and paratypes [Ishiga and Imoto, 1980, Plate 1, Figures 7, 8] from the Ashimiand Sasayama areas of Japan.

Pseudoalbaillella sp. G Ishiga (Plate 1, Figure 15). Pseudoalbaillella sp. G [Ishiga, 1982, pp. 334-335, Plate 1, Figures 1, 2; Ling et al., 1985, p. 358, Figures 3 N, 0]. Remarks: an overall conical outline and three weak, broad oblique bands in the pseudoabdominal segments chararcterize the present species.

Acknowledgments. Forsythe's fieldwork was supported by National Science Foundation grant EAR 82-06100. We thank Robert L. Bailey and Marcia A. Honz of Northern Illinois University for their technical assistance.

References

Cornell, W. C., and R. D. Simpson, New Permian albaillellid radiolarians from west Texas, Micropaleontology, 31, 271-279, 1985.

Dalziel, I. W. D., D. H. Elliot, D.L. Jones, J. W. Thomson, M. R. A. Thomson, N. A. Wells, and W. J. Zinsmeister, The geological significance of some Triassic microfossils from the South Orkney Islands, Scotia Ridge, Geol. Mag., 118, 15-25, 1981.

Davidson, J., C. Mpodozis, E. Godoy, F. Hervé, R. Pankhurst, and M. Brook, Late Paleozoic accretionary complexes on the Gondwana margin of southern Chile: Evidence from the Chonos Archipelago, this volume.

Deflandre, G., Albaillella nov. gen., Radiolaire fossile du Carbonifère inférieur, type d'une lignée aberrent eteinte, C. R. Hebd. Acad. Sci. 234, 872-874, 1952.

Deflandre, G., Radiolaires fossiles, in Traite de Zoologie, 1, edited by P. P. Grasse, pp. 389-436, Paris, Masson et Cie., Paris, 1953.

Douglass, R. C., and M. K. Nestell, K., Late Paleozoic Foraminifera from southern Chile, U.S. Geol. Surv. Prof. Pap., 858, 1-49, 1976.

Ehrenberg, C. G., Uber die Bildung der Kreidefelsen unde des Kreidemergels durch unsichtbare Orgamismen, Königlichen Akademie Wissenschaften Berlin, Abh., Jahrg., 1838, 59-147, 1838.

Forsythe, R. D., and C. Mpodozis, Geología del basamento Pre-Jurasico superior en el Archipielago Madre de Dios, Magallanes, Chile, Bol. Serv. Nacl. Geol. Miner., 39, 1-61, 1983.

Grunow, A., I. W. D. Dalziel, and D. V. Kent, Ellsworth-Whitmore mountains crustal block, western Antarctica: New paleomagnetic results and their tectonic significance, this volume.

Hattori, I., and M. Yoshimura, Lithofacies distribution and radiolarian fossils in the Nanjo area in Fukui Prefecture, central Japan, Proceedings First Radiolarian Symposium, Spec. Vol. 5, News of Osaka Micropaleontologist, pp. 103-116, Osaka, 1982.

Holdsworth, B. K., The relationship between the genus Albaillella Deflandre and the ceratoikiscid radiolaria, Micropaleontology, 15, 230-236, 1969.

Holdsworth, B. K., and D. L. Jones, Preliminary radiolarian zonation for Late Devonian through Permian time, Geology, 8, 281-283, 1980.

Ishiga, H., Late Carboniferous and Early Permian radiolarians from the Tamba Belt, southwest Japan, Chikyu Kagaku Chigaku Dantai Kenkyukai, 36, 333- 339, 1982.

Ishiga, H., and N. Imoto, Some Permian radiolarians in the Tamba District, southwest Japan, Chikyu Kagaku Chigaku Dantai Kenkyukai, 34, 333-345, 1980.

Ishiga, H., T. Kito, and N. Imoto, Late Permian radiolarian assemblages in the Tamba District and an adjacent area, southwest Japan, Chikyu Kagaku Chigaku Dantai Kenkyukai, 36, 10-22, 1982a.

Ishiga, H., T. Kito, and N. Imoto, Permian radiolarian biostratigraphy, Proceedings First Radiolarian Symposium, Spec. Vol. 5, News of Osaka Micropaleontologist, pp. 17-26, Osaka, 1982b.

Ishiga, H., T. Kito, and N. Imoto, Middle Permian radiolarian assemblages in the Tamba District

---

Plate 1. (Opposite) Pseudoalbaillellid radiolarians from southernmost Chile. Scale bar = 100 um.

Fig. 1. Pseudoalbaillella annulatus Ishiga. Sample B1.

Figs. 2-4. Pseudoalbaillella sp. cf. Ps. bulbosa Ishiga. Figure 2, sample B1; Figures 3, 4, sample B3.

Figs. 5-8. Pseudoalbaillella chilensis Ling and Forsythe, n. sp., Figure 5, sample IR 5, paratype, USNM 410835, Figure 6, sample IR 5, holotype, USNM 410836, Figure 7, sample IR 6, Figure 8, sample IR 6, paratype, USNM 410837.

Fig. 9. Pseudoalbaillella elegans Ishiga and Imoto. Sample B3

Figs. 10, 11. Pseudoalbaillella simplex Ishiga and Imoto. Figure 10, sample B1, Figure 11, sample IR 4.

Figs. 12-14. Pseudoalbaillella u-forma u-forma Holdsworth and Jones. Figure 12, sample IR 1, Figure 13, sample IR 3, Figure 14, sample IR 1.

Fig. 15. Pseudoalbaillella sp. G Ishiga. Sample B3.

and an adjacent area, southwest Japan, Chikyu Kagaku Chigaku Dantai Kenkyukai, 36, 272-281, 1982c.

Ishiga, H., N. Imoto, M. Yoshida, and T. Tanabe, Early Permian radiolarians from the Tamba Belt, southwest Japan, Chikyu Kagaku Chigaku Dantai Kenkyukai, 38, 44-52, 1984.

Ling, H. Y., R. D. Forsythe, and R. C. Douglass, Late Paleozoic microfaunas from southernmost Chile and their relation to Gondwanaland forearc development, Geology, 13, 357-360, 1985.

Miller, H., and P. Sprechmann, Eine devonische Faunula aus dem Chonos-Archipel Region Aisen, Chile, und ihre stratigraphische Bedeutung, Geol. Jahrb., Reihe B, 2, 37-45, 1978.

Müller, J., Über die Thalassicolen, Polycrstinen und Acantomentren des Mittelmeeres, Königlichen Akademie Wissenschaften Berlin, Abh., Jahrg., 1958, 1-62, 1958.

Nazarov, B. B., and A. R. Ormiston, Radiolaria from the late Paleozoic of the southern Urals, USSR and west Texas, USA, Micropaleontology, 31, 1-54, 1985.

Nazarov, B. B., and V. S. Rudenko, Some bilaterally symmetrical radiolarians of the late Paleozoic of the southern Urals (in Russian), Voprosy Mikropaleontologii, 24, 129-139, 1981.

Riedel, W. R., Subclass radiolaria, in The Fossil Record, edited by W. B. Harland et al., pp. 291-298, London Geol. Soc., London, 1967.

Rüst, D., Beitrage zur Kenntniss der fossilen radiolarien aus Gesteinen der Trias und der Paleozoischen schichten, Palaeontographica, 38, 107-200, 1892.

Ujiie, H., and Y. Hashimoto, Geology and radiolarian fossils in the inner zone of the Motobu Belt in Okinawa-jima and its environs, Chikyu, 5, 706-712, 1983.

Copyright 1987 by the American Geophysical Union.

THE LATE PALEOZOIC EVOLUTION OF THE GONDWANALAND CONTINENTAL MARGIN IN NORTHERN CHILE

C. M. Bell

School of Geography and Geology, College of St. Paul and St. Mary, Cheltenham, United Kingdom

Abstract. Tectonic activity on the Gondwanaland continental margin in northern Chile and northwestern Argentina has been continuous from the early Paleozoic to the present. Paleozoic accretion resulted from the buildup of accretionary and magmatic arc complexes, and possibly from the addition of exotic terranes. Paleozoic strata between 25°S and 29°S in northern Chile comprise two north-south elongated strips separated by a 100-km-wide graben infilled with younger rocks. The western strip consists of deep-sea turbidites and basic lavas of the Devonian or Early Carboniferous Las Tórtolas Formation. Subduction of these rocks during Carboniferous times produced the Chañaral mélange in the area south of 26°30'S. The mélange probably resulted from intrastratal movements of partly consolidated strata within an accretionary wedge. Further tectonic deformation of both the turbidites and the mélange was produced by northeast directed subduction. The subduction complex is bounded to the east by the Atacama strike-slip fault system. To the east of the graben are relatively undeformed Early Carboniferous lacustrine sedimentary rocks of the Chinches Formation. These were deposited in a deep, elongated basin, possibly of pull-apart type resulting from strike-slip movement parallel to the coastline. Late Carboniferous to Early Permian magmatic activity superimposed on both these sedimentary successions suggests seaward migration of the subduction zone. The development of the Mesozoic and Cenozoic Andean complex, which overlies the Paleozoic rocks with a marked unconformity, was not accompanied by the accretion of a further subduction complex.

Introduction

The late Paleozoic sedimentary and volcanic rocks in northern Chile form the geological basement of the Andean complex which developed as an ensialic magmatic arc during Mesozoic and Cenozoic subduction of an oceanic plate beneath the South American continental margin [James, 1971; Dalziel, 1986]. The first detailed field investigation of the Paleozoic rocks between 25°S and 27°S was undertaken by the author during a 3-month period in 1980-1981 [Bell, 1982, 1984]. Subsequent fieldwork on the Chañaral mélange, carried out in 1983, is reported here for the first time. The interpretation of the depositional environment, history of deformation, and tectonic setting presented here is based on these field investigations together with a survey of the geological literature of the region.

Paleozoic Geology of Northern Chile

The Paleozoic stratigraphy and tectonic development of northern Chile are poorly understood, partly due to the scattered and isolated exposures (Figure 1). Few of the rocks contain fossils of stratigraphic value, and most have been subjected to deformation and low- to medium-grade metamorphism. The only rocks of undisputed Precambrian age are the 1000 Ma schists at about 18°30'S [Pacci et al., 1980]. Damm et al. [1981] have reported 1730 Ma zircons from late Paleozoic granitoids at about 26°S. Mica schist, amphibolite, greenschist, phyllite, and marble of probable early Paleozoic age are known from several isolated occurrences (Figure 1). These rocks were subjected to metamorphism and deformation in an unknown tectonic setting, possibly during pre-Ordovician times.

The oldest sedimentary rocks to which a reliable age can be given are early Ordovician graptolite-bearing sediments at Sotoca [Cecioni, 1979] and on the Argentine border between 23°S and 24°S [Hervé et al., 1981]. These strata form part of a western, deepwater facies of the extensive early Paleozoic sediments of northwestern Argentina [Davidson et al., 1981]. Deposition was accompanied by widespread magmatic activity lasting from Middle to Late Ordovician times. The resultant calc-alkaline magmatic arc extended for 1300 km from 17°S in Bolivia to 29°S in Argentina, parallel with and to the east of the Chilean border. Plutonic rocks of this age have been reported from Sierra Moreno and south of Salar de Atacama in northern Chile. The magmatism was accompanied by the widespread Ocloyic tectonic event [Coira et al., 1982].

Sedimentary rocks of Devonian to early Carboniferous age are widespread between 21°S and 29°S. Lower Devonian shallow-marine strata occur at Cerro Lila and Sierra de Almeida [Ramírez and Gardeweg, 1982], and a few exposures are known from northwestern Argentina [Davidson et al., 1981]. Strata with fossils suggesting a Middle to Late Devonian age, in the coastal region between 21°S and 24°S, include the El Toco Formation,

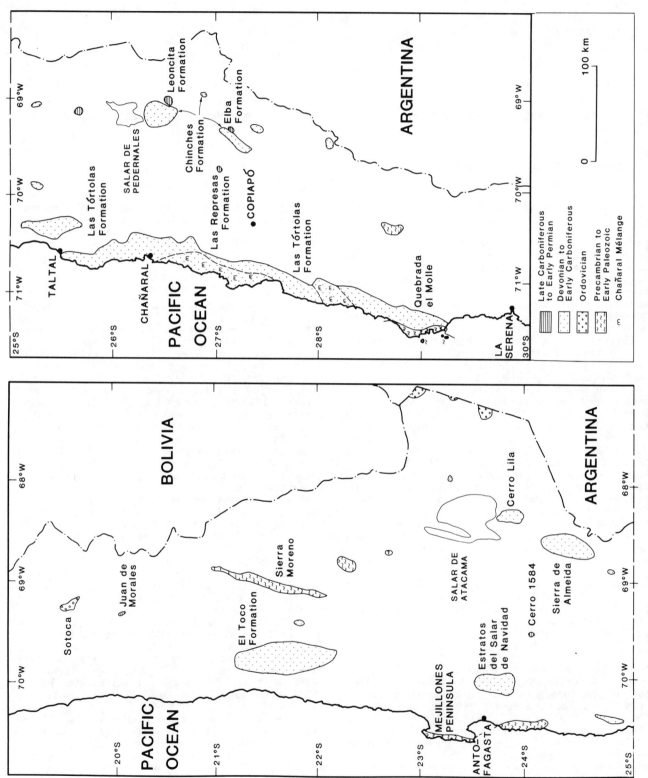

Fig. 1. The distribution of Paleozoic sedimentary and volcanic rocks in northern Chile (from Mapa Geologico de Chile, 1:1,000,000, 1982, Servicio Nacional de Geología y Minería).

deposited in a deltaic [Skarmeta and Marinovic, 1981] or turbidite environment [Breitkreuz and Bahlburg, 1985], and the shallow-marine Estratos del Salar de Navidad [Ferraris and Di Biase, 1978].

Extensive exposures of Devonian to early Carboniferous sedimentary and volcanic rocks, which form the subject of the present study, occur between 24°30'S and 29°S. In the west is the Las Tórtolas Formation and its deformed equivalent the Chañaral mélange [Bell, 1982, 1984], and in the east the Chinches Formation [Bell, 1985]. Contemporaneous calc-alkaline granitoids are known from isolated locations in northern Chile but are more abundant in Argentina [Coira et al., 1982]. The Late Devonian to Early Carboniferous tectonic and magmatic event has been named the Chanic phase. It also affected rocks of the central and southern Andes [Miller, 1984] where Early to mid-Carboniferous accretion of a subduction complex was associated with the development of an extensive magmatic arc [Gonzalez-Bonorino and Aguirre, 1970; Hervé et al., 1976, 1981].

Minor occurrences of Late Carboniferous to Early Permian shallow-marine and terrestrial sediments have been recorded between 20°S and 27°S (at Juan de Morales and Cerro 1584, and the Leoncita, Las Represas, and Elba Formations). The strata are less deformed and metamorphosed than the underlying Devonian to Early Carboniferous rocks. Deposition was followed by, and in places associated with, the development of a major magmatic belt which extended for approximately 4000 km from north to south (20°N to 40°S). These volcanic rocks of the Choiyoi Group together with associated calc-alkaline granitoids range in age from late Carboniferous to Early Triassic [Berg et al., 1983].

The most significant tectonic break recognized by most workers in Chile is the unconformity between the Paleozoic rocks and the overlying Andean complex [Dalziel, 1986; Coira et al., 1982]. Initiation of this complex was related to late Triassic development of a magmatic arc along much of the pacific margin of South America [James, 1971]. Despite the marked unconformity, recently published summaries of radiometric age dates [Berg et al., 1983; Aguirre, 1983] suggest that no significant break in magmatic activity occurred between the Paleozoic and Andean events.

Devonian to Early Carboniferous Sedimentary and Volcanic Rocks Between 25°S and 29°S

Devonian to early Carboniferous sedimentary and volcanic rocks form two north-south elongated strips separated by a 100-km-wide graben infilled with younger rocks of the Andean complex in the region between 25°S and 29°S (Figure 2). The low-grade metasedimentary rocks in each strip display evidence for distinct depositional and tectonic environments. In the western coastal region are the intensely deformed turbidites and minor basic volcanic rocks of the Las Tórtolas Formation [Bell, 1982, 1984]. In the south at Quebrada El Molle (Figure 2) is an isolated occurrence of possibly contemporaneous dacitic tuffs. To the east of the graben are the little-deformed fine-grained siliciclastic sedimentary rocks of the Chinches Formation [Mercado, 1982].

The Las Tórtolas Formation. Rocks of the Las Tórtolas Formation extend for at least 400 km from north to south (Figure 2). The basement of these strata is unknown, but in the south at Isla Gaviota the formation is faulted against probably early Paleozoic quartzites and garnet-actinolite-epidote schists. The Las Tórtolas Formation is unconformably overlain by Early to Middle Triassic volcanic and volcaniclastic rocks, and Jurassic marine sediments [Naranjo, 1978]. An upper age limit to the formation is provided by Lower Permian monzogranites (268 ± 8 Ma) which postdate the deformation. Bell [1982] suggested an Ordovician to Devonian age on the basis of a trace fossil assemblage. However, a conglomeratic limestone at Aguada de la Changa (Figure 3) contains upper Paleozoic spiriferaceans and bryozoa of probable Carboniferous or Permian age (P. D. Taylor, personal communication, British Museum of Natural History, 1984). This radiometric and paleontological evidence therefore suggests a Devonian or early Carboniferous age for the sediments.

The Las Tórtolas Formation consists essentially of a very thick and monotonous sequence of thinly bedded, fine-grained turbidites. Minor conglomerates and pebbly sandstones occur between Chañaral and Caleta Obispo. Paleocurrents in the turbidites, indicated by current scour marks, are directed toward the southeast in the area north of Chañaral [Bell, 1982]. Lithic clasts in the sandstones indicate a provenance of andesitic volcanic rocks, low-grade metasediments, and scarce granitoids. Rare limestone beds up to several meters thick have been recorded in the Las Tórtolas Formation between Pan de Azúcar and Carrizal Bajo. They consist of broken fragments of calcareous algae and have been interpreted as the products of deposition from turbidity currents. Basic volcanic rocks are interstratified with the turbidites between Pan de Azúcar and Bahía Copiapó. The most extensive exposures at Carrizal Bajo comprise a thick succession of pillow lavas, tuffs, and hyaloclastite breccias. These rocks have the geochemical characteristics of alkali basalts rather than normally depleted midocean ridge basalts [Bell, 1982].

Dacites at Quebrada El Molle. A 200-m-thick sequence of dacitic tuffs and agglomerates at Quebrada El Molle (Figure 2) probably originated as subaerially erupted ignimbrites. The steeply dipping strata are cut by a well-developed tectonic cleavage and are unconformably overlain by Upper Triassic to Jurassic sedimentary rocks. This stratigraphic position, together with the intense deformation, suggests that these rocks form part of the same late Paleozoic sequence as the Las Tórtolas Formation.

Chañaral mélange. The Chañaral mélange [Silver and Beutner, 1980] extends for approximately 70 km to the south of Chañaral (Figure 2). Other occurrences have been recorded farther south at Bahía Copiapó and Carrizal Bajo. Most of the mélange was produced by the fragmentation of strata of the

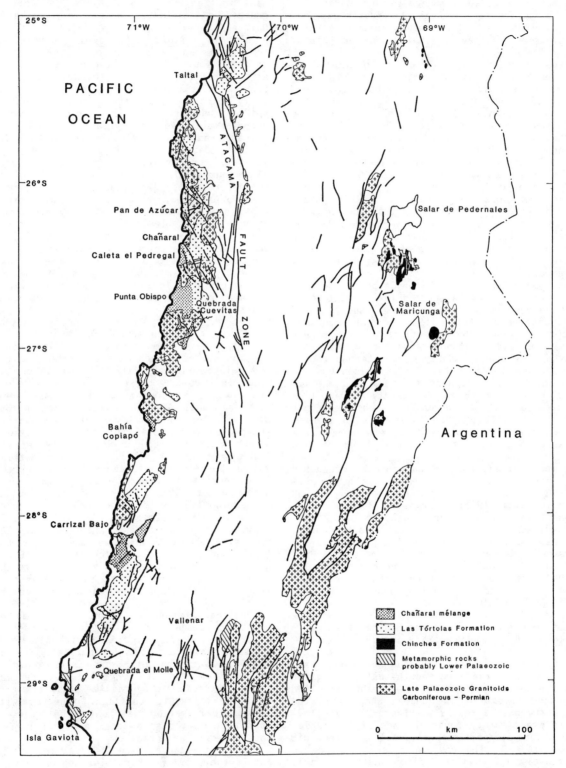

Fig. 2. The geology of Paleozoic sedimentary rocks in the region between 25°S and 29°S in northern Chile (from Mapa Geologico de Chile, 1:1,000,000, 1982, Servicio Nacional de Geología y Minería).

Las Tórtolas Formation, but it also contains a minor proportion of exotic blocks [Hsü, 1968] of rock types apparently not derived from this formation. Contacts between rocks of the Las Tórtolas Formation and the mélange are gradational, the turbidites becoming progressively more fragmented and mixed to form the mélange over distances ranging from centimeters to kilometers. In Quebrada Vegas Quemadas (Figure 4) and Quebrada Los Infieles the contact between the two sequences is slightly oblique to the stratification (Figure 3), and the turbidites become progressively more broken up in a westerly direction to form the mélange. Farther south, in Quebrada Cuevitas (Figure 2), the transition zone is about 10 km wide; bands of mélange up to 1 km wide are interlayered with and approximately parallel to the stratification. There are apparently no sedimentary contacts between mélange and undisturbed strata, and no stratigraphic truncation of the fragmented sequences has been observed.

Despite the intensity of the deformation, with complete destruction of bedding at the outcrop scale, the mélange between Bahía Las Animas and Quebrada Animas Viejas (Figure 3) retains elements of its original stratigraphy on a scale of hundreds of meters. The distinctive pillow lavas and tuffs form persistent horizons or trains of blocks, both in individual outcrops and over distances of up to 10 km along strike.

On the scale of meters to tens of meters, the mélange consists of sandstone blocks in a pelitic matrix. Blocks range from individual sand grains up to masses 150 m or more in diameter. These blocks are normally very poorly sorted with respect to size, and the proportion of blocks to matrix varies from interlocking masses of sandstone to small lenses of sandstone in a pelitic matrix. The distinction between blocks and matrix is not always clear, in places the contacts are sharp, but elsewhere gradational. Blocks and matrix are particularly difficult to distinguish in thin section. Many blocks are themselves internally fragmented, indicating recycling of the material [Hibbard and Williams, 1979]. Blocks are elongated and subrounded to angular in shape, and some form highly irregular lumps. Most are separated from one another by the matrix, but some are linked in sheets resembling irregular "chocolate-tablet" boudins.

Conglomerates and pebbly sandstones comprise a 100-m-wide zone in the mélange south of Caleta Obispo. Some of the matrix-supported pebble and granule conglomerates form distinct bands up to 5 m thick within the mélange, but most contacts between conglomerate and mélange are diffuse and poorly defined. In many places the conglomerates have been disaggregated, and the individual well-rounded detrital clasts are intimately mixed with other blocks in the mélange.

A distinctive breccia of angular to poorly rounded sandstone and siltstone fragments in a sandy or silty matrix forms 10- to 100-m-wide horizons within the mélange at Caleta Obispo and Caleta El Pedregal. The breccia is a relatively well-sorted mixture of individual detrital clasts and irregular fragments of sandstone and mudstone ranging from 1 mm to 50 cm in size. In places this breccia forms a matrix between other larger blocks in the mélange. In thin section the fragments have irregular, diffuse boundaries, and they show no evidence of strain of individual grains.

Volcanic rocks in the mélange. Volcanic rocks are abundant in the mélange south of Bahía Las Animas and at Bahía Copiapó (Figure 3). The basaltic pillow lavas, agglomerates, and tuffs were apparently interstratified with the turbidites prior to the disruption which produced the mélange. In thin section, most of the lavas are porphyritic with a variolitic texture of plagioclase laths. No primary ferromagnesian minerals are preserved, but secondary actinolite, epidote, biotite, chlorite, and zoisite indicate a greenschist facies of metamorphism.

Exotic blocks in the mélange. A single block of pyroxenite at Bahía Copiapó has tectonic contacts with the adjacent mélange. This ultrabasic plutonic rock could form part of the volcanic sequence or it may be an exotic block.

A band of six blocks of marble up to 150 m in diameter extends for 5 km from north to south-southeast of Caleta Obispo. The blocks have sharp tectonic contacts with the mélange and are rounded to slightly elongated. The marble is a breccia with a distinct planar fabric produced by flattening of the fragments. Small silicified echinoid fragments indicate a shallow marine origin.

A 30-m-wide block of bioclastic limestone, sandstone, and conglomerate at Aguada de la Changa contains poorly preserved fossils including spiriferaceans, bryozoa, crinoids, rugose corals, and bivalves. This fossil assemblage together with the lithology suggests a relatively high-energy, well-oxygenated marine depositional environment which contrasts strikingly with the deepwater depositional environment of the surrounding turbidites.

Deformation in the Las Tórtolas Formation
and the Chañaral Mélange

Origin of the Mélange

The first episode of deformation (designated here D_1) produced the Chañaral mélange by disruption of the strata of the Las Tórtolas Formation. In most localities, detailed evidence for this early episode (or episodes) of deformation has been masked by later tectonic and metamorphic events; but elsewhere, particularly in the 20 kilometers south of Chañaral, many D_1 structures have been preserved in the relatively competent arenaceous strata.

Much of the fragmentation which formed the mélange was accomplished by necking and boudinage of sandstone beds. The contacts between sandstone blocks and the more ductile pelitic matrix commonly exhibit irregular cusp and flamelike structures. In places, the matrix intrudes the blocks as branching veins and irregular masses. Most boudins are irregularly shaped, flattened oblate ellipsoids with abundant bulbous protrusions. They commonly form stacked or interlocking masses.

Blocks of finely laminated tuff and mudstone in

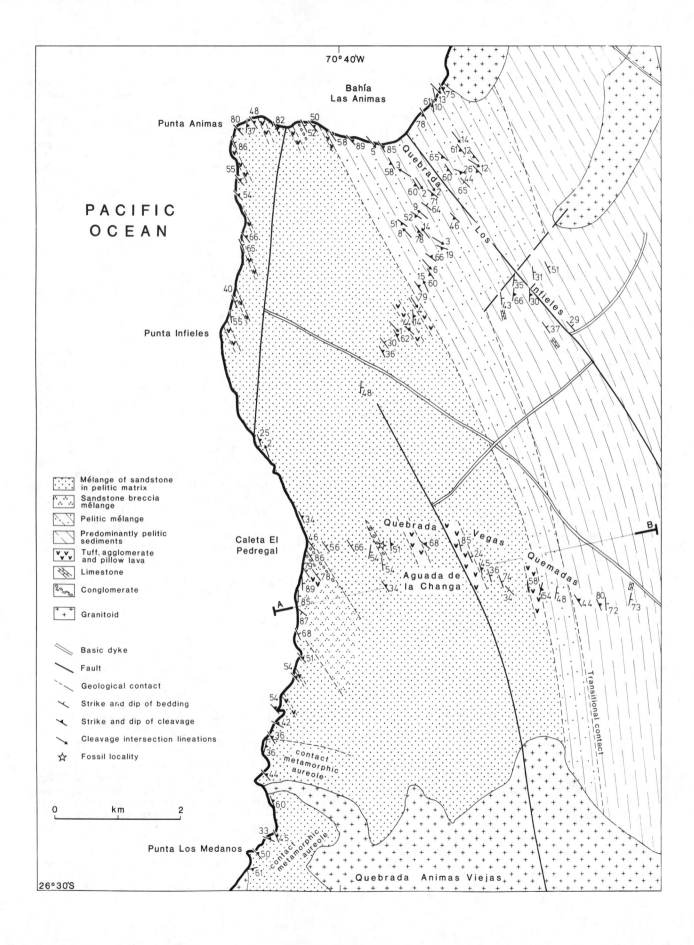

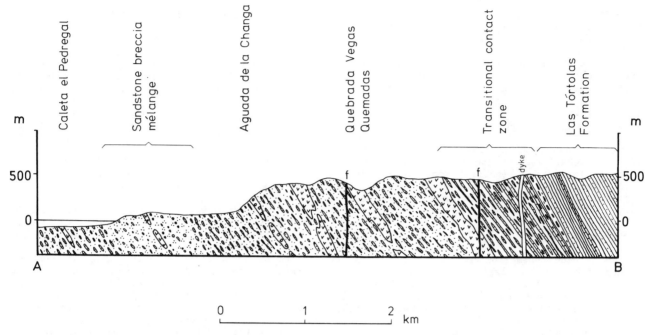

Fig. 4. Cross section through the Chañaral mélange along Quebrada Vegas Quemadas, showing the transitional contact between the mélange and the Las Tórtolas Formation.

the mélange show parallel sets of small-scale normal faults. Tight to isoclinal recumbent folds have also been recorded but are isolated and rare. In thin section there is no evidence to suggest that the D_1 deformation involved the strain of individual grains; by contrast most deformation appears to have resulted from movement between grains. Similarly, there is no evidence of cleavage, schistosity, or cataclasis associated with the formation of the mélange.

Subsequent Phases of Deformation

At many localities the D_1 structures have been overprinted by several subsequent phases of tectonic deformation. Two phases (designated here D_2 and D_3) were identified in the Las Tórtolas Formation north of Chañaral [Bell, 1984]. The D_2 event in the Las Tórtolas Formation produced tight chevron folds with a wavelength of about 100 m. Associated with these large-scale folds is an axial planar slaty cleavage or schistosity. In the mélange the slaty cleavage varies from strongly oblique to subparallel to trains of blocks. In most localities, D_3 produced structures and fabrics which are more pronounced than D_2. Folds vary in wavelength from millimeters up to 20 m. They are commonly asymmetrical and disharmonic; hinge zones are rounded with an interlimb angle of about 80°. The D_3 planar fabric is a crenulation cleavage of the D_2 schistosity in pelites, whereas in sandstones it usually forms a spaced fracture cleavage. The orientation of D_2 and D_3 structures is locally quite variable, but the majority of fold axes trend northwest and asymmetrical folds overturn toward the southwest. Fold axial planes and cleavage dip predominantly toward the northeast [Bell, 1984].

Two andesite dikes, 500 m east of Punta Animas, cut both D_1 and D_2 structures, but have themselves been deformed by D_3. These folded dikes indicate that despite their near-parallelism, the D_2 and D_3 structures are the products of distinct tectonic events separated by a phase of magmatic activity.

Origin of the Deformation

The origin of mélanges has been ascribed both to tectonic deformation and to large-scale slumping (these slumps, or olistostromes, may themselves be subsequently tectonized) [Hsü, 1968, 1974; Blake and Jones, 1974]. The Chañaral mélange shows abundant small-scale evidence for a soft-sediment origin. However, on a larger scale it occurs between, but grading into, undeformed strata. These contacts show none of the conformable depositional features of olistostromes [Boles and Landis, 1984]. The soft-sediment deformation therefore resulted not from superficial slumping [Woodcock, 1976] but from relative movements within and between packages of partly consolidated sediment. Boudins and cusps are suggestive of ductile extension of the more competent sandstones, accompanied by plastic flow of the less

Fig. 3. (Opposite) The geology of the region south of Bahía Las Animas, showing the relationship between the Las Tórtolas Formation and the Chañaral mélange.

competent pelitic layers. The "chocolate tablet" boudins and the small-scale normal faults indicate symmetrical layer-parallel extension [Cowan, 1974, 1982].

Some mélanges lack a primary cleavage [Hibbard and Williams, 1979], but most have a pervasive slaty cleavage which has been ascribed to deformation during the formation of the mélange [Hsü, 1968, 1974; Blake and Jones, 1974; Moore and Wheeler, 1978]. Doubts about this relationship between the fabric and the origin of mélanges have been expressed by Page [1978] and Cowan [1982]. At many localities in the Chañaral mélange the strongly foliated D_2 fabric is oblique to the D_1 structures (which have no associated tectonic foliation), and the cleavage is therefore not related to the formation of the mélange.

The Chinches Formation

The Chinches Formation consists essentially of fine-grained siliciclastic sedimentary rocks forming a north-south elongated strip at approximately 69°W and between 26°30'S and 27°30'S [Bell, 1985]. The strata are cut by late Carboniferous plutonic rocks [Sepulveda and Naranjo, 1982] and they contain fossil fish fragments of probable early Carboniferous age (D. L. Dineley, personal communication, University of Bristol, 1983).

The formation has a minimum thickness of 2500 m; the dominant sedimentary facies comprises monotonous sequences of laminated shales and parallel-bedded sandstones in units up to 1000 m thick. This facies is overlain by well-sorted, ripple-marked, and parallel-bedded siltstones and very fine-grained sandstones in sequences up to 25 m thick. Most of the ripples originated by wave activity, interference sets indicate frequent changes in wind direction. Fossils include rare plant impressions and the tracks of a large amphibian [Bell and Boyd, 1986]. The third facies in the upward coarsening sequences comprises cross-bedded sandstones in units up to 20 m thick, together with thin oolitic, pisolithic, and stromatolitic limestones. Fossils include fish scales and teeth, and thin-shelled mollusks. Clasts in the sandstones indicate a provenance of andesitic and rhyolitic volcanic rocks and low-grade metasediments. Rare horizons of rhyolitic tuff were recorded by Sepulveda and Naranjo [1982].

The rocks have been subjected to very low-grade metamorphism, and they are folded into open monoclinal folds with wavelengths of meters to kilometers. Tectonic cleavage is absent in most localities. This low degree of deformation contrasts markedly with that in the intensely deformed Las Tórtolas Formation.

The predominance of fine-grained sediments indicates a low-energy, partly or completely enclosed depositional basin. The fossils, sedimentary structures, facies, and sequences are all closely comparable with those found in both ancient and modern lake deposits [Picard and High, 1972; Link and Osborne, 1978]. The thick upward coarsening sequences are probably the products of the infilling of a lake which had an initial depth of 1000 m or more.

Summary and Tectonic Setting

The pre-Ordovician rocks of northern Chile include rare but widely distributed metamorphic rocks of unknown depositional environment and tectonic setting. In Ordovician times a marine sedimentary basin developed marginal to the Gondwana craton. Magmatism and deformation in a north-south trending belt has been interpreted as evidence for subduction, although some workers have suggested an intracratonic origin [Coira et al., 1982].

The Chañaral mélange originated by fragmentation and mixing of Devonian to early Carboniferous deep-sea turbidites, together with possible ocean island basalts, of the Las Tórtolas Formation. The initial deformation resulted from intrastratal movements of partly consolidated strata, probably by underthrusting and stacking of packages of sediment within an accretionary wedge. Subsequent phases of deformation resulted from continued deformation within the accretionary complex [Bell, 1984]. The orientations of structures suggest that this subduction was directed toward the northeast, probably oblique to the continental margin. Exotic blocks may have been mixed tectonically [Blake and Jones, 1974] or by diapirism [Williams et al., 1984].

The subduction which deformed the Las Tórtolas Formation was possibly contemporaneous with the deposition of the early Carboniferous Chinches Formation, suggesting that a north-south lacustrine basin developed parallel to the active continental margin. Independent evidence for active subduction is provided by the interbedded tuffs and the andesitic and rhyolitic detritus in the Chinches Formation, together with the dacitic tuffs at Quebrada El Molle. The great thickness of lacustrine sediments suggests deposition in a pull-apart basin, produced by strike-slip faulting [Steel and Gloppen, 1980], comparable with the Pliocene Ridge basin of California [Link and Osborne, 1978]. The presence of late Carboniferous to early Permian magmatic activity superimposed on these sedimentary successions suggests a seaward migration of the subduction zone. A possible modern-day equivalent to the Carboniferous tectonic setting of northern Chile is provided by the North Island of New Zealand where oblique subduction has produced an accretionary wedge together with a strike-slip fault system adjacent to a volcanic arc [Lewis, 1980].

Evidence for late Paleozoic accretion in other segments of the Andes has been recorded in southern Chile by Mpodozis and Forsythe [1983], and in Colombia by McCourt et al. [1984]. Helwig [1972] provided paleoclimatic evidence for Carboniferous accretion of exotic terranes in Bolivia. South American paleomagnetic data indicate that the area now forming northern Chile was in the south temperate zone during Devonian to Carboniferous times. The region apparently moved northward from about 75°S in the late Devonian to 40°S by the mid-Carboniferous [Smith et al., 1981]. However, the presence of large amphibians and stromatolites in the early Carboniferous Chinches Formation indicates a warm, possibly tropical climate. This paleoclimatic anomaly may possibly be explained by the accretion of an exotic terrane. The Paleozoic geology of northern Chile and northwestern Argentina is dominated by a series of north-south elongated magmatic arcs ranging in age from Ordovician to Triassic. The region was therefore an active continental margin for much of the Paleozoic. The calc-alkaline magmatic arcs, together with their associated sedimentary sequences, episodes of deformation, and metamorphism, can best be explained in terms of subduction of an oceanic plate beneath the Gondwana continental margin.

The paucity of Precambrian rocks in northern Chile, together with the presence of north-south elongated strips of Paleozoic strata (most of which exhibit distinctive stratigraphy, structures, and metamorphism) (Figure 1) indicates that accretion to the continental margin has resulted in continental growth. Litherland et al. [1985] have suggested that this accretion to the South American section of the Gondwana margin possibly dates back to 1300 Ma. By contrast with this evidence for Paleozoic accretion, the late Carboniferous to Triassic and the Triassic to Cenozoic (Andean) magmatic arc complexes were emplaced in an ensialic environment with no associated accretionary complexes.

Acknowledgments. I wish to acknowledge the field support provided by the Servicio Nacional de Geología y Minería of Chile. Research grants were received from the Royal Society and the Natural Environment Research Council of the United Kingdom.

References

Aguirre, L., Granitoids in Chile, Mem. Geol. Soc. Am., 159, 293-316, 1983.

Bell, C. M., The Lower Paleozoic metasedimentary basement of the coastal ranges of Chile between 25°30′ and 27°S, Rev. Geol. Chile, 17, 21-29, 1982.

Bell, C. M., Deformation produced by the subduction of a Palaeozoic turbidite sequence in northern Chile, J. Geol. Soc. London, 141, 339-347, 1984.

Bell, C. M., The Chinches Formation: an early Carboniferous lacustrine succession in the Andes of northern Chile, Rev. Geol. Chile, 24, 29-48, 1985.

Bell, C. M., and M. J. Boyd, A tetrapod trackway from the Carboniferous of northern Chile, Palaeontology, in press, 1986.

Berg, K., C. Breitkreuz, K-W. Damm, S. Pichowiak and W. Zeil, The north Chilean coast range--an example for the development of an active continental margin, Geol. Rundsch., 72, 715-731, 1983.

Blake, M. C., and D. L. Jones, Origin of Franciscan mélanges in Northern California, in Modern and Ancient Geosynclinal Sedimentation, Spec. Publ. 19, edited by R. H. Dott and R. H. Shaver, pp. 345-357, Society of Economic Paleontologists and Mineralogists, Tulsa, Okla., 1974.

Boles, J. R., and C. A. Landis, Jurassic sedimentary mélange and associated facies, Baja California, Mexico, Geol. Soc. Am. Bull., 95, 513-521, 1984.

Breitkreuz, C., and H. Bahlburg, Palaeozoic flysch series in the coastal Cordillera of northern Chile, Geol. Rundsch., 74, 1985.

Cecioni, A., El Tremadociano de Sotoca, I Región, norte de Chile, Actas Congr. Geol. Chileno 2nd, 3, pp. 159-164, 1979.

Coira, B., J. Davidson, C. Mpodozis, and V. Ramos, Tectonic and magmatic evolution of the Andes of northern Argentina and Chile, Earth Sci. Rev., 18, 303-332, 1982.

Cowan, D. S., Deformation and metamorphism of the Franciscan subduction zone complex northwest of Pacheco Pass, California, Geol. Soc. Am. Bull., 85, 1623-1634, 1974.

Cowan, D. S., Deformation of partly dewatered and consolidated Franciscan sediments near Piedras Blancas Point, California, in Trench-Forearc Geology, Spec. Publ. 10, edited by J. K. Leggett, pp. 439-457, Geological Society of London, 1982.

Dalziel, I. W. D., Collision and cordilleran orogenesis: An Andean perspective, in Collision Tectonics, Spec. Publ. 19, edited by M. P. Coward and A. C. Ries, pp. 389-404, Geol. Soc. London, 1986.

Damm, K-W., S. Pichowiak, and W. Zeil, The plutonism in the north Chilean coast range and its geodynamic significance, Geol. Rundsch., 70, 1054-1076, 1981.

Davidson, J., C. Mpodozis, and S. Rivano, El Paleozoico de Sierra de Almeida, al oeste de Monturaqui, Alta Cordillera de Antofagasta, Chile, Rev. Geol. Chile, 12, 3-23, 1981.

Ferraris, B., and F. Di Biase, Hoja Antofagasta, Carta Geologica de Chile, 1:250,000, No. 30, Instituto de Investigaciones Geologicas, Santiago, Chile, 1978.

Gonzalez-Bonorino, F., and L. Aguirre, Metamorphic facies series of the crystalline basement of Chile, Geol. Rundsch., 59, 979-994, 1970.

Helwig, J., Stratigraphy, sedimentation, paleogeography, and paleoclimates of Carboniferous ("Gondwana") and Permian of Bolivia, Am. Assoc. Petrol. Geol. Bull., 56, 1008-1033, 1972.

Hervé, F., F. Munizaga, M. Mantovani, and M. Hervé, Edades Rb/Sr neopaleozoicas del basamento cristalino de la Cordillera de Nahuelbuta, Actas Congr. Geol. Chileno 1st, 2, F19-F26, 1976.

Hervé, F., J. Davidson, E. Godoy, C. Mpodozis, and V. Covacevich, The late Paleozoic in Chile stratigraphy, structure and possible tectonic framework, An. Acad. Bras. Cienc., 53, 361-373, 1981.

Hibbard, J., and H. Williams, Regional setting of the Dunnage mélange in the Newfoundland Appalachians, Am. J. Sci., 279, 993-1021, 1979.

Hsü, K. J., Principles of mélanges and their bearing on the Franciscan-Knoxville paradox, Geol. Soc. Am. Bull., 79, 1063-1074, 1968.

Hsü, K. J., Mélanges and their distinction from olistostromes, in Modern and Ancient Geosynclinal Sedimentation, Spec. Publ. 19, edited by R. H. Dott and R. H. Shaver, pp. 321-333, Society of Economic Paleontologists and Mineralogists, Tulsa, Okal., 1974.

James, D. E., Plate tectonic models for evolution of the central Andes, Geol. Soc. Am. Bull., 82, 3325-3346, 1971.

Lewis, K. B., Quaternary sedimentation on the Hikurangi oblique-subduction and transform margin, New Zealand, in Sedimentation in Oblique-Slip Mobile Zones, Spec. Publ. 4, edited by P. F. Ballance and H. G. Reading, pp. 171-189, International Association of Sedimentologists, Blackwell, Oxford, England, 1980.

Link, M. H., and R. H. Osborne, Lacustrine facies in the Pliocene Ridge Basin Group: Ridge Basin, California, in Modern and Ancient Lake Sediments, Spec. Publ. 2, edited by A. Matter and M. E. Tucker, pp. 169-187, International Association of Sedimentologists, Blackwell, Oxford, England, 1978.

Litherland, M., B. A. Klinck, E. A. O'Connor, and P. E. J. Pitfield, Andean trending mobile belts in the Brazilian Shield, Nature, 314, 345-348, 1985.

McCourt, W. J., J. A. Aspden, and M. Brook, New geological and geochronological data from the Colombian Andes: continental growth by multiple accretion, J. Geol. Soc. London, 141, 831-845, 1984.

Mercado, M., Hoja Laguna del Negro Francisco, Region de Atacama, Carta Geologica de Chile, 1:100,000, No. 56, Servicio Nacional de Geología y Minería, Santiago, Chile, 1982.

Miller, H., Orogenic development of the Argentinian/Chilean Andes during the Palaeozoic, J. Geol. Soc. London, 141, 885-892, 1984.

Moore, J. C., and R. L. Wheeler, Structural fabric of a mélange, Kodiak Islands, Alaska, Am. J. Sci., 278, 739-765, 1978.

Mpodozis, C., and R. F. Forsythe, Stratigraphy and geochemistry of accreted fragments of the ancestral pacific floor in southern South America, Palaeogeogr., Palaeoclimatol., Palaeoecol., 41, 103-124, 1983.

Naranjo, J. A., Geologia de la zona interior de la Cordillera de la Costa entre los 26°00' y 26°20', Region de Atacama, Carta Geologica de Chile, 1:100,000, No. 34, Instituto de Investigaciones Geologicas, Santiago, Chile, 1978.

Pacci, D., F. Hervé, F. Munizaga, K. Kawashita and U. Cordani, Acerca de la edad Rb-Sr Precámbrica de rocas de la Formación Esquistos de Belén, Departamento de Parinacota, Chile, Rev. Geol. Chile, 11, 43-50, 1980.

Page, B. M., Franciscan mélanges compared with olistostromes of Taiwan and Italy, Tectonophysics, 47, 223-246, 1978.

Picard, M. D., and L. R. High, Criteria for recognizing lacustrine rocks, in Recognition of Ancient Sedimentary Environments, Spec. Publ. 16, edited by J. K. Rigby and W. K. Hamblin, pp. 108-145, Society of Economic Paleontologists and Mineralogists, Tulsa, Okal., 1972.

Ramírez, C. F., and M. Gardeweg, Hoja Toconao, Region de Antofagasta, Carta Geologica de Chile, 1:250,000, No. 54, Servicio Nacional de Geología y Minería, Santiago, Chile, 1982.

Sepulveda, P., and J. A. Naranjo, Hoja Carrera Pinto, Region de Atacama, Carta Geologica de Chile, 1:100,000, No. 53, Servicio Nacional de Geología y Minería, Santiago, Chile, 1982.

Silver, E. A., and E. C. Beutner, Melanges, Penrose Conference Report, Geology, 8, 32-34, 1980.

Skarmeta, J., and N. Marinovic, Hoja Quillagua, Region de Atacama, Carta Geologica de Chile, 1:250,000, No. 51, Instituto de Investigaciones Geologicas, Santiago, Chile, 1981.

Smith, A. G., A. M. Hurley, and J. C. Briden, Phanerozoic Paleogeographical World Maps, 102 pp., Cambridge University Press, New York, 1981.

Steel, R. J., and T. G. Gloppen, Late Caledonian (Devonian) basin formation, western Norway: Signs of strike-slip tectonics during infilling, in Sedimentation in Oblique-Slip Mobile Zones, Spec. Publ. 4, edited by P. F. Ballance and H. G. Reading, pp. 79-103, International Association of Sedimentologists, Blackwell, Oxford, England, 1980.

Williams, P. R., C. J. Pigram, and D. B. Dow, Mélange production and the importance of shale diapirism in accretionary terranes, Nature, 309, 145-146, 1984.

Woodcock, N. H., Ludlow Series slumps and turbidites and the form of the Montgomery Trough, Powys, Wales, Proc. Geol. Assoc., 87, 169-182, 1976.

Copyright 1987 by the American Geophysical Union.

PERMIAN TO LATE CENOZOIC EVOLUTION OF NORTHERN PATAGONIA:
MAIN TECTONIC EVENTS, MAGMATIC ACTIVITY, AND DEPOSITIONAL TRENDS

M. A. Uliana

Esso Exploration, Inc., Houston, Texas 77001

K. T. Biddle

Exxon Production Research Company, Houston, Texas 77001

Abstract. The late Paleozoic to late Cenozoic evolution of northern Patagonia was influenced significantly by events that occurred while the area was part of the South American sector of Gondwanaland. Late Paleozoic to Middle Triassic subduction along the edge of the supercontinent formed a broad convergent-margin system that is the underpinning of northern Patagonia. Deformation (Gondwanidian orogeny) associated with the subduction is recognized in both the forearc and the convergent backarc areas. Regional extension, accompanied by bimodal volcanism, began in the Late Triassic and led to the formation of a number of north-northwest trending rift basins in Patagonia, which generally followed the Gondwanidian basement grain. Continued extension in the Jurassic and Early Cretaceous led to the opening of the Rocas Verdes marginal basin in southern Chile and, ultimately, to the opening of the South Atlantic Ocean. Once oceanic crust began to form, faulting and volcanism declined in Patagonia. During the late Early Cretaceous to the Late Cretaceous, sags over the rift basins coalesced to form a broad backarc basin behind the volcanic arc to the west. These sags are suggestive of thermally driven subsidence. Subsidence of the evolving Atlantic margin allowed extensive marine transgressions to take place from the east. The stratigraphic record of northern Patagonia reflects these events. The upper Paleozoic to upper Mesozoic sedimentary sequences were deposited in basins directly associated with convergent activity along the margin of Gondwanaland or in rift basins created during its breakup. Even though the Tertiary evolution of Patagonia was dominated by events along the western margin of South America, the patterns of sediment transport, thickness, and general shoreline position were still influenced by the locations of the Mesozoic rifts formed during the breakup of Gondwanaland.

Introduction

We outline here the Carboniferous to Recent geological evolution of northern Patagonia. This area (Figure 1) was part of the South American sector of Gondwanaland from at least the late Paleozoic until the breakup of the supercontinent in the Mesozoic. Events that took place while this area was part of Gondwanaland have influenced either directly or indirectly much of the subsequent evolution of northern Patagonia.

Our approach is predominantly stratigraphic. We have used the distribution of rock units in time and space to reconstruct the major depositional, magmatic, and tectonic events that have affected the area. The stratigraphic record of Patagonia, parts of Chile, and formerly adjacent southern Africa is presented graphically as a chronostratigraphic chart (Plate 1) that runs roughly west to east at about 42°S latitude. Figure 2 is a cross section through the area covered by the chronostratigraphic chart and schematically shows the thicknesses and structural relationship between the major intervals discussed below. The information presented in these two figures is based on our previous studies, data compiled from the literature, and fieldwork by the senior author. The stratigraphic terms used here represent only a small fraction of those that have been applied to Patagonia. From the literature we have chosen names that either have precedence or are the best described.

By using the information shown in Plate 1 and Figure 2, the evolution of northern Patagonia can be divided conveniently into three parts. (1) a Carboniferous to Early Triassic interval when the area was part of Gondwanaland, (2) a Late Triassic to Early Cretaceous interval, during which the breakup of this part of Gondwanaland took place, and (3) an Early Cretaceous to Recent postbreakup interval.

Carboniferous to Early Triassic
(Prebreakup Phase)

Sedimentary rocks. Upper Paleozoic to Lower Triassic rocks are widespread in northern Patagonia. The outcrops of these rocks form two roughly linear belts: an eastern belt in Ventania and the Colorado Basin area and a western belt in the Patagonide Ranges that extends into south-

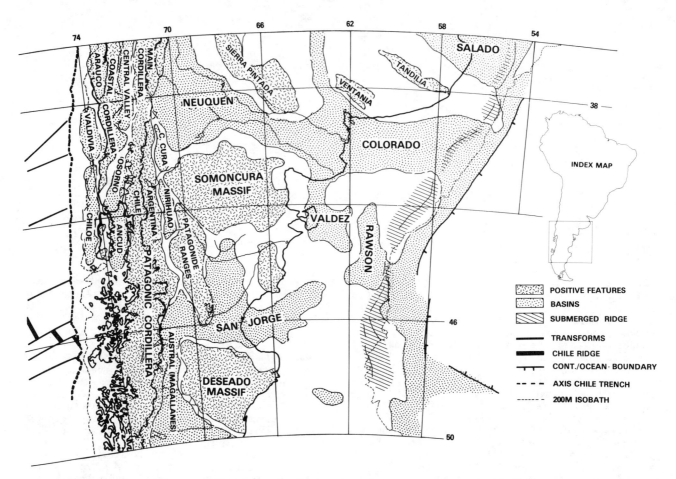

Fig. 1. Location map showing geologic provinces of northern Patagonia and surrounding areas.

central Chile [Lesta and Ferello, 1972; Hervé et al., 1981] and into the Deseado Massif (Figure 1 and Plate 1).

In the Patagonide Ranges, Upper Carboniferous deposits of the Tepuel Group (Plate 1) form a siliciclastic succession that is as much as 4,000 m thick [Page et al., 1984]. The lowest unit of this group, the Jaramillo Formation, is a sandy interval that was deposited in littoral environments. The middle unit, or Pampa de Tepuel Formation, is composed of pebbly mudstone, diamictite, and turbidites deposited in deeper water settings. The upper unit, the Mojon de Hierro Formation, is a progradational succession formed in shelfal to nonmarine environments. The Tepuel Group is overlain by the Lower Permian Rio Genoa Group [Cortiñas and Arbe, 1981]. This is a siliciclastic wedge as much as 800 m thick composed of stacked progradational/aggradational sequences deposited by constructive delta systems. Pebble composition [Ugarte, 1966] and paleocurrent analyses [Cortiñas and Arbe, 1981] indicate a source area in the Somoncura region to the northeast (Figure 1). Fossil plants suggest moderate to subtropical climate [Cúneo and Andreis, 1983].

The upper Paleozoic Panguipulli Formation of the Osorno Basin area of Chile (Figure 1 and Plate 1) is a rhythmically bedded succession of shales, siltstones, and sandstones with some olistostrome-like intervals. Clast composition indicates source areas in the present-day coastal ranges of Chile and from Patagonian basement rocks to the east [Thiele et al., 1976].

The eastern belt of upper Paleozoic sedimentary rocks crops out in the Ventania region of eastern Argentina (Figure 1). The succession there rests on a regional unconformity above the Devonian Lolén Formation [Harrington, 1970; Massabie and Rosello, 1984]. Tillites of the Sauce Grande Formation form the lowest part of the succession. These rocks are overlain by neritic to sublittoral shales and sandstones of the Bonete Formation, which are followed by sublittoral to nonmarine rocks of the Tunas Formation [Harrington, 1970, 1980]. This succession spans the entire Permian [Amos, 1972] and possibly includes some Upper

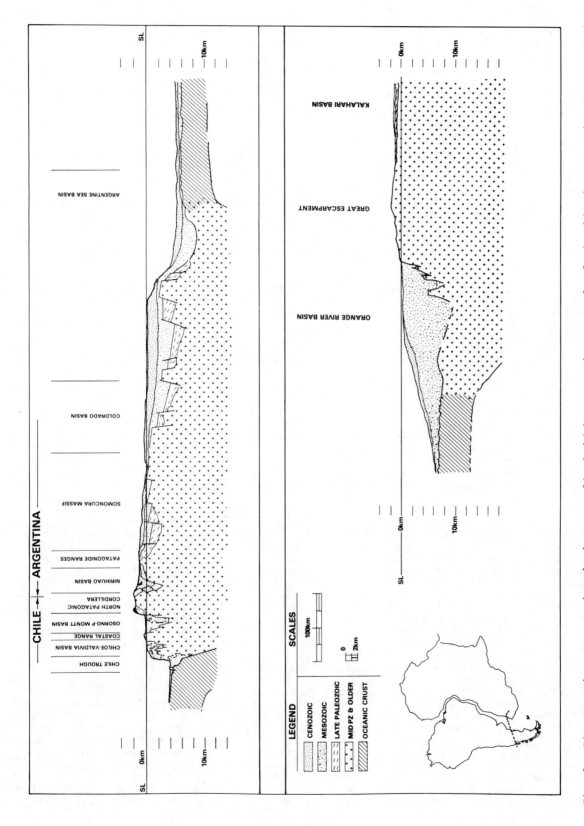

Fig. 2. Diagrammatic cross section showing generalized thicknesses and structural relationships between major tectonostratigraphic sequences discussed in the text.

Carboniferous [Harrington, 1972] and Lower Triassic rocks [Suero, 1957]. Similar rocks occur in the subsurface area of the Colorado Basin [Zambrano, 1972, 1980], in the Karoo Basin of southern Africa (Plate 1) [du Toit, 1937; Tankard et al., 1982], and in the Islas Malvinas (Falkland Islands) [Greenway, 1972].

Igneous rocks. Upper Paleozoic-lowermost Mesozoic igneous rocks occur throughout large areas of northern Patagonia and have been the subject of considerable work in the last 10 years (see summaries in the work of Toubes and Spikerman [1973], Halpern et al. [1975], Spikerman [1978]; Franchi and Page [1980], Lizuaín [1981], and Hervé et al. [1984]). These rocks are predominantly granites or rhyolites with associated rocks of intermediate composition and appear to be calc-alkalic [Llambías et al., 1984]. They are similar to coeval igneous rocks that occur in the Sierra Pintada (Figure 1) and the Frontal Cordillera of the southern La Pampa and Mendoza regions of Argentina between 38°N and 35°N. Radiometric ages of these rocks range from 335 to 230 Ma, but individual ages should be viewed with caution because of the complicated history of intrusion and lack of detailed information on dating techniques. These data and fieldwork in the Somoncura area (Figure 1) suggest two discrete magmatic events whose products are separated by a considerable gap in time [Llambías et al., 1984]. The older event is represented mainly by plutonic rocks and appears to be Early to middle Carboniferous in age [Llambías et al., 1984]. The younger event includes both plutonic and volcanic products and is Permian to Early Triassic in age [Llambías et al., 1984]. In addition, a group of late Paleozoic gabbroic intrusions has been recognized in the Patagonide Ranges [Lesta and Ferello, 1972; Ramos, 1983; Page et al., 1984]. The distribution of the major plutons indicates a broad igneous province as much as 500 km wide [Forsythe, 1982; Ramos, 1983] rather than a narrow linear batholithic belt.

Tectonics. The rocks described above define three northwest to north-northwest trending tectonostratigraphic belts that cut obliquely across northern Patagonia. The upper Paleozoic sedimentary rocks of the Patagonide Ranges and south central Chile form the westernmost belt and were deposited in a narrow north-northwest trending basin (or basins) that has (or have) been called the Cuenca Central Patagónica [Suero, 1962]. Several lines of evidence suggest that some of these rocks were deposited in a forearc setting [Hervé et al., 1981; Forsythe, 1982]. The local dominance of nonmarine deposits [DeGiusto et al., 1980], the obvious marine connection to the west [González, 1984], and the abrupt eastern margin of this belt close to the coeval magmatic province in northern Patagonia (Plate 1) [Forsythe, 1982] suggest a forearc to perhaps an intra-arc setting.

In the Osorno Basin area of Chile, the upper Paleozoic Panguipulli Formation (Plate 1) is deformed into tight, concentric folds cut by much younger normal faults [Thiele et al., 1976]. The deformation that produced the shortening has been dated as Permian to possibly early Triassic [Miller, 1979]. Upper Paleozoic sedimentary rocks of northern Patagonia also show evidence of late Paleozoic to early Mesozoic(?) shortening but are not as deformed or metamorphosed as rocks of the same age in Chile [Lesta and Ferello, 1972; Miller, 1976].

The upper Paleozoic-lowermost Mesozoic magmatic rocks form the second tectonostratigraphic belt. In northern Patagonia these rocks lie mostly to the east of sedimentary rocks discussed above (Plate 1) but locally overlap them [Ramos, 1983]. They have been variously interpreted as the products of westward dipping subduction beneath a separate Patagonian plate [Ramos, 1984], the results of pervasive intrusion across a collisional Hercynian-like orogen [Ramos, 1984, p. 320], or a signature of eastward dipping Andean-style subduction [Forsythe, 1982]. The first two interpretations do not satisfactorily explain the roughly age-equivalent high-pressure/low-temperature metamorphic rocks in coastal Chile [Saliot, 1969; Hervé et al., 1981; Forsythe, 1982]. A magmatic arc associated with eastward dipping subduction beneath the South American margin of Gondwanaland best accounts for the known data [Forsythe, 1982; Ramos, 1983]. Lock [1980] suggested that the subducted slab had a shallow dip during the late Paleozoic to explain the width of the magmatic arc and the occurrence of backarc deformation so far from the subducting margin (see below). The shallow slab model would necessitate a very low subduction angle and contact between lower and upper plates beneath much of Patagonia. This should have inhibited subduction related intrusion in Patagonia during this time unless the magmatism was associated with a migration of the arc during the transition from a steeply dipping to shallowly dipping state. Unfortunately, radiometric age control on the appropriate rocks is not yet refined enough to address this question.

The upper Paleozoic sedimentary rocks of the Ventania region occur to the northeast, or behind, the late Paleozoic-earliest Mesozoic magmatic arc and thus are interpreted as backarc basin deposits [Lock, 1980; Forsythe, 1982; Ramos, 1984; Gust et al., 1985]. This basin, or series of basins, was a segment of the Samfrau Geosyncline of du Toit [1937] and has a continuation in the Islas Malvinas and in southern Africa. In the Ventania area these rocks have been folded into a series of tight, northeast-vergent folds [Harrington, 1970, 1980]. This compressional deformation appears to have begun in the Permian and ended after the Early Triassic [Ramos, 1984]. This event is similar in both timing and style to deformation described in the Cape Fold Belt of southern Africa. There, folding and thrusting have been dated as occurring between 274 and 230 Ma [Halbich, 1983].

In summary, during late Paleozoic-Early Triassic, northern Patagonia appears to have been the site of a forearc to intra-arc area of clastic sedimentation and local gabbroic intrusions, a wide magmatic arc characterized by abundant calc-alkalic intrusions and their volcanic products, and a backarc area of sedimentation and deformation. These belts and the structural fabric created by late Paleozoic-earliest Mesozoic shortening across the area form the basement framework of northern Patagonia. The compressive ocean-continent convergent regime that led to the formation of these belts was replaced in the Mesozoic by an extensional one, but perhaps this occurred without the cessation of subduction.

Late Triassic to Early Cretaceous (Breakup Phase)

Sedimentary rocks. Subsurface studies in the oil-producing areas of southern South America and numerous field studies indicate that lower Mesozoic sedimentary rocks fill a number of narrow, northwest to north-northwest trending basins in central Chile and western Argentina [Yrigoyen and Stover, 1970; Charrier, 1979], Patagonia [Bracaccini, 1968; Lesta and Ferello, 1972; Cortés, 1981a], Tierra del Fuego [Natland et al., 1974], and across the southern Argentine shelf [Gust et al., 1985]. These rocks rest on a regional angular unconformity that is one of the most significant lithologic and structural discontinuities in Patagonia. Most commonly, these deposits are nonmarine and are consistently associated with volcanic and volcaniclastic rocks. Irregular distribution, abrupt and intricate facies changes, and the distribution of locally derived coarse-grained clastics composed of fragments of basement rocks [Robbiano, 1971; Cortés, 1981a; Gulisano and Pando, 1981] suggest that these rocks and the associated volcanic rocks were deposited in discrete fault-bounded basins.

Upper Triassic rocks appear to be poorly represented in Patagonia and adjoining Chile, but this may be an artifact of the difficulty encountered in dating the lower Mesozoic sedimentary rocks. Upper Triassic deposits are known from the Somoncura area (Figure 1 and Plate 1), where they are represented by the Garamilla and Los Menucos formations and the Dique Ameghino rhyolite [Llambías et al., 1984; Lapido et al., 1984; Cortés, 1981a]. In the Osorno-Puerto Montt Basin of Chile, the nonmarine Tralcán Formation appears to be of Late Triassic age [Charrier, 1979].

Sedimentary rocks of probable Early Jurassic age occur in the deepest parts of the Colorado and Valdez basins (Plate 1). These nonmarine rocks are unfossiliferous and have been penetrated by drilling in only a few places. We infer an Early Jurassic age for these rocks on the basis of a comparison with the Upper Triassic-Lower Jurassic section in the Somoncura area [Cortés, 1981a] and the presence of dated rocks as old as Bajocian in grabens and half grabens in southern Africa [Dingle et al., 1983]. In the Colorado Basin, Lower Jurassic(?) to Upper Jurassic rocks are known as the Fortín Formation (Plate 1). This is a red bed unit composed of conglomerates and coarse-grained sandstones that were deposited in alluvial fan and alluvial plain environments [Lesta et al., 1978]. Seismically constrained variations in thickness show that the Fortín Formation was deposited in three separate fault-bound depocenters [Zambrano, 1980].

The Fortín Formation is overlain by the Colorado Inferior Formation. This is a succession of nonmarine, red-brown to red-gray claystones, siltstones, and texturally immature sandstones. The deposits of the Colorado Inferior Formation are restricted to the axial parts of the Colorado Basin [Lesta et al., 1980a], but they are more widespread than the underlying Fortín Formation (Plate 1). The distribution of the formation as a whole indicates a gradual expansion of the area of deposition with time [Zambrano, 1980]. These rocks are also difficult to date, but are most likely of Early to middle Cretaceous age [Lesta et al., 1978]. They appear to be equivalent to the "Early Drift Sequence" of the Orange River Basin of southern Africa described by Gerrard and Smith [1982].

To the west, sedimentation was limited to areas west of the central Somoncura Massif (Figure 1) during the Jurassic. Here, grabens between the Somoncura Massif and the Patagonide Ranges contain Lower to Middle Jurassic nonmarine and locally marine units (among several others, the Puesto Lizarralde, Puntudo Alto, Lomas Chatas, Mulaguineu, and Piltriquitrón formations [Lesta et al., 1980a; Musacchio, 1981; Gabaldón and Lizuaín, 1983; Cortiñas, 1984]). During this time, marine incursions were from the north, presumably through the Neuquén Basin [Gabaldón and Lizuaín, 1983]. During the Late Jurassic global sea level highstand, marine influence extended well into the north Patagonian Andes through a possible connection with the Magallanes Basin to the south [Riccardi and Rolleri, 1980; Aguirre Urreta and Ramos, 1981]. Late Jurassic deposition in the Andean belt is represented by carbonates with local buildups and siliciclastic turbidites (Cotidiano and Tres Lagunas formations [Ramos, 1976; Ramos and Palma, 1983]), while in the grabens of the Patagonide Ranges and western Somoncura, the Late Jurassic depocenters contain a nonmarine (brackish?) sequence with black shales and stromatolitic limestones (Cañadon Asfalto Formation and Taquetrén Formation [Tasch and Volkheimer, 1970; Lesta and Ferello, 1972; Nullo and Proserpio, 1975]).

Early Cretaceous depositional patterns in western Patagonia have been discussed in several recent papers [Skarmeta, 1976; Haller and Lapido, 1982; Haller et al., 1981; Ramos and Palma, 1983]. The complex facies mosaic includes, from west to east, a belt of marine deposits with easterly derived pyroclastics (Isla Traiguén Formation [Skarmeta, 1976; Ramos, 1983]), a narrow marine trough along the Chile-Argentina border (Coyhaique Group [Haller et al., 1981]; Katterfield and Apeleg formations [Ramos and Palma, 1983]), with westerly derived volcanics and easterly derived siliciclastics and volcaniclastics [Aguirre Urreta and Ramos, 1981; Ramos and Palma, 1983], and a third belt of fluvial to lacustrine deposits located along the Patagonide Ranges and onlapping the Somoncura Massif (Las Heras Group; D129 Formation and equivalents [Lesta and Ferello, 1972; Cortiñas and Arbe, 1981]). By the end of the Hauterivian the areas under sedimentation had expanded to cover many of the regional highs located between the early Mesozoic graben troughs.

Igneous rocks. Volcanic rocks erupted during this time are very widespread in Patagonia. They are dominated by rhyolites and dacites in central and eastern Somoncura [Llambías et al., 1984] but include andesites and basalts in western Somoncura, the Patagonide Ranges, and the north Patagonian Andes [Lesta and Ferello, 1972; Haller and Lapido, 1982; Haller et al., 1981; Gust et al., 1985]. Geochemical analyses on some of the silicic volcanic rocks indicate anatexis of a crustal source and emplacement under conditions of regional extension [Bruhn et al., 1978; Gust et al., 1985]. Available stratigraphic and radiometric control show that the regional magmatism associated with extension lasted some 50 million years, from about 200 Ma to approximately 150 Ma, peak activity occurring around 165 to 155 Ma [Gust et al., 1985]. The timing of this igneous episode closely parallels that of the main Karoo volcanic event in southern Africa (Figure 3) [Bristow and Saggerson, 1983] (phases II and III of Dingle et al., 1983]).

After the middle Callovian, the area affected by igneous activity in Patagonia was greatly reduced. By the Late Jurassic most igneous activity was restricted to areas west of central Somoncura. There, the Upper Jurassic volcanic rocks are a calc-alkalic suite of andesites, dacites, and rhyolites [Haller et al., 1981; Haller and Lapido, 1982]. In the Patagonide Ranges and in western Somoncura, some Late Jurassic basalt flows and volcanic rocks with alkalic affinities, known as the Taquetrén Formation, occur [Coira, 1979; Cortés and Baldoni, 1984]. The geochemistry and significance of these rocks is poorly understood at present.

The Patagonic and Main Cordilleras of Chile and westernmost Argentina (Figure 1) are dominated by the uplifted and eroded Andean batholith. Intrusive rocks in this area range in age from as old as Early Devonian [Lizuaín, 1981] to as young as late Miocene [González Díaz and Valvano, 1979], but most of the plutonic rocks are of either Late Jurassic or Cretaceous age [Malvicini and Llambías, 1982]. Most of the available radiometric ages fall between 135 and 80 Ma, the majority grouping between 100 and 85 Ma [Haller and Lapido, 1982]. Thus, although most of these plutonic rocks of the Andean batholith postdate the time interval discussed in this section, clearly some were formed at this time. These rocks are described as granodiorites, tonalites, diorites, and gabbros [e.g., González Díaz and Valvano, 1979; Haller et al., 1981]. Geochemical studies of the batholith near 45°S show that these rocks are calc-alkalic [Wells, 1979; Ramos et al., 1982]. They form large plutons with metamorphic aureoles and commonly show some evidence of cataclastic deformation.

Tectonics. Crustal extension, which began in northern Patagonia by Late Triassic, continued throughout the Jurassic and culminated in the opening of the South Atlantic Ocean in the earliest Cretaceous. This extension formed a number of northwest to north-northwest trending fault-bounded basins in southern South America. The Colorado Basin is the best described example of these basins in northern Patagonia. The time of initiation of extension is difficult to document because of the uncertainty of dating nonmarine sections and because of the depth of burial of most of the early basin fill. The basins created by extension are grabens and half grabens. They are best seen on seismic data from the Colorado, San Jorge, and Magallanes basins, but can also be seen on the Deseado Massif [Lesta et al., 1978; Biddle et al., 1986]. Here, about 1000 m of nonmarine Triassic section bound by north-northwest striking normal faults are exposed [DeGiusto et al., 1980].

Plate 1. Chronostratigraphic chart that illustrates stratigraphic relationship, nomenclature, and dominant rock composition across northern Patagonia and southern Africa. Key references by region are: Chiloe Basin - Mordojovich, 1974, 1981; Cecioni, 1980. Coastal Cordillera - Saliot, 1969; Hervé et al., 1981; Valenzuela Ayala, 1982. Osorno-Puerto Montt Basin - Katz, 1963, 1970, 1971; Illies, 1967; García, 1968; Moreno and Parada, 1976; Hervé et al., 1976. Patagonic Cordillera - Cazau, 1980; Ramos, 1976, 1978, 1979, 1983; Ramos et al., 1982; Ramos and Palma, 1983; Pesce, 1978; González Bonorino, 1979; Hervé et al, 1979; Thiele and Hein, 1979; Haller and Lapido, 1982; Haller et al., 1981; González Diaz and Nullo, 1980; Rapala et al., 1983, 1984. Patagonide Ranges - Lesta and Ferello, 1972; Lesta et al., 1980a; Spikermann, 1978; Codignotto et al., 1979, Franchi and Page, 1980; Musacchio, 1981; Cortiñas and Arbe, 1981; Cortiñas, 1984; Page et al., 1984. Somoncara Massif - Stipanicic and Methol, 1972, 1980; Núñez et al., 1975; Coira, 1979; Lapido and Page, 1979; Ardolino, 1981; Llambías et al., 1984; Lapido et al., 1984; Franchi et al., 1984. Arroyo Verde - Haller, 1979; Lizuaín and Sepúlveda, 1979; Cortéz, 1981a, b; Mendia and Bayarsky, 1981; Llambías et al., 1984; Franchi et al., 1984. Valdes Basin - Masiuk et al., 1976; Haller, 1979; Spiegelman and Busteros, 1979; Lesta et al., 1980b. Colorado Basin - Harrington, 1970, 1980; Zambrano, 1972, 1980; Jordi and Lehner, 1973; Rolleri, 1973; Yrigoyen, 1975; Valera and Cingolani, 1976; Lesta et al., 1978, 1980b; Ludwig et al., 1979. Orange River Basin - Siesser and Dingle, 1981; Gerrard and Smith, 1982; Jaunich, 1983; Dingle et al., 1983; Bristow and Saggerson, 1983. Namibia-West Karoo Basin - Martin, 1973; Rust, 1975; Tankard et al., 1982; Dingle et al., 1983. Cape-Karoo Basin - Dingle, 1973; Dingle et al., 1983; Lock et al., 1975; Lock, 1980; du Toit, 1979; Dunlevey and Hiller, 1979; Tankard et al., 1982.

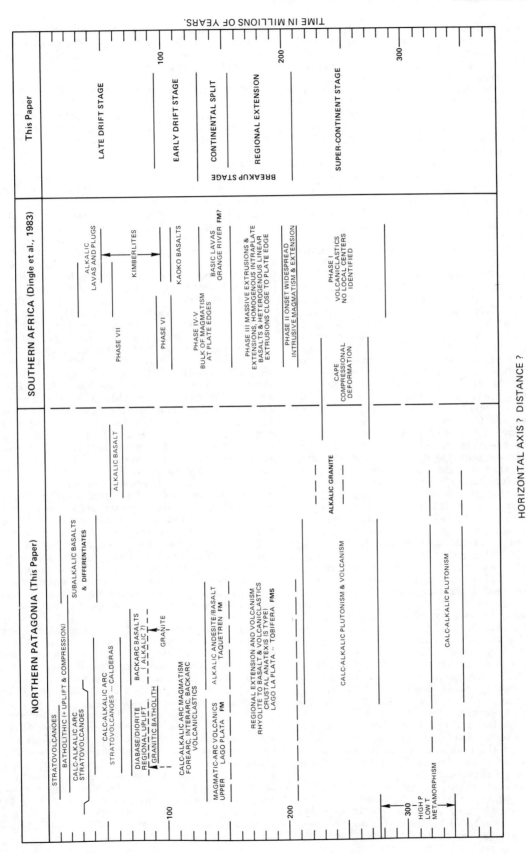

Fig. 3. Temporal distribution of key magmatic events in northern Patagonia and southern Africa. The horizontal axis runs from west (left) to east (right), but is not to scale. For relative position of corresponding units see Figure 2 and Plate 1.

During the early stages of extension the fault-bounded basins were small and isolated as shown by the limited distribution of the early basin fill in Plate 1. During the Late Jurassic and earliest Cretaceous the areas affected by normal faulting enlarged and became interconnected to some extent.

The igneous activity associated with the extension reached a maximum during the Middle Jurassic and began to decrease after the Callovian (Figure 3). In northern Patagonia, volcanism was widespread and appears to have involved considerable crustal anatexis [Bruhn et al., 1978; Gust et al., 1985]. Intrusion and volcanism that is likely subduction related also occurred to the west near the Chilean-Argentine border during the Middle and Late Jurassic [Ramos, 1983]. Evidence of Triassic subduction-related igneous activity, however, is equivocal.

To the south and west of Patagonia, extension continued until about the latest Jurassic when a piece of continental crust was detached from the western margin of southernmost South America, and a small basin floored by oceanic crust had formed [Bruhn et al., 1978; Saunders et al., 1979; Dalziel, 1981]. At about 43°50'S, within the Patagonide Ranges, a number of noritic and gabbroic (Gabro Cresta de los Bosques) sills and small intrusions with olivine tholeiite composition that intrude the marine Lias section have been interpreted to have marginal basin affinities and may represent the local expression of this event [Page, 1984]. Remnants of the oceanic basin are preserved south of 50°S as the Rocas Verdes ophiolitic assemblage of Chile [Dalziel, 1981].

Evidence on the Atlantic side of Patagonia suggests that the onset of the spreading was somewhat younger there than in the Rocas Verdes marginal basin. The oldest identified magnetic anomaly in the southern Atlantic Ocean is M11, just off the tip of southern Africa [Larson and Ladd, 1973]. However, recently released multichannel seismic data suggest that this anomaly occurs within extended continental crust [Gerrard and Smith, 1982; Jaunich, 1983]. The oldest, clearly oceanic magnetic anomaly is M6 (I. Norton, personal communication, 1985). If the time scale of Harland et al. [1982] is used, seafloor spreading would be older than about 128 Ma in this area. The oldest identified magnetic anomaly on the Argentine side of the South Atlantic is M4 [Rabinowitz and LaBreque, 1979], an indication that oceanic crust had formed there by about 127 Ma. This is about 25 million years after the peak of bimodal volcanic activity in Patagonia and about 10 million years after oceanic crust began to form in the Rocas Verdes marginal basin. Thus, we view the extension and volcanism that began in the Late Triassic and continued through the Jurassic as the precursor of events that led to the creation of oceanic crust in the Rocas Verdes basin during the Late Jurassic and in the South Atlantic Ocean during the Early Cretaceous.

Once seafloor spreading began in the South Atlantic Ocean, normal faulting declined in Patagonia. Subsidence caused by lithospheric thinning and faulting was replaced by thermally driven subsidence. Areas of deposition above extensional basins, such as the Colorado Basin, expanded (Plate 1), and passive-margin deposits accumulated along both the Argentina and southern Africa continental margins.

Early Cretaceous to Recent (Postbreakup Phase)

Sedimentary rocks. Early Cretaceous and younger sedimentary units are much more widely distributed in northern Patagonia than sedimentary rocks deposited before or during the breakup of Gondwanaland (Plate 1). Lower Cretaceous sedimentary rocks are entirely nonmarine in large areas of northern Patagonia. Depositional patterns and thicknesses in the central and eastern parts of Patagonia were likely controlled by thermally induced sags over the Mesozoic rift basins [e.g., Zambrano, 1980]. Between these areas, Lower Cretaceous rocks are thin or absent (Plate 1).

In the Colorado Basin, successive isopach maps of the Upper Cretaceous Colorado Superior Formation and overlying Cenozoic depositional sequences [Zambrano, 1980, Figure 5] illustrate the change through time toward more areally extensive and thinner deposits. During the Late Cretaceous a marginal basement ridge at the edge of the Patagonian segment of South America (Figure 1) [Urien and Zambrano, 1973; Rolleri, 1973] subsided enough to allow transgressions from the east to occur [Camacho, 1967; Charrier and Malumián, 1975; Lesta et al., 1978]. After this time the magnitude of coastal onlap and marine encroachment attained during each transgressive event closely follows the global trend of coastal onlap defined by Vail et al. [1977]. The same general effect has been recorded across the Atlantic, in southern Africa [Siesser and Dingle, 1981].

Maastrichtian-Paleocene deposits overlie a regional disconformity and represent a distinctive episode of marine transgression and subsequent regression. The peak of marine flooding occurred during the Maastrichtian [Bertels, 1969], when most of Patagonia was drowned (Plate 1). Nonmarine conditions returned to the Somoncura area and areas to the west after the Danian. During the Cenozoic, successive transgressions affected smaller and smaller areas of Patagonia (Plate 1 and paleogeographic sketches by Franchi et al. [1984]).

After the Maastrichtian, the eastern section of northern Patagonia was dominated by shallow- to deep-shelfal marine deposition. These rocks form a package of stacked prograding wedges on a wide continental shelf and adjacent coastal plain and are rich in fine-grained pyroclastic debris [e.g., Spiegelman and Busteros, 1979].

In general, in the Patagonian region only limited material was derived from erosion of Mesozoic and older rocks from the Late Cretaceous on. This was due to subdued relief throughout most of Patagonia at this time. As a result, many Upper Cretaceous and Cenozoic sequences are rich in volcanic material erupted to the west and contain only modest volumes of other rock fragments or detrital quartz. Dominant lithofacies are fluvial with abundant paleosols and shallow lacustrine diatomite deposits.

West of the Somoncura Massif, deposition was associated with a subduction-related magmatic arc and occurred in a series of deep backtroughs with variable amounts of connection to the ocean. A discussion of the complex Cenozoic depositional patterns in the western third of Patagonia is beyond the scope of this paper. However, the subject has been analyzed by González Bonorino and González Bonorino [1978], Cazau [1980], Ramos et al. [1982], Katz [1970, 1971], Garcia [1968], Cecioni [1980], and Franchi et al. [1984].

Igneous rocks. Volumetrically, the most important postbreakup igneous rocks are associated with the Andean arc along the western edge of South America. The locations of calc-alkalic eruptive fields changed with time but were confined to the western third of Patagonia (Plate 1). These fields produced thick piles of flows and proximal pyroclastic deposits that accumulated in fault-bound depressions and are intricately associated with pyroclastics deposited in subaerial, lacustrine, paralic, and marine environments [González Bonorino and González Bonorino, 1978; Rapela et al., 1983, 1984; Franchi et al., 1984].

In the north Patagonic Andes (Figure 1), field relationships and radiometric dates indicate that most of the Andean batholith was emplaced during the Cretaceous between 135 and 80 Ma [Toubes and Spikerman, 1973; González Díaz and Valvano, 1979; Lizuaín, 1981]. The main plutonic episode occurred between 100 and 85 Ma and has been correlated to an episode of global high seafloor spreading rates [Haller and Lapido, 1982]. The volcanic equivalents of the Late Cretaceous batholith are poorly preserved in the north Patagonic Cordillera, but Eocene and younger volcanic and volcaniclastic rocks are common.

There is also a Miocene episode of granite intrusion in the north Patagonic Andes [Malvicini and Llambías, 1982]. K/Ar ages of these rocks in Argentina range from 16 ± 1 Ma to 9 ± 1 Ma [González Díaz and Valvano, 1979]; in Chile, ages of granodiorites of 12.1 ± 0.4 Ma have been obtained [Hervé et al., 1979].

An eruptive complex of basalt flows, basaltic breccias, ignimbrites, and tuffs of Late Cretaceous age occurs along the eastern flank of the Andes in northern Chubut (Tres Picos Prieto Formation [Franchi and Page, 1980]). Although remnants of this unit are scarce, it can be locally as much as 500 m thick and is thought to have extended at least 500 km from southern Neuquén to the central Chubut [Franchi and Page, 1980; Baker et al., 1981]. Basalts of this age include both quartz and olivine tholeiites in central Patagonia [Baker et al., 1981].

Away from the Andes, the Tertiary volcanic rocks are most common in the Somoncura area. There, olivine basalts of the Somoncura, La Mesada, and Puesto Muñoz formations extend over wide areas to make the Somoncura Plateau [Yllanez and Lema, 1979; Ardolino, 1981]. Eruptive centers are not commonly recognized, and the basalts are interpreted as the products of fissure type eruptions. They range in age from 33 ± 3 Ma to 15 ± 1 Ma [Ardolino, 1981] and presumably are derivatives of partial melt of peridotites in the upper mantle [Corbella, 1984]. The Quiñelaf Formation also is found in the Somoncura area. This formation consists of basanites, alkalic basalts, trachybasalts, and trachytes related to large eruptive centers [Corbella, 1982a, b]. They range in age from 37 to 11 Ma [Ardolino, 1981] and might be associated with crustal thinning [Corbella, 1984].

In the Colorado Basin area, Paleogene alkalic basalts with basic and ultrabasic xenoliths, tuffs, and hyaloclastites occur. These rocks were erupted along the normal faults that bound the southern flank of the basin [Lesta et al., 1978].

Tectonics. After the breakup of Gondwanaland and the beginning of seafloor spreading in the south Atlantic area, the tectonics of western Patagonia were dominated by events that occurred along the western margin of South America and resulted in voluminous intrusion and volcanism in the western third of the continent. During much of this time some extension appears to have occurred within the arc. During the Miocene, however, regional uplift of the Andes took place and Cenozoic rocks were slightly shortened [Thiele and Hein, 1979; González Bonorino, 1979].

The postbreakup structural style of the north Patagonic Andes is dominated by block faulting [González Bonorino, 1979; Thiele and Hein, 1979; Thiele et al., 1979]. The faults strike north or north-northwest but are locally associated with west-northwest and northwest striking faults. This segment of the Andes lacks the thin-skinned type of deformation [González Bonorino, 1979; Thiele et al., 1979] seen to the north and to the south in the Neuquén and Magallanes basins. Evidence of east-west shortening is limited to a narrow belt 50 to 60 km wide. Here, high-angle basement involved reverse faults, and a few associated folds that deform beds that contain mid-Miocene mammalian fossils occur. Seismic sections indicate that these structures are very young and involve reactivation (inversion) of older normal faults.

During the late Paleogene and early Miocene a number of fault-bounded basins formed in western Argentina and Chile. These include the Collon Cura and Rio Chico troughs [González Bonorino, 1979; González Díaz and Nullo, 1980], the Ñirihuao Basin [Cazau, 1980], the Lolco-Bio Bio Basin, and the Arauco, Osorno-Puerto Mont-Golfo de Ancud basins, and Golfo de Penas-Chiloe-Pucatrihue-Valdivia basins of Chile [Garcia, 1968; Katz, 1970, 1971; Cecioni, 1980; Mordojovich, 1981; Forsythe and Nelson, 1985]. These basins trend to the north or north-northwest and appear to be extensional to transtensional in origin. Locally, the fill of these basins has been uplifted and deformed by compression. The Nirihuau Basin is a good example of this [González Bonorino, 1979].

Throughout the rest of northern Patagonia, Late Cretaceous and younger rocks are subhorizontal. They are deformed only by small displacement normal faults. Regional thermally driven subsidence probably continued to occur over the Mesozoic rift basins and along the Atlantic margin, but at decreasing rates. This was enough, however, to localize drainage systems and sediment transport paths over the basins formed during the breakup of Gondwanaland [Franchi et al., 1984].

Conclusions

1. The Permian to late Cenozoic tectonic and stratigraphic evolution of northern Patagonia can be divided conveniently into three phases: (1) a prebreakup of Gondwanaland phase that lasted from at least the late Paleozoic to the Early Triassic; (2) a Late Triassic to Early Cretaceous phase dominated by extension during the breakup of Gondwanaland; and (3) a postbreakup phase from the Early Cretaceous to the present.

2. The late Paleozoic to Early Triassic prebreakup phase was dominated by subduction-related tectonics. Patagonia was the site of an Andean-type margin with a wide magmatic arc-backarc system and associated compressional deformation.

3. During the Late Triassic to the Early Cretaceous, breakup of this sector of Gondwanaland, widespread extension, and associated volcanism took place in Patagonia. Extensional basins, such as the Colorado Basin, formed at this time.

4. The postbreakup phase of evolution, lasting from the Early Cretaceous to the present, has been dominated by subduction-related activity along the western margin of South America and by decreasing thermally driven subsidence over the Mesozoic extensional basins. Continued subsidence of the Atlantic margin, coupled with eustatic sea level changes, led to maximum flooding of Patagonia in the Maastrichtian.

5. Features formed before and during the breakup of Gondwanaland have strongly influenced the subsequent evolution of Patagonia. For example, the prebreakup basement grain may have controlled the location of some of the extensional features formed during the breakup of this part of Gondwanaland. The location of the breakup stage basins controlled postbreakup subsidence and hence sediment transport systems and thickness of postbreakup sedimentary units. Along the southern flank of the Colorado Basin, breakup stage faults have localized, and acted as conduits for, Tertiary volcanism. Faults formed during either the prebreakup or breakup phase of evolution have been reactivated as reverse faults along the eastern edge of the Andes during the Tertiary and have influenced the location of some deformation associated with the Andean Mountain belt.

Acknowledgments. Early versions of this paper were reviewed by R. D. Forsythe, I. O. Norton, F. L. Wehr, and an anonymous reviewer. Their efforts have improved this paper, and we thank them for their constructive comments. We also thank Exxon Production Research Company for permission to publish this contribution.

References

Aguirre Urreta, M. B., and V. A. Ramos, Estratigrafía y paleontología de la alta cuenca del Rio Roble, Provincia de Santa Cruz, Actas Congr. Geol. Argent. 8th, 3, 101-138, 1981.

Amos, A., Las cuencas carbónicas y pérmicas de Argentina, An. Acad. Bras. Cienc., 44, suppl., 21-36, 1972.

Ardolino, A., El vulcanismo Cenozoico del borde occidental de la Meseta de Somuncura, Provincia del Chubut, Actas Congr. Geol. Argent. 8th, 3, 7-23, 1981.

Baker, P. E., W. J. Rea, J. Skarmeta, R. Caminos, and D. C. Rex, Igneous history of the Andean Cordillera and Patagonian Plateau around latitude 46°S, Philos. Trans. R. Soc. London, Ser. A, 303(1474), 105-149 1981.

Bertels, A., Estratigrafía del límite Cretacico-Terciario de la Patagonia Septentrional, Rev. Asoc. Geol. Argent., 24(1), 41-54, 1969.

Biddle, K. T., M. A. Uliana, R. M. Mitchum, M. G. Fitzgerald, and R. C. Wright, The stratigraphic and structural evolution of the central and eastern Magallanes Basin, southern South America, in Foreland Basins, edited by P. A. Allen and P. Homewood, Spec. Publ. Int. Assoc. Sedimentol. 8, Blackwell Scientific, Oxford, in press, 1986.

Bracaccini, O. I., Panorama general de geología patagónica, Actas Congr. Geol. Argent. 3rd, 1, 17-47, 1968.

Bristow, J. W., and G. P. Saggerson, A general account of Karoo vulcanicity in southern Africa, Geol. Rundsch., 72, 1015-1060, 1983.

Bruhn, R. L., C. R. Stern, and M. J. DeWitt, Field and geochemical data bearing on the development of a Mesozoic volcano-tectonic rift zone and backarc basin in southernmost South America, Earth Planet. Sci. Lett., 41, 32-46, 1978.

Camacho, H. H., Las transgresiones del Cretácico superior y Terciario de la Argentina, Rev. Asoc. Geol. Argent., 22(4), 257-279, 1967.

Cazau, L. B., Cuenca Ñirihuau-Ñorquinco-Cushamen, Simp. Geol. Reg. Argent. 2nd, 2, 1149-1171, 1980.

Cecioni, G., Darwin's Navidad embayment, Santiago region, Chile, as a model of the southeastern Pacific shelf, J. Pet. Geol., 2(3), 309-321, 1980.

Charrier, R., El Triásico en Chile y regiones adyacentes de Argentina--Una reconstruccion paleogeográfica y paleoclimática, Comun. Univ. Chile Fac. Cienc. Fis. Mat. Dep. Geol., 26, 1979.

Charrier, R., and N. Malumián, Orogénesis y epirogénesis de la región austral de America del sur durante el Mesozoico y el Cenozoico, Rev. Asoc. Geol. Argent., 30(2), 193-207, 1975.

Codignotto, J., F. Nullo, J. Panza, and C. Proserpio, Estratigrafía del grupo Chubut entre Paso de Indios y Las Plumas, Provincia del Chubut, Argentina, Actas Congr. Geol. Argent. 7th, 1, 471-480, 1979.

Coira, B. L., Descripción geológica de la Hoja 40d, Ingeniero Jacobacci, Provincia de Rio Negro, Serv. Geol. Nac. Bol., 168, 1979.

Corbella, H., Complejo volcánico alcalino Sierra Negra de Telsen, Patagonia extraandina, Argentina, Actas Congr. Geol. Latinoam. Geol. 5th, 2, 225-238, 1982a.

Corbella, H., Naturaleza del Complejo Alcalino Sierra de Queupuniyeu Patagonia extraandina norte, Argentina, Actas Congr. Geol. Latinoam. 5th, 2, 197-208, 1982b.

Corbella, H., El vulcanismo de las Altiplanicie del Somuncura, Geol. Prov. Buenos Aires Relat. Congr. Geol. Argent. 9th, 1(10), 267-300, 1984.

Cortés, J. M., El substrato precretácico del extremo noreste de la Provincia del Chubut, Rev. Asoc. Geol. Argent., 36(3), 211-235, 1981a.

Cortés, J. M., Estratigrafía cenozoica y estructura al oeste de la Península de Valdez, Chubut. Consideraciones tectónicas y paleogeográficas, Rev. Asoc. Geol. Argent., 36(4), 424-445, 1981b.

Cortés, J. M., and A. M. Baldoni, Plantas fósiles jurásicas al sur del Rio Chubut Medio, Actas Congr. Geol. Argent. 9th, 4, 432-443, 1984.

Cortiñas, J. A., Estratigrafía y facies del Jurásico entre Nueva Lubecka, Ferrarotti y cerro Colorado. Su relación con los depositos coetaneos del Chubut Central, Actas Congr. Geol. Argent. 9th, 2, 283-299, 1984.

Cortiñas, J., and H. Arbe, El Cretácico continental de la región comprendida entre los cerros Guadal y Ferrarotti, Departamento Tehuelches, Provincia del Chubut, Actas Congr. Geol. Argent. 8th, 3, 359-372, 1981.

Cúneo, R., and R. R. Andreis, Estudio de un bosque de licófitas en la Formación Nueva Lubecka, Pérmico de Chubut, Argentina. Implicancias paleoclimáticas y paleogeográficas, Ameghiniana, Rev. Asoc. Paleontol. Argent., 20(1-2), 132-140, 1983.

Dalziel, I. W. D., Back-arc extension in the southern Andes: A review and critical reappraisal, Philos. Trans. R. Soc. London., Ser. A, 300, 319-335, 1981.

DeGiusto, J. M., C. A. DiPersia, and E. Pezzi, Nesocratón del Deseado, Simp. Geol. Reg. Argent. 2nd, 2, 1389-1430, 1980.

Dingle, R. V., Mesozoic palaeogeography of the southern Cape, South Africa, Palaeogeogr. Palaeoclimatol. Palaeoecol., 13, 203-213, 1973.

Dingle, R. V., W. G. Siesser, and A. R. Newton, Mesozoic and Tertiary Geology of Southern Africa, 374 pp., A. A. Balkema, Rotterdam, 1983.

Dunlevey, J. N., and N. Hiller, The Witteberg-Dwyka contact in the southwestern Cape, Trans. Geol. Soc. S. Afr., 82, 251-256, 1979.

du Toit, A. L., Our Wandering Continents, 366 pp., Oliver and Boyd, Edinburgh, 1937.

du Toit, S. R., The Mesozoic history of the Agulhas Bank in terms of the plate-tectonic theory, in Some Sedimentary Basins and Associated Ore Deposits of South Africa, edited by A. M. Anderson and W. J. van Biljon, pp. 197-204, Spec. Publ. 6, Geol. Soc. S. Afr., 1979.

Forsythe, R., The late Paleozoic to early Mesozoic evolution of southern South America: A plate tectonic interpretation, J. Geol. Soc. London, 139, 671-682, 1982.

Forsythe, R., and E. Nelson, Geological manifestations of ridge collision: Evidence from the Golfo de Penas-Taitao Basin, southern Chile, Tectonics, 4, 477-495, 1985.

Franchi, M., and R. F. N. Page, Los basaltos cretácicos y la evolución magmática del Chubut occidental, Rev. Asoc. Geol. Argent., 35(2), 208-229, 1980.

Franchi, M. R., F. E. Nullo, E. G. Sepúlveda, and M. A. Uliana, Las sedimentitas terciarias, Geol. Prov. Buenos Aires Relat. Congr. Geol. Argent. 9th, 1(9), 215-266, 1984.

Gabaldón, V., and A. Lizuaín, Estratigrafía y sedimentología del Liásico del noroeste del Chubut, Argentina, Actas Congr. Geol. Latinoam. 5th, 3, 509-528, 1983.

García, A. F., Estratigrafía del Terciario de Chile central, in El Terciario de Chile, Zona Central, pp. 25-57, Sociedad de Geologíe de Chile, Santiago, 1968.

Gerrard, I., and G. C. Smith, Post Paleozoic succession and structure of the southwestern African continental margin, Mem. Assoc. Pet. Geol., 34, 49-74, 1982.

González, C. R., Rasgos paleogeográficos del Paleozoico superior de Patagonia, Actas Congr. Geol. Argent. 9th, 1, 191-205, 1984.

González Bonorino, F., Esquema de la evolución geológica de la Cordillera Norpatagónica, Rev. Asoc. Geol. Argent., 34(3), 184-202, 1979.

González Bonorino, F., and G. González Bonorino, Geología de la región de San Carlos de Bariloche: Un estudio de las formaciones Terciarias del Grupo Nahuel Huapi, Rev. Asoc. Geol. Argent., 33(3), 175-210, 1978.

González Díaz, E. F., and F. E. Nullo, Cordillera Neuquina, Simp. Geol. Reg. Argent. 2nd, 2, 1099-1147, 1980.

González Díaz, F., and J. Valvano, Plutonitas Cretácicas y neoterciarias entre el sector norte del lago Nahuel Huapi y el Lago Traful (Provincia del Neuquén), Actas Congr. Geol. Argent. 7th, 1, 227-242, 1979.

Greenway, M. E., The geology of the Falkland Islands, Br. Antarct. Surv. Rep., 76, 43 pp., 1972.

Gulisano, C., and G. A. Pando, Estratigrafía y facies de los depósitos jurásicos entre Piedra del Aguila y Sañicó, Depto Collon Cura, Prov. del Neuquén, Actas Congr. Geol. Argent. 8th, 3, 553-577, 1981.

Gust, D. A., K. T. Biddle, D. W. Phelps, and M. A. Uliana, Associated Middle to Late Jurassic volcanism and extension in southern South America, Tectonophysics, 116, 223-253, 1985.

Halbich, I. W., Geodynamics of the Cape Fold Belt in the Republic of South Africa, a summary, in Profiles of Orogenic Belts, Geodyn. Ser., vol. 10, edited by N. Rast and F. M. Delany, pp. 21-29, AGU, Washington, D. C., 1983.

Haller, M. J., Estratigrafía de la región al poniente de Puerto Madryn, Provincia del Chubut, Republica Argentina, Actas Congr. Geol. Argent. 7th, 1, 285-297, 1979.

Haller, M. J., and O. R. Lapido, El Mesozoico de la Cordillera Patagónica Central, Rev. Asoc. Geol. Argent., 35(2), 230-247, 1979.

Haller, M. J. and O. R. Lapido, The Jurassic-Cretaceous volcanism in the Patagonian Septentrional Andes, Earth Sci. Rev., 18, 395-410, 1982.

Haller, M. J., O. R. Lapido, A. Lizuaín, and R. F. N. Page, El martititono-neocomiano en la evolución de la Cordillera Norpatagonica, in Cuencas Sedimentarias del Jurásico y Cretácico de America del Sur, vol. 1, pp. 221-237, Comité Sudamericano del Jurásico y Cretácio, Buenos Aires, 1981.

Halpern, M., P. N. Stipanicic, and R. O. Toubes, Geocronología (Rb/Sr) en los Andes Australes Argentinos, Rev. Asoc. Geol. Argent., 30(2), 180-192, 1975.

Harland, W. B., A. V. Cox, P. G. Llewellyn, G. A. G. Pickton, A. G. Smith, and R. Walters, A Geologic Time Scale, 131 pp., Cambridge University Press, New York, 1982.

Harrington, H. J., Las Sierras Australes de Buenos Aires: Cadena aulacogénica, Rev. Asoc. Geol. Argent., 35(2), 151-181, 1970.

Harrington, H. J., Sierras Australes de Buenos Aires, Simp. Geol. Reg. Argent. 2nd, 2, 395-407, 1972.

Harrington, H. J., Sierras Australes de la Provincia de Buenos Aires, Simp. Geol. Reg. Argent. 2nd, 2, 967-983, 1980.

Hervé, F., F. Munizaga, M. Mantovani, and M. Hervé, Edades Rb/Sr neopaleozoicas del basamento cristalino de la Cordillera de Nahuebuta, Actas Congr. Geol. Chil. 1st, 2, F19-F26, 1976.

Hervé, F., E. Araya, J. Fuenzalida, and A. Solano, Edades radiométricas y tectónica neógena en el sector costero de Chiloe continental, Xa region, Actas Congr. Geol. Chil. 2nd, 1, F1-F18, 1979.

Hervé, F., J. Davidson, E. Godoy, C. Mpodozis, and V. Covacevich, The late Paleozoic in Chile: Stratigraphy, structure and possible tectonic framework, An. Acad. Bras. Cienc., 53(2), 361-373, 1981.

Hervé, F., M. Suárez, and A. Puig, The Patagonian batholith south of Tierra del Fuego: Timing and tectonic implications, J. Geol. Soc. London, 141, 909-917, 1984.

Illies, H., Randpazifische Tektonik und Vulkanismus im sudlichen Chile, Geol. Rundsch., 57, 81-101, 1967.

Jaunich, S., The southwestern African continental margin, AAPG Stud. Geol., 15(2), 2.2.3-70-2.2.3-122, 1983.

Jordi, H. A., and P. Lehner, Regional seismic profiles across the Atlantic margin of South America and South Africa, An. Congr. Bras. Geol. 27th, 3, 67-90, 1973.

Katz, H. R., Erdoel Geologische Untersuchungen im Chilenischen Langstalerdoel und Kohle, Erdol, Kohle Erdgas Petrochem, 16, pp. 1089-1094, 1963.

Katz, H. R., Rundpazifische Bruchtektonik am beispiel Chiles und Neusseelands, Geol. Rundsch., 59(3), 598-926, 1970.

Katz, H. R., Continental margin in Chile--Is tectonic style compressional or extensional? Am. Assoc. Pet. Geol. Bull., 55(10), 1753-1758, 1971.

Lapido, O. R., and R. F. N. Page, Relaciones estratigráficas y estructura del Bajo de la Tierra Colorada (Provincia del Chubut), Actas Congr. Geol. Argent. 7th, 1, 299-313, 1979.

Lapido, O. R., A. Lizuaín, and E. Núñez, La Cobertura sedimentaria mesozoica, Geol. Prov. Buenos Aires Relat. Congr. Geol. Argent. 9th, 1(6), 139-162, 1984.

Larson, R. L., and J. W. Ladd, Evidence for opening of the south Atlantic in the Early Cretaceous, Nature, 246, 209-212, 1973.

Lesta, P., and R. Ferello, Región extraandina de Chubut y norte de Santa Cruz, in Geología Regional Argentina, edited by A. F. Leanza, pp. 61-683, Academia Nacional de Ciencias, Córdoba, 1972.

Lesta, P. J., M. A. Turic, and E. Mainardi, Actualización de la información estratigráfica en la Cuenca del Colorado, Actas Congr. Geol. Argent. 7th, 1, 701-713, 1978.

Lesta, P., R. Ferello, and G. Chebli, Chubut extraandino, Simp. Geol. Reg. Argent. 2nd, 2, 1307-1380, 1980a.

Lesta, P., E. Mainardi, and R. Stubelj, Plataforma continental Argentina, Simp. Geol. Reg. Argent. 2nd, 2, 1577-1602, 1980b.

Lizuaín, A., Características y edad del plutonismo en los alrededores del Lago Puelo, Provincia del Chubut, Actas Cong. Geol. Argent. 8th, 3, 607-616, 1981.

Lizuaín, A., and E. Sepúlveda, Geología del gran Bajo del Gualicho (Provincia de Rio Negro), Actas Congr. Geol. Argent. 8th, 3, 407-422, 1979.

Llambías, E. J., R. Caminos, and C. Rapela, Las plutonitas y vulcanitas del ciclo eruptivo gondwánico, Geol. Prov. Buenos Aires Relat. Congr. Geol. Argent. 9th, 1(4), 85-117, 1984.

Lock, B. E., Flat-plate subduction and the Cape Fold Belt of South Africa, Geology, 8, 35-39, 1980.

Lock, B. E., R. Shone, A. T. Coates, and T. J. Broderick, Mesozoic Newark type sedimentary basins within the Cape Fold Belt of southern Africa, Int. Sedimentol. Congr. 9th, 4, 217-226, 1975.

Ludwig, W. J., J. I. Ewing, C. C. Windisch, A. G. Lonardi, and F. F. Rios, Structure of Colorado Basin and continental-ocean crust boundary off Bahia Blanca, Argentina, Mem. Am. Assoc. Pet. Geol., 19, 113-124, 1979.

Malvicini, L., and J. Llambías, El magmatismo Mioceno y las manifestaciones metalíferas asociadas en la Argentina, Congr. Geol. Latinoam. 5th, 3, 547-566, 1982.

Martin, H., The Atlantic margin of southern Africa between latitude 17°S and the Cape of Good Hope, in The Ocean Basins and Margins, vol 1, edited by A. E. M. Nairn and F. G. Stehli, pp. 277-300, Plenum, New York, 1973.

Masiuk, V., D. Becker, and A. García Espiasse, Micropaleontología y sedimentología del Pozo YPF (Ch. PV.es-1), Península de Valdés, provincia del Chubut, Republica Argentina, in Importancia y Correlaciones, vol. 24, 22 pp., A.R.P.E.L., YPF, Buenos Aires, 1976.

Massabie, A. C., and E. A. Rosello, La discordancia pre-formación Sauce Grande y su entorno estratigráfico, Sierras Australes de la provincia de Buenos Aires, Actas Congr. Geol. Argent. 9th, 1, 337-352, 1984.

McLachlan, I. R., and I. K. McMillan, Microfaunal biostratigraphy, chronostratigraphy and history of Mesozoic and Cenozoic deposits on the coastal margin of South Africa, Spec. Publ. 6, Geol. Soc. S. Afr., 161-181, 1979.

Mendía, J. E., and A. Bayarsky, Estratigrafía del Terciario en el valle inferior del Rio Chubut, Actas Congr. Geol. Argent. 8th, 3, 593-606, 1981.

Miller, H., El basamento de la Provincia de Aysén (Chile) y sus correlaciones con las rocas premezosoicas de la Patagonia Argentino, Actas Congr. Geol. Argent. 6th, 1, 125-141, 1976.

Miller, H., Unidades estratigráficas y estructurales del basamento andino en el Archipielago de Los Chonos, Aisen, Chile, Actas Congr. Geol. Chil. 2nd, 1(A), 103-120, 1979.

Mordojovich, C., Geology of a part of the Pacific margin of Chile, in Geology of Continental Margins, edited by C. A. Burk and C. L. Drake, pp. 591-598, Springer-Verlag, New York, 1974.

Mordojovich, C., Sedimentary basins of Chilean Pacific offshore, AAPG Stud. Geol., 12, 63-82, 1981.

Moreno, H., and M. A. Parada, Esquema geologico de la Cordillera de los Andes entre los paralelos 39°00' y 41°30'S, Actas Congr. Geol. Chil. 1st, 1(A), 213-A226, 1976.

Musacchio, E. A., Estratigrafía de la Sierra Pampa de Agnia en la región extraandina de la provincia de Chubut, Argentina, Actas Congr. Geol. Argent. 8th, 3, 343-357, 1981.

Natland, M. L., E. Gonzalez, A. Cañón, and M. Ernst, A system of stages for correlation of Magallanes basin sediments, Mem. Geol. Soc. Am., 139, 126 pp., 1974.

Nullo, F., and C. Proserpio, La Formación Taquetrén en Cañadón del Zaino (Chubut) y sus relaciones estratigráficas en el ámbito de la Patagonia de acuerdo a la flora, República Argentina, Rev. Asoc. Geol. Argent., 30, 133-150, 1975.

Núñez, E., E. W. Bachmann, I. Ravazzoli, A. Britos, M. Franchi, A. Lizuaín, and E. Sepúlveda, Rasgos geológicos del sector oriental del Macizo de Somuncura, Provincia de Rio Negro, República Argentina, Congr. Iberoamer. Geol. Econ. 2nd, 4, 247-266, 1975.

Page, R. F. N., C. O. Limarino, O. López Gamundi, and S. Page, Estratigrafía del grupo Tepuel en su perfil tipo y en la región El Molle, Provincia de Chubut, Actas Congr. Geol. Argent. 9th, 1, 619-632, 1984.

Page, S., Los gabros bandeados de la Sierra de Tepuel, cuerpos del sector suroeste, Provincia del Chubut, Actas Congr. Geol. Argent. 9th, 2, 584-599, 1984.

Pesce, A., Estratigrafía de la Cordillera Patagónica entre los 43°30' y 44° de latitud sur y sus areas mineralizadas, Actas Congr. Geol. Argent. 7th, 1, 257-270, 1978.

Rabinowitz, P. D., and J. LaBrecque, The Mesozoic South Atlantic Ocean and evolution of its continental margins, J. Geophys. Res., 84, 5973-6002, 1979.

Ramos, V. A., Estratigrafía de los Lagos La Plata y Fontana, Provincia del Chubut, Actas Congr. Geol. Chil. 1st, 1(A), 43-64, 1976.

Ramos, V. A., El vulcanismo del Cretácico inferior de la Cordillera Patagonica de Argentina y Chile, Actas Congr. Geol. Argent. 7th, 1, 423-435, 1978.

Ramos, V. A., Tectónica de la región del río y lago Belgrano, Cordillera Patagónica Argentina, Actas Congr. Geol. Chil. 2nd, 1(B), 1-32, 1979.

Ramos, V. A., Evolución tectónica y metalogénesis de la cordillera Patagónica, Actas Congr. Nac. Geol. Econ. 2nd, 1, 107-124, 1983.

Ramos, V. A., Patagonia: Un continente Paleozoico a la deriva?, Actas Cong. Geol. Argent. 9th, 2, 311-325, 1984.

Ramos, V. A., and M. A. Palma, Las lutitas pizarrenas fosilíferas del cerro Dedo, Lago la Plata, Provincia del Chubut, Rev. Asoc. Geol. Argent., 38(2), 148-160, 1983.

Ramos, V. A., H. Niemeyer, J. Skarmeta, and J. Muñoz, The magmatic evolution of the Austral Patagonian Andes, Earth Sci. Rev., 18(3-4), 411-443, 1982.

Rapela, C., L. Spalletti, J. Merodio, and E. Aragón, Evolución magmática y geotectónica de la "Serie Andesítica" andina (Paleoceno-Eoceno) en la Cordillera Norpatagónica, Rev. Asoc. Geol. Argent., 38, 469-484, 1983.

Rapela, C. W., L. Spalletti, J. C. Merodio, and E. Aragón, El vulcanismo Paleoceno-Eoceno de la Provincia volcánica Andino-Patagónica, Geol. Prov. Buenos Aires Relat. Congr Geol. Argent. 9th, 1(8), 189-213, 1984.

Riccardi, A., and E. O. Rolleri, Cordillera Patagónica austral, Simp. Geol. Reg. Argent. 2nd, 2, 1173-1306, 1980.

Robbiano, J. A., Contribución al conocimiento estratigráfico de la Sierra del Cerro Negro, Pampa de Agnia, Provincia del Chubut, Argentina, Rev. Asoc. Geol. Argent., 24(1), 41-56, 1971.

Rolleri, E. O., Acerca de la Dorsal del Mar Argentino y su posible significado geológico, Actas Congr. Geol. Argent 5th, 4, 203-220, 1973.

Rust, I. C., Tectonic and sedimentary framework of Gondwana basins in southern Africa, in Gondwana Geology, edited by K. S. W. Campbell, pp. 537-564, Australian National University Press, Canberra, 1975.

Saliot, P., Etude géologique dans l'Ile de Chiloe (Chili), Bull. Soc. Geol. Fr., XI(7), 388-399, 1969.

Saunders, A. D., J. Tarney, C. R. Stern, and I. W. D. Dalziel, Geochemistry of Mesozoic marginal basin floor igneous rocks from southern Chile, Geol. Soc. Am. Bull., 190, 237-258, 1979.

Siesser, W. G., and R. V. Dingle, Tertiary sea-level movements around southern Africa, J. Geol., 89, 83-96, 1981.

Skarmeta, J., Evolución tectónica y paleogeográfica de los Andes Patagónicos de Aisén (Chile) durante el Neocomiano, Actas Congr. Geol. Chil. 1st, 1(B), 1-15, 1976.

Spiegelman, A. T., and A. G. Busteros, Caracterización litoestratigráfica de las sedimentitas terciarias en las localidades de Barrancas Blancas (Puerto Madryn), Bahia Cracker e Isla Escondida (Punta Lobos), Provincia del Chubut, República Argentina, Actas Congr. Geol. Argent. 7th, 2, 659-671, 1979.

Spikerman, J. P., Contribución al conocimiento de la intrusividad en el Paleozoico de la región extraandina del Chubut, Rev. Asoc. Geol. Argent., 33(1), 17-35, 1978.

Stipanicic, P. N., and E. J. Methol, Macizo de Somun Cura, in Geología Regional Argentina, edited by A. F. Leanza, pp. 581-599, Academia Nacional de Ciencias, Córdoba, 1972.

Stipanicic, P. N., and E. J. Methol, Comarca Nordpatagonica, Simp. Geol. Reg. Argent. 2nd, 2, 1071-1097, 1980.

Suero, T., Geología de la Sierra de Pillahuincó (Sierras Australes de la Provincia de Buenos Aires), Prov. Buenos Aires, Laboratorio de Ensayo de Materiales e Investigaciones Technologicas, 2, Ministerio de Obras Publicas, La Plata, Buenos Aires, 1957.

Suero, T., Paleogeografía del Paleozoico superior de la Patagonia (Republica Argentina), Rev. Asoc. Geol. Argent., 16(1-2), 35-42, 1962.

Tankard, A. J., M. P. A. Jackson, K. A. Eriksson, D. K. Hobday, D. R. Hunter, and W. E. C. Minter, Crustal Evolution of Southern Africa, 3.8 Billion Years of Earth History, 502 pp., Springer-Verlag, New York, 1982.

Tasch, P., and W. Volkheimer, Jurassic conchostracans from Patagonia, Paleontol. Contrib. Pap., 50, 1-23, Univ. Kans., 1970.

Thiele, R., and R. Hein, Posición y evolución tectónica de los Andes Nord-Patagonicos, Actas Congr. Geol. Chil. 2nd, 1(B), 33-46, Santiago, 1979.

Thiele, R., F. Hervé, and M. A. Parada, Bosquejo geologico de la isla Huapi, lago Ranco, Provincia de Valdivia: Contribucion al conocimiento de la formacion Panguipulli (Chile), Actas Congr. Geol. Chil. 1st, 1(A), 115-136, 1976.

Thiele, R., J. C. Castillo, R. Hein, G. Romero, and M. Ulloa, Geología del sector fronterizo de Chiloe continental entre los 43°00' y 43°45' latitud sur, Chile (Comunas de Futaleufú y Palena), Actas Congr. Geol. Argent. 7th, 1, 577-591, 1979.

Toubes, R. O., and J. P. Spikerman, Algunas edades K/Ar y Rb/Sr de plutonitas de la Cordillera Patagónica entre los paralelos 40° y 44° de latitud sur, Rev. Asoc. Geol. Argent., 28(4), 382-396, 1973.

Ugarte, F., La cuenca compuesta Carbonífera-Jurásica de la Patagonia meridional, An. Univ. Patagonia "San Juan Bosco," Geology, 1(1), 37-68, 1966.

Urien, C. M., and J. J. Zambrano, The geology and the tectonic framework in the Argentine Continental Margin and Malvinas Plateau, in The Ocean Basins and Margins, edited by A. E. M. Nairn and F. G. Stehli, pp. 135-170, Plenum, New York, 1973.

Vail, P. R., et al., Seismic stratigraphy and global changes of sea level, in Seismic Stratigraphy--Applications to Hydrocarbon Exploration, edited by Charles E. Payton, pp. 49-212, Mem. Am. Assoc. Pet. Geol., 26, 1977.

Valenzuela Ayala, E., Estratigrafía de la boca occidental del Canal de Chacao, X region, Chile, Actas Congr. Geol. Chil. 3rd, 1(A), 343-376, 1982.

Varela, R., and C. A. Cingolani, Nuevas edades radimétricas del basa-mento aflorante en el perfil del Cerro Pan de Azucar-Cerro del Corral y consideraciones sobre la evolución geocronológica de las rocas ígneas de las Sierras Australes, Provincia de Buenos Aires, Actas Congr. Geol. Argent. 6th, 1, 543-556, 1976.

Wells, P., The geochemistry of the Patagonian batholith between 45°S and 46°S latitude, M.Sc. thesis, University of Birmingham, England, 50 pp., 1979.

Yllanez, E., and H. A. Lema, Estructuras anulares y geología del noreste de Telsen (Provincia del Chubut), Actas Congr. Geol. Argent. 7th, 1, 445-454, 1979.

Yrigoyen, M. R., Geología del subsuelo y plataforma continental, Geol. Prov. Buenos Aires Relat. Congr. Geol. Argent. 6th, 139-168, 1975.

Yrigoyen, M. R., and L. W. Stover, La palinología como elemento de correlación del Triásico en la Cuenca Cuyana, Actas Congr. Geol. Argent. 2nd, 4, 427-447, 1970.

Zambrano, J. J., La Cuenca del Colorado, in Geología Regional Argentina, edited by A. F. Leanza, pp. 419-438, Academia Nacional de Ciencias, Córdoba, 1972.

Zambrano, J. J., Comarca de la cuenca Cretácica de Colorado, Simp. Geol. Reg. Argent. 2nd, 2, 1033-1070, 1980.

Copyright 1987 by the American Geophysical Union.

PETROLOGY AND FACIES ANALYSIS OF TURBIDITIC SEDIMENTARY ROCKS OF THE PUNCOVISCANA TROUGH
(UPPER PRECAMBRIAN-LOWER CAMBRIAN) IN THE BASEMENT OF THE NW ARGENTINE ANDES

P. Ježek and H. Miller[1]

Geologisch-Paläontologisches Institut der Westfälischen Wilhelms-Universität
Münster, Federal Republic of Germany

Abstract. The late Precambrian to Early Cambrian metagraywacke/metapelite sequence of the NW Argentine Andes (Puncoviscana Formation) and its higher-grade metamorphic equivalents are interpreted as mainly gravity current deposits of a large submarine fan system. The petrologic composition of 31 graywackes points to a recycled orogen provenance. The analysis of facies and paleocurrent directions indicates a westward prograding fan association upon a continuously subsiding inactive continental margin as the most appropriate model for the deposition of the Puncoviscana Formation and equivalent rocks. Intracontinental orogenic belts (Brazilides, about 650 Ma) in the interior of the Brazilian Shield are assumed to be the main source areas.

Introduction

The northern part of the upper Precambrian to Lower Cambrian basement of the NW Argentine Andes is composed of metagraywacke/metapelite sequences (Puncoviscana Formation). These low-grade to very low-grade metamorphic rocks can be traced into medium-grade to high-grade metamorphic equivalents to the south. The basement outcrops (Figure 1) extend over an area of 1200 km by 300 km from the Bolivian border (type locality of the Puncoviscana Formation) [Turner, 1960] southward to Córdoba in central Argentina. Age determinations are based on Early Cambrian trace fossil content [Aceñolaza, 1978] and Early to Middle Cambrian Rb/Sr data of a first orogenic event [Bachmann et al., 1986]. North of Tucumán, Upper Cambrian to Lower Ordovician shallow water sedimentary units (Mesón Group) discordantly overlie the Puncoviscana Formation rocks. In the Puncoviscana Formation sequences of the area between Tucumán and Salta (Figure 1), petrological and sedimentary facies analyses, including sedimentological profiles and paleocurrent data, have been provided. As these rocks and the higher-grade metamorphic equivalents represent the oldest component of the NW Argentinian Andes, the provenance of the sediments and the paleogeographic setting are of great importance for understanding the early history of the western continental margin of South America.

[1] Now at Institut für Allgemeine und Angewandte Geologie der Universität München, D-8000 München, Federal Republic of Germany.

Petrology of the Puncoviscana Formation Rocks

More than 100 thin sections were used to characterize the composition of the different rock types. For the identification of the very fine fraction several scanning electron microscope (SEM) analyses were realized.

The fine-grained to medium-grained sandstones are composed of monocrystalline and polycrystalline quartz, feldspar, mainly plagioclase with minor potassic feldspar, a variety of predominantly sedimentary to low-grade metamorphic lithic fragments, mica, and heavy minerals. The grain size of the sandstones generally ranges between a few micrometers to approximately 0.8 mm. The monocrystalline quartz occurs as clear subangular grains of the entire size range. In some thin sections, rounded grains with opaque inclusions and diagenetic overgrowth rims have been observed. The monocrystalline quartz shows both undulose and straight extinction. The polycrystalline varieties include very fine grained cherts and siliceous siltstones, medium-grained sutured and polygonal metaquartzitic fragments, and elongated quartz mylonites.

Subangular plagioclase grains, generally of albite to oligoclase composition, are abundant. No zoned or euhedral grains were found, but twinning is common. Small albite particles are usually untwinned and were recognized by SEM. Locally (e.g., Figure 1, localities 1, 3, and 5), in some graywackes, microcline or microperthite clasts of approximately 0.5 mm in diameter are present.

The lithic fragment association is composed of a high amount of sedimentary detritus. Pelithic intraclasts and siltstone fragments are most abundant. A smaller portion of low-grade metamorphic phyllite and slate particles is also present. Detrital carbonate grains are very rare (Figure 1, localities 3 and 4, west of locality 2). Detrital muscovite chlorite and muscovite-chlorite intergrowths are present. A few biotite and biotite-chlorite grains in some samples are of local importance only (Figure 1, localities 3 and 4, west of locality 2). In general the lithics are restricted to the coarser grain size range of 0.1 mm to 0.8 mm and show a higher degree of roundness compared to monocrystalline quartz and feldspar.

Among the accessory minerals, fragments and euhedral crystals of zircon are most common.

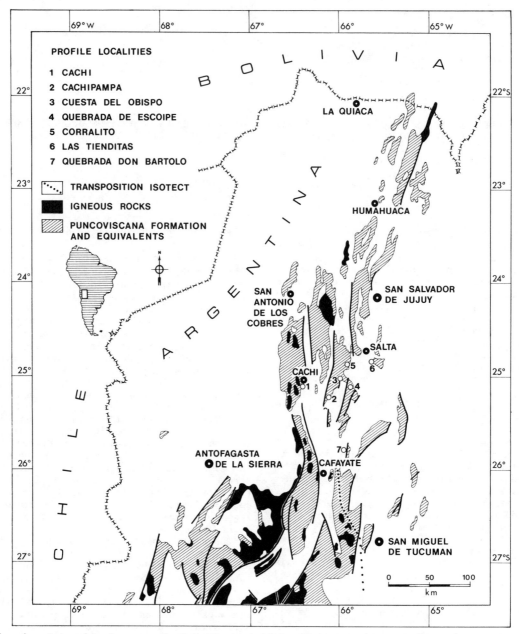

Fig. 1. Generalized map of the NW Argentine upper Precambrian to Lower Cambrian metasedimentary basement, including lower Paleozoic intrusions, showing the locations of sedimentological profiles within the low-grade to very low-grade metamorphic zone between Tucumán and Salta.

Tourmaline, epidote, apatite, sphene, rutile, and opaques (generally pyrite, ilmenite, hematite, magnetite, and titanomagnetite) complete the heavy mineral assemblage. The largest zircon crystals reach up to 0.4 mm in length. Many of the opaques are presumed to be of diagenetic origin. Besides the detrital accessory minerals, sphene is enriched in the lower tectonic levels that are exposed to the south (Figure 1, south of locality 7) and has been produced partly by metamorphic processes. The epidote concentration rises to the west adjacent to plutonic domes.

The matrix, in some cases visible due to the bimodal grain size distribution of the sandstones with low ratio of clasts to matrix (Figure 1, localities 2 and 3), develops primary flow orientation of tiny phyllosilicate sheets around larger grains. Graded bedding in such rocks is caused by an upward decrease in the ratio of large to small grains and suggests that the matrix is at least partially of sedimentary origin. Commonly, however, a continuous grain size spectrum within a clast-supported fabric and a southward increase in the effects of pressure solution and crystallization of white mica obscure the identification of the matrix.

According to the ratios of quartz, feldspar, and the lithic fragments, the sedimentary material includes fine-grained lithic graywackes and lithic arenites. A few specimens correspond to feldspathic graywackes and sublitharenites [Pettijohn et al., 1972].

Two types of pelites rhythmically alternating with the psammitic beds are distinguished by color, structural features, and lithostratigraphic relationships. Green to gray pelites with a weakly graded lamination of centimeter scale and structureless, red to dark purple pelites were distinguished. Although the chemical composition of both is very similar [Willner et al., 1985], petrologic differences exist. The green pelite units contain detritus-rich laminae of light components (quartz and feldspar) in the lower parts and parallel-oriented phyllosilicates (chlorite and illite) in the upper parts of the sedimentary units. No bedding structures have been observed in the red pelites. The randomly arranged network of phyllosilicates in the red pelites is composed of loosely distributed coarser angular grains of quartz and feldspar. The main characteristic feature for distinguishing both types is a submicroscopic hematite pigmentation, which, exceeding 1.5% Fe_2O_3, causes the red color of the rock [Franke and Paul, 1980]. The green pelitic layers are directly related to the underlying psammitic beds. They show horizontal lamination and graded bedding and rarely exceed a few centimeters per sedimentary unit. The red pelites, on the other hand, are cyclically intercalated with the green to gray graywacke/pelite alternating successions, lack any sedimentary structure, and often reach several meters in thickness.

Two major, locally restricted, oligomict conglomerate occurrences of more than 100 m of outcrop extension are present in the Puncoviscana Formation and in the equivalent Suncho Formation sequences. They contain pebbles of varying abundance, size, and composition. They are mostly well rounded and from 3 to 20 cm in diameter. The larger one (the Corralito Formation) [Salfity et al., 1975] consists of metagraywacke, metapelite, and vein quartz, and, rarely, chalcedony and carbonate pebbles. South of Cachi (Figure 1, 50 km south of locality 1 at Seclantás), rhyolite clasts, and at Suncho (150 km southwest of Tucumán), basic volcanic clasts (F. R. Durand, personal communication, 1982) are present. The matrix consists of silty to fine-grained psammitic material. Further information is given later in the section on facies analysis and paleocurrents.

Locally, carbonates also exist within the Puncoviscana Formation rocks. The carbonate rocks are generally brecciated. They form mostly isolated massive bodies, in some places transitionally grading into alternating clastic sequences [Omarini, 1983]. The dark color of the rocks gives evidence of a high organic carbon content. Because of intense internal cataclasis, recrystallization, and mainly tectonic contacts against the adjacent rocks, poor information has been obtained about the original petrologic composition and microfacies so far.

Provenance of the Puncoviscana Formation Sediments

Geochemical analyses of 120 rock samples have been previously provided from both very low-grade to low-grade metamorphic terranes in the north of Tucumán (Puncoviscana Formation) and medium-grade to high-grade metamorphic terranes to the south [Willner et al., 1985]. The results point to widespread, poorly recycled graywackes and pelites as the predominant source rocks for the quartz-rich to quartz-intermediate metagraywackes [Crook, 1974] of the Puncoviscana Formation and its equivalents. The petrologic study was expected to yield a similarly uniform picture.

The detrital clasts can be divided into three groups according to the parent rock type. The most abundant contains reworked sedimentary fragments, i.e., cherts, siltstones, and pelitic intraclasts. The second assemblage is characterized by the presence of low-grade to very low-grade metamorphic detritus such as strained monocrystalline quartz, metaquartzites with both sutured and polygonal subgrain boundaries, elongated quartz mylonites, slates and phyllites, muscovite, muscovite-chlorite, and chlorite and biotite flakes. The third group is of local importance only and is composed of clasts of plutonic and hydrothermal origin, myrmekitic quartz, microcline and microperthite (K), quartz-feldspar and quartz-feldspar-mica fragments, vein quartz intergrown with hematite, and accessory apatite.

For the purpose of petrologic provenance interpretation, applying the methods of Dickinson and Suczek [1979] and Dickinson et al. [1983], 31 thin sections of graywackes, from more than 100 samples from the area between Tucumán and Humahuaca (Figure 1) were selected for point count analysis. From each specimen originating from the base of a sandstone bed, 500 points of detrital sand-sized clasts, except for monocrystalline detrital mica, carbonate, and heavy mineral grains, were counted. The sum equals 100% quartz + feldspar + lithic (QFL).

The results are presented in four triangular diagrams (Figure 2) showing the framework proportions of total quartz (44-80%) composed of the monocrystalline (15-60%) and the polycrystalline variety ($Q = Qm + Qp$), the feldspars (5-20%) composed of plagioclase (6-19%) and potassic feldspar (up to 4%; $F = P + K$), and unstable lithics (9-44%) of volcanic and sedimentary to low-grade metamorphic origin ($L = Lv + Ls$). The percentages refer to the sum of sand-size detrital quartz, feldspar, and lithic fragments [Dickinson and Suczek, 1979]. Heavy minerals, micas, carbonate grains, and the matrix are not included in this scheme. Despite the large geographical distribution of the samples (more than 300 km in a

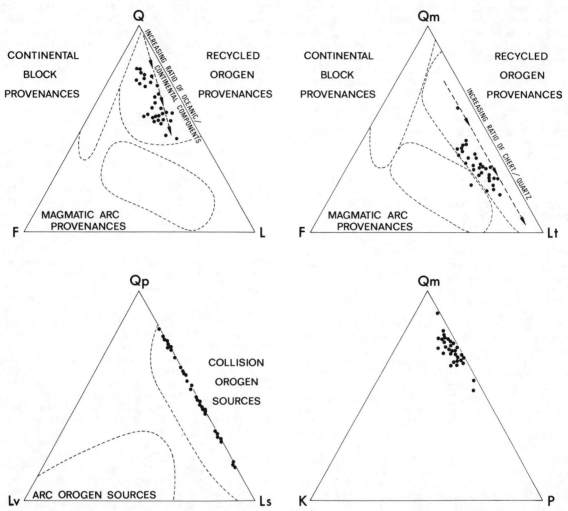

Fig. 2. Triangular plots [Dickinson and Suczek, 1979] of the petrologic composition of 31 Puncoviscana Formation graywackes from the area between Tucumán and Humahuaca. Q is total quartzose grains; F, total feldspar grains; l, total lithic fragments; P, plagioclase; Qp, polycrystalline quartzose; Qm, monocrystalline quartzose; Lv, total unstable volcanic rock fragments; Ls, unstable sedimentary rock fragments; and Lt, total lithic rock fragments.

north-south direction) all of them are closely spaced within the recycled orogen fields in the QFL and QmFLt diagrams (monocrystalline-Qm, total feldspar grains-F, and total polycrystalline lithic fragments-Lt). They roughly follow a trend from the foreland uplift provenance composition toward that of recycled detritus of collision orogens and related terranes showing a very low ratio of oceanic to continental components.

Furthermore, in the QpLvLs and PmPK diagrams, (polycrystalline-Qp, total volcanic-metavolcanic rock fragments-Lv, unstable sedimentary-metasedimentary rock fragments-Ls, quartz grains-Qm, plagioclase feldspar grains-P, and K-feldspar grains-K) (Figure 2), the two composition fields reflecting the difference between magmatic arc and recycled orogen provenance are enhanced. The Puncoviscana Formation sample plots are clearly distant from the Lv and K edges, thus again representing recycled orogen derivation with a relatively high proportion of mature and stable detritus.

Samples collected from profiles of the area west of Salta tend to show a successive vertical enrichment in monocrystalline quartz at the expense of the lithics (Figure 1, e.g., locality 1). However, for the reconstruction of stratigraphic and regional trends the evaluation of additional data is indispensable.

Thus a sedimentary to weakly metamorphic recycled orogen terrane including granitic rocks is interpreted as the predominant provenance area of the Puncoviscana Formation rocks and its equivalents. A continuous reworking of the basin floor also influences their petrologic composition.

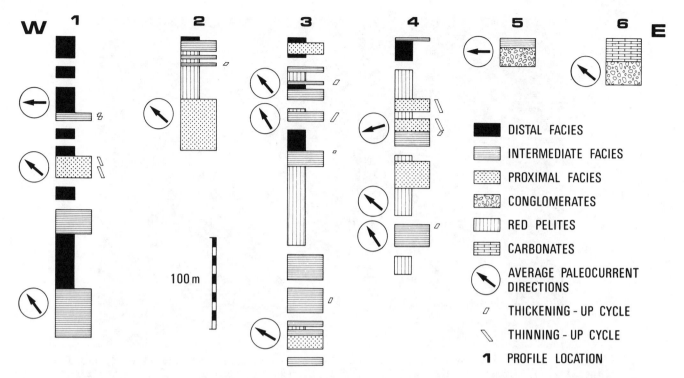

Fig. 3. The series of facies profiles of one to six intersects of the study area west of Salta approximately along latitude 25°S, parallel to the sediment transport (from E to W). Lateral correlation is impossible due to discontinuous outcrop, lack of guide horizons, and vertical block displacements. However, obviously more-proximal facies types, dominating in the east, grade into more distal ones toward the west. Gaps in the profiles represent small tectonic offsets.

Facies Analysis and Paleogeography

The sedimentary rocks described above represent a submarine fan facies association [Ježek et al., 1985; Ježek and Miller, 1986]. On the basis of sedimentological characteristics of the studied sections (Figure 3), bed thickness, psammite/pelite ratio, bed contacts, surface and internal bedding structures, grain size, and lithology, four types of clastic sedimentary submarine fan related facies have been differentiated. The facies types are represented by distinct continuous bed sequences, indicating different transport and depositional mechanisms. Four types of gravity currents are attributed to the conglomerate, proximal, intermediate, and distal facies type respectively: debris flow currents, transition of grain flow and fluidized flow to high-velocity/high-density turbidity currents, high-velocity/high-density turbidity currents, and low-velocity/low-density turbidity currents.

The two studied occurrences of coarse clastic rocks north of Tucumán (Figure 1, locality 5 and 50 km south of locality 1), with mostly disorganized internal textures, represent debris flow deposits within proximal distributary channels off submarine canyons.

The proximal facies is most abundant in the Tucumán area. It consists of upward thinning cycles of thick-bedded sandstone sequences with scarce internal structures. Average sandstone bed thicknesses exceed 30 cm, and the psammite/pelite ratio is greater than 5. These sequences are interpreted as deposits of transitional grain flow to high-velocity/high-density turbidity currents on midfan intrachannel areas or proximal lobes.

The intermediate facies dominating the zone between Tucumán and Salta (Figure 1, e.g., locality 7) is characterized by medium-bedded sandstone/pelite alternations predominantly forming upward thickening cycles with abundant surface structures (ripple marks and flute casts) and mostly incomplete Bouma sequences (AE, ACE, and ACDE). These sediments are assumed to have been deposited in outer fan lobes or on midfan interchannel sites. The average thicknesses of the sandstone beds range between 30 cm and 10 cm, and the psammite/pelite ratio varies between 5 and 1.

Distal thin-bedded pelitic successions with parallel lamination become common west of Salta (Figure 1, e.g., localities 1 and 4). They contain the upper Bouma intervals (CDE and DE), thus representing the low-velocity/low-density ends of turbidity currents. Medusoid imprints and trace fossil assemblages are most common in this facies, which is interpreted as an outer fan-fringe sediment.

Fig. 4. A typical intermediate facies sequence of the Puncoviscana Formation (Figure 1, locality 7) overlain to the left (E) by red pelites. The psammatic beds to the right (W) are arranged in upward thickening cycles with increasing maximum bed thickness and decreasing number of beds per cycle. The outcrop shown is 50 m wide.

Massive red pelites form common cyclic intercalations within the intermediate facies sequences (Figure 3, profile 2). They are supposed to mark periods of quiescence and an oxidizing environment in the bottom sediment [Faupl and Sauer, 1978]. The rare but thick bodies of dark fractured carbonate rocks (Figure 1, locality 6) are interpreted as being derived from locally developed rises.

From well-preserved sedimentary structures (ripple marks, flute casts, slump folds, and imbricated clasts) more than 800 paleocurrent data were obtained by one of us (P.J.). In general, they show a unidirectional pattern of sediment transport from east to west (Figure 3). Other minor directions may indicate the activity of oceanic currents.

Bed sequences of the intermediate facies are commonly arranged in upward thickening cycles of several meters, representing short periods of fan lobe progradation (Figure 4). In sections west of Salta (Figure 3, profiles 1, 3, and 4), repeated upward fining facies successions of hundreds of meters scale indicate a long-term basin subsidence.

Conclusions

The Puncoviscana Formation sedimentary rocks and their higher-grade metamorphic equivalents were deposited within a large submarine fan system. According to the transport and depositional mechanisms and sedimentological properties, six types of facies are distinguished: the channel, proximal, intermediate, and distal submarine fan facies, a pelagic facies, and a carbonate facies.

The arrangement of facies successions on a scale of hundreds of meters indicates a continuous basin subsidence in the area west of Salta, with periods of fan-lobe progradation documented by meter-scale upward thickening cycles. The nearly unidirectional paleocurrent distribution points to a source area to the east. High tectonic levels of recycled intracontinental orogenic belts (Brazilides) in the interior of the Brazilian Shield, probably bordered by a lowland plain, are assumed to be the main sources of the Puncoviscana Formation rocks and their higher-grade metamorphic equivalents. Most of the history of the Brazilian cycle had ended before the Phanerozoic, as is indicated by an unfolded or only weakly folded Eocambrian sedimentary cover [Aceñolaza and Miller, 1982]. Additional material was contributed by synsedimentary basin reworking.

The sedimentological, geochemical, and petrological data indicate the existence of an inactive continental margin setting along the Pacific edge of the Brazilian Shield in latest Precambrian to Early Cambrian time.

Acknowledgments. This work was executed in the course of the scientific cooperation convention signed between the Universities of Münster and Tucumán. It was supported by the Deutsche Forschungsgemeinschaft, the Deutscher Akademischer Austauschdienst and the Stiftung Volkswagenwerk. We are indebted to the reviewers for their valuable comments.

References

Aceñolaza, F. G., El Paleozoico inferior de Argentina según sus trazas fósiles, Ameghiniana, 15(1/2), 15-64, 1978.

Aceñolaza, F. G., and H. Miller, Early Paleozoic orogeny in southern South America, Precambrian Res., 17(2), 133-146, 1982.

Bachmann, G., B. Grauert, and H. Miller, Isotopic dating of polymetamorphic metasediments from NW-Argentina, Zentralbl. Geol. Palaentol., Teil 1, 1985, 1257-1268, 1986.

Crook, K. A. W., Lithogenesis and geotectonics: The significance of compositional variation in flysch arenites (greywackes), in Modern and Ancient Geosynclinal Sedimentation, edited by R. H. Dott and R. M. Shaver, pp. 304-310, Society of Economic Paleontologists and Mineralogists Tulsa, Okla., 1974.

Dickinson, W. R., and C. A. Suczek, Plate tectonics and sandstone composition, Am. Assoc. Pet. Geol. Bull., 63(12), 2164-2182, 1979.

Dickinson, W. R., L. S. Beard, G. R. Brakenridge, J. L. Erjavec, R. C. Ferguson, K. F. Inman, F. A. Lindberg, and P. T. Ryberg, Provenance of North American Phanerozoic sandstones in relation to tectonic setting, Geol. Soc. Am. Bull., 94, 222-235, 1983.

Faupl, J., and R. Sauer, Zur Genese roter Pelite in Turbiditen der Flyschgosau in den Ostalpen (Oberkreide-Alttertiär), Neues Jarhb. Geol. Palaeontol. Monatsh., 1978(2), 65-86, 1978.

Franke, W., and J. Paul, Pelagic redbeds in the Devonian of Germany--Deposition and diagenesis, Sediment. Geol., 25, 2312-2356, 1980.

Ježek, P., and H. Miller, Deposition and facies distribution of turbidic sediments of the Puncoviscana Formation (upper Precambrian-Lower Cambrian) within the basement of the NW-Argentine Andes, Zentralbl. Geol. Palaeontol., Teil 1, 1985, 1235-1244, 1986.

Ježek, P., A. P. Willner, F. G. Aceñolaza, and H. Miller, The Puncoviscana trough--A large basin of late Precambrian to Early Cambrian age on the Pacific edge of the Brazilian Shield, Geol. Rundsch., 75(3), 573-584, 1985.

Omarini, R. H., Caracterizatión litholǵica, diferenciación y génesis de la Formación Puncoviscana entre el Valle de Lerma y la Faja Eruptiva de la Puna, Doctoral thesis, 202 pp., Universidad Nacional de Salta, Buenos Aires, 1983.

Pettijohn, F. J., P. E. Potter, and R. Siever, Sand and Sandstone, 618 pp., Springer-Verlag, New York, 1972.

Salfity, J. A., R. H. Omarini, B. Baldis, and W. J. Gutiérrez, Consideraciones sobre la evolución geológica del Precámbrico y Paleozoico del Norte Argentino, Actas Congr. Ibero Am. Geol. Econ. II, 4, 341-361, 1975.

Turner, J.C., Estratigrafía de la Sierra de Santa Victoria y adyacencias, Bol. Acad. Nac. Cience. Córdoba, 41(2), 163-206, 1960.

Willner, A. P., H. Miller, and P. Ježek, Geochemical features of an upper Precambrian-Lower Cambrian greywacke/pelite sequence (Puncoviscana trough) from the basement of the NW-Argentine Andes, Neues Jarhb. Geol. Palaeontol. Monatsh., 1985(8), 498-512, 1985.

ASPECTS OF THE STRUCTURAL EVOLUTION AND MAGMATISM IN WESTERN NEW SCHWABENLAND, ANTARCTICA

G. Spaeth

Geologisches Institut, Rheinisch-Westfälische Technische Hochschule
Aachen, Federal Republic of Germany

Abstract. Data and observations on rock units of the northern Kraul Mountains (Vestfjella) and the Ahlmann Ridge in western New Schwabenland (western Queen Maud Land) are presented. The numerous dolerite dikes of both regions, which are apparently of Mesozoic age, indicate crustal extension. The majority have been intruded parallel to observed or assumed major fracture zones which dissect this part of the western edge of the East Antarctic Shield. These dolerite dikes and related sills are chemically similar to the basaltic lava flows of the Kraul Mountains. For this reason and because of the relationship between the dikes and the Permo-Carboniferous sedimentary rocks (Beacon Supergroup), they are regarded as Mesozoic and equivalent to the Jurassic volcanic rocks of the Transantarctic Mountains and not upper Precambrian, as indicated on the Geologic Map of Antarctica, (1:5,000,000, Craddock (1972)). Paleomagnetic and radiometric dating studies are in progress. The results of recent geophysical investigations also suggest dissection of the crust of the Antarctic continent and shelf at the eastern edge of the Weddell Sea. This dissection, shown by fracture tectonics, may be related to the opening of the Weddell Sea and a failed rift during the fragmentation of Gondwana. Geochemical, petrographic, and structural investigations on presumed Mesozoic dolerite dikes and the basaltic-andesitic lavas of late Precambrian age on the Ahlman Ridge demonstrate the difference in age of these magmatic events. The phase of fracture tectonics, as shown by surface morphology and by the emplacement of the dolerite dikes, was preceded by at least two phases of deformation of the upper Precambrian platform sedimentary and volcanic rocks. One is inferred to be a compressive strain possibly connected with the Ross Orogeny.

Introduction

Western New Schwabenland, comprising the westernmost part of Queen Maud Land, lies between two significant rift zones of Antarctica. To the east is the Jutulstraumen rift, which extends in a south southwest-north northeast direction, and to the west, in the Weddell Sea, a large rift system is assumed [Elliot, 1975]. The geological structure of western New Schwabenland is therefore of great importance to the understanding of the geotectonic setting of this part of Antarctica and the age and the evolutionary history of these rift zones.

For another reason, that of geotectonic interpretation, the geology of this region is of considerable importance. In many graphic representations in the literature, the Ross Orogen has been placed in numerous different positions in the Weddell Sea sector, and even the question of its northern extension is still open and cannot as yet be answered. The existence and the position of the Ross Orogen at the edge of the Weddell Sea is of great importance for the reconstruction of Gondwana, especially regarding the former connection between Antarctica and Africa.

The ice-free areas of western New Schwabenland (Figure 1) are not very extensive compared with other coastal regions of Antarctica, and they are generally not as accessible. In spite of this, several publications have described the geology of these regions in some detail. The geology of the Heimefront Range has been described by Juckes [1972], and that of the Kraul Mountains (Vestfjella) by Hjelle and Winsnes [1972] and Furnes and Mitchell [1978]. Roots [1953] has dealt with the geology of the area as a whole. Several publications from Neethling [e.g., Neethling, 1969, 1972] are significant for the Ahlmann Ridge, Borg Massif, and Kirwan Escarpment. The results of these and other works of South African geologists have recently been compiled and presented in a synthesis by Wolmarans and Kent [1982]. From these publications it emerges that the age relations of several lithological units in these areas are questionable because of the lack of radiometric age determinations and because of uncertainty regarding the structural relations. This is also partly true for the Kraul Mountains.

During two expeditions in the austral summer of 1982-1983 and of 1983-1984 the author carried out structural investigations in the northern Kraul Mountains (Figure 2) [Behr et al., 1983] and the nunataks of the northern Ahlmann Ridge [Spaeth and Peters, 1984] (Figure 1). Approximately 100 rock samples were chemically analyzed. The average compositions of certain rock groups are presented and discussed here with regard to their importance in clarifying age relationships. Radiometric,

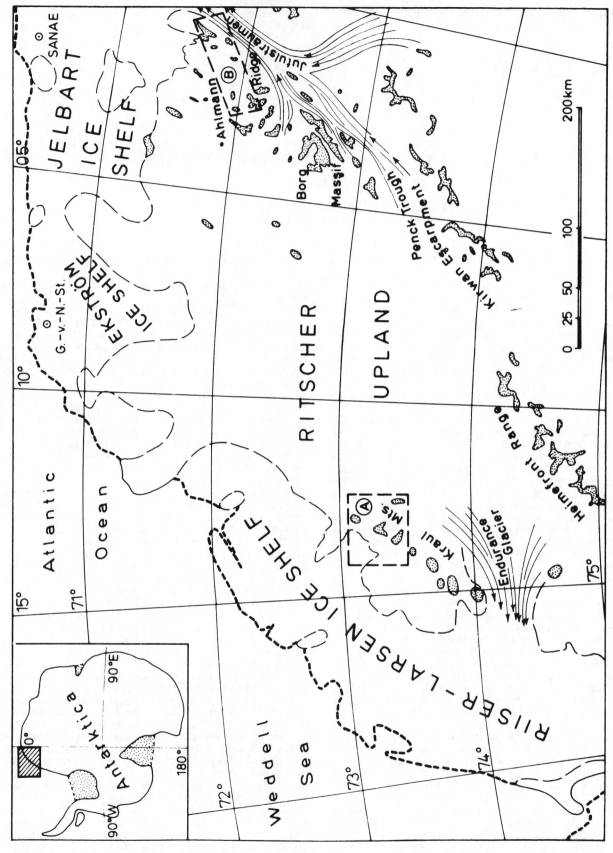

Fig. 1. Sketch map of western New Schwabenland, Antarctica. Areas with exposures are stippled. Area A: area of fieldwork in 1983. Area B: area of fieldwork in 1984.

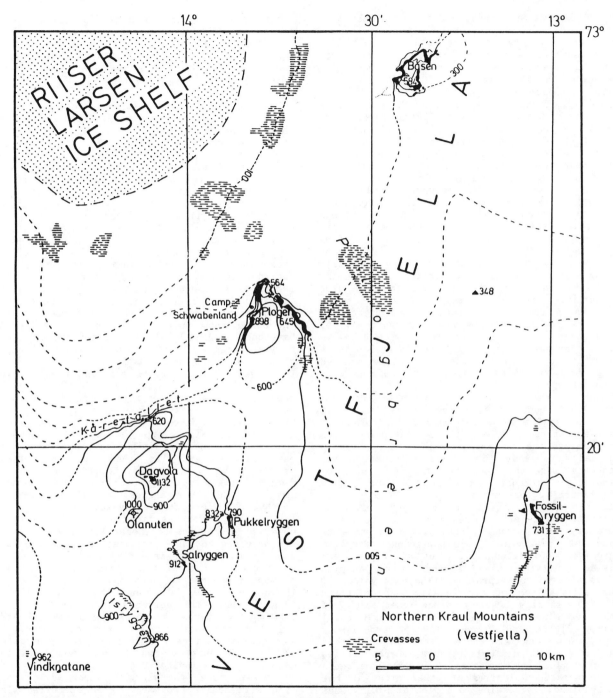

Fig. 2. Map of the northern Kraul Mountains (Vestfjella) showing area A of Figure 1 (after Norwegian topographical map of Queen Maud Land 1:250,000, sheet Vestfjella Aust).

paleomagnetic, and geochemical investigations were also carried out by other expedition members, but these studies are still in progress and only some preliminary results are mentioned in this paper.

The Northern Kraul Mountains

Lithologic units. The Kraul Mountains (Vestfjella) are in the western part of New Schwabenland and extend for approximately 130 km in a north northeast-south southwest direction. The six nunataks or nunatak groups of the northern half of this mountain range (Basen, Plogen, Fossilryggen, Pukkelryggen, Salryggen, and Dagvola, (Figure 2)) which were studied by the author consist mainly of volcanic rocks. The small nunatak group Fossilryggen, which lies to the east of the main mountain range (Figure 2), is the exception.

It is composed of sandstones and mudstones with plant fossils of Early Permian age; these sediments are intruded by dolerite dikes and sills. At the southern end of the Kraul Mountains in the nunataks of Utpostane there is an olivine gabbro intrusive [Hjelle and Winsnes, 1972].

The basaltic flows are well exposed in more than 400 m of vertical exposure of the massif-like mountains of Basen and Plogen. Total thickness of the pile of basaltic flows is estimated to be 800 m to 1000 m. Hjelle and Winsnes [1972] estimated a total thickness of up to 2000 m in the southern Kraul Mountains. The thickness of the individual flows, as judged by vesicular layers and the appearance of ropy surfaces, amounts to only a few meters for the Plogen massif and rarely exceeds 10 m. At the Basen massif there are numerous flows with thicknesses of approximately 1 m or less. The sequence of basaltic flows, which are often highly amygdaloidal, has minor intercalations of tuff layers and is intruded by dolerite sills having a thickness of a few meters. Intercalations of thin layers of sandy sediments are also present but only very rarely. Prominent mafic dike swarms cut the above-mentioned rock units. The dikes are generally between 1 m and 5 m wide; one quarter of them are wider than 5 m, and only a few are narrower than 1 m. Their strike length is estimated to be in the range of a few hundred meters to a few kilometers, for their limits were seldom observed.

None of these dikes could definitely be described as feeder dikes for the flows. The basalt flows are quite strongly altered, and even in fresh-appearing samples, where plagioclase and pyroxene are the main constituents and ore minerals and hornblende the auxiliary constituents, secondary minerals such as chlorite, calcite, epidote, and prehnite are present. These secondary minerals, together with quartz, also constitute the main minerals filling amygdales. By contrast, the dolerite dikes and sills are fresh. These are mostly fine to medium grained and composed of plagioclase, clinopyroxene (augite, pigeonite, and titanaugite), orthopyroxene, and olivine. Olivine is not present in all samples and is altered to serpentine. The mean values of chemical analyses of the basalts and dolerites (Table 1, samples 1 and 2) agree reasonably well with the values reported by Furnes and Mitchell [1978], Hjelle and Winsnes [1972], and Juckes [1968]. The rocks are characterized by relatively low K_2O and SiO_2 contents (Figure 3, open circles and triangles) as compared with volcanic rocks and associated dolerites of other areas in New Schwabenland. According to a plot of SiO_2 versus Zr/TiO_2 [Winchester and Floyd, 1977], used here because of the rock alteration, the volcanic rocks can, in most cases, be classified as subalkaline basalts (Figure 4). K/Rb and K/Ba ratios are also listed in Table 1; the K/Rb ratios suggest that they are continental basalts [Gunn, 1965].

Structures. The basaltic layers from Plogen, the largest mountain in the northern Kraul Mountains, dip at low angles (approximately 10°) toward the south and southwest. At Basen, the second-largest mountain in the region, the flows are more or less horizontal or dip slightly toward the north and northwest. The sedimentary beds at Fossilryggen dip shallowly in a westerly direction. Hjelle and Winsnes [1972] also reported general westerly dips in the entire Kraul Mountains, resulting from a slight tilting of faulted blocks.

Hjelle and Winsnes [1972] assumed the existence of large faults between the larger mountains and the nunatak groups. However, such faulting was not observed in the field. The present author identified several small normal faults at Plogen and Basen (Figure 5a). Two distinct groups can be identified on the basis of strike direction; one set strikes northwest-southeast (f_3 and f_4 in Figure 5a), the other northeast-southwest (f_1 and f_2 in Figure 5a). The latter is the most frequent. Vertical displacement along the faults is between 1 m and 30 m.

The above-mentioned dolerite dikes, especially those that are steeply dipping or vertical, are regarded as providing evidence for extension tectonics, and because of their great abundance, information concerning their strike and thickness is of great importance for structural analysis. The dikes occur in all nunataks and massifs that have been studied by the author but are most abundant at Plogen and in the nunatak group Pukkelryggen. The dikes are conspicuous as a result of their grayish-brown color and characteristic columnar jointing. A total of 150 dikes were observed; strikes and dips were measured, and thicknesses were estimated or measured. Figure 5b shows that the dikes are relatively tightly oriented and that two groups are present. The dominant set (120 dikes) strikes approximately northeast-southwest and dips steeply to the southeast or northwest. A second, subordinate set (30 dikes) strikes approximately northwest-southeast and dips at lower angles to the northeast or southwest. Where the two systems occur together, the steeper northeast trending dikes were always observed to cut the southeast trending dikes. It was observed in many instances that the shallower dipping dikes merged into sills.

Thickness of the dikes ranges from 0.1 m to 35 m; most frequent are those from 1 m to 10 m. In the approximately 6-km-long northeastern cliff of Plogen, where the northeast striking dikes are common, the cumulative thickness of the 66 dikes is 275 m, suggesting an extension of the crust in a northwest-southeast direction of approximately 5%. For the nunatak group Pukkelryggen, an extension of 4.7% is indicated.

In many cases the steeply dipping dikes coincide with faults, movement having occurred along contact zones as indicated by slickensides. Many of the dikes have been sheared into large lens-shaped bodies cut by slickensided surfaces. From this it can be deduced that the fracturing outlasted or postdated the emplacement of the dikes.

Age relationships. The age of the dolerite intrusives in the sediments of Fossilryggen is interpreted, on structural evidence, to be post-Permian. Hjelle and Winsnes [1972] reported an age of 220 Ma for these dolerites on the basis of a preliminary K-Ar dating made by Krylow [see

TABLE 1. Average Analyses of Volcanic Rocks (Flows) and Dolerites (Dikes) from Western New Schwabenland

	Sample 1		Sample 2		Sample 3		Sample 4	
	Mean	Range	Mean	Range	Mean	Range	Mean	Range
Oxides, %								
LOI	2.50	0.50 - 4.25	1.68	0.48 - 3.67	2.01	0.77 - 5.45	3.15	1.75 - 4.63
SiO_2	51.55	47.68 - 61.46	48.76	46.78 - 52.99	48.09	43.11 - 55.54	53.56	48.84 - 60.14
Al_2O_3	13.56	11.46 - 14.91	13.37	10.12 - 15.37	10.61	6.87 - 14.43	13.64	11.02 - 15.93
CaO	7.75	5.06 - 10.56	10.43	7.49 - 13.44	9.18	5.25 - 13.18	7.48	4.68 - 16.05
MgO	5.26	1.75 - 7.33	6.96	4.52 - 11.23	10.51	5.05 - 19.15	4.97	1.36 - 7.51
Na_2O	3.16	2.40 - 4.34	2.38	1.32 - 3.16	2.02	1.09 - 3.49	2.63	0.09 - 3.69
K_2O	0.93	0.15 - 1.83	0.53	0.14 - 1.06	0.59	0.12 - 0.97	1.76	0.02 - 4.10
TiO_2	1.75	0.79 - 4.81	1.78	1.24 - 4.00	2.59	0.60 - 5.51	1.46	0.81 - 2.92
Fe_2O_3	11.69	6.88 - 15.45	13.61	11.97 - 19.71	14.61	10.97 - 16.81	12.50	9.58 - 15.69
MnO	0.16	0.10 - 0.23	0.18	0.14 - 0.20	0.19	0.17 - 0.22	0.18	0.14 - 0.22
P_2O_5	0.26	0.10 - 0.78	0.19	0.10 - 0.51	0.28	0.06 - 0.73	0.19	0.09 - 0.68
Total	98.79		99.65		98.92		98.47	
Trace Elements, ppm								
Ba	598.25	187 - 1654	396.90	124 - 1101	215.96	101 - 641	626.50	138 - 1513
Sr	361.00	91 - 753	269.45	131 - 378	291.07	36 - 757	159.46	28 - 492
Rb	27.42	37 - 53	19.83	21 - 34	21.93	21 - 47	76.61	23 - 165
Zr	190.83	95 - 509	137.32	85 - 304	178.69	62 - 503	167.07	92 - 253
Y	30.50	22 - 51	25.18	20 - 57	22.48	20 - 53	26.71	20 - 41
K/Rb	304.51	260.22 - 357.06	241.70	114.48 - 338.38	226.11	131.42 - 489.49	192.89	80.09 - 293.40
K/Ba	12.99	5.28 - 21.47	10.79	6.02 - 18.22	23.90	10.56 - 76.98	24.71	6.67 - 36.94

Sample 1 is volcanic rocks, flows of Kraul Mountains, average of 22 analyses; Sample 2 is dolerites, dikes of Kraul Mountains, average of 12 analyses; Sample 3 is dolerites, dikes of northern Ahlmann Ridge, average of 29 analyses; Sample 4 is volcanic rocks, flows of Straumsnutane region, Straumsnutane Formation, average of 28 analyses. All iron as Fe_2O_3. LOI is loss on ignition.

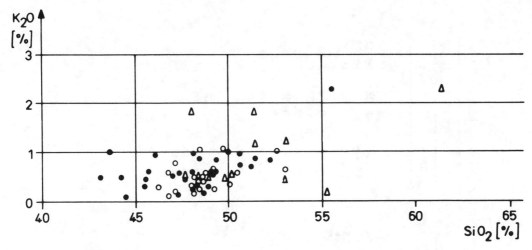

Fig. 3. Plot of K_2O versus SiO_2 for basalts and dolerites from western New Schwabenland, Antarctica (triangles are volcanic rocks of the Kraul Mountains; open circles, dolerites of the Kraul Mountains; solid circles, dolerites of the northern Ahlmann Ridge).

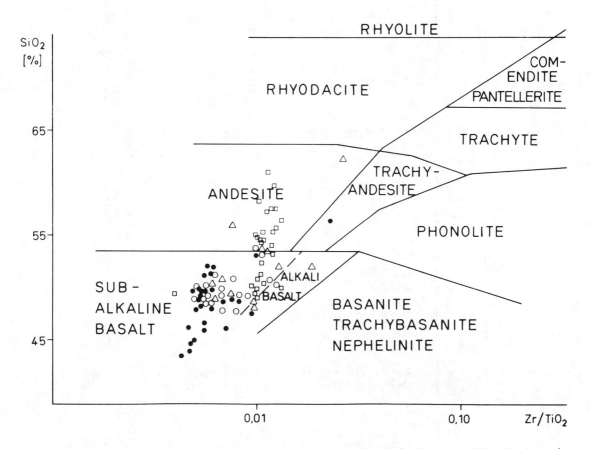

Fig. 4. SiO_2, versus Zr/TiO_2 diagram [after Winchester and Floyd, 1977] showing the delimited fields for volcanic rocks (squares are volcanic rocks of the northern Ahlmann Ridge; other symbols as for Figure 3).

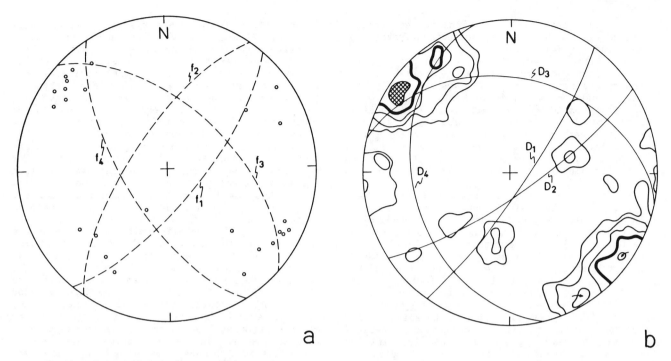

Fig. 5. Pole diagrams with structural data from the northern Kraul Mountains, Antarctica. Schmidt net, lower hemisphere. (a) Poles of 25 normal faults. (b) Poles of 150 dolerite dikes; contours, 1%, 3%, 5%, 7%, 10%, per 1% area.

Hjelle and Winsnes, 1972]. Also, a Jurassic age has been obtained [Rex, 1972] for a dolerite intrusive in the vicinity of the Kraul Mountains. In view of the large geochemical and petrographical similarities between the dikes and sills on the one hand and the lava flows on the other, Hjelle and Winsnes [1972] assumed that both lithological units were of roughly the same age, implying a Jurassic age for the basaltic flows. The preliminary K-Ar age dating of 440 Ma after Krylow [see Hjelle and Winsnes, 1972] of a basaltic flow in the northern Kraul Mountains is, in their and our opinion, not reasonable. Based on this doubtful age determination, the rocks of the Kraul Mountains are shown in the 1:5,000,000 geological map [Craddock, 1972] as late Precambrian.

The chemical analyses that we have carried out confirm the suggested similarities between the dikes and the lava flows and lend credence to the suggestion of a Jurassic magmatic event. Further, there are several new but as yet unpublished K-Ar radiometric age determinations (M. Peters, personal communication, 1985) on plagioclase which yielded ages between 160 and 180 Ma for several dolerite dikes and sills of the northern Kraul Mountains, comparable to those age determinations published by Furnes and Mitchell [1978] for dikes of the southern Kraul Mountains; further K-Ar dating of basaltic flows, carried out on plagioclase separates, has indicated ages of 150 and 250-280 Ma. The latter ages were, however, obtained from a rather altered rock and may not be reliable. Furthermore, it should be mentioned that Furnes and Mitchell [1978] reported Triassic ages of four basaltic lavas in the southern Kraul Mountains.

Paleomagnetic investigations on numerous samples from the flows and intrusives also point to a Jurassic age (M. Peters, personal communication, 1985). These data, although of a preliminary nature, suggest that the magmatic event of the Kraul Mountains can be placed into and around the Early Jurassic period.

The Northern Ahlmann Ridge

Lithologic units. The numerous nunataks and massifs of the Ahlmann Ridge, in the northeast of western New Schwabenland, were investigated during several expeditions by South African geologists [Wolmarans and Kent, 1982]. From the Straumsnutane region in the northeastern Ahlmann Ridge, where the present author carried out a structural study, a more detailed geological description is already available [Watters, 1972].

The mountains of the Ahlmann Ridge and the Borg Massif are built up of rock sequences of a sedimentary-volcanogenic platform cover that compose the Ritscherflya Supergroup of probable Proterozoic age [Wolmarans and Kent, 1982]. Into this sequence were intruded the extensive, mainly mafic, Borgmassivet Intrusives, also of Proterozoic age. Mafic dikes of probably Mesozoic age have invaded these Proterozoic units. The rock sequence of the Ritscherflya Supergroup has an estimated thickness of at least 3500 m and is subdivided into the predominantly sedimentary Ahlmannryggen Group and the predominantly volcanogenic Jutulstraumen Group.

The Straumsnutane Formation belongs to the latter group and comprises all the nunataks of the Straumsnutane region along the western side of the Jutulstraumen Rift. This formation, with a minimum thickness of 860 m [Watters, 1972], consists mainly of lavas, often amygdaloidal. In some nunataks there are minor thin sedimentary layers intercalated in the lava sequence, and some flows are pillowed. Thickness of individual lava flows generally amounts to a few meters but can exceed 50 m.

All these mafic volcanic rocks underwent a weak metamorphism; the most frequent metamorphic minerals are chlorite and epidote. The Borgmassivet Intrusives are missing in the Straumsnutane region, and the only intrusions are basaltic and doleritic dikes which are practically unaltered. Petrographically, the dike rocks of the Ahlmann Ridge are very similar to the dolerites of the Kraul Mountains. However, in a few cases, olivine dolerites with higher contents (15-20%) occur, and several dike rocks also show a basaltic rather than doleritic texture. The average chemical composition (Table 1, sample 3) and the plots of these dolerites in Figures 3 and 4 indicate a close agreement in chemical composition between the dolerite dikes of both areas. The chemical composition of the volcanic rocks of the Straumsnutane Formation is, however, very different. Petrographically, these have been described as porphyritic andesites [Watters, 1972] and are clearly somewhat more acidic (Table 1, sample 4). According to Figure 4 they are basaltic-andesites and andesites which form a fairly well-defined grouping in Figure 4 and can in this way be distinguished from the other rock groups.

Structures. Structural measurements were carried out on most nunataks of the Straumsnutane region, particularly on the Snökallen, Snökjerringa, Bolten, Utkikken, and Trollkjelpiggen.

The layering of the lava beds is mostly horizontal to weakly inclined to the southeast and northwest. Steep dips also occur; this is the case in shear zones of assumed downthrown faults and on the western margin of the Jutulstraumen. The weakly dipping lava layers form synclinal structures, which were sometimes observed in the field.

Shearing is a conspicuous feature in the volcanic rocks of Straumsnutane. Generally, the shearing increases from west to east and is most intensive near the Jutulstraumen. Further, alteration of the rocks is at its strongest where the shearing is most intensive. Where the shearing is weakly developed, the shear planes have intervals of a few millimeters to a few centimeters; however, along the margin of the Jutulstraumen the strong shearing has produced a schistosity. Measurements of the shear planes and of the schistosity (s) indicated a strike between north-northeast and northeast with a generally steep dip (about $70°$) to the east-southeast and southeast (Figure 6a).

Another notable structural feature of the Straumsnutane region is a system of small overthrusts. These are very frequent and have a relatively regular distribution. Figure 6b shows that these constitute a conjugate system of overthrust planes (p_1 and p_2) that strike approximately northeast-southwest and dip at approximately $25°$ to the southeast or northwest. The northwest dips are less frequent than those to the southeast, and scatter is insignificant. Thrust planes are always thickly coated with epidote and are frequently exposed over many square meters and occasionally over hundreds of square meters. For each plane, two to four measurements were carried out in different positions; strikes and dips from 58 of these small overthrusts were recorded, but many more of them were observed. Well-developed slickensides, filled feather joints, and the offsets of key lithologic units made it possible to determine the sense and the direction of the displacement due to shearing. The offset is, as far as it could be measured, not very large, amounting to a maximum of a few meters. Both sets of thrust planes are probably of the same age, and where two differently inclined planes intersected, one frequently was observed to converge with the other by bending. Occasionally, these small overthrusts die out in contacts between lava flows. The overthrust planes are not cut by the steep shear planes and the schistosity as indicated by the unsheared state of the thick epidote coatings on the former. The dolerite dikes, however, do cut the overthrust planes and are not affected by overthrusts. A few of these small overthrusts were also found on the nunataks of Grunehogna, in the central part of the Ahlmann Ridge, which consist of the Ahlmannryggen Group sediments and Borgmassivet Intrusives.

The dolerite dikes were not present to the same extent in the Ahlmann Ridge as in the Kraul Mountains; while only 38 dikes were observed and measured, it is probable that more are present. These also were recorded in the central part of the Ahlmann Ridge. Most of the dikes strike in a north-northeast or northeast direction (d_1 and d_2 in Figure 6c) and are either vertical or steeply dipping. A few strike west-northwest to east-southeast or west to east and partly dip at lower angles. The dikes mostly have thicknesses between 1 and 5 m, the widest observed having a width of 25 m. Except for the ubiquitous columnar jointing, the dikes are devoid of structural features.

Figure 7 illustrates the typical setting of the above-mentioned structural features. The relations between the different structures are easily identifiable, and situations similar to that illustrated can be found in nearly all nunataks of the Strausnutane region.

Age relations. Only a few reliable radiometric age determinations (mainly Rb-Sr and K-Ar method on whole rock) are presently available for this area [Wolmarans and Kent, 1982]. Ages in excess of 1700 Ma are probable for the sediments of the Ahlmannryggen Group. The Borgmassivet Intrusives have been dated at approximately 1700 Ma. An age of approximately 820 Ma [Eastin et al., 1970] was determined for the volcanics of the Straumsnutane

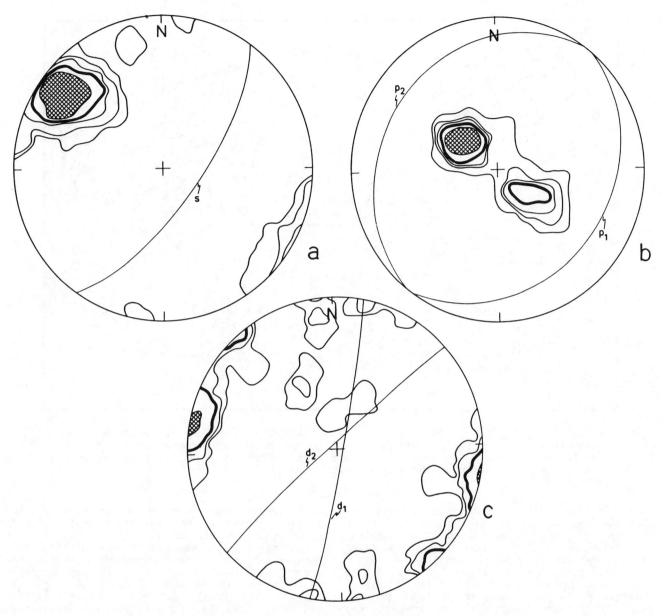

Fig. 6. Pole diagrams with structural data from the northern Ahlmann Ridge, Antarctica. Schmidt net, lower hemisphere. (a) Narrow-spaced shear planes and schistosity, 258 poles; contours, 1%, 3%, 5%, 7%, 10%, per 1% area. (b) Small overthrusts, 171 poles from 58 thrust planes; contours, 1%, 3%, 5%, 7%, 10%, per 1% area. (c) Dolerite dikes, 38 poles; contours, 2.5%, 5%, 10%, 25%, per 1% area.

Formation, and olivine dolerite from the central Ahlmann Ridge has yielded an age of 192 Ma [Aucamp, 1972].

Radiometric determinations on samples taken by M. Peters during our fieldwork have so far yielded two results of about 526 Ma (K-Ar method on sericite) for sheared Straumsnutane volcanics and one age of 190 Ma (K-Ar method on plagioclase for a dolerite dike at Grunehogna nunatak in the central Ahlmann Ridge) (M. Peters, personal communication,

1985). Paleomagnetic measurements on several dolerite dikes of the northern Ahlmann Ridge again confirm a Mesozoic, probably Early Jurassic, age (M. Peters, personal communication, 1985).

Discussion and Conclusions

The close geochemical and petrographical relationship between basaltic lava flows and dolerite dikes of the Kraul Mountains, supported by radio-

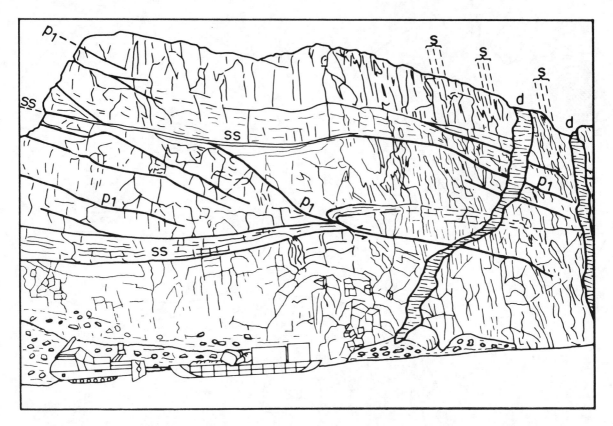

Fig. 7. Sketch from the south wall of Snökallen nunatak, northeastern Ahlmann Ridge, Antarctica, with the typical setting of structures. Height of the exposure is approximately 150 m. ss is layering in the lavas and sedimentary layers; s, narrow-spaced shear planes; p_1, small overthrusts; d, dolerite dikes.

metric evidence, supports the above mentioned Mesozoic, probably Early Jurassic, age. Such an age for the dolerite dikes cannot be disputed, because of structural relationships to Lower Permian units. In view of their age and field relationships these igneous rock units of the Kraul Mountains can be correlated with the Ferrar Supergroup of the Transantarctic Mountains, as named by Kyle et al. [1981]. The basaltic lava flows may be equated to the Kirkpatrick Basalt Group, and the dolerite dikes and sills to the Ferrar Dolerite Group.

As indicated by our fieldwork and publications by the authors mentioned in the introduction of this paper, mafic dike swarms are found throughout western New Schwabenland. However, the dikes appear to be more frequent in the Kraul Mountains than in the Ahlmann Ridge because fewer dikes were found in the latter region, while nearly the same number of nunataks of similar dimensions were investigated in both regions. Possibly, a rifting during the Jurassic was concentrated in the area of the Kraul Mountains. The Ahlmann Ridge mafic dikes of Mesozoic age occur together with Precambrian dolerite intrusives and can be distinguished from the latter by having insignificant alteration and distinctly different chemical composition. An Early Jurassic age is presently indicated by only a few radiometric determinations supported by paleomagnetic age indications.

The majority of the dolerite dikes in the Kraul Mountains have a northeast-southwest direction and are oriented parallel to most of the small normal faults and to the suggested large-scale tensional fractures (Figure 8). The latter have been inferred mainly from the subglacial bedrock topography [Hjelle and Winsnes, 1972]. The close relation between the orientation of the dikes and the predominant fracture system is also true for the Ahlmann Ridge. In addition, the orientations of the dikes and of the Mesozoic and post-Mesozoic fractures seem to be controlled by Precambrian and possibly early Paleozoic tectonic features (shear zones). This emerges from the comparison of the orientation diagrams for the northern Ahlmann Ridge seen in Figure 8. It can be assumed that weak zones in the crust guided the Mesozoic rifting, as Kyle et al. [1981] supposed for other Antarctic areas.

The crustal extension, as indicated by the dikes and fracturing, was oriented northwest-southeast. A subordinate extension perpendicular to this direction also occurred. However, this rifting in western New Schwabenland, although very

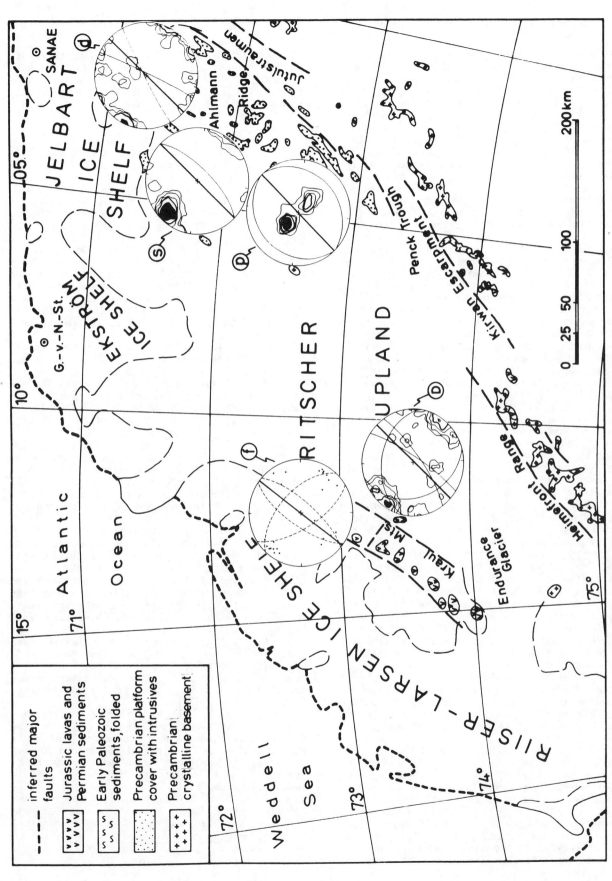

Fig. 8. Geological sketch map of western New Schwabenland, Antarctica, with pole diagrams of the structural data relating to both areas of fieldwork. Straight bold lines in the diagrams indicate main orientations. f is normal fault; D, dolerite dikes of the Kraul Mountains; s, shear planes and schistosity; p, small overthrusts; d, dolerite dikes in the northeastern Ahlmann Ridge.

intensive in some localities (Jutulstraumen and the Kraul Mountains), did not lead to an oceanic spreading center. The question that remains is, Were these rift systems independent, or did the rifting wander from the east to the west to develop more fully into an ocean floor spreading regime in the Weddell Sea area? A very distinct fracturing of the shelf with subsidence of a thick rock sequence, interpreted as a stack of basaltic lava flows, is indicated by geophysical research in the eastern Weddell Sea [Hinz and Krause, 1982]. It should be pointed out that the Mesozoic magmatic events and the post-Mesozoic block faulting have resulted in a strong dissection of the continental margin of western New Schwabenland.

With regard to the influence and trend of the Ross Orogen in the Weddell Sea sector, there seem to be only uncertain indications. Aucamp et al. [1972] placed a folded sedimentary complex of the southern Kirwan Escarpment (Figure 8), the Urfjell Group, into the lower Paleozoic; however, there are no radiometric age data or fossil evidence to support such an age. Neethling [1970] suggests that during the Ross Orogeny, movements between the crystalline basement and the platform cover of the area west to the Jutulstraumen-Pencksökket Rift occurred as a result of strong compressional forces. This opinion is supported by a radiometric age of 590 Ma for a diaphtoritic rock. Further radiometric data of about 480 Ma from the crystalline basement of New Schwabenland have been reported [Wolmarans and Kent, 1982], obviously pointing to a thermal event causing rejuvenation.

The numerous small overthrusts, which in this paper have been described from the northern Ahlmann Ridge and which seem to have an even wider distribution, are also seen as having developed during the Ross Orogeny. These are compressional structures, the direction and age of which fit well into the framework of a tectonic strain acting during the Ross Orogeny. They must be younger than 526 Ma, an age which has been determined on minerals from a steep shear zone in the northeastern Ahlmann Ridge, but older than the Early Jurassic, since the overthrust planes do not cut the dolerite dikes.

In light of the data presented here, it is suggested that western New Schwabenland was at least the near foreland of the Ross Orogen. In the area of the Kraul Mountains there are certainly no rocks originating from the Ross cycle, but it is not clear whether such rocks occur in the southern and southwestern parts of western New Schwabenland. This is a task for further research in the Kirwan Escarpment and in the Heimefront Range.

Acknowledgments. I wish to thank the Deutsche Forschungsgemeinschaft for financial support of these studies. I also thank the Alfred-Wegener-Institut, Bremerhaven, and the Council for Scientific and Industrial Research, South Africa, which provided logistical support for the expeditions. Furthermore, I am grateful to J. Krynauw, team leader in the South African Earth Science Programme 1983-1984, B. Watters, and M. Peters for discussions and advice. I also owe my gratitude to G. Friedrich and W. L. Plüger, Institut für Mineralogie und Lagerstättenlehre, RWTH Aachen, for making available their laboratory facilities and to P. Schüll for geochemical and petrographical investigations of rock samples.

References

Aucamp, A. P. H., The geology of Grunehogna, Ahlmannryggen, western Dronning Maud Land, S. Afr. J. Antarct. Res., 2, 23-31, 1972.

Aucamp, A. P. H., L. G. Wolmarans, and D. C. Neethling, The Urfjell Group, a deformed (?) early Palaeozoic sedimentary sequence, Kirwanweggen, western Dronning Maud Land, in Antarctic Geology and Geophysics, edited by R. J. Adie, pp. 557-562, Universitetsforlaget, Oslo, 1972.

Behr, H. J., H. Kohnen, M. Peters, G. Spaeth, and K. Weber, Die geologische Expedition zu den Kraul-Bergen, westliches Neuschwabenland/Antarktika--Bericht über ihren Verlauf und erste Ergebnisse, Ber. Polarforsch. 13/1983, pp. 13-26, Alfred-Wegener-Inst. für Polarforsch., Bremerhaven, 1983.

Craddock, C., Geologic map of Antarctica, scale 1:5,000,000, American Geographical Society, New York, 1972.

Eastin, R., R. Faure, and D. C. Neethling, The age of the Trolljellrygg Volcanics of western Queen Maud Land, Antarct. J. U. S., 5, 157-158, 1970.

Elliot, D. H., Tectonics of Antarctica: A review, Am. J. Sci., 275-A, 45-106, 1975.

Furnes, H., and J. G. Mitchell, Age relationships of Mesozoic basalt lava and dykes in Vestfjella, Dronning Maud Land, Antarctica, Skr. Nor. Polarinst., 169, 45-68, 1978.

Gunn, B. M., K/Rb and K/Ba ratios in Antarctic and New Zealand tholeiites and alkali basalts, J. Geophys. Res., 70, 6241-6247, 1965.

Hinz, K., and W. Krause, The continental margin of Queen Maud Land/Antarctica: Seismic sequences, structural elements and geological development. Geol. Jahrb., Reihe E, 23, 17-41, 1982.

Hjelle, A., and T. Winsnes, The sedimentary and volcanic sequence of Vestfjella, Dronning Maud Land, in Antarctic Geology and Geophysics, edited by R. J. Adie, pp. 539-546, Universitetsforlaget, Oslo, 1972.

Juckes, L. M., The geology of Mannefallknausane and part of Vestfjella, Dronning Maud Land, Br. Antarct. Surv. Bull., 18, 65-78, 1968.

Juckes, L. M., The geology of northeastern Heimefrontfjella, Dronning Maud Land, Br. Antarct. Surv. Sci. Rep., 65, 44 pp., 1972.

Kyle, P. R., D. H. Elliot, and J. F. Sutter, Jurassic Ferrar Supergroup tholeiites from the Transantarctic Mountains, Antarctica, and their relationship to the initial fragmentation of Gondwana, in Gondwana Five, edited by M. M. Cresswell and P. Vella, pp. 283-287, A. A. Balkema, Rotterdam, 1981.

Neethling, D. C., Pre-Gondwana sedimentary rocks of Queen Maud Land, Antarctica, in Gondwana Stratigraphy, International Union of Geological Sciences Symposium, Buenos Aires, 1967, edited by A. J. Amos, pp. 1153-1162, UNESCO, Paris, 1969.

Neethling, D. C., South African earth science exploration of western Queen Maud Land, Antarctica, Ph.D. thesis, Univeristy of Natal, Pietermaritzburg, 1970.

Neethling, D. C., Age and correlation of the Ritscher Supergroup and other Precambrian rock units, Dronning Maud Land, in Antarctic Geology and Geophysics, edited by R. J. Adie, pp. 547-556, Universitetsforlaget, Oslo, 1972.

Rex, D. C., K-Ar age determinations on volcanic and associated rocks from the Antarctic Peninsula and Dronning Maud Land, in Antarctic Geology and Geophysics, edited by R. J. Adie, pp. 133-136, Universitetsforlaget, Oslo, 1972.

Roots, E. F., Preliminary note on the geology of western Dronning Maud Land, Nor. Geol. Tidsskr., 32, 18-33, 1953.

Spaeth, G., and M. Peters, Geologische Untersuchungen im nördlichen Ahlmann-Rücken, mittleres Neuschwabenland/Antarktika--Bericht über die Teilnahme an der Geländekampagne 1983/84 im "South African Antarctic Earth Sciences Programme": Ablauf, Logistik, Geologie des Arbeitsgebietes und erste Ergebnisse, Ber. Polarforsch., 19/1984, pp. 174-185, Alfred-Wegener-Inst. für Polarforsch., Bremerhaven, 1984.

Watters, B. R., The Straumsnutane volcanics, western Dronning Maud Land, Antarctica, S. Afr. J. Antarct. Res., 2, 23-31, 1972.

Winchester, J. A., and P. A. Floyd, Geochemical discrimination of different magma series and their differentiation products using immobile elements, Chem. Geol., 20, 325-343, 1977.

Wolmarans, L. G., and L. E. Kent, Geological investigations in western Dronning Maud Land--A synthesis, S. Afr. J. Antarct. Res., Suppl. 2, 1-93, 1982.

Copyright 1987 by the American Geophysical Union.

PLATE TECTONIC DEVELOPMENT OF LATE PROTEROZOIC PAIRED METAMORPHIC COMPLEXES
IN EASTERN QUEEN MAUD LAND, EAST ANTARCTICA

Kazuyuki Shiraishi,[1] Yoshikuni Hiroi,[2] Yoichi Motoyoshi,[3] and Keizo Yanai[1]

Abstract. Two Proterozoic metamorphic complexes of the East Antarctic Shield are exposed in eastern Queen Maud Land: Yamato-Belgica and Lützow-Holm to the southwest and northeast, respectively. In the Yamato-Belgica Complex, igneous activity is widespread. The metamorphic portions of this complex consist mainly of amphibolite facies rocks. Granulite facies rocks occur locally and are partly included in a syenitic intrusive. The granulite facies rocks belong to the low-pressure type. On the other hand, the regional metamorphism of the dominantly metasedimentary Lützow-Holm Complex is of the medium-pressure type, grading progressively from upper amphibolite facies to granulite facies toward the southwest. There is textural evidence of prograde recrystallization. The prograde P-T-time(t) paths of the rocks from different metamorphic zones differ from each other and also from the regional metamorphic geotherm based on the mapping of metamorphic zones. High P/T-type metamorphism in the early stage of regional metamorphism is suggested by garnet porphyroblasts rimmed with symplectic intergrowth of spinel, orthopyroxene, and plagioclase in troctolitic rocks. We thus infer that the Yamato-Belgica and Lützow-Holm complexes form paired metamorphic complexes and that the latter was formerly situated in a subduction zone. Moreover, chemical compositions of metamorphosed ultrabasic and basic rocks occurring mainly in the western part of the Lützow-Holm Complex resemble those of cumulate rocks associated with ophiolites. Therefore, they are possibly fragments of oceanic crust which were tectonically fractured and emplaced into the sedimentary pile prior to regional metamorphism. Consequently, a plate tectonics model would be the most suitable interpretation for the development of the late Proterozoic metamorphic complexes in the region.

Introduction

The East Antarctic Shield has often been omitted from discussions on Gondwana geology because of insufficient data. Quite recently, however, geologic and petrologic studies as well as geochronologic studies, especially in the Indian Ocean sector, indicate that the East Antarctic Shield is not uniform but is composed of various geologic units [e.g., Grew, 1982; James and Tingey, 1983]. In Enderby Land (45°-60°E longitude), two complexes, the Archean Napier Complex, and the Proterozoic Rayner Complex, are recognized (Figure 1) [Kamenev, 1972; Sheraton et al., 1980; Ellis, 1983]. The Napier Complex, which cratonized at 2.5 Ga, is characterized by extremely high-temperature conditions of metamorphism, whereas the 1.0 Ga Rayner Complex is a mobile zone metamorphosed under the upper amphibolite to lower granulite facies conditions at the margin of the Napier Complex [Ellis, 1983].

In this paper we will describe two Proterozoic metamorphic complexes in eastern Queen Maud Land (30°-45°E longitude) to the west of Enderby Land, and discuss contrasting features between the two complexes. Then we will attempt to construct a model of the tectonic development of eastern Queen Maud Land on the basis of the geologic and petrologic evidence. Detailed petrography and petrochemistry of the complexes are given elsewhere.

Two Complexes in Eastern Queen Maud Land

Eastern Queen Maud Land is composed of the Prince Olav Coast, Lützow-Holm Bay region (Soya Coast), Yamato Mountains, and Belgica Mountains (Figure 1). Geological surveys have been conducted by Japanese geologists, and the results have been published by the National Institute of Polar Research, Tokyo, as the Antarctic Geological Map Series. The high-grade metamorphic rocks exposed in the Lützow-Holm Bay region and the Yamato Mountains were originally termed the Lützow-Holm Bay System [Tatsumi and Kizaki, 1969]. Yoshida [1978, 1979] divided the rocks exposed in the Lützow-Holm Bay region and the western part of the Prince Olav Coast into three units, from northeast to southwest: the Okuiwa, Ongul, and Skallen groups. He considered that the Ongul and Skallen groups were polymetamorphic sequences, and that the Okuiwa group, which was subjected to amphibolite facies metamorphism, was younger than the other two groups.

However, recent geologic and petrologic studies revealed that the Prince Olav Coast and the Lüt-

---

[1]National Institute of Polar Research, Tokyo 173, Japan.
[2]Department of Earth Sciences, Chiba University, Chiba 260, Japan.
[3]Department of Geology and Mineralogy, Hokkaido University, Sapporo 060, Japan.

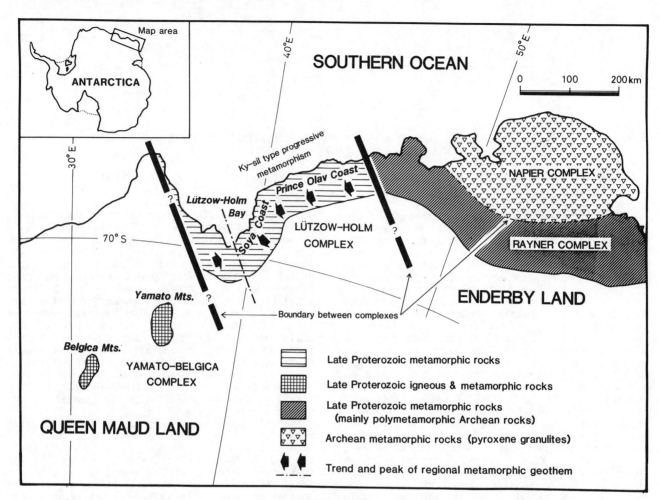

Fig. 1. Map showing the distribution of and relationship among the four Precambrian metamorphic complexes in eastern Queen Maud Land and Enderby Land.

zow-Holm Bay region have lithologic, deformational, and metamorphic features in common, and that the metamorphic grade increases gradually from northeast to southwest [Hiroi et al., 1983a, b; Shiraishi et al., 1984; Motoyoshi et al., 1985]. Therefore, the rocks from the Prince Olav Coast and Lützow-Holm Bay region form a single geological unit which is designated as the Lützow-Holm Complex (Figure 1).

On the other hand, the inland mountain ranges, Yamato and Belgica mountains, are characterized by widespread igneous activity such as intrusion of granites and syenites. Metamorphic rocks are composed mainly of amphibolite facies rocks with subordinate granulite facies rocks. The contact between the two facies is tectonic, and the relationship between them is not yet clear [Shiraishi et al., 1983b]. The basic type of the granulite facies metamorphic rocks is of low-pressure type and is distinct from that of the Lützow-Holm Complex (medium-pressure type) as described below. Thus, the plutonometamorphic complex in the inland region is distinct from the Lützow-Holm Complex and is termed the Yamato-Belgica Complex.

The boundary between the Lützow-Holm Complex and the Yamato-Belgica Complex is covered by continental ice (Figure 1). East of the Lützow-Holm Complex is the Proterozoic Rayner Complex [Kamenev, 1972; Grew, 1978]. Although the boundary between them is not exposed, it is most likely tectonic as suggested by the presence of the large Sinnan Glacier (45°E longitude).

Granite and pegmatite are widely distributed throughout eastern Queen Maud Land. They form stocks, sheets, dikes, and veins. About 50 mineral and whole-rock K-Ar, Rb-Sr, and U-Pb radiometric ages ranging from 350 to 560 Ma have been reported on the granites, pegmatites, and metamorphic rocks [Yanai and Ueda, 1974; Kojima et al., 1982]. Moreover, Shibata et al. [1985] showed Rb-Sr mineral isochron ages of 469 Ma, biotite K-Ar ages of 469 to 483 Ma, and a hornblende K-Ar age of 502 Ma from the granulite facies rocks in eastern Queen Maud Land. It is conceivable that the early Paleozoic granite and pegmatite activity reset the radiometric ages of the metamorphic rocks throughout the study area, as suggested by Grew [1982] for the Indian Ocean sector of East Antarctica.

Geological and petrographical features of the

TABLE 1. Yamato-Belgica Complex Showing Typical Mineral Assemblages in the Major Rock Types

	Granulite Facies Rocks	Amphibolite Facies Rocks
Pelitic gneisses	Op + Bi + Pl + Q ± Kf	Gt + Bi + Pl + Kf + Q
Intermediate to basic gneisses	Op + Cp + Bi + Kf + Pl + Q ± Hb Cp + Hb + Bi + Pl + Q ± Kf	Hb + Bi + Kf + Pl + Q Cp + Hb + Bi + Sc + Pl + Q
Calc-silicate rocks	Wo + Cp + Sc ± Pl + Sph (+ Gt + Q)	Sp + Hum + Do + Cc + Phl + Fo Cp + Phl + Hb + Sc + Cc Gt + Cp + Sc + Cc + Q

Parentheses indicate retrograde mineral. Mineral abbreviations: And, andalusite; Anth, anthrophyllite; Bi, biotite; Cc, calcite; Cor, corundum; Crd, cordierite; Cp, clinopyroxene; Do, dolomite; Ged, gedrite; Gr, grossular; Gt, garnet; Hb, hornblende; Hum, humite; Kf, K-feldspar; Ky, kyanite; Mus, muscovite; Ol, olivine; Op, orthopyroxene; Pl, plagioclase; Phl, phlogopite; Q, quartz; Sapp, sapphirine; Sc, scapolite; Sill, sillimanite; Sp, spinel; Sph, sphene; Sta, staurolite; Wo, wollastonite.

Yamato-Belgica and Lützow-Holm complexes are described in the following sections.

Yamato-Belgica Complex. Granulite facies rocks in the Yamato-Belgica Complex are exposed only in the Yamato Mountains. They are composed of two-pyroxene biotite gneiss, two-pyroxene amphibolite, orthopyroxene biotite gneiss, and calc-silicate gneiss. Amphibolite facies rocks comprise quartzofeldspathic (granitic) gneisses, biotite-hornblende gneiss, biotite amphibolite, and marble and skarn with minor garnet-biotite gneiss. The metamorphic facies series of the amphibolite facies metamorphism has not been revealed. Migmatitic gneiss which has paleosomes of granulite facies rocks occurs locally in the Yamato Mountains. No ultrabasic rocks have been found in this complex.

There are two interpretations for the relationship between the rocks of the two facies. According to one interpretation, both facies belong to the same metamorphic sequence and occupied different crustal levels prior to the tectonic movement [Asami and Shiraishi, 1983]. According to the second interpretation, the granulite facies rocks were metamorphosed in an older stage than the amphibolite facies metamorphism which is contemporaneous with the early Paleozoic granite and pegmatite activity [Kizaki, 1965]. Widespread syenitic intrusives characterize the complex. The syenite suite is divided into three lithologic groups [Shiraishi et al., 1983a]: two-pyroxene syenite, clinopyroxene quartz monzosyenite, and clinopyroxene syenite. The close field relationship between the two-pyroxene syenite and the granulite facies rocks suggests that emplacement of the syenitic rocks took place during the granulite facies metamorphism [Shiraishi et al., 1983b]. Other syenites are younger than the two-pyroxene syenite and are cut by metamorphosed basic dikes, granite, and pegmatite [Yanai et al., 1982].

Typical mineral assemblages in the Yamato-Belgica Complex are shown in Table 1. No Al_2SiO_5 mineral has been found in this complex, probably as none of the bulk compositions are sufficently aluminous.

For the granulite facies rocks, two-pyroxene geothermometry yields metamorphic temperatures up to 750°C [Asami and Shiraishi, 1983]. The calc-silicate gneiss in the granulite facies rocks contains an association of grossular garnet + quartz + wollastonite + anorthite. The textural relationships show the anorthite + wollastonite assemblage was stable at the peak of the metamorphism, and the grossular garnet and quartz association is a reaction product formed during the retrograde metamorphism. The experimentally determined equilibrium relation of the reaction, grossular + quartz = anorthite + wollastonite, shows that the anorthite and wollastonite assemblage is stable under the lower-pressure condition at a constant temperature [Windom and Boettcher, 1976]. The wollastonite is close to pure composition, and the anorthite contains only 7 mol % albite, while the grossular garnet contains a considerable amount of andradite component (Ad_{21-18}) [Asami and Shiraishi, 1985]. Low albite contents in plagioclase may not shift the equilibrium curve very far to the high-pressure side [Windom and Boettcher, 1976]. On the other hand, the stability field of the grossular garnet with the considerable andradite component expands toward the high-temperature side [Huckenholz et al., 1974, 1981]. Thus, a significantly low-pressure condition compared with that of the Lützow-Holm Complex is suggested, and the metamorphic facies series is of the low-pressure type.

A Rb-Sr whole-rock isochron age of 718.4 ± 33.7 Ma with an initial $^{87}Sr/^{86}Sr$ ratio 0.70899 ± 0.00026 was obtained for the granulite facies rocks [Shibata et al., 1986]. The Rb-Sr whole-rock isotopic system is supposed to have been extensively affected and homogenized during the granulite facies metamorphism [Shibata et al., 1986]. Therefore, the age is considered to date the granulite facies metamorphism.

Lützow-Holm Complex. The complex is mainly composed of well-layered pelitic and intermediate gneisses with subordinate amounts of calcareous and basic to ultrabasic rocks. Granitic to granodioritic migmatite of anatectic origin is abundant. The intermediate gneisses are biotite-hornblende gneiss and charnockitic gneiss in the eastern and western part of the complex, respectively. Metamorphosed basic to ultrabasic rocks which are now amphibolite, hornblendite, pyroxenite, and peridotite occur as concordant layers and as isolated blocks within the well-layered gneisses

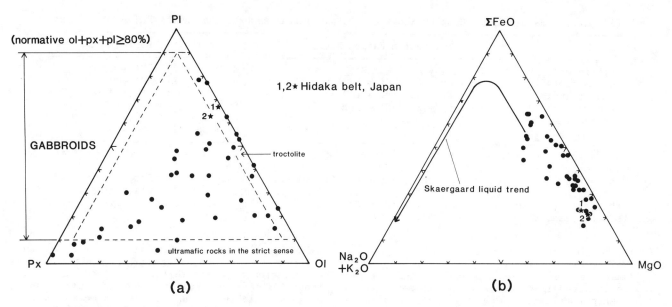

Fig. 2. Diagrams showing characteristics of chemical compositions of the basic and ultrabasic rocks from the Lützow-Holm Complex. Troctolites from the ophiolite complex of the Hidaka belt in Japan [Miyashita et al., 1980] are also shown for comparison. (a) Plot of normative plagioclase, pyroxenes, and olivine. (b) AFM diagram.

mainly in the western part of the complex. Figure 2 illustrates characteristics of the chemical compositions of the basic to ultrabasic rocks. Some of the ultrabasic rocks may be derived from troctolites whose mineral assemblage of plagioclase + olivine suggests crystallization of basic magma at low pressures. The wide range in chemical composition suggests that these rocks are derived from various parts of layered gabbro.

At least two episodes of deformation are recognized throughout the Lützow-Holm Complex: earlier isoclinal to tight folds with axial planes trending northwest-southeast to north-south, and later, open to tight folds with axial planes of east-west to northeast-southwest trends [Yoshida, 1978; Matsumoto et al., 1982; Hiroi et al., 1983c; Shiraishi et al., 1985]. The earlier folds are contemporaneous with the main metamorphism, since mineral lineations shown by dimensional preferred orientation of amphiboles, sillimanite, and biotite are parallel to the earlier fold axes. The migmatitic rocks show dome structures, which are generally a few hundred meters in diameter in the eastern part of the complex [Hiroi et al., 1983c; Shiraishi et al., 1985]. The later fold episode probably occurred immediately after the main metamorphism and clearly before the intrusion of thermally metamorphosed andesitic and basaltic dikes. Local mylonitization occurred before the emplacement of the early Paleozoic granite and pegmatite.

Plutonic rocks are not widespread except for the migmatite of anatectic origin and the early Paleozoic granite and pegmatite which form dikes and small stocks. Plutonic charnockite was reported from the southwestern Lützow-Holm Bay region, where the highest metamorphic temperatures were attained [Yoshida, 1978]. Minor andesitic and basaltic dikes in the central part of the Prince Olav Coast are thermally metamorphosed by the early Paleozoic granite and pegmatite.

Since Maegoya et al. [1968] first reported an Rb-Sr isochron age of 1110 Ma on K-feldspar, most investigators have considered that the main regional metamorphism occurred at that time [e.g., Yoshida, 1978, 1979; Yoshida and Aidawa, 1983]. However, Shibata et al. [1986] report a 683.1 ± 13.2 Ma Rb-Sr whole-rock isochron age with an initial $^{87}Sr/^{86}Sr$ ratio of 0.70471 ± 0.00011 and a 683.0 ± 85.0 Ma age with an initial $^{87}Sr/^{86}Sr$ ratio of 0.70564 ± 0.00075 for the granulite facies and the amphibolite facies rocks, respectively. It is noteworthy that the whole-rock isochron ages between the Yamato-Belgica Complex and the Lützow-Holm Complex are almost identical, but the initial $^{87}Sr/^{86}Sr$ ratio of the Yamato-Belgica Complex is higher.

In the Lützow-Holm Complex, three zones of different metamorphic facies are recognized from northeast to southwest: amphibolite facies, transitional, and granulite facies (Figure 3) [Hiroi et al., 1983a, b]). Typical mineral assemblages in the three zones are shown in Table 2. The sillimanite + K-feldspar assemblage is stable throughout the complex. In addition, metastable kyanite occurs as inclusions within garnet and plagioclase in most of the sillimanite-bearing rocks regardless of the metamorphic grade, suggesting prograde recrystallization of the rocks from the kyanite to sillimanite stability fields.

Calcium-poor amphiboles (cummingtonite and anthophyllite) are common in the amphibolite facies rocks, while orthopyroxene is ubiquitous in rocks from the granulite facies zone (Figure 3). In the transitional zone, orthopyroxene occasionally occurs in rocks with appropriate bulk chemical com-

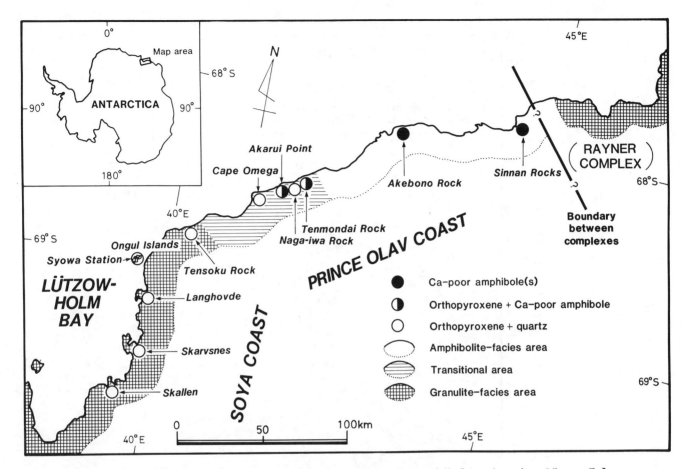

Fig. 3. Distribution of orthopyroxene and Ca-poor amphiboles in the Lützow-Holm Complex.

TABLE 2. Lützow-Holm Complex Showing Typical Mineral Assemblages in the Main Rock Types

	Granulite Facies Zone	Transitional Zone	Amphibolite Facies Zone
Pelitic gneisses	Sp+Gt+Sill(Ky)±Pl Op+Gt+Bi+Pl+Kf+Q Sill+Gt+Bi+Pl+Kf+Q	Sp+Gt+Sill(Ky)±Pl Op+Gt+Bi+Pl+Q+±Hb Crd+Sill+Bi+Kf+Q Sill+Gt+Bi+Pl+Kf+Q Ged+Gt+Bi+Pl+Q	Sta±Pl±Gt Cor+Kf+Mus+Pl+Bi±Sill Sill(Ky)+Gt+Bi+Pl+Q±Crd Anth+Gt+Bi+Pl+Q
Intermediate-basic gneisses	Op+Cp+Pl+Kf+Q±Hb±Bi Gt+Op+Cp+Hb+Bi+Pl±Q Cp+Gt±Q	Op+Cp+Hb+Bi+Pl+Kf+Q Gt+Op+Hb+Bi+Pl+Q	Cum+Hb+Bi+Pl±Q Gt+Hb+Bi+Pl±Q±Kf Cp+Hb+Pl±Sc±Bi±Q
Ultrabasic rocks	Op+Cp+Hb+Phl+Pl Sp+Op+Hb+Pl Op+Cp+Gt+Pl Op+Sp+Pl*	Sp+Op+Hb+Bi+Pl Ol+Sp+Op+Hb+Bi Gt+Op+Pl+Sp*	
Calc-silicate rocks	Hb+Phl+Ol+Cp+Cc±Sp Cp+Sc+Pl+Cc Cp+Sc+Gt+Q+Cc	Ol+Phl+Hb+Cp+Cc Sc+Phl+Cc+Cp+Pl+Q Gt+Sc+Cc+Pl	Gt+Cp+Hb+Pl+Q Gt+Cp+Cc+Q

Parentheses indicate relict mineral. Mineral abbreviations are shown in Table 1.
*Symplectite.

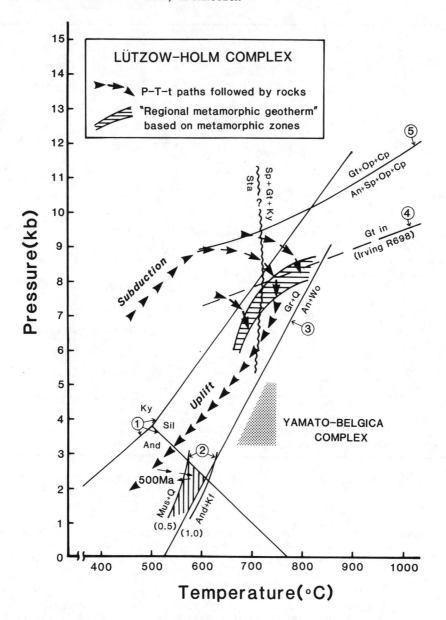

Fig. 4. Metamorphic geotherm of Lützow-Holm Complex and the P-T condition of the Yamato-Belgica Complex. Arrows indicate the prograde P-T paths of the granulite facies, transitional, and amphibolite facies zones. (1) Holdaway [1971], (2) Kerrick [1972], (3) Windom and Boettcher [1976], (4) Irving [1974], and (5) Herzburg [1978].

positions, and calcium-poor amphibole also occurs [Shiraishi et al., 1984].

In the eastern amphibolite facies zone, staurolite is present as inclusions in garnet and plagioclase, but staurolite is never in direct contact with quartz. On the other hand, a spinel + garnet + sillimanite (or kyanite) association, which is a product of staurolite breakdown in the absence of quartz, is found in the western part of the amphibolite facies zone to the granulite facies zone [Motoyoshi et al., 1985; Hiroi et al., 1983a, b; Y. Hiroi, in preparation, 1986].

In the granulite facies zone, systematic examination of the Fe-Mg distribution between orthopyroxenes and clinopyroxenes in the charnockitic gneisses shows a distinct thermal gradient to the southwest [Motoyoshi, 1986]. All this petrographical evidence supports the progressive metamorphism from east to west in the Lützow-Holm Complex. Moreover, the thermal axis of the regional metamorphism is found at the southern end of the Soya Coast. West of this point, the metamorphic grade decreases (Figure 1) [Motoyoshi, 1986].

Thermal metamorphism associated with the early

TABLE 3. Comparison of Yamato-Belgica and Lützow-Holm Complexes

	Yamato-Belgica Complex	Lützow-Holm Complex
Source rocks	Various sedimentary rocks and basic intrusive rocks No ultrabasic rocks Rare aluminous pelitic rocks High initial $^{87}Sr/^{86}Sr$ ratio (0.709)	Various sedimentary rocks, basic intrusive rocks, and ultrabaisc rocks (cumulate complex?) Low initial $^{87}Sr/^{86}Sr$ ratios (0.705-0.706)
Metamorphism	Upper amphibolite facies and low-pressure type granulite facies (relation uncertain)	Upper amphibolite to granulite facies (medium-pressure type, progressive)
Age of metamorphism	~700 Ma	~700 Ma
Plutonism	Extensive syenite intrusions Posttectonic mafic dikes Early Paleozoic granite-pegmatite	Local intrusive charnockite Posttectonic mafic dikes Early Paleozoic granite-pegmatite

Paleozoic granite and pegmatite resulted in the local formation of andalusite in the kyanite- and sillimanite-bearing rocks [Hiroi et al., 1983a].

P-T-Time Path of the Lützow-Holm Complex

The P-T conditions of the regional metamorphism of the Lützow-Holm Complex have been discussed by many authors [Yoshikura et al., 1979; Yoshida, 1979; Yoshida and Aikawa, 1983; Suzuki, 1979, 1982, 1983, 1984; Kanisawa and Yanai, 1982]. Hiroi et al. [1983b] estimated the P-T conditions of 6-7 kbar and 680°C for the amphibolite facies rocks and 6.5-7.5 kbar and 750°C for the transitional rocks. Motoyoshi [1986] estimated the metamorphic conditions of the granulite facies to be as high as 850°C and 10 kbar. Figure 4 illustrates the regional metamorphic geotherm based on the metamorphic zones and P-T estimates.

Observed textures of minerals in pelitic and ultrabasic rocks indicate several lines of evidence of prograde recrystallization of the rocks. For example, in pelitic rocks, metastable kyanite occurs as inclusions within garnet and plagioclase, as mentioned above. Moreover, the product of staurolite breakdown includes kyanite and not sillimanite in the granulite facies and transitional zones [Hiroi et al., 1983a; Y. Hiroi, in preparation, 1986]. This means that the staurolite breakdown in the rocks of granulite facies and transitional zones has occurred in the kyanite stability field (Figure 4). In the ultrabasic granulite of troctolite composition in the transitional zone (Figure 2), garnet, including hornblende, spinel, and sapphirine, shows a breakdown texture to the symplectic intergrowth of spinel + orthopyroxene + plagioclase. Experimentally and theoretically determined subsolidus phase equilibria for ultrabasic granulites by Irving [1974], Herzberg [1978], and Obata and Thompson [1981] suggest the following crystallization and recrystallization history for the ultrabasic rocks (Figure 4): (1) crystallization from magma under low-pressure conditions, (2) regional metamorphic recrystallization under high-pressure and relatively low-temperature conditions to form garnet, and (3) subsequent recrystallization under conditions of lower pressures and higher temperatures, resulting in the formation of symplectic intergrowth of orthopyroxene, spinel, and plagioclase around garnet.

This evidence implies that the P-T paths followed by rocks from each zone are different from the regional metamorphic geotherm implicit in the mapped metamorphic zonation (Figure 4). The P-T paths followed by rocks suggest an initial pressure increase more substantial than the associated temperature increase in the early stage of metamorphism, implying rapid burial. Thus, the medium-pressure-type metamorphic rocks of the Lützow-Holm Complex experienced an early high-P/T-type metamorphism.

Tectonic Interpretation of Eastern Queen Maud Land

Lützow-Holm Complex in Gondwanaland

The position of the eastern Queen Maud Land region in relation to other fragments of Gondwanaland is one of reconstruction problems [Lawver and Scotese, this volume]. It is interesting that Sri Lanka with India fit into a socket of East Antarctica near the Lützow-Holm Bay area and the coast of Enderby Land. Amphibolite facies Vijayan Complex and granulite facies Highland Group in Sri Lanka were attributed to the Rayner and Napier complexes, respectively, by Grew and Manton [1979]. However, if the model of Lawver and Scotese [this volume] is valid, it is possible to correlate the Lützow-Holm Complex with the Vijayan Complex in Sri Lanka. General structural trends of northwest-southeast in the Lützow-Holm Complex seem to support the model. Despite the 1.0 to 1.5 Ga ages reported for the Vijayan Complex, the age of the regional metamorphism in Sri Lanka is still controversial [e.g., Perera, 1984]. The type of metamorphism as well as geochronological data of the Vijayan Complex is an interesting subject. On

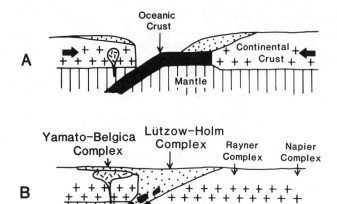

Fig. 5. Plate tectonic model of late Proterozoic metamorphic complexes in eastern Queen Maud Land. See text for details.

the other hand, more rigorous geochronological studies in eastern Queen Maud Land are much needed.

Formation of Gondwanaland

The contrasting features of the two metamorphic complexes in eastern Queen Maud Land are shown in Table 3. Despite the discrepancy between ~1100 Ma and newly reported 700 Ma ages, it can safely be assumed that the main regional metamorphism in the Lützow-Holm Complex occurred in the late Proterozoic and was contemporaneous with that of the Yamato-Belgica Complex. The relatively high initial $^{87}Sr/^{86}Sr$ ratio of 0.709 obtained from the granulite facies rocks of the Yamato-Belgica Complex is close to the 0.711 value which Grew [1978] reported from the Rayner Complex. However, the rocks of the Lützow-Holm Complex with lower initial ratios near 0.705-0.706 may have been derived from material younger than the precursors to the Yamato-Belgica Complex. A possible interpretation for the low initial ratios of the rocks from the Lützow-Holm Complex is contamination by rocks with much lower initial ratios, such as younger igneous rocks.

The low-pressure-type character of the Yamato-Belgica Complex and high- to medium-pressure-type character of the Lützow-Holm Complex suggest paired metamorphism in the late Proterozoic. As discussed by Miyashiro [1973], the plate tectonic model would be the most suitable for interpretation of the tectonic development of the paired metamorphic terrane. Moreover, the wide range of chemical compositions and mode of occurrence in the field of the metamorphosed ultrabasic and basic rocks suggest that the rocks originated from cumulate rocks associated with ophiolites [Coleman, 1977] and that they were tectonically fractured and emplaced in the metasedimentary gneisses. In addition, these rocks are distributed much more in the western part than in the eastern part of the Lützow-Holm Complex. Therefore, we infer the presently missing oceanic crust to the west of the Lützow-Holm Complex. More detail about the metamorphosed ultrabasic and basic rocks is given in a separate paper [Hiroi et al., 1986].

Figure 5 shows a model of the tectonic development of eastern Queen Maud Land mainly based on the petrological evidence. Source rocks of the Lützow-Holm Complex might deposit on the margin of the craton in Enderby Land, whereas those of the Yamato-Belgica Complex might deposit on the craton on the other side, beyond an ocean intervening between the two cratons. Stage A represents the initiation of subduction of the oceanic crust toward the southwest. Subsequently, at stage B, oceanic crust might be torn off and tectonically introduced into the sedimentary pile, and eventually the oceanic crust is entirely exhausted. The occurrence of the isolated blocks of the ultrabasic rocks mainly in the western part of Lützow-Holm Complex may be explained by such a mechanism. The descending slab might pull the rocks of the Lützow-Holm Complex to deeper levels before the thermal peak of regional metamorphism. In the Yamato-Belgica Complex, extensive igneous activity including syenite intrusion may have occurred during the subduction. Thus, we propose the scheme of continent-continent collision for development of the East Antarctic Shield, which is one of the fragments of Gondwanaland. It may follow that Gondwanaland formed in such a manner.

Acknowledgments. We thank all members of the Japanese Association of Geologists for Antarctic Research for valuable discussions. We are also indebted to anonymous reviewers for their comments on the early version of the manuscript.

References

Asami, M., and K. Shiraishi, Mineral parageneses of basic to intermediate metamorphic rocks in the Yamato Mountains, East Antarctica, Mem. Natl. Inst. Polar Res. Spec. Issue Jpn., 28, 183-197, 1983.

Asami, M., and K. Shiraishi, Retrograde metamorphism in the Yamato Mountains, East Antarctica, Mem. Natl. Inst. Polar Res. Spec. Issue Jpn., 37, 147-163, 1985.

Coleman, R. G., Ophiolites, 229 pp., Springer-Verlag, New York, 1977.

Ellis, D. J., The Napier and Rayner complexes of Enderby Land, Antarctica: Contrasting styles of metamorphism and tectonics, in Antarctic Earth Science, edited by R. J. Oliver, P. R. James, and J. B. Jago, pp. 20-24, Australian Academy of Science, Canberra, 1983.

Grew, E. S., Precambrian basement at Molodezhnaya Station, East Antarctica, Geol. Soc. Am. Bull., 89, 801-813, 1978.

Grew, E. S., The Antarctic margin, in The Ocean Basins and Margins, vol. 6, edited by A. E. M. Nairn and F. G. Stehli, pp. 697-755, Plenum, New York, 1982.

Grew, E. S., and W. I. Manton, Archean rocks in Antarctica, 2.5-billion-year uranium-lead ages of pegmatites in Enderby Land, Science, 206 (4417), 443-445, 1979.

Herzberg, C. T., Pyroxene geothermometry and

geobarometry: Experimental and thermodynamic evaluation of some subsolidus phase relations involving pyroxenes in the system $CaO-MgO-Al_2O_3-SiO_2$, Geochim. Cosmochim. Acta, 42, 945-957, 1978.

Hiroi, Y., K. Shiraishi, K. Yanai, and K. Kizaki, Aluminum silicates in the Prince Olav and Soya coasts, East Antarctica, Mem. Natl. Inst. Polar Res. Spec. Issue Jpn., 28, 115-131, 1983a.

Hiroi, Y., K. Shiraishi, Y. Nakai, T. Kano, and S. Yoshikura, Geology and petrology of Prince Olav Coast, East Antarctica, in Antarctic Earth Science, edited by R. L. Oliver, P. R. James, and J. B. Jago, pp. 32-35, Australian Academy of Science, Canberra, 1983b.

Hiroi, Y., K. Shiraishi, and Y. Yoshida, Geological map of Sinnan Rocks, scale 1:25,000, Antarct. Geol. Map Ser., Sheet 14, (with explanatory text 7 pp.), Natl. Inst. of Polar Res., Tokyo, 1983c.

Hiroi, Y., K. Shiraishi, Y. Motoyoshi, S. Kanisawa, K. Yanai, and K. Kizaki, Mode of occurrence, bulk chemical compositions, and mineral textures of ultramafic rocks in the Lützow-Holm Complex, East Antarctica, Mem. Natl. Inst. Polar Res. Spec. Issue Jpn., 43, 62-84, 1986.

Holdaway, M. J., Stability of andalusite and the aluminum silicate phase diagram, Am. J. Sci., 271, 97-131, 1971.

Huckenholz, H. G., W. Lindhuber, and J. Springer, The join $CaSiO_3-Al_2O_3-Fe_2O_3$ of the $CaO-Al_2O_3-Fe_2O_3-SiO_2$ quaternary system and its bearing on the formation of granditic garnets and fassaitic pyroxenes, Neues Jahrb. Miner. Abh., 121, 160-207, 1974.

Huckenholz, H. G., W. Lindhuber, and K. T. Fehr, Stability relationships of grossular + quartz + wollastonite + anorthite, I, The effect of andradite and albite, Neues Jahrb. Mineral Abh., 142, 223-247, 1981.

Irving, A. J., Geochemical and high pressure experimental studies of garnet pyroxenite and pyroxen granulite xenoliths from the Delegate basaltic pipes, Australia, J. Petrol., 15, 1-40, 1974.

James, P. R., and R. J. Tingey, The Precambrian geological evolution of the East Antarctic metamorphic shield--A review, in Antarctic Earth Science, edited by R. L. Oliver, P. R. James, and J. B. Jago, pp. 5-10, Australian Academy of Science, Canberra, 1983.

Kamenev, E. N., Geological structure of Enderby Land, in Antarctic Geology and Geophysics, edited by R. J. Adie, pp. 579-583, Universitetsforlaget, Oslo, 1972.

Kanisawa, S., and K. Yanai, Metamorphic rocks of the Cape Hinode district, East Antarctica, Mem. Natl. Inst. Polar Res. Spec. Issue Jpn., 21, 71-85, 1982.

Kerrick, D. M., Experimental determination of muscovite + quartz stability with $P_{H_2O} < P_{total}$, Am. J. Sci., 272, 946-958, 1972.

Kizaki, K., Geology and petrography of the Yamato Sanmyaku, East Antarctica, JARE Sci. Rep., Ser. C (Geol.), 3, 27 pp., 1965.

Kojima, H., K. Yanai, and T. Nishida, Geology of the Belgica Mountains, Mem. Natl. Inst. Polar Res. Spec. Issue Jpn., 21, 32-46, 1982.

Lawver, L. A., and C. R. Scotese, A revised reconstruction of Gondwanaland, this volume.

Maegoya, T., S. Nohda, and I. Hayase, Rb-Sr dating of the gneissic rocks from the East Coast of Lützow-Holm Bay, Antarctica, Mem. Fac. Sci. Kyoto Univ. Ser. Geol. Mineral., 35, 131-138, 1968.

Matsumoto, Y., T. Nishida, K. Yanai, and H. Kojima, Geology and geologic structure of the northern Ongul Islands and surroundings, East Antarctica, Mem. Natl. Inst. Polar Res. Spec. Issue Jpn., 21, 47-70, 1982.

Miyashiro, A., Metamorphism and Metamorphic Belts, 492 pp., Allen & Unwin, London, 1973.

Miyashita, S., M. Komatsu, and S. Hashimoto, Sapphirine from metamorphic layered complex of Mt. Poroshiri, Hidaka metamorphic belt, Hokkaido, Proc. Jpn. Acad., Ser. B, 56, 108-113, 1980.

Motoyoshi, Y., Prograde and progressive metamorphism of the granulite facies Lüzow-Holm Bay region, East Antarctica, D.Sc. Thesis, 238 pp., Hokkaido Univ., Sapporo, 1986.

Motoyoshi, Y., S. Matsubara, H. Matsueda, and Y. Matsumoto, Garnet-sillimanite gneisses from the Lützow-Holm Bay region, East Antarctica, Mem. Natl. Inst. Polar Res. Spec. Issue Jpn., 37, 82-94, 1985.

Obata, M., and A. B. Thompson, Amphibole and chlorite in mafic and ultramafic rocks in the lower crust and upper mantle: A theoretical approach, Contrib. Mineral. Petrol., 77, 74-81, 1981.

Perera, L. R. K., Co-existing cordierite-almandine --A key to the metamorphic history of Sri Lanka, Precambrian Res., 25, 349-364, 1984.

Sheraton, J. W., L. A. Offe, R. J. Tingey, and D. J. Ellis, Enderby Land, Antarctica--An unuual Precambrian high-grade metamorphic terrain, J. Geol. Soc. Aust., 27, 1-18, 1980.

Shibata, K., K. Yanai, and K. Shiraishi, Rb-Sr mineral isochron ages of metamorphic rocks around Syowa Station and from the Yamato Mountains, East Antarctica, Mem. Natl. Inst. Polar Res. Spec. Issue Jpn., 37, 164-171, 1985.

Shibata, K. K. Yanai, and K. Shiraishi, Rb-Sr whole-rock ages of metamorphic rocks from eastern Queen Maud Land, East Antarctica, Mem. Natl. Inst. Polar Res. Spec. Issue Jpn., 43, 133-148, 1986.

Shiraishi, K., M. Asami, and H. Kanaya, Petrochemical character of the syenitic rocks from the Yamato Mountains, East Antarctica, Mem. Natl. Inst. Polar Res. Spec. Issue Jpn., 28, 183-197, 1983a.

Shiraishi, K., M. Asami, and Y. Ohta, Geology and petrology of the Yamato Mountains, in Antarctic Earth Science, edited by R. L. Oliver, P. R. James, and J. B. Jago, pp. 50-53, Australian Academy of Science, Canberra, 1983b.

Shiraishi, K., Y. Hiroi, and H. Onuki, Orthopyroxene-bearing rocks from the Tenmondai and Nagaiwa rocks in the Prince Olav Coast, East Antarctica: First appearance of orthopyroxene in progressive metamorphic sequence, Mem. Natl. Inst. Polar Res. Spec. Issue Jpn., 33, 126-144, 1984.

Shiraishi, K., Y. Hiroi, K. Moriwaki, K. Sasaki, and H. Onuki, Tenmondai Rock, Scale 1:25,000, Antarct. Geol. Map Ser., Sheet 19 (with explanatory text 7 pp.), Natl. Inst. Polar Res., Tokyo, 1985.

Suzuki, M., Metamorphism and plutonic rocks in the Cape Omega area, East Antarctica, Mem. Natl. Inst. Polar Res. Spec. Issue Jpn., 14, 128-139, 1979.

Suzuki, M., On the association of orthopyroxene-garnet-biotite found in the Lützow-Holmbukta region, East Antarctica, Mem. Natl. Inst. Polar Res. Spec. Issue Jpn., 28, 86-102, 1982.

Suzuki, M., Preliminary note on the metamorphic conditions around Lützow-Holm Bay, East Antarctica, Mem. Natl Inst. Polar Res. Spec. Issue Jpn., 28, 132-143, 1983.

Suzuki, M., Reexamination of metapelites in the Cape Omega area, westernmost part of Prince Olav Coast, East Antarctica, Mem. Natl. Inst. Polar Res. Spec. Issue Jpn., 33, 145-154, 1984.

Tatsumi, T., and K. Kizaki, Geology of the Lützow-Holm Bay region and the "Yamato Mauntains" (Queen Fabiola Mountains), Antarct. Map Folio Ser., folio 12, sheets 9 and 10, Am. Geogr. Soc., New York, 1969.

Windom, K. E., and A. L. Boettcher, The effect of reduced activity of anorthite on the reaction grossular + quartz = anorthite + wollastonite: A model for plagioclase in the earth's lower crust and upper mantle, Am. Mineral., 61, 889-896, 1976.

Yanai, K., and Y. Ueda, Absolute ages and geological investigations on the rocks in the area around Syowa Station, East Antarctica (in Japanese), Nankyoku Shiryo (Antarct. Rec.), 48, 70-81, 1974.

Yanai, K., T. Nishida, H. Kojima, K. Shiraisi, M. Asami, Y. Ohta, K. Kizaki, and Y. Matsumoto, Geological map of the Central Yamato Mountains, Massif B and Massif C, 1:25,000, Antarct. Geol. Map Ser., Sheet 28 (with explanatory text 10 pp.), Natl. Inst. Polar Res., Tokyo, 1982.

Yoshida, M., Tectonics and petrology of charnokites around Lützow-Holmbukta, East Antarctica, J. Geosci. Osaka City Univ., 21, 65-152, 1978.

Yoshida, M., Metamorphic conditions of the polymetamorpic Lützow-Holmbukta region, East Antarctica, J. Geosci. Osaka City Univ., 22, 97-139, 1979.

Yoshida, M., and N. Aikawa, Petrography of a discordant metabasite from Skallen, Lützow-Holmbukta, East Antarctica, Mem. Natl. Inst. Polar Res. Spec. Issue Jpn., 28, 144-165, 1983.

Yoshikura, S., Y. Nakai, and T. Kano, Petrology and geothermometry of the clinopyroxene-garnet rock from Cape Ryûgû, East Antarctica, Mem. Natl. Inst. Polar Res. Spec. Issue Jpn., 14, 172-185, 1979.

TECTONIC POSITION OF KAROO BASALTS, WESTERN ZAMBIA

Raphael Unrug

Department of Geological Sciences, Wright State University, Dayton, Ohio 45435

Abstract. Widespread subsurface occurrences of Karoo basalts are present along the eastern margin of the Barotse basin trending north-northwest in western Zambia and in adjacent parts of Angola. To the south these basalts are continuous with the basalts underlying neighboring parts of Namibia and Botswana and with the Batoka basalts traversed by the Zambezi River gorges east of Victoria Falls. The Barotse basin is a half graben with a step fault zone at its eastern margin that follows a line offsetting structural and geophysical trends in the Pan-African Damaran-Lufilian mobile belt. Faulting and basalt effusions occurred along a zone of structural weakness that possibly represents a Pan-African transcurrent fault. Trends of basalt occurrences in eastern Namibia, northern Botswana, and southern Zambia also follow Pan-African dislocation zones. Karoo volcanism in the center of the African plate is closely related to preexisting lines of structural weakness associated with the Pan-African tectonothermal event.

Introduction

Karoo rocks fill the north-northwest trending Barotse basin that underlies western Zambia and central Angola. To the south, the Barotse basin merges with the large Kalahari basin which extends over parts of Zimbabwe, Botswana, and Namibia. A major part of the Barotse basin is covered by younger deposits including Cretaceous sediments and the Kalahari Group of Tertiary to Recent age (Figure 1).

The Karoo sequence of the Barotse basin consists of terrigenous clastics with coal at the base and basalts at the top [Money, 1972]. Two grabens with downfaulted Karoo rocks extend east from the Barotse basin. These are the Kafue graben and the mid-Zambezi graben (Figure 2). In Zambia the Karoo basalts are present in the southern part of the Barotse basin and in the two eastern grabens.

Occurrences of Basalts

Subsurface occurrences of basalts in the Barotse basin are known in several boreholes drilled by the Geological Survey of Zambia [Money, 1972], and one surface occurrence is present in the northeastern part of the basin [Thieme and Johnson, 1981]. That basalts rest over various locally recognized formations of the Karoo Supergroup, indicates a period of erosion and possibly faulting preceding the lava effusions. The thickness of the basalts increases westward, from 30 m at the eastern margin of the basin. The greatest thickness traversed is 390 m in the southernmost of the three boreholes situated along the Zambezi River (Figure 2). The other boreholes in this area were stopped in basalts.

The basalts are tholeiites [Ridgway and Money, 1981]. Variation of textural characteristics permits determination of the boundaries of individual flows. Twenty-three flows are present in the zone of maximum known thickness of the basalts [Money, 1972].

At the junction of the Barotse basin and the mid-Zambezi River graben, basalts are present over a large area at the surface, or under a thin cover of aeolian Kalahari sand. The Victoria Falls and the Zambezi River gorge below the falls are carved in these basalts. Several flows are visible in the walls of the Zambezi River gorge, which is 100 m deep, but the total thickness of the basalts is not known.

Surface occurrences of basalts are present at the eastern end of the Kafue River graben. Four flows are present there, producing a typical trap morphology. More basalts are probably buried under swamps of the Kafue River valley.

Minor surface occurrences of basalts in the mid-Zambezi River graben are located on the southern shore of the Kariba Reservoir (Figure 2).

Tectonics of the Barotse Basin

The cover of the Cretaceous and Kalahari sediments that mask much of the Barotse basin make it necessary to base tectonic studies of the basin on geophysical data and regional considerations. Satellite imagery also provides information on the structure of the basin.

A system of fractures in the basalts exposed at the surface is visible on Landsat imagery along the Zambezi River gorge downstream of Victoria Falls. The fractures are accompanied by north-northeast and south-southwest trending faults that produce fault blocks tilted to the south. The intersections of flow boundaries with the surface are seen as arcuate lines in these blocks [Mallick et al., 1981].

The markedly angular drainage pattern in the

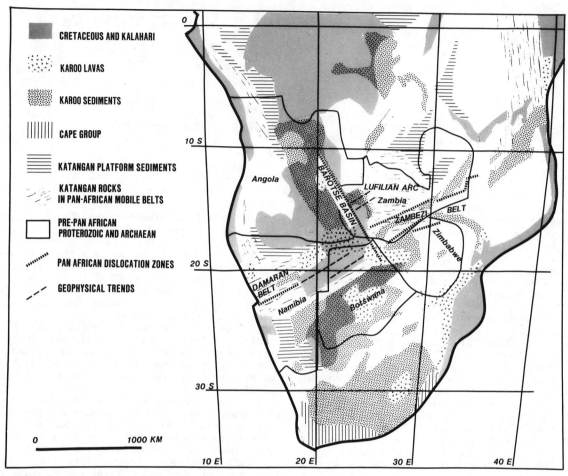

Fig. 1. Geologic map of southern Africa showing the regional position of the Barotse basin.

Barotse basin is undoubtedly a surface expression of a fracture system similar to that seen at the surface along the Zambezi River gorge. The angular drainage pattern is evident over a range of fluvial channel sizes, from first order streams to major rivers including the Zambezi River. The drainage, developed on a thin veneer of Kalahari sand and Karoo clastics that overlies granitic terranes east of the Barotse basin, does not show the angular pattern; a dendritic drainage pattern is present there. Farther west the angular drainage pattern appears abruptly, marking the location of the eastern boundary of the Barotse basin.

Phototone lineaments seen on Landsat imagery have greater length than the drainage lineaments. They are parallel to some drainage lineaments and also tend to coincide with regular kinks in the course of major valleys and drainage lineaments (Figure 2). The phototone lineaments are interpreted as surface expressions of major fault zones. Some of the west-southwest trending parallel phototone lineaments occurring in groups coincide with gravity anomalies as shown by Mazac [1972] and with the trends of magnetic anomalies [Saviaro, 1979].

Boreholes of the Geological Survey of Zambia provide constraints on the subsurface occurrence of basalts (Figure 2). This evidence and the assumption that phototone lineaments indicate major fault trends have been used to delineate major fault zones (Figure 2). The faulting was a late event in the evolution of the basin, since it affects the sedimentary fill. It is speculated that faults at the eastern margin of the basin functioned as conduits for linear effusions of basaltic lavas. In this interpretation, faulting in the Barotse basin was active during the lava effusions in late Karoo time (Jurassic). Tilting of lava flows in fault blocks along the Zambezi River gorge suggests that faulting outlasted the volcanic activity.

Tectonics of the Kafue and Mid-Zambezi Grabens

The two northeast trending grabens containing the midcourses of the Kafue River and the Zambezi River are, at least partly, late Karoo structures. An erosional outlier of condensed lower Karoo succession present on the basement block separating the two grabens [Drysdall et al., 1972] suggests

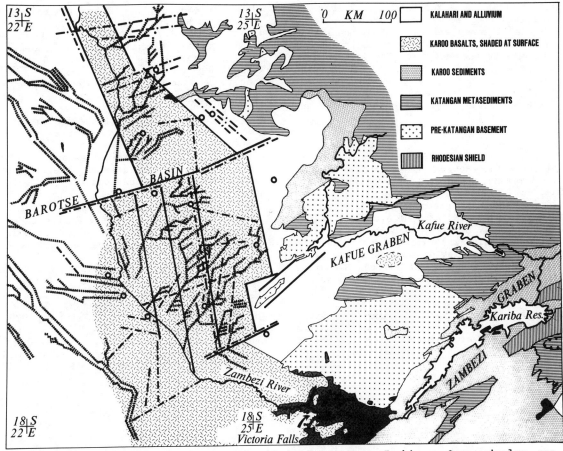

Fig. 2. Geologic map of the Barotse basin in western Zambia. Open circles are boreholes; dotted lines, drainage lineaments; and dash-dot lines, phototone lineaments. Solid lines are faults.

continuity of the sedimentary basin during the late Paleozoic. However, coarse-grained lithologies of upper Karoo rocks along fault scarps bordering the Kafue graben indicate fault activity during the later part of Karoo sedimentation [Barr and Brown, 1968]. Faults bordering the mid-Zambezi graben displace the entire sedimentary succession of the Karoo Supergroup and are late structures. They may be coeval with faults affecting the basalts at the southwestern end of the mid-Zambezi graben.

Regional Tectonics

Data on the structure of the Barotse basin in Zambia discussed in this paper pertain only to the southeastern part of the entire basin. Little is known of the northwestern part of the Barotse basin in Angola. However, the fault zone recognized in western Zambia on the basis of subsurface information and indirect surface evidence continues to the northwest into Angola, and it is shown with more basalts in the Karoo Supergroup in the tectonic map of Africa [Choubert, 1968].

The Barotse basin is interpreted as a half graben with a step-faulted eastern margin. This fault zone is coincident with a line that offsets structural and geophysical trends in the Damaran-Lufilian mobile belt and was interpreted as a transcurrent fault related to the Pan-African tectonothermal event [Unrug, 1983].

The distribution of Karoo volcanics in Southern Africa is thought to follow ancient structural lines [Cox, 1970; Pretorius, 1979]. The Karoo basalts of western Zambia are no exception to this rule, as their distribution coincides with the Pan-African structural plan (Figure 1).

Karoo basalts of the Kafue graben and the mid-Zambezi graben are associated with the Mwembeshi dislocation zone and the Zambezi dislocation zone, respectively [de Swardt et al., 1965]. Both these transcurrent fault zones are late Pan-African structures formed as a result of the collision of the Congo and the Kalahari cratons [Unrug, 1983].

Conclusions

Results of this work can be summarized as follows:

1. The existence of a faulted eastern margin of the Barotse basin has been established.

2. Faulting of the Karoo basins in Zambia occurred in late Karoo time.

3. The fault zones affecting Karoo basins in Zambia follow older lines of structural weakness formed during late phases of the Pan-African tectonothermal event.

Acknowledgments. This work was supported by Wright State University research grants 241117 and 216011.

References

Barr, M. W. C., and A. G. Brown, Karroo rocks of part of the Kafue trough between Mapanza Mission and Ndundumwense, Rec. Geol. Surv. Zambia, 11, 29-50, 1968.

Choubert, G., (Ed.), International tectonic map of Africa, 1:5,000,000, Ass. of Afr. Geol. Surv. and UNESCO, Rabat, Morocco, 1968.

Cox, K. G., Tectonics and vulcanism of the Karroo period and their bearing on the postulated fragmentation of Gondwanaland, in African Magmatism and Tectonics, edited by T. N. Clifford and I. G. Gass, pp. 211-235 Oliver and Boyd, Edinburgh, 1970.

de Swardt, A. M. J., P. Garrard, and J. G. Simpson, Major zones of transcurrent dislocation and superposition of orogenic belts in part of central Africa, Geol. Soc. Am. Bull., 76, 89-102, 1965.

Drysdall, A. R., R. L. Johnson, T. A. Moore, and J. G. Thieme, Outline of the geology of Zambia, Geol. Mijnbouw, 51, 265-276, 1972.

Mallick, D. I. J., F. Hapgood, and A. C. Skinner, A geological interpretation of Landsat imagery and air photography of Botswana, Overseas Geol. Miner. Resour., 56, 36 pp., 1981.

Mazac, O., Reconnaissance gravity survey of Zambia, Tech. Rep. 76, 40 pp., Geol. Surv. of Zambia, Lusaka, 1972.

Money, N. J., An outline of the geology of western Zambia, Rec. Geol. Surv. Zambia, 12, 103-123, 1972.

Pretorius, D. A., The aeromagnetic delineation of the distribution patterns of Karroo volcanics and consequent implications for the tectonics of the subcontinent, Bull. Botswana Geol. Surv. Dep., 22, 93-139, 1979.

Ridgway, J., and N. J. Money, Karroo basalts from western Zambia and geochemical provinces in central and southern Africa, Geol. Rundsch., 70, 868-873, 1981.

Saviaro, K., Preliminary analysis of airborne magnetic surveys in Zambia, Bull. Botswana Geol. Surv. Dep., 22, 159-179, 1979.

Thieme, J. G., and R. L. Johnson, Geological map of the Republic of Zambia, 1:1,000,000, Geol. Surv. of Zambia, Lusaka, 1981.

Unrug, R., The Lufilian arc: A microplate in the Pan-African collision zone of the Congo and the Kalahari cratons, Precambrian Res., 21, 181-196, 1983.

SYMPOSIUM PARTICIPANTS

1. B. Stait
2. K. Kenna
3. L. Beltan
4. G.C. Young
5. W. Buggish
6. Z. Yang
7. Mrs. S. Unrug
8. P.D. Toens
9. S. Chatterjee
10. C. Tripathi
11. J.M. Dickins
12. S.C. Shah
13. S. Ramanathan
14. M.R.A. Thomson
15. J.L. Smellie
16. B. Saldukas
17. D. Fütterer
18. H.M. Kapoor
19. V. Raiverman
20. U. Rosenfeld
21. K. Pedersen
22. M. Ernesto
23. S. Bergström
24. S. Defauw
25. P. English
26. A. Sahni
27. S.L. Jain
28. E.H. Colbert
29. W.R. Hammer
30. A. Inderbitzen
31. E. Stump
32. S.C. Borg
33. H.J. Rakotoarivelo
34. R. Frish
35. R.P. Agarwal
36. M. Manassero
37. S. Archangelski
38. G.J. Retallack
39. G. Spaeth
40. D.I.M. Macdonald
41. J. Utting
42. T.N. Taylor
43. J. Splettstoesser
44. G.D. McKenzie
45. P.K. Dutta
46. T.S. Laudon
47. V.D. Choubey
48. G. Webers
49. D.A. Coates
50. Y. Kristoffersen
51. B.C. Storey
52. M. Miller
53. R.R.B. Von Frese
54. L.D. McGinnis
55. M.A. Uliana
56. E. Zeller
57. G. Dreschoff
58. C. Macellari
59. C.F. Burrett
60. M. Abed
61. R.L. Leary
62. J.M.G. Miller
63. B. Oelefsen
64. M.R. Johnston
65. H. Kreuzer
66. M. Morales
67. S.L. Ettreim
68. C. Chonglakmani
69. C. Craddock
70. T. Singh
71. R.J. Pankhurst
72. J. Krishna
73. H.Y. Ling
74. A. Meneilly
75. S.R. Rooney
76. J.A. Crame
77. J. Yang
78. B. Waugh
79. G.P. Gravenor
80. J.M. Zawiskie
81. R.O. Utgard
82. L. Krissek
83. I.P. Martini
84. V. von Brunn
85. R. Unrug
86. A.P. Willner
87. A. Rocha-Campos
88. A.K. Cooper
89. B.R. Watters
90. J.N.J. Visser
91. Mrs. C. Wright
92. R. Askin
93. W.J. Zinsmeister
94. H. Miller
95. N. Hotton
96. L.D.B. Herrod
97. D.D. Blankenship
98. J.C. Crowell
99. A. Grunow
100. I.W.D. Dalziel
101. P. Fitzgerald
102. D.H. Elliot
103. L. Lavver
104. T.O. Wright
105. D. Beike
106. F. Tessensohn
107. G. Kleinschmidt
108. M. Venkateswarlu
109. I.O. Norton
110. A.J. Rowell
111. C.M. Bell
112. E. Smoot
113. S. Prakash
114. C.R. Scotese
115. K.K. Venkataramani
116. J.W. Collinson
117. R.P. Wright
118. M.R. Banks
119. S.M. Casshyap
120. C.F. Burrett
121. R.D. Powell
122. K. Pigg
123. C. Mason
124. M.E. Brookfield
125. R.D. Forsythe
126. S. Wu
127. P.N. Webb

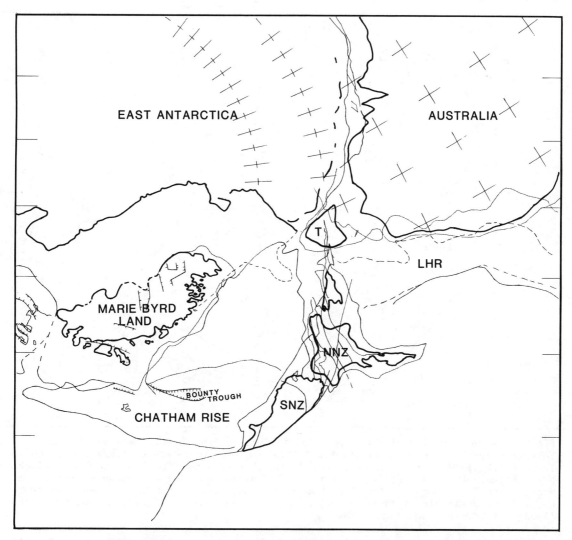

Fig. 2. Reconstruction of New Zealand-Marie Byrd Land-East Antarctica-Australia modified from Grindley and Davey [1982]. Continental outlines are the same as in Figure 1. The dashed light line is the presumed continental part of Lord Howe Rise (LHR). T is Tasmania; SNZ, southern New Zealand; NNZ, northern New Zealand.

assuming that they are not allochthonous. Previous reconstructions of Gondwana have resorted to three different solutions with regard to the fit of the pieces of West Antarctica. The first solution has been to ignore the problem and to simply show the Antarctic Peninsula overlapping the Falkland Plateau [Dietz and Holden, 1970; Smith and Hallam, 1970; Norton and Sclater, 1979]. An alternative solution has been to place the Antarctic Peninsula on the Pacific side of South America [Barron et al., 1978]. While there are no marine magnetic data to dispute this conclusion, geological data seem to imply that the westward facing tip of South America fronted an active subduction zone [Dalziel, 1980] during the period 120-165 Ma, the period for which an overlap must be considered a problem [Lawver et al., 1985]. The third solution has involved movement of the peninsula with respect to East Antarctica [de Wit, 1977; Barker et al., 1976; Dalziel and Elliot, 1982]. From data presented by Longshaw and Griffiths [1983] and Grunow et al. [this volume] concerning the paleomagnetic position for the pieces of West Antarctica, constraints can be placed on the amount of rotation as well as the paleolatitude that both the Ellsworth-Whitmore Mountains crustal block and the Antarctic Peninsula could have undergone if the paleomagnetic results are strictly adhered to. While our revised version of Gondwana does not use the new poles determined by Grunow et al. [this volume], we do use their conclusion that the Antarctic Peninsula and the Ellsworth Mountains-Whitmore Mountains were roughly in the same position relative to each other during the Jurassic as they are now. We have moved the Ellsworth Mountains block and the Whitmore Moun-

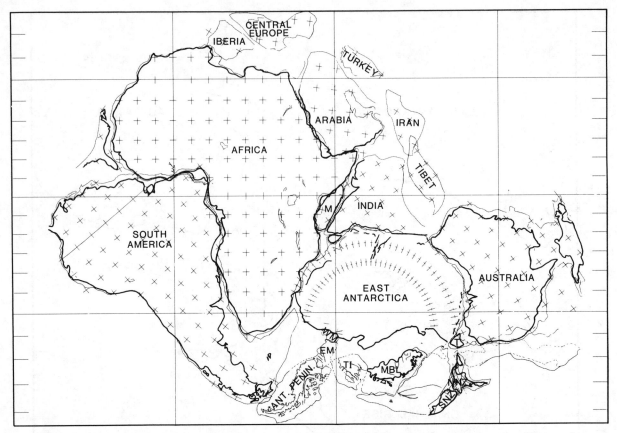

Fig. 3. Reconstruction of Gondwana. Continental outline is the same as in Figure 1. Poles for reconstruction to Africa fixed in a present-day reference frame are given in Table 1. WM is Whitmore Mountains; EM, Ellsworth Mountains; TI, Thurston Island block; NNZ, northern New Zealand; SNZ, southern New Zealand attached to Campbell Plateau; M, Madagascar; T, Tasmania. The rotation of the Antarctic Peninsula/Ellsworth-Whitmore Mountains crustal blocks is based on the paleomagnetic work of Longshaw and Griffiths [1983] and Grunow et al. [this volume].

tains block slightly closer together on the basis of the recent work of Doake et al. [1983] and Garrett et al. [this volume]. The prebreakup locations of the pieces of West Antarctica with respect to Gondwana, with the exception of Marie Byrd Land, are tentative, so they are shown with a light line to distinguish them from the pieces that are constrained by marine magnetic anomalies. Other pieces, particularly those north of the main body of Gondwana, are also shown with a light line, since marine magnetic anomaly identifications are again not available to constrain them in a prebreakup configuration.

Our revised reconstruction is shown in Figure 3. We have kept Africa fixed and plotted the reconstructed pieces in a Mercator projection. The advantages of plotting the pieces fixed to Africa and using a Mercator projection are twofold. First, the continents remain recognizable to those accustomed to seeing the world plotted in a Mercator projection. Second, the pieces that are not controversial are plotted without much enlargement, while the more poorly positioned pieces of New Zealand and West Antarctica are enlarged so that some of the controversial areas are easier to decipher. We feel that the major problems in our revised reconstruction of Gondwana are the placement of the pieces of West Antarctica and the relative position of the whole assemblage. Table 1 lists the poles of rotation that we used to produce the reconstruction shown in Figure 3.

Since we choose to have Africa remain fixed, the poles are either listed as rotations of the pieces to Africa or are listed as being rotated to another plate, fixed to that plate, and then the package of plates rotated to Africa. To change the location of Gondwana with respect to the present reference frame, it would only be necessary to change the rotation pole for Africa.

Du Toit's [1937] reconstruction of Gondwana (Figure 4) stands up remarkably well with time; the only obvious change is the location of Antarctica with respect to Africa. Otherwise, the location of Madagascar with respect to Africa has not changed, although its location has produced debate during the last 20 years, and India, Sri

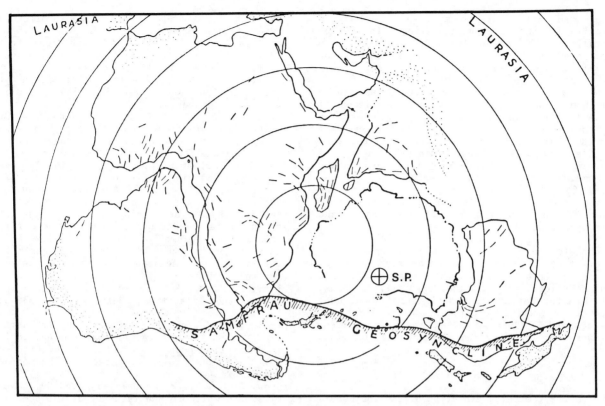

Fig. 4. Reproduction of Figure 7 in Our Wandering Continents [du Toit, 1937], showing the reassembly of Gondwana during the Paleozoic era. The space between the various portions was then mostly land. Short lines indicate the Precambrian or Early Cambrian "grain." Diagonal rules show the "Samfrau" Geosyncline of the late Paleozoic. Stippled marks indicate our regions of Late Cretaceous and Tertiary compression, according to Lambert's equal area polar projection.

TABLE 1. Rotation Poles for Jurassic Gondwana Fit

Plate	Latitude	Longitude	Angle	With Respect to Plates
Africa	90.0	0.0	0.0	Held Fixed
South America	45.5	-32.2	58.2	Africa
India	-4.44	16.74	-92.77	East Antarctica
Sri Lanka	-13.67	31.11	-107.14	East Antarctica
Australia	-1.58	39.02	-31.29	East Antarctica
Madagascar	-3.41	-81.70	19.73	Africa
Antarctic Peninsula	73.87	108.59	-41.79	East Antarctica
Ellsworth Mountains	72.64	100.37	-37.44	East Antarctica
Whitmore Mountains	72.55	97.64	-39.73	East Antarctica
Thurston Island	62.27	21.84	13.27	East Antarctica
Marie Byrd Land	62.27	21.84	13.27	East Antarcitca
North New Zealand	24.19	-19.91	44.61	Australia
South New Zealand	65.14	-52.00	62.38	Marie Byrd Land
Chatham Rise	41.00	-15.90	7.47	South New Zealand
East Antarctica	-7.78	-31.42	58.0	Africa
Florida	57.00	-20.80	88.90	Africa
Central Europe	64.3	147.3	20.6	Africa
Iberia	50.0	3.3	-27.0	Central Europe
Arabia	26.5	21.5	-7.6	Africa
Iran/Turkey	2.02	36.9	-24.6	Arabia
Tibet	-43.0	0.5	8.0	India

Lanka, and East Antarctica have remained in approximately the same configuration. The fit of Australia to East Antarctica and South America to Africa are easily recognizable. It is interesting that the reconstruction of Gondwana has not been greatly revised in the past 50 years.

Our revised reconstruction of Gondwana places constraints on the area in which the pieces of West Antarctica can be placed. The paleomagnetic results of Grunow et al. [this volume] indicate that there may be a major break between the peninsula and Marie Byrd Land, since our reconstruction indicates that Marie Byrd Land cannot have been rotated in the simple fashion that is suggested for the peninsula and the Ellsworth and Whitmore Mountains. Further work, particularly in the region between the Antarctic Peninsula and Marie Byrd Land, is needed.

Acknowledgments. This work was supported by DPP grant 84-05968 to the Institute for Geophysics and by contributions to the Paleoceanographic Mapping Project. This is the University of Texas, Institute for Geophysics Contribution number 666. We wish to thank I. Dalziel, D. Sandwell, J. Sclater, and T. Shipley as well as one anonymous reviewer for reading and commenting on this manuscript.

References

Barker, P. F., et al., Evolution of the southwestern Atlantic Ocean basin: Results of Leg 36, Deep Sea Drilling Project, Initial Rep. Deep Sea Drill. Proj. 1969, 36, 993-1014, 1976.

Barron, E. J., C. G. A. Harrison, and W. W. Hay, A revised reconstruction of the southern continents, EOS Trans. AGU, 59, 436-449, 1978.

Bergh, H. W., Mesozoic seafloor off Dronning Maud Land, Antarctica, Nature, 269, 686-687, 1977.

Cooper, R. A., J. B. Jago, A. J. Rowell, and P. Braddock, Age and correlation of the Cambrian-Ordovician Bowers Supergroup, north Victoria Land, Antarctic Earth Sciences, edited by R. L. Oliver, P. R. James, and J. B. Jago, pp. 128-131, Australian Academy of Sciences, Canberra, 1983.

Dalziel, I. W. D., Comments on Mesozoic evolution of the Antarctic Peninsula and the southern Andes, Geology, 8, 260-261, 1980.

Dalziel, I. W. D., and D. H. Elliot, West Antarctica: Problem child of Gondwanaland, Tectonics, 1, 3-19, 1982.

de Wit, M. J., The evolution of the Scotia Arc as a key to the reconstruction of southwestern Gondwanaland, Tectonophysics, 37, 53-81, 1977.

Dietz, R. S., and J. D. Holden, Reconstruction of Pangea: Breakup and dispersion of continents, Permian to Present, J. Geophys. Res., 75, 4939-4956, 1970.

Doake, C. S. M., R. D. Crabtree, and I. W. D. Dalziel, Subglacial morphology between Ellsworth Mountains and Antarctic Peninsula: New data and tectonic significance, in Antarctic Earth Sciences, edited by R. L. Oliver, P. R. James, and J. B. Jago, pp. 270-273, Australian Academy of Sciences, Canberra, 1983.

Drewry, D. J., and S. R. Jordan, The bedrock surface geology of Antarctica, in Sheet 3 of Antarctica: Glaciological and Geophysical Folio, edited by D. J. Drewry, Scott Polar Research Institute, Cambridge, England, 1983.

du Toit, A. L., Our Wandering Continents, an Hypothesis of Continental Drifting, 366 pp., Oliver and Boyd, Edinburgh, 1937.

Garrett, S. W., L. D. B. Herrod, and D. R. Mantripp, Crustal structure of the area around Haag Nunataks, West Antarctica: New aeromagnetic and bedrock elevation data, this volume.

Grew, E., The Antarctic margin, in The Ocean Basins and Margins, edited by A. E. M. Nairn and F. G. Stehli, pp. 697-775, Plenum, New York, 1982a.

Grew, E., Sapphirine-bearing rocks in Antarctica, south India, Madagascar, and southern Africa (abstract), Volume of Abstracts, for Fourth International Symposium on Antarctic Earth Sciences, pp. 73, Adelaide University, Adelaide, Australia, 1982b.

Griffiths, J. R., Revised continental fit of Australia and Antarctica, Nature, 249, 336-338, 1974.

Grindley, G. W., and F. J. Davey, The reconstruction of New Zealand, Australia, and Antarctica, in Antarctic Geoscience, edited by C. Craddock, pp. 15-29, University of Wisconsin Press, Madison, 1982.

Grunow, A. M., I. W. D. Dalziel, and D. V. Kent, Ellsworth-Whitmore Mountains crustal block, West Antarctica: New paleomagnetic results and their tectonic significance, this volume.

Hayes, D. E., et al., Initial Rep. Deep Sea Drill. Proj. 1969, 28, pp. 3-942, 1975.

Katz, M. B., Sri Lanka in Gondwanaland and the evolution of the Indian Ocean, Geol. Mag., 115, 237-244, 1978.

Katz, M. B., and C. Premoli, India and Madagascar in Gondwanaland based on matching Precambrian lineaments, Nature, 297, 312-315, 1979.

Ladd, J. W., South Atlantic sea-floor spreading and Caribbean tectonics, Ph.D. thesis, 251 pp., Columbia University, New York, 1974.

Lawver, L. A., J. G. Sclater, and L. Meinke, Mesozoic and Cenozoic reconstructions of the South Atlantic, Tectonophysics, 114, 233-254, 1985.

Longshaw, S. K., and D. H. Griffiths, A paleomagnetic study of Jurassic rocks from the Antarctic Peninsula and its implications, J. Geol. Soc. London, 140, 945-954, 1983.

McKenzie, D., and J. G. Sclater, The evolution of the Indian Ocean since the Late Cretaceous, Geophys. J. R. Astron. Soc., 25, 437-528, 1971.

Molnar, P., T. Atwater, J. Mammerickx, and S. M. Smith, Magnetic anomalies, bathymetry, and the tectonic evolution of the South Pacific since the Late Cretaceous, Geophys. J. R. Astron. Soc., 40, 383-420, 1975.

Norton, I. O., and P. Molnar, Implications of a revised fit between Australia and Antarctica for the evolution of the eastern Indian Ocean, Nature, 267, 338-339, 1977.

Norton, I. O., and J. G. Sclater, A model for the evolution of the Indian Ocean and the breakup of